A collection of twenty original essays in the history of science and mathematics, written in honour of D. T. Whiteside. The topics covered embrace the main themes of Whiteside's scholarly work, emphasizing Newtonian topics.

The investigation of difficult things

TOM WHITESIDE

(Photograph by Philippa Whiteside)

The investigation of difficult things

Essays on Newton and the history of the exact sciences
in honour of D. T. Whiteside

Edited by
P. M. HARMAN
Lancaster University
ALAN E. SHAPIRO
University of Minnesota

CAMBRIDGE
UNIVERSITY PRESS

Published by the Press Syndicate of the University of Cambridge
The Pitt Building, Trumpington Street, Cambridge CB2 1RP
40 West 20th Street, New York, NY 10011-4211, USA
10 Stamford Road, Oakleigh, Victoria 3166, Australia

First published 1992

Printed in Great Britain at the University Press, Cambridge

A catalogue record of this book is available from the British Library

Library of Congress cataloguing in publication data available

ISBN 0 521 37435 9 hardback

Contents

Contents

Contributors

ZEV BECHLER
School of History
Tel-Aviv University
Ramat Aviv 69 978
Tel-Aviv, Israel

DOMENICO BERTOLONI
MELI
Jesus College
Cambridge CB5 8BL, England

HENK J. M. BOS
Mathematical Institute
University of Utrecht
Budapestlaan 6
3508 TA Utrecht, The Netherlands

J. BRUCE BRACKENRIDGE
Department of Physics
Lawrence University
Appleton, Wisconsin 54912, USA

JED Z. BUCHWALD
Institute for the History and
   Philosophy of Science and
   Technology
Victoria College, University of
   Toronto
Toronto, Ontario M5S 1K7,
   Canada

I. BERNARD COHEN
Department of History of Science
Science Center 235
Harvard University
Cambridge, Massachusetts 02138,
   USA

LENORE FEIGENBAUM
Department of Mathematics
Tufts University
Medford, Massachusetts 02155,
   USA

EMIL A. FELLMANN
Euler-Archiv
Arnold Bröcklin-Strasse, 37
CH-4051 Basel, Switzerland

KARIN FIGALA
Institut für Geschichte der Technik
Technische Universität München
8 München 26, Germany

D. H. FOWLER
Mathematics Institute
University of Warwick
Coventry CV4 7AL, England

ALAN GABBEY
5 Lough Road
Upper Ballinderry
Lisburn
Co. Antrim BT28 2PQ, UK

BERNARD R. GOLDSTEIN
Department of History and
   Philosophy of Science
University of Pittsburgh, 2604 CL
Pittsburgh, Pennsylvania 15260,
   USA

RONALD GOWING
RICHST
The Royal Institution
21 Albemarle St.
London W1X 4BS, England

Contributors

JEREMY GRAY
Department of Mathematics
The Open University
Milton Keynes MK7 6AA,
    England

A. RUPERT HALL
14 Ball Lane
Tackley
Oxford OX5 3AG, England

P. M. HARMAN
Department of History
Lancaster University
Lancaster LA1 4YG, England

†JOHN HARRISON

ULRICH PETZOLD
Institut für Geschichte der Technik
Technische Universität München
8 München 26, Germany

ALAN E. SHAPIRO
Program in History of Science and
    Technology
Tate Laboratory of Physics
116 Church Street SE
University of Minnesota
Minneapolis, Minnesota 55455,
    USA

P. E. SPARGO
School of Education
University of Capetown
Rondebosch 7700, South Africa

N. M. SWERDLOW
Department of Astronomy and
    Astrophysics
University of Chicago
5640 South Ellis Avenue
Chicago, Illinois 60637, USA

CURTIS A. WILSON
St John's College
P.O. Box 2800
Annapolis, Maryland 21404, USA

Preface

This collection of essays written in honour of D. T. Whiteside on the occasion of his sixtieth birthday celebrates his unique contribution to scholarship: *The Mathematical Papers of Isaac Newton* is a towering intellectual achievement, undoubtedly one of the landmarks of twentieth-century scholarship.

Tom Whiteside's arrival in Cambridge in 1956 as a graduate student was a singularly fortunate event for the history of science. He was 'fresh' from a tour of duty with the tank corps in the North African desert. Before that he had completed his bachelor's degree at Bristol University in French and Latin, with some mathematics. Now he was to embark on his career as a historian of mathematics and science. Newton's scientific papers, which had remained essentially unread since his death in 1727, were still languishing in the University Library seventy-five years after the fifth Earl of Portsmouth had donated them to the University. In the course of the research for his doctoral degree Tom had his initial encounter with Newton's mathematical papers. His thesis, which – he recalls – he wrote in twenty-nine days and nights in the spring of 1959, was subsequently published in the inaugural volume of the *Archive for History of Exact Sciences*; it soon established itself as a classic paper, 'Patterns of mathematical thought in the later seventeenth century.' Enticed by the 'depth and luxuriance' of Newton's unexplored mathematical manuscripts, he set out, almost immediately after receiving his degree, to publish an edition of them. He accomplished this herculean task in a little over twenty years, with the eight volumes appearing between 1967 and 1981.

As all Newton's scholars know too well, Newton's manuscripts are in an extremely disordered state. Not only do they lack chronological sequence, but frequently the sheets of a single work are separated and scattered throughout the collection. Tom's first and very considerable task was to arrange the chaotic mass of these manuscripts into a coherent chronological sequence representing Newton's lifelong mathematical investigations. His next, far more demanding and creative task was to gather the papers together in such a way that they could enlighten the scholarly community. Bringing an extraordinary historical sensitivity to the papers, he was able to bring Newton's mathematical thinking to life once again and place it solidly within the context of the seventeenth-century mathematical tradition. The introductions, notes, and commentary are a rich source for understanding Newton's mathematics and, indeed, broad areas of the history of early mathematics and of seventeenth-century science. Until the publication of the *Mathematical Papers* Newton's reputation as a mathematician was based on only a few published works and claims he had made in his controversy with Leibniz over the discovery of the calculus. The *Mathematical Papers* enabled the scholarly community to grasp fully Newton's stature as a mathematician. At the same time, it set new standards of excellence for all editions of scientific papers.

In contributing to this volume, Tom Whiteside's friends and colleagues do not merely wish to add further to the list of honours which this monument of editorial scholarship has brought him, though we should mention some of the more notable ones: the Koyré medal of the International Academy of the History of Science in 1968, the Sarton Medal of the History of Science Society in 1977, election as a Fellow of the British Academy in 1975, and the award of an honorary doctorate by Lancaster University in 1987. Rather these essays have been undertaken in appreciation of his informal and personal role in shaping work on Newton and the history of mathematics and the exact sciences during the past thirty years. Tom has been extremely generous with his assistance. Recipients of his extraordinary letters have garnered from a treasure trove which he has readily distributed. He has played a particularly important role in encouraging and guiding the early efforts of younger scholars, many of them post-doctoral visitors to Cambridge. Yet, as their copious acknowledgements to him attest, he has also provided valuable assistance to more senior scholars. The obvious excellence of his own work has been rendered less forbidding by his warm and genuine interest in the work of others.

All who know Tom Whiteside as editor of Newton's *Mathematical Papers*, will know of his appreciation of Adolf Prag for his friendship and his assistance in that work. Wherever Tom Whiteside is lauded as editor of Newton's mathematical papers, the contribution of this most self-effacing of scholars should be honoured.

The association between Tom Whiteside and Cambridge University Press has had a very special quality. By unreservedly committing itself to the heady aspirations of an unknown young scholar and then translating that into the magnificent publishing achievement of *The Mathematical Papers of Isaac Newton*, the Press has earned the gratitude of the world of scholarship. It is therefore fitting that Cambridge should publish this volume of essays, and we wish to thank the Syndics for their ready support. We are especially grateful to Richard Ziemacki for his friendly enthusiasm in helping us to bring this project to fruition.

We take the title of this collection of essays from the penultimate paragraph of the last, and most quoted, of the Queries to the *Opticks* where Newton alludes to 'the investigation of difficult things' in mathematics and natural philosophy. Like Newton, Tom has not shirked the investigation of difficult things. Indeed, he has encouraged and cajoled others into investigating them too.

Publications of D. T. Whiteside

'Wren the mathematician'. *Notes and Records of the Royal Society of London*, 15(1960):107–11. Reprinted in *The Royal Society. Its Origins and Founders*, ed. Sir Harold Hartley (London: The Royal Society, 1960), pp. 107–11.

'Bibliography [of John Wallis]'. *Notes and Records of the Royal Society of London*, 15(1960):66–7. Reprinted in *The Royal Society*, ed. Hartley, pp. 66–7.

'Brouncker's mathematical papers'. *Notes and Records of the Royal Society of London*, 15(1960):157. Reprinted in *The Royal Society*, ed. Hartley, p. 157.

'Patterns of mathematical thought in the later seventeenth century'. *Archive for History of Exact Sciences*, 1(1961):179–388.

'Henry Briggs: the binomial theorem anticipated'. *Mathematical Gazette*, 45(1961):9–12.

'Newton's discovery of the general binomial theorem'. *Mathematical Gazette*, 45(1961):175–80.

'The expanding world of Newtonian research'. *History of Science*, 1(1962):16–29.

'Kepler the mathematician [review of Johannes Kepler, *Gesammelte Werke Band IX: Mathematische Schriften*, ed. and with notes by Franz Hammer (Munich: C. H. Beck, 1960)]'. *History of Science*, 1(1962):86–90.

'After the *Principia* [review of *The Correspondence of Isaac Newton. Volume III: 1688–1694,* ed. H. W. Turnbull (Cambridge University Press, 1961)]'. *History of Science*, 1(1962):96–100.

'Scientific papers of Newton [review of *Unpublished Scientific Papers of Isaac Newton*, ed. A. R. Hall and M. B. Hall (Cambridge University Press, 1960)]'. *History of Science*, 2(1963):125–30.

The Mathematical Works of Isaac Newton. Volume I. New York/London: Johnson Reprint Corporation, 1964.

'Newton's early thoughts on planetary motion: a fresh look'. *The British Journal for the History of Science*, 2(1964):117–37.

'Isaac Newton; birth of a mathematician'. *Notes and Records of the Royal Society of London*, 19(1964):53–62.

[Review of *Les Coniques d'Apollonius de Perge: Oeuvres traduites pour la première fois du grec en français, avec une introduction et des notes*, ed. Paul Ver Eecke, new edn (Paris: Albert Blanchard, 1963)]. *History of Science*, 3(1964):124–7.

'Newtonian dynamics [review of John Herivel, *The Background to Newton's 'Principia'* (Oxford: Clarendon Press, 1966)]'. *History of Science*, 5(1966):104–17.

'Newton's marvellous year: 1666 and all that'. *Notes and Records of the Royal Society of London*, 21(1966):32–41.

'Sources and strengths of Newton's early mathematical thought'. *The Texas Quarterly*, 10(1967):69–85. Reprinted in *The Annus Mirabilis of Sir Isaac Newton 1666–1966*, ed. R. Palter (Cambridge, Mass./London: MIT Press, 1970), pp. 69–85.

The Mathematical Papers of Isaac Newton. Volume I: 1664–1666. With the assistance in publication of M. A. Hoskin. Cambridge University Press, 1967.

The Mathematical Works of Isaac Newton. Volume II. New York/London: Johnson Reprint Corporation, 1967.

'A face-lift for Newton: current facsimile reprints'. *History of Science*, 6(1967):59–68.

The Mathematical Papers of Isaac Newton. Volume II: 1667–1670. With the assistance in publication of M. A. Hoskin. Cambridge University Press, 1968.

The Mathematical Papers of Isaac Newton. Volume III: 1670–1673. With the assistance in publication of M. A. Hoskin and A. Prag. Cambridge University Press, 1969.

The Mathematical Principles underlying Newton's 'Principia Mathematica'. Being the ninth Gibson Lecture in the History of Mathematics delivered within the University of Glasgow on 21st October 1969. University of Glasgow, 1970.

'Before the *Principia*: The maturing of Newton's thoughts on dynamical astronomy, 1664–1684'. *Journal for the History of Astronomy*, 1(1970):5–19.

'The mathematical principles underlying Newton's *Principia Mathematica*'. *Journal for the History of Astronomy*, 1(1970):116–38.

'Barrow, Isaac'. *Dictionary of Scientific Biography.* Edited by Charles Coulston Gillispie. 16 vols. New York: Scribner's, 1970–80. Vol. 1, pp. 473–6.

The Mathematical Papers of Isaac Newton. Volume IV: 1674–1684. With the assistance in publication of M. A. Hoskin and A. Prag. Cambridge University Press, 1971.

'Collins, John'. *Dictionary of Scientific Biography*, 3(1971):348–9.

The Mathematical Papers of Isaac Newton. Volume V: 1683–1684. With the assistance in publication of M. A. Hoskin and A. Prag. Cambridge University Press, 1972.

'Gregory, David'. *Dictionary of Scientific Biography*, 5(1972):520–2.

'Gregory, James'. *Dictionary of Scientific Biography*, 5(1972):524–30.

'Newton's mathematical method'. *Bulletin of the Institute of Mathematics and its Applications*, 8(1972):173–8.

The Unpublished First Version of Isaac Newton's Cambridge Lectures on Optics, 1670–1672: a Facsimile of the Autograph, now Cambridge University Library MS Add. 4002. Cambridge: University Library, 1973.

The Mathematical Papers of Isaac Newton. Volume VI: 1684–1691. With the assistance in publication of M. A. Hoskin and A. Prag. Cambridge University Press, 1974.

'Keplerian planetary eggs, laid and unlaid, 1600–1605'. *Journal for the History of Astronomy*, 5(1974):1–21.

'Mercator, Nicolaus'. *Dictionary of Scientific Biography*, 9(1974):310–12.

[Review of *Internationales Kepler-Symposium Weil der Stadt 1971. Referate und Diskussion*, eds. F. Krafft, K. Meyer, and B. Sticker. (Hildesheim: H. A. Gerstenberg, 1973)]. *Studies in History and Philosophy of Science*, 4(1974):387–92.

'Astronomical eggs – laid and unlaid – in Kepler's planetary theories, 1600–1605'. *Vistas in Astronomy*, 18(1975):553–5.

'In search of Thomas Harriot [Review of *Thomas Harriot: Renaissance Scientist*, ed. J. W. Shirley (Oxford: Clarendon Press, 1974)]'. *History of Science*, 13(1975):61–70.

'Newton's lunar theory: from high hope to disenchantment'. *Vistas in Astronomy*, 19(1976):317–28.

The Mathematical Papers of Isaac Newton, Volume VII: 1691–1695. With the assistance in publication of A. Prag. Cambridge University Press, 1976.

[Review of *The Correspondence of Isaac Newton. Volume V: 1709–1713*, ed. A. R. Hall and L. Tilling (Cambridge University Press, 1975)]. *Annals of Science*, 33(1976):319–22.

[Review of *The Gresham Lectures of John Flamsteed*, ed. Eric G. Forbes (London: Mansell, 1975)]. *Annals of Science*, 33(1976):417–20.

[Review of *The Gresham Lectures of John Flamsteed*, ed. Eric G. Forbes (London: Mansell, 1975)]. *Archives internationales d'histoire des sciences* 27(1977):317–20.

'From his claw the Greene Lyon [Review of B. J. T. Dobbs, *The Foundations of Newton's Alchemy: or "The Hunting of the Greene Lyon"* (Cambridge University Press, 1975)]'. *Isis*, 68(1977):116–21.

[Review of *The Correspondence of Isaac Newton. Volume VII: 1718–1727*, ed. A. R. Hall and L. Tilling (Cambridge University Press, 1977)]. *Annals of Science*, 36(1979):539–42.

'Kepler, Newton and Flamsteed on refraction through a "regular aire": the mathematical and the practical'. *Centaurus*, 24(1980):288–315.

The Mathematical Papers of Isaac Newton. Volume VIII: 1697–1722. With the assistance in publication of A. Prag. Cambridge University Press, 1981.

'Newton the mathematician'. *Contemporary Newtonian Research*. Edited by Zev Bechler. Dordrecht/Boston/London: Reidel, 1982, pp. 109–27.

'Newtonian motion [Review of Richard S. Westfall, *Never at Rest: A Biography of Isaac Newton* (Cambridge University Press, 1980)]'. *Isis*, 73(1982):100–7.

The Making of Newton's 'Principia', 1664–1687. An exhibition in the University Library [Cambridge]. June/July 1987. Cambridge: University Library, 1987.

The Preliminary Manuscripts for Isaac Newton's 1687 'Principia' 1684–1685. Facsimiles of the Original Autographs, now in Cambridge University Library, with an Introduction. Cambridge University Press, 1989.

'The latest on Newton...'. *Notes and Records of the Royal Society of London*, 44(1990):111–17.

'The prehistory of the "Principia" from 1664 to 1686'. *Notes and Records of the Royal Society of London*, 45(1991):11–61.

[This bibliography does not include shorter reviews.]

I

Mathematics and astronomy before Newton

1 Lunar velocity in the Ptolemaic tradition

BERNARD R. GOLDSTEIN

Acknowledgements

Juliane Lay, John Roche, John North, and José Luis Mancha kindly supplied me with readings from manuscripts in Paris and Oxford. Dr Mancha, Dr Alan C. Bowen, and Dr José Chabás read drafts of this paper, and I have greatly benefited from their comments.

Introduction

In ancient and medieval astronomy lunar velocity is used to compute the duration of eclipses and the time from mean syzygy to true syzygy.[1] In Ptolemy's *Almagest*, a procedure is given for computing 'instantaneous' velocity without justification, based on the simple lunar model. A different rule is given by Regiomontanus (d. 1476), also without justification, that seems to reflect a recognition of the effect of the second lunar inequality, known as the evection, on the 'instantaneous' lunar velocity at syzygy. It is perhaps curious that, although instantaneous velocity could not be defined in this period, some astronomers could compute it correctly (i.e., according to the models for lunar motions they held to be true). Indeed, even the concept of uniform velocity was problematic.

Thus, for example, in Archimedes' *On Spirals* uniform velocity is described in terms of the proportionality of ratios formed between like quantities (e.g., length to length, or time to time). This appears in the first two propositions: Prop. (1) 'If a point moves at a uniform rate along a line, and two lengths be taken on it, they will be proportional to the times describing it'; and Prop. (2) 'If each of two points on different lines respectively move along them at a uniform rate, and if lengths be taken, one on each line, forming pairs, such that each pair are described in equal times, the lengths will be proportional'.[2] In other words, a ratio of length to time (or an angle to time) was not considered legitimate in ancient and medieval mathematics.[3]

---

[1] The term 'syzygy' is used for conjunction, or opposition, of the Sun and the Moon in celestial longitude.

[2] T. L. Heath (trans.), *The Works of Archimedes*, Cambridge University Press, 1897 (Dover reprint: n.d.), p. 155.

[3] Cf. E. J. Dijksterhuis, *The Mechanization of the World Picture*, Oxford University Press, 1961, pp. 191–2.

3

It is possible that Regiomontanus' rule was used to compute lunar velocity tables before he stated it; yet, no earlier source has been found in which this procedure is discussed.[4] In this paper we shall consider the rule for computing lunar velocity according to Ptolemy (*fl. ca.* 150) based on his simple lunar model (note that he gives no table of lunar velocity),[5] the table for lunar velocity by al-Battānī (d. 929) computed from Ptolemy's simple lunar model,[6] the two tables for lunar velocity by Levi ben Gerson (d. 1344),[7] and some of the tables for lunar velocity in the Alfonsine corpus (composed in the fourteenth century),[8] as well as the rule stated by Regiomontanus.[9]

[4] Dr J. L. Mancha informs me that in one of the manuscripts with a lunar velocity table ascribed to John of Genoa, we find 'Hic sciendum quod in tabula motus lune est computatum illud quod contingit lune propter equationem centri' (Paris, Bibliothèque Nationale, abbr. BN, lat. 7282, 129v), that is, 'One should understand here that what has been computed in this table for the motion of the Moon is that which affects the Moon due to the equation of center'. A similar remark appears in a manuscript containing John of Montfort's version of this lunar velocity table: Paris, BN lat. 7283, 44v. The expression 'equation of center' for the Moon in the Alfonsine tables refers to the entries in column 2 of the lunar correction table (whose argument is the double elongation of the Moon from the Sun) that does not have any role in Ptolemy's simple lunar model. These passages suggest that those who computed these lunar velocity tables knew what they were doing, but they do not reveal the mathematical argument that led to the method of calculation. For a discussion of two alternative explanations for the method of calculation, see Section IV.

[5] Ptolemy's simple lunar model is described in the *Almagest* IV. For a discussion of Ptolemy's lunar models see O. Neugebauer, *The Exact Sciences in Antiquity* (Providence: Brown University Press, 1957), pp. 192–9; and O. Pedersen, *A Survey of the Almagest* (Odense: University Press, 1974), pp. 159–202.

[6] Al-Battānī's table of lunar velocity appears in C. A. Nallino, *Al-Battānī sive Albatenii Opus Astronomicum*. 3 vols. (Milan 1899–1907), vol. 2, p. 88. For a discussion of the way in which it was computed, see B. R. Goldstein, *The Astronomical Tables of Levi ben Gerson* (New Haven: Connecticut Academy of Arts and Sciences, 1974), pp. 108–13. For the presence of this table among the Toledan tables, see G. J. Toomer, 'A survey of the Toledan tables', *Osiris*, 15 (1968), pp. 5–174, on p. 86.

[7] See Goldstein, *Astronomical Tables*, Table 22; see also B. R. Goldstein, *The Astronomy of Levi ben Gerson (1288–1344)*, Studies in the History of the History of Mathematics and the Physical Sciences, 11 (New York, Berlin: Springer-Verlag, 1985).

[8] See E. Poulle, *Les Tables Alphonsines* (Paris: Editions du CNRS, 1984), pp. 210–13; cf. E. Poulle, 'The Alfonsine tables and Alfonso X of Castille', *Journal for the History of Astronomy*, 19 (1988), pp. 97–113. There are three versions of a table for lunar velocity that differ from al-Battānī's table, and they are ascribed to John of Lignères (*fl. ca.* 1330), John of Genoa, and John of Montfort. See E. Poulle, 'John of Lignères', *Dictionary of Scientific Biography*, 7 (1973), pp. 122–8. Little is known about John of Genoa and John of Montfort, although it is clear that they are somehow associated with the Parisian astronomers in the 1330s who compiled the Latin version of the Alfonsine tables (see P. Duhem, *Le Système du monde* [Paris: Hermann, 1954],

I

Ptolemy's procedure for computing the lunar velocity at syzygy (in degrees per hour) for a given lunar anomaly, α, expressed algebraically, is the following:[10]

$$v(\alpha) = 0;32,56 + 0;32,40 \cdot \Delta \qquad (1)$$

where $0;32,56°/^h$ is the hourly mean motion of the Moon in longitude ($\bar{v}$), $0;32,40°/^h$ is the hourly mean motion in anomaly ($\bar{v}_\alpha$), and Δ is the difference in the correction for anomaly, c, in the simple model from α to $\alpha + 1°$:

$$\Delta = c(\alpha + 1°) - c(\alpha) \qquad (2)$$

where α is an integer.[11] This yields reasonably good agreement with al-Battānī's table for lunar velocity.[12] Toomer comments that, since (as Ptolemy himself states) there is no effect of Ptolemy's second lunar inequality on the lunar position at syzygy, the second inequality can be ignored in computing the velocity there; but this statement is clearly wrong.[13] When the value of a trigonometric function is 0, its derivative is often 1: hence, we should expect the maximum effect on velocity when the second lunar inequality is 0°. Pedersen demonstrated (using partial derivatives: see Appendix 1) that the second lunar inequality may not be ignored and showed that of the three terms that define the lunar velocity according to Ptolemy's complete lunar model [see eq. (a1)], the first is identical to the first term in eq. (1), the second vanishes at syzygy, and the third term corresponds to the second term in eq. (1). However, in the second term in eq. (1) Ptolemy used the wrong value for the mean anomaly by disregarding the effect of the second inequality. As we shall see, Regiomontanus computed what Pedersen proved for the second term in eq.

vol. 4, pp. 74–5). As Dr Mancha informs me, John of Montfort's tables are dated January 1332 (MS Paris, BN lat. 7283, 44r) and, surprisingly, John of Genoa's tables are also dated 1332 in the same MS (Paris, BN lat. 7283, 45r). For the presence of al-Battānī's table of lunar velocity in the Alfonsine corpus, see Poulle, *Les Tables Alphonsines*, p. 210. A variant on this table is also found in the Alfonsine corpus: see B. R. Goldstein, 'Solar and lunar velocities in the Alfonsine tables', *Historia Mathematica*, 7 (1980), pp. 134–40.

9   Regiomontanus, *Epitome of the Almagest*: *Epytoma Joannis de Monte Regio in Almagestum Ptolomei* (Venice: J. Hamman-Hertzog, 1496); reprinted in Regiomontanus, *Opera Collectanea*, ed. F. Schmeidler (Osnabrück 1972). The rule for computing lunar velocity occurs in VI, 4.

10   Almagest VI,4 (G. J. Toomer [trans.], *Ptolemy's Almagest*, Berlin, New York: Springer-Verlag, 1984, p. 282).

11   Note that the values for $c(\alpha)$ are tabulated at degree intervals in al-Battānī's tables, ed. Nallino, 1907, vol. 2, pp. 78–83, col. 3.

12   Ed. Nallino, *al-Battānī*, vol. 2, p. 88.

13   See Toomer, *Almagest*, p. 282 n15 where he cites the discussion in Pedersen 1974, p. 226.

(1), above, and we suggest that Regiomontanus (or his predecessor) formulated this rule for computing lunar velocity at syzygy based on his understanding of Ptolemy's procedure.

To justify Ptolemy's rule (i.e., ignoring the second lunar inequality), let us consider λ, the lunar longitude, and $\bar{\lambda}$, the mean lunar longitude, where:

$$\lambda = \bar{\lambda} + c. \tag{3}$$

For λ_1 and λ_2 this means that

$$\lambda_1 = \bar{\lambda}_1 + c_1$$

and

$$\lambda_2 = \bar{\lambda}_2 + c_2.$$

Thus,

$$\Delta\lambda = \lambda_2 - \lambda_1 = \Delta\bar{\lambda} + \Delta c$$

and

$$v = \frac{\Delta\lambda}{\Delta t} = \frac{\Delta\bar{\lambda}}{\Delta t} + \frac{\Delta c}{\Delta t}. \tag{4}$$

Now

$$\Delta\alpha = \bar{v}_\alpha \cdot \Delta t$$

or

$$\Delta t = \Delta\alpha / \bar{v}_\alpha,$$

and

$$\frac{\Delta\bar{\lambda}}{\Delta t} = \bar{v}.$$

Hence, eq. (4) becomes

$$v = \bar{v} + \frac{\Delta c}{\Delta\alpha / \bar{v}_\alpha}$$

or

$$v = \bar{v} + \bar{v}_\alpha \cdot \Delta c, \tag{5}$$

where

$$\Delta\alpha = 1° \tag{6}$$

as in eq. (2), above. Equation (5) corresponds to Ptolemy's instructions for computing the lunar velocity (see eq. (1), above: $\Delta c = \Delta$ when $\Delta\alpha = 1°$); hence, given his assumption that the second inequality can be ignored, his formula is correct. Note that eq. (5) can also be obtained from eq. (4) by using the relation:

$$\frac{\Delta c}{\Delta t} = \frac{\Delta c}{\Delta\alpha} \cdot \frac{\Delta\alpha}{\Delta t}$$

which, with the condition in eq. (6), leads directly to eq. (5) because

$$\bar{v}_\alpha = \frac{\Delta\alpha}{\Delta t}.$$

6

Table 1. *Lunar velocity with al-Battānī's lunar corrections*

I argument (°)	II Δ (°)	III v: comp. (°/h)	IV v: text (°/h)	V v: comp. (°/h)
0	−0; 4,50	0;30,18	0;30,18	0;29,34
30	−0; 4,18	0;30,36	0;30,35	0;29,56
60	−0; 2,47	0;31,25	0;31,25	0;31, 0
90	−0; 0,24	0;32,43	0;32,41	0;32,39
120	0; 2,24	0;34,14	0;34,14	0;34,36
150	0; 4,47	0;35,32	0;35,31	0;36,16
180	0; 5,45	0;36, 4	0;36, 4	0;36,56

The entries in cols. II and IV have been derived from al-Battānī's tables. The entries in col III have been computed from the entries in col. II according to Ptolemy's rule, whereas the entries in col. V have been computed from the entries in col. II according to Regiomontanus' rule.

Thus, in eq. (4)

$$v = \frac{\Delta\bar{\lambda}}{\Delta t} + \frac{\Delta c}{\Delta t}$$

or

$$v = \bar{v} + \frac{\Delta c}{\Delta\alpha} \cdot \frac{\Delta\alpha}{\Delta t}$$

$$= \bar{v} + \bar{v}_\alpha \cdot \Delta c,$$

where

$$\Delta\alpha = 1°.$$

In Table 1, col. II, we display al-Battānī's values for Δ; in col. III the lunar velocities computed from them according to eq. (1); and in col. IV the entries in al-Battānī's table for lunar velocity. The agreement is very good [i.e., T(ext) − C(omputation) = 0″ or −1″] except for α = 90° (where T − C = −2″). In order to get the computed velocity equal to the velocity in the text for 90°, the value for Δ would have to be −0;0,28° instead of −0;0,24°, but with −0;0,25° the velocity would be 0;32,42 which is only 1″ greater than the value in the text (this would require that c(91°) = −5;0,27° instead of −5;0,26°, the value that appears in al-Battānī's table). Another method proposed for computing these values (see Appendix 2) yields exact agreement for α = 90° (as well as for α = 0° and α = 180°) and nowhere do the differences between text and computation exceed 1″. Since al-Battānī does not say how the entries in his table were computed, we cannot determine his method with certainty. But, if

7

we take agreement with recomputation as a guide, it seems slightly more likely that he used the method described in Appendix 2 according to which the progress of the Moon is computed for an interval of an hour.

II

According to Regiomontanus' *Epitome of the Almagest* VI,4, the lunar velocity at syzygy may be computed from the following formula:

$$v(\alpha') = 0;32,56 + 0;41,49 \cdot \Delta. \tag{7}$$

This formula differs from eq. (1) in significant ways. Firstly, the values $0;32,56°/h$ and $0;41,49°/h$ represent the Moon's hourly mean motion in longitude ($\bar{v}$), and the corrected hourly mean motion in anomaly at syzygy (v_α), respectively. Secondly, the argument is the true lunar anomaly, α', that appears in Ptolemy's complete lunar model, rather than the mean anomaly, α. The expression, Δ, is defined as before in eq. (2). We can express eq. (7) more generally as

$$v(\alpha') = \bar{v} + v_\alpha \cdot \Delta c \tag{8}$$

which has the advantage that we can substitute other values for the basic parameters.

It has been suggested that v_α, the corrected hourly mean motion in anomaly at syzygy in Regiomontanus' formula, was computed as follows:[14]

$$v_\alpha = 0;32,40 + 1;1 \cdot 0;9 = 0;41,49 \tag{9}$$

where $0;32,40°/h$ is the hourly mean motion in anomaly ($\bar{v}_\alpha$), $1;1°/h$ is the hourly mean motion in double elongation ($\bar{v}_{2\eta}$), and $0;9°$ is the difference between c' (the correction to the mean anomaly found in column 4 of al-Battānī's lunar correction tables) for arguments of double elongation $0°$ and $1°$. Again, we can express (9) more generally as

$$v_\alpha = \bar{v}_\alpha + \bar{v}_{2\eta} \cdot \Delta c' \tag{10}$$

where $\Delta c'$ is restricted as stated above.

The parameter derived in (9) can be justified by a procedure similar to the one we invoked to justify eq. (1). We define

$$\alpha' = \alpha + c'.$$

[14]   See N. M. Swerdlow and O. Neugebauer, *Mathematical Astronomy in Copernicus's De Revolutionibus*, Studies in the History of Mathematics and Physical Sciences, 10 (New York, Berlin: Springer-Verlag, 1984), pp. 274–6.

Thus,

$$\Delta\alpha' = \Delta\alpha + \Delta c'$$

and

$$v_\alpha = \bar{v}_\alpha + \frac{\Delta c'}{\Delta t}.$$

Now

$$\bar{v}_{2\eta} = \Delta(2\eta)/\Delta t$$

or

$$\Delta t = \Delta(2\eta)/\bar{v}_{2\eta}.$$

Hence,

$$v_\alpha = \bar{v}_\alpha + \bar{v}_{2\eta} \cdot \Delta c' \tag{11}$$

where

$$\Delta(2\eta) = 1°. \tag{12}$$

Let us compute the hourly lunar velocity (as suggested by Regiomontanus) by the formula in eq. (7). Using the table for corrections due to the lunar anomaly for Ptolemy's simple model in al-Battānī's tables, we compute the velocities displayed in Table 1, col. V, that are certainly different from the velocities based on eq. (1) displayed in Table 1, col. III.

III

We will now investigate the evidence for the use of Ptolemy's complete lunar model for computing lunar velocity in the fourteenth century, more than 100 years before Regiomontanus wrote his *Epitome*, restricting our attention to the two lunar velocity tables in the astronomical works of Levi ben Gerson and some of the lunar velocity tables in the corpus of tables associated with the name of Alfonso, king of Castile (reigned: 1252–84).

It has been argued that one of Levi's lunar velocity tables was computed in the same way as al-Battānī's lunar velocity table except that Levi used his own lunar model at syzygy rather than Ptolemy's model.[15] As a result, there are some small differences between their respective tables. Levi's other lunar velocity table is more puzzling. There seems to be no indication in the text of his *Astronomy* that he changed his mind on the method of computing these velocities, yet it is also clear that no simple lunar model could account for the entries in this table. In 1974 I proposed a 'skew-equant' model to account for this table of lunar velocity.[16] This 'skew-equant' model adequately accounted

---

[15]   See Goldstein, *Astronomical Tables*, pp. 108–13.

[16]   In this 'skew-equant' model the distances from the center of the deferent to the equant and to the observer are not equal: the distance from the equant to the center of the deferent was set equal to 1;22 and the distance from the center of the deferent to the observer was 5;20 (see Goldstein, *Astronomical Tables*, pp. 114–15).

Table 2. *Lunar velocity according to Levi ben Gerson*

I argument (°)	II Levi Δ (°)	III 'Regiom.'/Levi v: comp. (°/h)	IV Levi v: text (°/h)	V 'Regiom.'/al-B. v: comp. (°/h)
0	−0; 4,50	0;29,34	0;29,35	0;29,34
30	−0; 4,20	0;29,55	0;29,57	0;29,56
60	−0; 2,53	0;30,55	0;31, 0	0;31, 0
90	−0; 0,24	0;32,39	0;32,40	0;32,39
120	0; 2,31	0;34,41	0;34,36	0;34,36
150	0; 4,52	0;36,20	0;36,15	0;36,16
180	0; 5,45	0;36,56	0;36,56	0;36,56

The entries in col. II have been derived from Levi's tables (ed. Goldstein, table 35, col. V). The entries in col. III have been computed from those in col. II according to Regiomontanus' rule. The entries in col. IV are taken from Levi's second table of lunar velocity (ed. Goldstein, table 22, col. IV). For purposes of comparison, we display entries in col. V (copied from Table 1, col. V) that are computed according to Regiomontanus' rule based on al-Battānī's correction table.

for the velocities but it had the unfortunate consequence that the maximum lunar correction at syzygy would reach 6;24° instead of Ptolemy's 5;1°. However, it now seems that Levi may not have had recourse to a new model at all, but that he was taking into account the second lunar inequality (represented in Ptolemy's complete lunar model as well as in Levi's complete lunar model). If Levi used Regiomontanus' rule, we should then determine whether he used al-Battānī's values for Δ, or the values in his own table of lunar corrections.[17]

When we compute the values for Δ from Levi's table for lunar anomaly according to his own lunar model for syzygy and apply them in eq. (7), we find the results displayed in Table 2.[18]

It is clear that Levi's table for lunar velocities shows better agreement with the velocities based on al-Battānī's table for the lunar correction (Table 2, col. V, which is copied from Table 1, col. V, above) than with those in Table 2 based on the entries in Levi's own table for the lunar correction ($|T-C| \leqslant 1''$ vs. $|T-C| \leqslant 5''$). Moreover, this method yields better agreement than the

[17] Note that there is a scribal error in the entry for 61° of anomaly in al-Battānī's table: read 4;11,53° instead of 4:11,33°. In the Hebrew version of al-Battānī's tables by Abraham Bar Ḥiyya (twelfth century), we find 4;11,53° (cf. MS Paris, BN heb. 1038, 36b) and this is the value we used in this computation.

[18] For Levi's table of lunar anomaly, see Goldstein, *Astronomical Tables*, Table 35, col. V.

method based on the 'skew-equant' model where $|T-C| \leqslant 3''$. As was true for al-Battānī, Levi does not explain how he computed the entries in his table of velocities. But using the criterion of agreement with recomputation, it would seem that he computed his table with al-Battānī's corrections for lunar anomaly and the formula described by Regiomontanus.

We now turn to the tables of lunar velocity in the corpus of the Alfonsine tables, most of which have not yet been examined in detail:[19]

(i)   in a table at degree intervals in MS Oxford, Bodlian Library, abbr. Bodl., Canon Misc. 499, 41v–42r (ascribed to John of Lignères in a later hand),[20] the lunar velocity ranges from $0;29,37°^{/h}$ to $0;36,53°^{/h}$ (see Table 3, col. IV);

(ii) in a table at 6° intervals ascribed to John of Genoa (MS Paris, Bibliothèque Nationale, abbr. BN, lat. 7282, 129r–129v)[21] but that might belong to John of Lignères (cf. MS Wolfenbüttel 2401, f. 311) the lunar velocity ranges from of $0;29,37,13°^{/h}$ to $0;36,58,54°^{/h}$, and from $0;11,50,53°^{/mn}$ to $0;14,47,33°^{/mn}$, where the unit ($°^{/mn}$) is degrees per sixtieth of a day (see Table 3, cols. V and VI);

(iii) in a table at degree intervals ascribed to John of Montfort (MSS Paris, BN lat. 7283, 43r–44r, and Oxford, Bodl. Canon. Misc. 499, 151v–152v),[22] the lunar velocities range from $0;11,51,9,11°^{/mn}$ to $0;14,47,8°^{/mn}$ (see Table 3, col. VII), corresponding to a minimum of $0;29,37,52,57°^{/h}$ and a maximum of $0;36,51,57,50°^{/h}$.

[19] I wish to thank Dr Donald W. Olsen for drawing my attention to the passage in Poulle, *Les Tables Alphonsines*, pp. 210–11, where these lunar velocity tables are listed, and for supplying me with some preliminary calculations based on a modern analysis of Ptolemy's lunar models showing the effect of the second inequality on the lunar velocity.

[20] I am grateful to Dr John Roche and to Dr Mancha for transcribing this table for me. Dr Mancha also collated this table with another copy that appears on folios 154v–155r in the same manuscript (there were few variants). Note that a second foliation of this manuscript increases by ten the folio numbers used here. This table also appeared in a printed edition: *Tabulae Astronomicae, quas vulgo, ..., Resolutas vocant...* per Ioannem Schonerum (Norimbergae apud Io. Petreium 1536). The entry for 150° is the correct value, 0;36,14, in the edition I consulted: *Opera Mathematica* Ioannis Schoneri (Norinbergae in officina I. Montani & U. Neuberi, 1551), part III, 41r.

[21] I am grateful to Mme Juliane Lay for transcribing the two relevant columns from this MS and from MS Paris, BN lat. 7295A, 137r, I also wish to thank Dr Mancha for providing me with a collation of the hourly lunar velocity table in this MS with the corresponding tables in MSS Paris, BN lat. 7286C, 56v; 7432, 255r; and 7286A, 117r (trivial variants in these MSS have not been recorded in the notes to Table 3). Cf. Goldstein, 'Alfonsine tables', p. 139, where the same maximum and minimum values were reported, based on MS Paris, BN lat. 7295A, 137r.

[22] I am again grateful to Mme Lay for transcribing the Paris copy of this table for me.

Table 3. *A comparison of lunar velocities in the Alfonsine corpus*

I argument (°)	II Δ (°)	III v: comp. (°/h)	IV J. Lignères (°/h)	V John of Genoa (°/h)	VI John of Genoa (°/mn)	VII John of Montfort (°/mn)	VIII v: comp. (°/mn)
0	−0; 4,46	0;29,37,13	—	0;29,37,13	0;11,50,53	—	0;11,50,53
1	4,45	29,37,55	0;29,37	—		0;11,51, 9,11*	11,52,10
6	4,43	29,39,18	29,39	29,39,19	11,51,43	11,51,42,38	11,51,43
12	4,39	29,42, 5	29,41	29,42, 6	11,52,50	11,52,49,31	11,52,50
18	4,33	29,46,16	29,45	29,46,17	11,54,31	11,54,29,51	11,54,30
24	4,25	29,51,51	29,51	29,51,51	11,56,44	11,56,43,38	11,56,44
30	−0; 4,14	0;29,59,30	0;29,59	0;29,59,31	0;11,59,48	0;11,59,47,35	0;11,59,48
36	4, 1	30, 8,34	30, 9	30, 8,34	12, 3,25	12, 3,24,58	12, 3,25
42	3,45	30,19,43	30,20	30,19,43	12, 7,53	12, 7,52,31	12, 7,53
48	3,27	30,32,15	30,32	30,32,15	12,12,54	12,12,53,31	12,12,54
54	3, 6	30,46,53	30,47	30,46,53	12,18,45	12,18,44,41	12,18,45
60	−0; 2,43	0;31, 2,54	0;31, 3	0;31, 2,55	0;12,25,10	0;12,25, 9,18	0;12,25,10
66	2,20	31,18,56	31,19	31,18,55	12,31,34	12,31,33,55	12,31,34
72	1,53	31,37,44	31,38	31,37,45	12,39, 6	12,39, 5,25	12,39, 6
78	1,25	31,57,15	31,57	31,57,15	12,46,54	12,46,53,38	12,46,54
84	0,57	32,16,45	32,17	32,16,45	12,54,42	12,54,41,51	12,54,42
90	−0; 0,24	0;32,39,44	0;32,40	0;32,39,45	0;13, 3,54	0;13, 3,53,41	0;13, 3,54
96	+0; 0,13	33, 5,31	33, 5	33, 8,31*	13,14,13	13,14,12,25	13,14,12
102	0,39	33,23,38	33,24	33,30,36*	13,21,28	13,21,27,11	13,21,27
108	1,16	33,49,24	33,49	33,49,21*	13,31,46	13,31,45,55	13,31,46
114	1,46	34,10,18	34,10	34,10,19	13,40, 8	13,40, 7,35	13,40, 7
120	0; 2,22	0;34,35,23	0;34,36	0;34,35,24	0;13,50,10	0;13,50, 9,35	0;13,50, 9

126	2,55	34,58,23	34,58	13,59,58*	13,59,21,25	13,59,21
132	3,24	35,18,35	35,19	35,18,25*	14, 7,26,21	14, 7,25
138	3,53	35,38,47	35,39	35,38,47	14,15,31,18	14,15,31
144	4,20	35,57,36	35,58	35,57,36	14,23, 2,48	14,23, 2
150	0; 4,43	0;36,13,37	0;36,18*	0;36,13,37	0;14,30, 0,51*	0;14,29,27
156	5, 4	36,28,15	36,28	36,28,15	14,35,18,35	14,35,18
162	5,18	36,38, 0	36,38	36,38, 0	14,39,12,41	14,39,12
168	5,30	36,46,22	36,46	36,46,22	14,42,33,21	14,42,33
174	5,37	35,51,14	36,51	36,53,15*	14,44,30,25	14,44,30
180	+0; 5,40	0;36,53,20	0;36,53	0;36,58,54*	0;14,44,47, 8*	0;14,45,20

The entries in col. IV have been copied from MS Oxford, Canon Misc. 499, 41v–42r; in cols. V and VI from MS Paris, BN lat. 7282, 129r–129v; and in col. VII from MS Paris, BN lat. 7283, 43r–44r.

Col. IV, 150. Read: 0;36,**14**.
Col. V, 96. Read: 33,**5**,31 (with MS Paris, BN lat. 7286A).
Col. V, 102. Read: 33,**23**,37 (with MS Paris, BN lat. 7286A).
Col. V, 108. Read: 33,**49**,35 (with MSS Paris, BN lat. 7432 and 7295A).
Col. V, 132. Read: 35,**18**,35 (with MSS Paris, BN lat. 7432 and 7295A).
Col. V, 174. See the discussion at the end of Section III.
Col. V, 180. See the discussion at the end of Section III.
Col. VI, 126. Read: 13,59,**21**.
Col. VI, 132. Read: 14,**7,27**.
Col. VI, 174. See the discussion at the end of Section III.
Col. VI, 180. See the discussion at the end of Section III.
Col. VII, 1. See the discussion at the end of Section III.
Col. VII, 150. The discrepancy here seems to be due to the computation and not to a copyist's error.
Col. VII, 180. See the discussion at the end of Section III.

Since c (and hence Δ) is only given to seconds in the Alfonsine tables, the introduction of sexagesimal thirds and fourths in tables of lunar velocity is unwarranted. In Table 3, col. II, we display the values for Δ computed from the entries for the lunar corrections in the Alfonsine tables.[23] To compute the velocities according to the Alfonsine tables, we first compute a value for v_α in eq. (10), above, based on the Alfonsine parameters: $\bar{v}_\alpha = 0;32,39,44,54°/h$, $\bar{v}_{2\eta} = 1;0,57,13,28°/h$, and $\Delta c' = 0;9°$.[24] Thus,

$$v_\alpha = \bar{v}_\alpha + \bar{v}_{2\eta} \cdot \Delta c'$$

$$= 0;32,39,44,54 + 1;0,57,13,28 \cdot 0;9 = 0;41,48,19,55$$

$$\approx 0;41,48.$$

Then we compute v from eq. (8), above, where $\bar{v} = 0;32,56,27,33°/h$:[25] the results in degrees per hour are displayed in Table 3, col. III, and the corresponding entries in degrees per sixtieth of a day are displayed in col. VIII.

It is clear that the agreement of the computed values with the entries in all versions of these Alfonsine lunar velocity tables is generally quite good, and that there are only a few problematic cases (these are marked with an asterisk in Table 3). Unfortunately, these problematic cases affect the minimum and maximum values that one might well consider as characterizing these tables. Moreover, while it is clear that the three versions of these tables are very similar, their exact relationship has not yet been determined.

In Table 3, col. IV, a problem occurs at 150°, but this is an isolated error in one copy only; otherwise $|T-C| \leqslant 1''$. In cols. V and VI, the entries in the text corresponding to 174° and 180° agree with each other, but not with recomputation: I have no explanation to offer for this.[26] In col. VII, there is excellent agreement with the recomputed values (i.e. $|T-C| \leqslant 1'''$, where the entries in the table are rounded to thirds) except for 150° and at the beginning and end of the table: there is certainly confusion near 0° and 180° (in fact, near minimum the entries for 1° and 3° are identical and the entries for 2°, 4° and 5° are all 0;11,51,25,55, while near maximum the entries for 174° to 176° are all identical as are the entries for 177° to 180°). So, here the maximum and minimum values are less informative than one might expect.

[23] See Poulle, *Les Tables Alphonsines*, pp. 148–53, col. 5.

[24] See *ibid.*, pp. 136, 139.   [25] See *ibid.*, p. 135.

[26] A similar problem afflicts one of the other lunar velocity tables in the Alfonsine corpus: cf. Goldstein, 'Alfonsine tables', I am informed by Dr Mancha that in MS Paris, BN lat. 7286A, 117r, in a table for lunar velocity not ascribed to John of Genoa but significantly related to his table, the entry for 174° is 0;36,51,15, which is much closer to the recomputed value. It is possible that a scribal error occurred in the copy of the Alfonsine table of lunar anomaly used to compute these entries.

IV

We have argued that the second lunar inequality was used in computing lunar velocity by Levi ben Gerson. Moreover, despite the discrepancies among the three versions of lunar velocity tables in the Alfonsine corpus discussed above, the agreement found with recomputation is sufficient to allow us to conclude that the entries in all three versions also take into account the second lunar inequality. Since Levi ben Gerson and the Alfonsine astronomers were active in the 1330s, it is not possible to assign priority and, on the basis of sources currently available, one cannot say if they knew of each other's work.

From the preceding remarks, it would seem that Regiomontanus' rule was already known in the fourteenth century. Yet, in the absence of direct testimony, a note of caution should be added. Was there another way to compute these lunar velocities available at the time? We have already mentioned the usefulness of the method described in Appendix 2 for computing lunar velocities according to Ptolemy's simple lunar model. Obviously, this method can be modified to take into account any other lunar model. In fact, there is a fourteenth-century text which seems to tell us to do something very much like that: John of Saxony's canons for the Alfonsine tables, composed in 1327.[27] In chapter 22, concerning the time from mean syzygy to true syzygy, there is an instruction for computing lunar velocity as the progress of the Moon in 1/60 of a day (i.e., $0;24^h$).[28] John of Saxony does not specify which lunar model should be used for computing these positions 1/60 of a day apart but, in presenting a worked example, Poulle reasonably assumed that the complete lunar model was intended.[29] The same argument would hold for the lunar velocity tables of Levi ben Gerson as well.

Thus, we have two alternative methods for explaining the computation of lunar velocity tables in the fourteenth century, and more evidence may be needed to decide between them.[30]

[27] Poulle, *Les Tables Alphonsines*, p. 17.

[28] *Ibid.*, pp. 84–5. Dr Mancha informs me that an almost identical method is reported in John of Genoa's *Canones eclipsium* dated 1332 and preserved in at least 3 MSS (Paris, BN lat. 7322 and 7281; Oxford, Digby 97). John of Genoa gives rules for calculating positions of the Moon at an interval of 1 hour rather than 1/60 day as in John of Saxony's canons.

[29] Poulle, *Les Tables Alphonsines*, pp. 217–18.

[30] Recently, Dr Mancha has found evidence to suggest that the computation of lunar velocity as the progress of the Moon in a given time interval may go back at least to John of Sicily's commentary on the canons for the Toledan tables, composed *ca.* 1290 (see Paris, BN lat. 7281, ff. 46r–138r, esp. f. 111r), and this remains to be explored.

Appendix 1

Pederson presents the following argument (translated into our notation) to support his claim that the second lunar inequality cannot be ignored in computing the lunar velocity at syzygy.[31] Let

$$\lambda = \bar{\lambda} + e(2\eta, \alpha')$$

where e is the total correction (or 'equation') due to both the first and the second inequalities. Then

$$v = \bar{v} + \frac{de(2\eta, \alpha')}{dt}$$

or

$$v = \bar{v} + \frac{\partial e}{\partial(2\eta)} \cdot \bar{v}_{2\eta} + \frac{\partial e}{\partial\alpha'} \cdot v_{\alpha}. \qquad (a1)$$

At syzygy the second term vanishes. As for the third term,

$$v_{\alpha} = \bar{v}_{\alpha} + \frac{dc'}{dt}$$

or

$$v_{\alpha} = \bar{v}_{\alpha} + \frac{dc'}{d(2\eta)} \cdot \bar{v}_{2\eta}. \qquad (a2)$$

Therefore, $v_{\alpha} = \bar{v}_{\alpha}$ only when the second term in eq. (a2) is 0: this is true for Ptolemy's simple model, but not for his complete model (even at syzygy).

Appendix 2

We present the following method for computing the lunar velocity.[32] To find the hourly lunar velocity at anomaly α, let

$$\alpha_1 = \alpha - (\bar{v}_{\alpha}/2)$$

and

$$\alpha_2 = \alpha + (\bar{v}_{\alpha}/2)$$

i.e., α_1 is the anomaly at a half hour before α, and α_2 is the anomaly at a half hour after α. Then

$$\lambda_1 = \bar{\lambda}_1 + c(\alpha_1)$$

and

$$\lambda_2 = \bar{\lambda}_2 + c(\alpha_2).$$

[31]   Pedersen, *Almagest*, p. 226.   [32]   Goldstein, *Astronomical Tables*, pp. 111–12.

But

$$\bar{\lambda}_2 = \bar{\lambda}_1 + \bar{v}.$$

Hence,

$$\lambda_2 - \lambda_1 = \bar{v} + c(\alpha_2) - c(\alpha_1).$$

and this is the progress of the Moon in an hour which we assign to anomaly α. This method yields the same results, to sexagesimal thirds, as the precise determination by means of the modern formula for the instantaneous lunar velocity at anomaly α in Ptolemy's simple model; but it avoids introducing the concept of 'instantaneous' velocity.

2 Shadow measurement: the *Sciametria* from Kepler's *Hipparchus* – a translation with commentary

N. M. SWERDLOW

For many years Kepler intended to write a book called *Hipparchus, or On the Sizes and Distances of the Three Bodies of the Sun, the Moon, and the Earth.* The title was in honor of Hipparchus who had himself written a lost work on this subject referred to by Ptolemy and Pappus. The distances and sizes of the sun and moon were naturally of interest for cosmology, but Kepler's purpose was also practical, namely, for the determination of the parallaxes and apparent diameters of the sun and moon, and the apparent diameters of the shadows of the moon and earth, all required for the computation of eclipses. Indeed, the only section of the *Hipparchus* that he brought close to a finished state is a series of eighteen theorems and sixteen problems called *Sciametria* or 'Shadow measurement' concerned with the various relations between these quantities and their application to eclipses. The little tract is quite ingenious and contains some of Kepler's most elegant geometry although no philosophy. It is here translated with an introduction and commentary.

Kepler first expressed his intention of writing a book called *Hipparchus* in a brief peroration to the *Astronomiae pars optica* of 1604. He had just treated subjects from Books 4, 5, and 6 of Ptolemy's *Almagest* pertinent to optics, such as the apparent diameters of the sun and moon, the illumination and shadows of the earth and moon, parallax, and eclipses. If God prolong his life and strength, Kepler says, he will write another small book that may be called a second part or an appendix to the present work in which he will improve the three books of Ptolemy with ingenious and highly agreeable problems, and will show how to investigate the same subjects treated by Ptolemy more briefly and with fewer and easily comparable observations. This work is all the more necessary since the parts of Tycho Brahe's *Progymnasmata* concerning the *lunaria*, subjects related to the moon, appeared without the benefit of demonstrations. And because the principal object of the little book will be to investigate the sizes and distances of the sun, the moon, and the earth, the same topic that Hipparchus had examined in his own book of that title, as is clear from Theon, therefore that it may be fruitful and fortunate, let this little book be named *Hipparchus*.[1]

Alas, it was not to be fruitful or fortunate, and in Kepler's lifetime it was not to be at all. For twenty-five years he worked on the *Hipparchus* from time to time, wrote out a number of pages, including a title page, in something close to a finished copy, and often referred to the book in his publications and correspondence: After the *Astronomia nova* he promised (1607); but I progress very slowly (1608); Galileo's telescope will be of great assistance (1610); the theory of demonstrating the proportions of the sun, moon, and earth, although very ingenious, is very uncertain in application (1612); astronomers will patiently bear a short delay until either my *Epitome* or my *Hipparchus* is published (1617); I remember how much time I expended in vain, the problems are very beautiful, the work itself impassable (1619); after the *Rudolphine Tables* are published if God prolong my life and strength (1624); it will be the next publication if God grant me life (1627); something fatal happened to my *Hipparchus*, for among its theorems (although not yet published nor polished with final care) one is spurious (1629, the last reference).[2]

Fragments of the *Hipparchus* from Kepler's manuscripts were first published by Christian Frisch in 1860 and more completely by Volker Bialas and Friederike Boockmann as recently as 1988.[3] To judge from Kepler's references to the work and from the published fragments, the *Hipparchus* was to contain the *lunaria*, the subjects of Books 4–6 of the *Almagest* excepting lunar theory itself, although its methods could be applied to deriving the elements of lunar theory from observation. The principal subjects were to be everything pertinent to eclipses, that is, the apparent diameters of the sun and moon, the shadow and penumbra of the moon and earth, the variation of these quantities due to change of distance, the parallaxes and distances of the sun and moon, the true and apparent hourly motion of the moon, and the magnitudes and durations of eclipses. On Kepler's title page, under what appears to be the more general heading *Memoranda of the Restorations of the Lunaria*, the title is: *Many very*

[1] *Johannes Kepler Gesammelte Werke*, abbr: *KGW* (Munich: C. H. Beck, 1937–), vol. 2, p. 378. The reference to Theon is to a passage in what is now identified as Pappus's commentary on *Almagest* V (*Commentaires de Pappus et de Théon d'Alexandrie sur l'Almageste*, ed. A. Rome, *Studi e Testi* 54 (Rome, 1931), pp. 67–8) that has been translated and discussed in N. Swerdlow, 'Hipparchus on the distance of the sun', *Centaurus* 14(1969):287–305, and G. J. Toomer, 'Hipparchus on the distances of the sun and moon', *Archive for History of Exact Sciences* 14(1974):126–42.

[2] Quoted or paraphrased from the citations collected in *KGW*, vol. 20.1 (*Manuscripta Astronomica* I) pp. 551–6.

[3] *Joannis Kepleri Astronomi Opera Omnia*, abbr: *KOO* (Frankfurt and Erlangen, 1858–71), vol. 3, pp. 520–49 and a few fragments in other volumes; *KGW*, vol. 20.1, pp. 183–268.

beautiful demonstrations and a related book to which I have given the name Hipparchus, or On the Sizes and Distances of the Three Bodies of the Sun, the Moon, and the Earth. The work, he says, was begun many years ago in Prague and to a great extent worked on continuously in Linz, especially in the last year, 1616, which presumably means that the title page was written the following year.[4]

Kepler had begun to study these problems in the *Astronomia pars optica*, and indeed they form the principal subjects of chapters 6–11 in which some of Kepler's innovations in the theory of eclipses were first described, among them the penumbra and the circle of illumination of the earth and moon. But what remained to be done in the intended 'second part' or 'appendix' was a systematic study of the *lunaria* and the derivation of accurate numerical values from observation. The former Kepler was well on the way to accomplishing, at least in the *Sciametria*; the latter seemed ultimately to defeat him, as shown by the discouraging remarks quoted earlier about his difficulties and delays. Nevertheless, in the *Epitome astronomiae Copernicanae* and the *Rudolphine Tables* he explained some of the procedures from the *Hipparchus*, and the tables for the computation of eclipses were based upon parameters that he had found in some way, perhaps using some of those very procedures, if not to find the parameters, then to check their consistency with each other. Still, it was probably the obstacles in applying the 'very beautiful demonstrations' that prevented Kepler from ever completing the work, and it will be fairly obvious that, because of the difficulty of making sufficiently precise measurements, the *Sciametria* is better in principle than in practice.

The fragments presumed to belong to the *Hipparchus* are found principally in Volume 1 of Kepler's manuscripts, now in the Archive of the Academy of Sciences in St Petersburg. The largest sections are (with page numbers in *KGW* 20.1) first the *Sciametria* itself (185–205), then three numbered chapters on: I. The apparent diameters of the sun and moon found from solar eclipses (206–7); II. The lunar eccentricity at syzygy and the variation of the distance of the moon, considering Hipparchus, Ptolemy, Copernicus, and Tycho (207–12); III. The hourly motion of the moon (i.e. its true velocity or elongation from the sun) at syzygy examined both theoretically and observationally, with an extended consideration of the physical theory of the motion of the moon (212–27). A heading *Second Part of the Book, On the Theory of Eclipses* is followed by a Chapter XI *On the criticized method*, mostly a criticism of Ptolemy's procedures for finding the apparent diameters of the

---

[4]   *KGW*, vol. 20.1, p. 183. 'Haec Pragae inchoata a multis annis...Lincij verò magna parte continuata praesertim ultimo anno 1616.' An unnumbered Problem at the end of the *Sciametria* (*ibid.*, p. 204) uses lunar eclipses of 26 June and 20 December of 1620, and Theorem XII must date from after the publication of Books V–VII of the *Epitome astronomiae Copernicanae* in 1621.

moon and shadow and other quantities necessary to the theory of eclipses (253–9), followed by an outline of Kepler's method (259–60), which is presumably an outline of the intended contents of the *Hipparchus* pertinent to eclipses. There are smaller sections on: computation of parallax in the circle of altitude (234–5); an analemma for solar eclipses (236–9); variation of the length and semidiameter of the shadow of the earth under minimum and maximum assumed solar distances (240–2); parallax and distance of the moon found from ancient solar eclipses (246–8); various configurations of the moon's penumbra and shadow projected on to the disc of the earth (265–8, from vol. XX of Kepler's MSS.). In addition there are several lists of subjects Kepler has treated or will treat. A list of seven chapters (250), the fourth of which is '*Sciametria*, completed by six theorems', appears to concern chapters of the *Hipparchus*. A *Catalogue of Problems to be Proposed* (250–2), apparently dating from early 1601/02,[5] contains much that fits the subject matter of the *Hipparchus*. Far more comprehensive is the *Contents in the Memoranda of the Restorations of the Lunaria* (228–33), a guide to the contents of a manuscript of at least 280 folios on all aspects of lunar theory including those in the *Hipparchus*.[6] There is a list (260–1), divided into three parts, of quantities that may be discovered by observation (it appears of eclipses), that are given by theory or computatión, and that are to be found. Finally, there are a number of other fragments that I really do not know how to describe, including a short list of corrections or additions to the *Sciametria* (261).

The *Sciametria*

The title *Sciametria* means 'Shadow measurement'. The work does contain a good deal concerning the shadow of the earth and the shadow and penumbra of the moon, and one purpose is surely to determine the size of the shadow of the earth, for which no less than four methods are given. But there is far more, namely, the various relations between the distances and parallaxes of the sun and moon, the apparent diameters of the sun, the moon, the earth's shadow, the moon's shadow and penumbra, and the magnitudes of eclipses. Kepler called the *Hipparchus* and its theorems and problems *ingeniosissima* (1612), *pulcherrima* (1619, 1627), and *iucundissima* (1627), and it is clear that the references are to the *Sciametria*.[7] After reading the little tract, it is hard to disagree with his judgment. Further, the *Sciametria* is entirely original,

---

5   *KGW*, vol. 20.1, p. 502.

6   There are references to a lunar eclipse of 26 Aug 1616 and computations for as late as 18 July 1620. I do not known how much of this material survives. Under the titles *Lunaria* and *Restitutionum Lunarium Adversaria* (*KGW*, vol. 20.1, pp. 269–320, 321–92), Bialas arranges much material on lunar theory, a part of which was previously published by Frisch.   7   See the citations referred to in note 2.

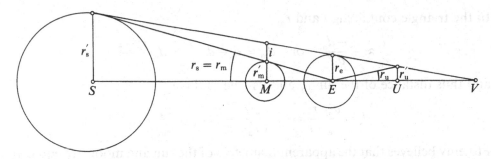

Figure 1

examining the relation between these quantities in a way at once more ingenious and more direct than in any previous treatment, in particular Ptolemy's in Books 5 and 6 of the *Almagest*, the foundation of all later work. The tract consists of eighteen theorems and sixteen problems in addition to some that were cancelled or not numbered, and it does not appear to have been completed although most of it is in something like a final form.

The background to the *Sciametria*, and the subject of its first part, is what came to be called the 'Hipparchus diagram' used in *Almagest* V,15 to demonstrate the distance of the sun from a known lunar distance, although it was originally used by Hipparchus to find the distance of the moon from an assumed solar parallax and distance.[8] The principle of Ptolemy's demonstration is shown in Figure 1 in which S is the sun and M the moon contained by the shadow cone of the moon, and E the earth and U the earth's shadow contained by the shadow cone of the earth with vertex V. It is assumed that the points of tangency lie on diameters of the bodies of the sun, moon, and earth. Let the distances of the sun and moon at which their apparent diameters are equal be $d_s = SE$ and $d_m = ME$ in units of the radius of the earth r_e; at these distances let the equal apparent radii of the sun and moon be r_s and r_m; and let the apparent radius of the earth's shadow at the same distance $d_m = EU = ME$ be r_u. We are given d_m, $r_m = r_s$, and the ratio r_u/r_m, and we wish to find d_s. Now, since r_m and r_u are small angles, the true radii of the moon and shadow, r'_m and r'_u, are respectively

$$r'_m \approx d_m \sin r_m, \quad r'_u \approx \frac{r_u}{r_m} \cdot r'_m,$$

and considering the parallel lines $r'_m + i$ and r'_u equidistant from r_e,

$$r'_m + i = 2r_e - r'_u, \quad r_e - i = r'_m + r'_u - r_e.$$

8 See the articles of Swerdlow and Toomer in note 1 and also A. Van Helden, *Measuring the Universe, Cosmic Dimensions from Aristarchus to Halley* (University of Chicago Press, 1985), pp. 10–13, 17–19. I have found Van Helden's excellent book very helpful, particularly the treatment of Kepler in chapters 6 and 8.

In the triangle containing i and r_e,

$$\frac{i}{r_e} \approx \frac{d_s - d_m}{d_s} \quad \text{or} \quad d_m = d_s - d_s \cdot \frac{i}{r_e} = d_s \cdot \frac{r_e - i}{r_e},$$

and thus distance of the sun d_s and its true radius r'_s,

$$d_s = \frac{d_m \cdot r_e}{r_e - i} = \frac{d_m \cdot r_e}{r'_m + r'_u - r_e}, \quad r'_s = \frac{d_s}{d_m} \cdot r'_m.$$

Ptolemy believes that the apparent diameters of the sun and moon are equal at the greatest distance of the moon, and thus in the numerical solution, in accordance with his earlier demonstrations,

$$d_m = 64;10 r_e, \quad r_m = r_s = 0;15,40°, \quad r_u / r_m = 2;36,$$

from which

$$r'_m = 0;17,33 r_e \quad r'_u = 0;45,38 r_e, \quad i = 0;56,49 r_e,$$

$$r_e - i = 0;3,11 r_e, \quad d_s \approx 1210 r_e, \quad r'_s \approx 5\tfrac{1}{2} r_e.$$

Ptolemy's demonstration is consistent in that the result is a solar distance that must have appeared plausible, with a horizontal parallax of about 0;3°, a value that remained canonical until the early seventeenth century. But consistency is not necessarily easy to achieve. The quantity $r_e - i = r'_m + r'_u - r_e$, is quite small, and the slightest reduction in r'_m or r'_u can make it smaller still and cause d_s to increase greatly. For example if $r_m = 0;15°$ instead of $0;15,40°$, and everything else remains the same, we shall have

$$r'_m = 0;16,48 r_e, \quad r'_u = 0;43,41 r_e, \quad r_e - i = 0;0,29 r_e,$$

from which $d_s \approx 8000 r_e$, more than six times as great. Further, Ptolemy's belief that the apparent diameters of the sun and moon are equal at the greatest lunar distance would make annular eclipses impossible and is simply untrue. But reducing the lunar distance at which the apparent diameters are equal would likewise cause the distance of the sun to increase. Later attempts to apply Ptolemy's procedure, such as those of al-Battānī and Copernicus, ran into difficulties of this kind, and by late in the sixteenth century Tycho had given up the procedure and simply assumed a solar distance of $1150 r_e$, both because it gave the expected parallax of 0;3° and also because expressed in diameters of the earth it was close to the 'mystical number' 576.[9]

Kepler wished to do better, but initially his accomplishment was largely negative and critical. In the *Astronomia nova* XI he examined observations of Mars near opposition made by Tycho in 1582/83 that were believed to show a parallax in excess of the sun's – since at opposition Mars is at about one-half the distance of the sun, one would expect about 0;6° – and observations of his

---

[9] On this entire subject see Van Helden, *Measuring the Universe*, pp. 31, 45–6, 49–50. For what it is worth $576 = 24^2 = 9 \cdot 64 = 3^2 \cdot (2^2)^3$.

own near Mars's first station in February of 1604.[10] He found that the planet showed no measurable parallax, and concluded that its parallax was very small, not exceeding 0;4°, the range of instrumental error since the observations were taken as accurate to $\pm 0;2°$. This implies that the solar parallax is less than 0;2°, corresponding to a distance of more than $1700r_e$, but Kepler does not draw this conclusion. Rather, he says, the sun is not closer than $230r_e$, where its size would be equal to the earth's, nor at an infinite distance.[11] But although no definite number has yet been demonstrated, it lies between $700r_e$ and $2000r_e$ (the former in his *Mysterium Cosmographicum*, the latter in observations of eclipses) 'as I shall demonstrate in my *Hipparchus*'. What Kepler has in mind here is not clear, and neither is the source of these distances, but in his *Ephemerides* for 1617 he assumed a distance of $1800r_e$, corresponding to a parallax of a little under 0;2°. He explains that he likes the number $1800r_e$ because it is about thirty times the lunar distance of nearly $60r_e$, so that the proportion of the distance of the sun from the earth to the distance of the moon from the earth is the same as the proportion of the revolutions of the moon and earth about the center of the earth, that is, the ratio of the month to the day.[12]

In the *Epitome astronomiae Copernicanae* IV,i,4 Kepler further reduced the solar parallax to about 0;1° by another, more elaborate 'archetypal' argument, which, very briefly, goes like this:[13] For many excellent reasons – theological, archetypal, harmonical, numerological – it was suitable that the apparent diameter of the sun seen from the earth be $\frac{1}{720}$ of a circle or $\frac{1}{2}°$. Hence, the apparent radius of the sun $r_s = \frac{1}{4}°$ and the distance of the sun from the earth in true solar radii r_s' must be $d_s = r_s'/\tan\frac{1}{4}° = 229.18166 r_s'$. Now because the earth was to be the home of the 'measuring creature', it is suitable that the body of the earth be the measure of bodies, i.e. volumes, and the radius of the earth the measure of distances. Thus, the ratio of the body of the sun to the body of the earth ought to be equal to the ratio of the distance of the sun to the radius of the earth, or, since bodies are as the cubes of their radii,

$$r_s'^3/r_e^3 = d_s/r_e \quad \text{or} \quad d_s = r_s'^3/r_e^2.$$

Letting $d_s/r_s' = 229.18166 = k$, we set $r_s' = d_s/k$, from which

$$d_s = d_s^3/k^3 r_e^2 \quad \text{or} \quad d_s = k^{\frac{3}{2}} r_e = 3469\tfrac{1}{2} r_e.$$

10   *KGW*, vol. 3, pp. 120–9. This is the famous measurement that Tycho used to test the Ptolemaic and Copernican theories. I hesitate to say more about it because I do not believe that the subject has received adequate investigation.

11   *KGW*, vol. 3, p. 129. See Van Helden, *Measuring the Universe*, p. 61. In Figure 1 $r_s'/d_s = \tan r_s$. Hence if $r_s \approx \frac{1}{4}°$ and $r_s' = r_e$, $d_s = r_e/\tan\frac{1}{4}° \approx 230 r_e$. We shall see Kepler doing something very much like this again.

12   Van Helden, *Measuring the Universe*, p. 79.

13   *KGW*, vol. 7, pp. 277–81. See Van Helden, *Measuring the Universe*, pp. 83–4.

The true radius

$$r'_s \sim d_s^{\frac{1}{3}} = k^{\frac{1}{2}}r_e = 15;8r_e.$$

Kepler uses more steps and finds $d_s = 3469\frac{1}{3}r_e$ and $r'_s > 15r_e$, but in any case the parallax is about $0;1°$, that is, $\sin^{-1}(r_e/d_s) = 0;0,59,45° \approx 0;1°$.

And there is more, for the apparent diameter of the moon should also be $\frac{1}{720}$ of a circle, both because of the number itself as before, and also because of solar eclipses – ordained by the Creator in order to teach the 'observing creature' the measure of the motion of the stars – which require that the apparent diameters of the sun and moon be equal. Now a similar relation holds between the bodies and distance of the moon and earth as between the sun and earth, namely,

$$r_e^3/r'^3_m = d_m/r_e \quad \text{or} \quad d_m = r_e^4/r'^3_m.$$

As in the relation for the sun, the distance of the moon in radii of the moon is $d_m = r'_m/\tan\frac{1°}{4} = 229.18166r'_m = kr'_m$, so that $r'_m = d_m/k$, and thus

$$d_m = k^3r_e^4/d_m^3 \quad \text{or} \quad d_m = k^{\frac{3}{4}}r_e = 58;54r_e.$$

The true radius

$$r'_m \sim d_m^{-\frac{1}{3}} = k^{-\frac{1}{4}}r_e = 0;15,25r_e.$$

Kepler says that $d_m < 59r_e$ and $r_e < 4r'_m$. Now the distance is a very significant result, first because $59r_e$ is about the lunar distance at apogee found by Tycho, and second because it follows, 'by way of a long, circuitous demonstration, which see in my *Hipparchus*', that the distance of the moon is a mean proportional between the distance of the sun and the radius of the earth, that is, between the semidiameter of the earth's sphere and the semidiameter of the earth's body. Kepler is saying that

$$d_s/d_m = d_m/r_e \quad \text{or} \quad d_m = (d_s r_e)^{\frac{1}{2}},$$

and indeed $d_m = (3469\frac{1}{3})^{\frac{1}{2}}r_e = 58;54r_e$, which appears very remarkable. However, although Kepler appears unaware of what lies behind this relation – unless that is what he means by the 'long, circuitous demonstration' in the *Hipparchus* – it is not exactly a miracle, for since, as we have seen, $d_s = k^{\frac{3}{2}}r_e$ and $d_m = k^{\frac{3}{4}}r_e$, it follows that

$$\frac{d_s}{d_m} \cdot r_e = \frac{k^{\frac{3}{2}}}{k^{\frac{3}{4}}} \cdot r_e = k^{\frac{3}{4}}r_e = d_m,$$

and thus $d_s = d_m^2$. And what is more, $r'_m = r'^{-\frac{1}{2}}_s$.

Kepler's remark that the relation between the solar and lunar distances is demonstrated in the *Hipparchus* tells us something about the contents of that book. It happens that in Theorem XI of the *Sciametria* Kepler uses a solar parallax of $0;1°$, the value derived here, and thus the implication of the remark is that this section of the *Epitome* rests upon more detailed demonstrations in, or intended for, the *Hipparchus*, and conversely that the *Hipparchus* had, or

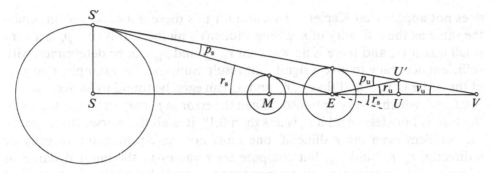

Figure 2

was intended to have, archetypal arguments of the same sort found in the *Epitome* (and indeed throughout Kepler's works). In fact, as we shall note in the commentary, several parts of the *Sciametria* are at the foundation of procedures or results described in both the *Epitome* and the *Rudolphine Tables*, and the same may be true of other parts of the *Hipparchus* no longer surviving, so even if Kepler never completed or published the *Hipparchus*, there can be no doubt that it left its mark on two of his principal works.

The principal subject of the *Hipparchus* was supposed to be the sizes and distances of the sun, moon, and earth, and in this investigation, even if the solar distance had to be found through archetypal reasons, Kepler did not abandon the 'Hipparchus diagram'. In fact, in the *Sciametria* he used it in an entirely new way that is simpler and more elegant than its previous application.[14] The modified diagram, in which only angles need be considered, is shown in Figure 2, in which the parallaxes of the sun and moon are p_s and p_m, and their apparent radii $r_s = r_m$. The apparent radius of the shadow is r_u, the parallax of the shadow at the distance at which the moon crosses it $p_u = p_m$, and the angle of half the vertex of the shadow is v_u. By considering that the exterior angle of a triangle is equal to the sum of the interior and opposite angles, the following relations appear:

(1) in triangle $S'VE$, $r_s = p_s + v_u$,

(2) in triangle $U'VE$, $p_u = r_u + v_u$,

and combining these

(3) in triangle $S'U'E$, $p_s + p_u = r_s + r_u$,

from which, since $p_m = p_u$,

(3a) $p_s = r_s + r_u - p_m$.

These relations appear in Theorems I–III and Problem XV of the *Sciametria*, and the last is, in principle, a way to find the parallax of the sun. However, it

[14]   This has been noted by Van Helden, *Measuring the Universe*, pp. 61–2.

does not appear that Kepler did so, and for this there is a good reason, much the same as the difficulty of applying Ptolemy's method. It is that p_s is a very small quantity, and there is no way that r_s, r_u, and p_m can be determined with sufficient accuracy to give a significant result. Suppose, for example, that p_s is of the order of $0;1°$ and that r_s, r_u, and p_m can each be found to an accuracy of $\pm 0;0,20°$, which is very optimistic. Then the error in p_s may well be $\pm 0;1°$, and the task is hopeless. And if p_s is less than $0;1°$, it is all the worse. But to make the problem even more difficult, one must not merely measure, directly or indirectly, r_s, r_u, and p_m, but compute their values at the lunar distance at which $r_s = r_m$, which necessitates measuring r_m and also makes the control of distance in the lunar theory crucial. And it is precisely distance – the subject, after all, of Part IV of the *Astronomia nova* – that is the most uncertain part of planetary and, even more so, lunar theory.

On the other hand, Problem II shows how to find the radius of the shadow from the relation $r_u = p_s + p_m - r_s$, and this Kepler does apply in the *Epitome* VI,v,7 where, as an example, $p_s = 0;1,1°$, $p_m = 1;2,15°$, $r_s = 0;15,30°$, from which $r_u = 0;47,46°$.[15] One may also use this relation to check the consistency of values used by others. For example, at the solar distance $1210r_e$ and the lunar distance $64;10r_e$ at which $r_m = r_s$, Ptolemy sets $p_s = 0;2,51°$, $p_m = 0;53,34°$, $r_m = r_s = 0;15,40°$. Hence $r_u = 0;40,45°$, and indeed Ptolemy himself takes $r_u = 2;36r_m = 0;40,44°$, which is so close that one might almost imagine that Ptolemy also knew this relation, although the consistency of his demonstration of the solar distance is really sufficient in itself to assure the consistency of this relation. One might also guess that the values used by al-Battānī and Copernicus would not turn out so well, and this guess would be correct.

Kepler did not have the last word on the application of the Hipparchus diagram, for it was soon considered in two other treatises. The first was Philipp Lansberg's *Uranometriae Libri Tres*,[16] a work that, like his tables, may well have eclipsed Kepler's through the middle of the seventeenth century, but has now fallen into a certain obscurity. In the *Prolegomena* Lansberg uses the diagram, which he calls the 'diagram of Aristarchus', to examine and rudely criticize the consistency of the demonstrations of solar distance and the underlying parameters of Ptolemy, al-Battānī, Copernicus, Tycho, Longomontanus, and Kepler. Although his understanding is primitive by comparison with Kepler's, he had learned a few things about the application of the diagram from the *Epitome*, which he does not directly acknowledge in his demonstrations in Book II. Using one procedure from the *Epitome*, corresponding to relation (1) above, when the sun is at apogee $r_s = 0;16,47°$

---

[15] *KGW*, vol. 7, p. 491. Since p_s is very small compared to r_u, its accuracy is not so crucial.

[16] Middelburg, Zeeland, 1631; I have used a printing dated 1652.

and $v_u = 0;14,34°$, from which he finds $p_s = r_s - v_u = 0;2,13°$ and $d_s = 1550;52r_e$. It is of interest to note that he derives r_s at apogee from the eclipse of 585 BC, often called the eclipse of Thales, which he believes to be the same eclipse, total at the Hellespont, used by Hipparchus.[17] The parallax he confirms by no less than twenty-one solar eclipses extending from 585 BC to AD 1612. One of these, the eclipse reported by 'Julius Capitolinus' as an omen for the short rule of Gordian III, he takes to be that of 12 April 237, which in turn he uses to redate the consulships of Pius and Pontianus cited by Censorinus from 238 to 237, thereby removing the linchpin of ancient chronology.[18] There are other curiosities in this work.

Lansberg's poor reputation in general is due to the acute criticism of Jeremiah Horrocks in the collection of treatises published by John Wallis as *Astronomia Kepleriana defensa et promota*, one of which, *Disputatio V. De diagrammate Hipparchicho*, is a worthy successor to the *Sciametria*.[19] Horrocks, it need hardly be said, was far more able than Lansberg in his use of the diagram, which he named the 'diagram of Hipparchus', although he too says it was first invented by Aristarchus, something he must have learned from Lansberg since he had not seen Aristarchus's work (and Aristarchus does not have the same diagram). His first chapter contains thirteen theorems and seven problems using a complex version of the diagram based upon Kepler's figure in the *Epitome*, but going beyond what is in the *Epitome* and thus independently rediscovering at least a part of the *Sciametria*. The next three chapters are devoted to a devastating examination of Lansberg and of Lansberg's criticism of his predecessors. Finally, in the fifth chapter Horrocks shows 'the foolishness (*vanitas*) of those who hope to demonstrate the distance of the sun securely and accurately from the diagram of Hipparchus'. His reasons are first, that none of the other parameters is known with sufficient accuracy to give a quantity as small as the solar parallax – although use of the diagram does show the solar parallax to be insensible – and second, that the parallax is far smaller, and thus the distance of the sun far larger, than is commonly believed.

Nevertheless, the task is not hopeless, for Horrocks's has another way of finding the parallax and distance of the sun, and it is this: The semidiameters of the planets, he says, are proportional to their distances from the sun. This does not agree with Kepler, who believed that the volumes of the planetary bodies were proportional to their distances, but it is confirmed by telescopic

17   *Ibid.*, pp. 29–36.

18   *Ibid.*, pp. 52–3. The source is *Hist. Aug.*, Gord. XXIII, 2; that he prefers this to Censorinus says something of his judgment, but his identification of the eclipse may well be correct.

19   *Jerimiae Horroccii ... Opera Posthuma* (London, 1673), pp. 102–74.

observation.[20] For, from measurements of the apparent diameters of the planets with the telescope, it is concluded that their apparent diameters as seen from the sun are all 0;0,30° more or less, and at greatest distance are exactly 0;0,30°. The conclusion is obvious. 'Who', he asks, 'will deny it to be in the highest degree likely (*maxime verisimile*) that the earth, our planet, also appears 0;0,30° [in diameter] at the sun, so that the parallax of the sun is 0;0,15°?' (This means, by the way, that the solar parallax is 0;0,15° at every planet.) The corresponding distance is $13751r_e$, and after some further discussion he decides that the mean distance of the earth from the sun is $14000r_e$ and the solar parallax about 0;0,14°.[21] I need not point out to readers of this volume that it would be anachronistic to regard Horrocks's solar parallax as progress. Nevertheless, his work is very ingenious, and one wishes it would at last receive some competent study.

In this introduction we have considered only the part of the *Sciametria* concerned with the distance of the sun, the diagram of Hipparchus, and some related matters. The remainder will be considered in the commentary to the translation. The work as a whole falls into three principal sections.

1.   Theorems 1–7, Problems 1–2, 15–16: the relations of the quantities in the Hipparchus diagram, i.e. the parallaxes and apparent radii of the sun, moon, and the earth's shadow, and half the vertex angle of the shadow, an auxiliary quantity.

2.   Definitions 1–3, Theorems 8–13, Problems 3–7: the relations, including those to the previous quantities, of the radii of the circle of illumination, the shadow and penumbra of the moon, and the disc of the earth, and the projection of the motion of the moon on to the disc of the earth.

3.   Theorems 14–18, Problems 8–14: the relations, in eclipses, of the radii of the sun, moon, and shadow to the magnitude of the eclipse and the arcs of immersion, half-delay, half-duration, and latitude.

The translation that follows was originally made from Frisch's edition (*KOO*), and then revised and augmented in accordance with the new edition by Bialas and Boockmann (*KGW*), which contains additional material and a number of better readings.[22] The object of the translation has been above all

[20]   Kepler's argument for the proportionality of volume to distance is in the *Epitome* IV,i,4 (*KGW*, vol. 7, pp. 281–2), and the point is essential to his physical explanation of the three-halves power law in IV,ii,4 (*ibid.*, pp. 306–8.)

[21]   Horrocks, *Opera Posthuma*, pp. 160–4. Much the same argument appears in Horrocks's *Venus in Sole Visa*, published by Hevelius in 1662, of which there is a good discussion in Van Helden, *Measuring the Universe*, pp. 108–12.

[22]   *KOO*, vol. 3, pp. 520–33; *KGW*, vol. 20.1, pp. 185–205.

clarity. It is not that Kepler himself is not clear – in this text he generally is – but a literal rendering would sometimes be cumbersome, and thus where it seemed appropriate I have converted phrases and sentences into mathematical notation. The original is, in any case, readily available for comparison. Likewise, although elegant variation may have been a virtue in Neo-Latin, it is not in contemporary English, and therefore I have not attempted to search out, discover or inquire into equivalents of Kepler's numerous variations on 'to find' and a few other words. A number of emendations to the text in *KGW* are noted, and since the text is based upon an autograph, these are either places that Kepler made slips of the pen that he would have caught in preparing the work for publication, or perhaps errors in *KGW*.[23] A few more serious errors that could not be fixed so easily are marked [!] and mentioned in the commentary, while some minor slips in grammar or punctuation are translated according to the obvious sense of the text without specific notice. Kepler provided five figures for the entire text, of which the first two, each used for many demonstrations, are fairly complicated. These two, along with the third, have been divided into simpler figures which are numbered in the translation in parentheses, and have been rotated from a vertical to a horizontal orientation to economize space in printing.[24] Finally, some cancelled and unnumbered theorems and problems that have not been translated are briefly described in an appendix.

Notation in the commentary

The following list contains the notation used most frequently in the commentary along with the extremal values of quantities either taken directly or computed from those given by Kepler in the *EAC*.

APO   *Astronomiae pars optica* (1604)

EAC   *Epitome astronomiae Copernicanae* (1618–21)

KGW   *Johannes Kepler Gesammelte Werke* (1937–)

TR   *Tabulae Rudolphinae* (1627)

Δ finite difference

---

[23] Frisch not infrequently made silent emendations to the text – there is no apparatus – some of which, marked F, I have adopted, although many more I have not. I assume that the new edition, which generally makes good sense, is intended as an uncorrected transcription of Kepler's autograph.

[24] The originals may be found in *KGW*, vol. 20.1, pp. 185, 189, 195. In his useful notes Bialas has likewise drawn a number of simpler figures for each theorem (ibid. 506–13). Frisch too has annotated the text with special attention to the computations in Problems XI, XIII, and XIV (*KOO*, vol. 3, pp. 718–21).

x' alteration of any quantity $x' = x \pm \Delta x$

d_m distance of moon from earth $53;59r_e$ $58;54r_e$

d_s distance of sun from earth $3469\frac{1}{3}r_e$

d_v distance to vertex of earth's shadow $236r_e$ $246r_e$

m magnitude of an eclipse

p_m horizontal parallax of moon $0;58,22°$ $1;3,22°$

p_s horizontal parallax of sun $0;0,59°$ $0;1,1°$

p_u horizontal parallax of earth's shadow at place of transit of the moon $= p_m$

r_d apparent radius of the disc of the earth as seen from moon $= p_m$

r_e radius of the earth   860 German miles ($1° = 15$ German miles)

r_i apparent radius of 'illumination' from center of earth at place of transit of moon $1;12,22°$ $1;17,55°$

r_m apparent radius of moon from center of earth $0;15°$ $0;16,22°$

r_p apparent radius of moon's penumbra at center of earth as seen from moon, limits including r_{um} $0;30,30°$ $0;32,25°$

r_s apparent radius of sun from center of earth $0;15°$ $0;15,33°$

r_{st} true radius of sun $15;8r_e$

r_u apparent radius of earth's shadow from center of earth at place of transit of moon $0;43,50°$ $0;49,40°$

r_{um} apparent radius of moon's shadow at center of earth as seen from moon $0°$ $0;1,22°$

v_u half the vertex angle (semivertex) of earth's shadow $0;14°$ $0;14,33°$

Translation and commentary

Theorem I

The angle of parallax of the upper limb of the sun when located in the horizon and the angle of half the vertex of the shadow of the earth added together equal the angle of the semidiameter of the sun seen from the center of the earth.

With center A and radius AD (Figure 3), let the circle of the body of the sun be described. Now let the distance of the center of the earth from the center of the sun be AB, and with center B and radius BE less than AD, let the circle of the body of the earth be described. Let straight line DE be drawn touching both circles in points D, E, and let it be continued in the direction of E. Then let centers A, B be joined, and AB produced will intersect DE in point C. Further, let B the center of the earth be joined with the point of tangency D, and A the center of the sun with the point of tangency E.

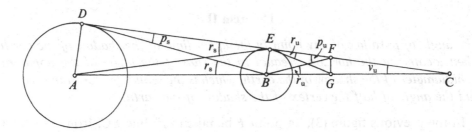

Figure 3

Accordingly, in triangle *BDC* straight line *BC* intersects the two lines *DB*, *DC*; therefore the exterior angle *DBA* is equal to the sum of the interior and opposite angles *BDC*, *BCD*. But *DBA* is the angle of the semidiameter of the sun *DA* seen at the center of the earth *B*. And angle *CDB*, that is *EDB*, is the angle of the horizontal parallax of point *D*. For the line of sight *ED* touches the circle of the earth *BE* at *E*, and thus when the observer is located at *E*, point *D* is seen in the horizon. Finally, *DCB*, that is *ECB*, is the angle of half the vertex of the shadow. Therefore the parallax of the sun and half the vertex of the shadow added together equal the semidiameter of the sun seen at the center of the earth, q.e.d.

Commentary. In triangle *BDC* exterior angle $r_s = p_s + v_u$.

Problem I

Given the apparent semidiameter of the sun and its horizontal parallax, to find the angle of [half the vertex of] the shadow of the earth.

Subtract the horizontal parallax of the sun from the apparent semidiameter of the sun, and you will form half the vertex of the shadow of the earth.

Corollary.[25] As much as the apparent semidiameter of the sun increases, by just so much also increases the sum assembled from both the semivertices of the shadow of the earth and the parallaxes of the sun.

Comm. From Th. I, $v_u = r_s - p_s$. Corollary: $\Delta(p_s + v_u) = \Delta r_s$. In *EAC* VI,v,7 (*KGW* 7, 490) this rule is given with the example: $r_s = 0;15°$, $p_s = 0;1°$, thus $v_u = 0;14°$. It follows that the length of the shadow cone, the distance from the earth to the vertex, $d_v = r_e/\tan v_u \approx 246 r_e$. Kepler uses the relation $d_v = d_s r_e/(r_{st} - r_e)$, and where $d_s = 3469 r_e$ and $r_{st} = 15 r_e$, he finds $d_v > 247 r_e$ (corr. $247\frac{3}{4} r_e$).

[25]   186:6 *om. I* F.

Theorem II

The angle of parallax of any point in the upper limb of the shadow of the earth when located in the horizon is equal to the sum of the angle of the apparent semidiameter of the shadow of the earth, which is the semidiameter at that point, and the angle of half the vertex of the shadow of the earth.

In the previous figure (3), let point F be taken in[26] line EC, from which let FG fall perpendicular to BC, and let F be joined with B and G with E. Now in triangle FBC, when side CF is produced, the exterior[27] angle EFB will be equal to the sum of the interior and opposite angles FBC, FCB. But angles EFB and EGB are, to all acuteness of perception, equal. Therefore also EGB will be equal to the sum of ECB and GBF. But EFB is the parallax of point F in the upper limb of the rising or setting shadow, because FE touches the bodies of the earth and sun at E, D, so that when the observer is located at E the same line FE is the horizon. And FBC, that is FBG, is the angle at the center of the earth of the apparent semidiameter of the shadow FG, drawn through point F, just as FEG is the angle of the same shadow seen from point E of the horizon, again to every perception equal to angle FBG. Finally, FCB, that is ECB, is the angle of half the vertex of the shadow. Therefore, etc., q.e.d.

Briefly: The horizontal parallax of the moon is equal to the sum of the semidiameter of the shadow of the earth and half the vertex of the shadow.

Comm. In triangle FBC exterior angle $p_u = r_u + v_u$, and since $p_m = p_u$, $p_m = r_u + v_u$

Theorem III

The sum of the parallaxes of the sun and of a particular point of the shadow of the earth [when both are] situated in the horizon is equal to the sum of the apparent semidiameters of the sun and the shadow of the earth at that point.

In the figure (3), because DBA or DEA, the semidiameter of the sun, is greater than BDE, the parallax of the sun, by angle BCE, which is the semivertex of the shadow of the earth when the sun is located at A, by Theorem I; and because FBG or FEG, the semidiameter of the shadow in the transit of the moon FG, is less than EFB, the horizontal parallax of the moon in point F, by the same angle BCE by Theorem II; therefore the excess of the former and the defect of the latter are equal. Therefore $DEA + FEG = BDE + EFB$, q.e.d.

Comm. $r_s - p_s = v_u$ (Th. I, Pr. I); $p_u - r_u = v_u$ or $p_m - r_u = v_u$ (Th. II); thus $r_s - p_s = p_u - r_u$ or $p_s + p_u = r_s + r_u$. The 'semivertex' (*semimucro*) is half the vertex angle of the shadow. The 'transit of the moon' (*transitus lunae*), a

---

[26]   186:14 linea/in linea.   [27]   186:16 exactior/exterior F.

standard term, is the distance of the moon when it crosses through the earth's shadow in a lunar eclipse.

Problem II

Given the horizontal parallaxes of the luminaries and the apparent semidiameter of the sun, to find the semidiameter of the shadow of the earth.

From the sum of the horizontal parallaxes of the sun and the moon, subtract the apparent semidiameter of the sun; the remainder is the semidiameter of the shadow of the earth, as great as the semidiameter is in that place which the moon shows.

Comm. Since $p_m = p_u$, from Th. III $(p_s + p_m) - r_s = r_u$. This rule is given in *TR* XXXI, Precept 141 (*KGW* 10, 207) and in *EAC* VI,v,7 (*KGW* 7, 491) with the example: $r_s = 0;15,30°$, $p_s = 0;1,1°$, $p_m = 1;2,15°$, and thus $r_u = 0;47,46°$.

Theorem IV

If from the sum of the semidiameters of the sun and the shadow of the earth, the parallax of the moon less the sun, as much as it can be in the horizon, be subtracted, the remainder is twice the parallax of the sun.

For by Theorem III the sum of the whole horizontal parallax of the moon and the horizontal parallax of the sun is equal to the sum of the apparent semidiameters of the sun and the shadow of the earth in the point that the moon shows by its transit. But the whole parallax of the moon reduced by the parallax of the sun is called the 'parallax of the moon less the sun'. Therefore twice the parallax of the sun plus the parallax of the moon less the sun is equal to the said semidiameters. So when the parallax of the moon less the sun is subtracted from the sum of the semidiameters of the sun and the shadow, the remainder is twice the parallax of the sun, q.e.d.

Comm. The *parallaxis lunae a sole*, a conventional term here translated 'parallax of the moon less the sun', is the difference of the lunar and solar parallax, $p_m - p_s$, that was regularly used and tabulated for the computation of solar eclipses. Its maximum value occurs in the horizon. Since $p_m = p_u$, by Th. III,

$$(r_s + r_u) - (p_m - p_s) = (p_m + p_s) - (p_m - p_s) = 2p_s.$$

Theorem V

The difference of the semivertices of the shadow formed at varying distances of the sun from the earth is equal, although in the opposite direction, to the difference of the semidiameters of the shadow of the earth in any one place of the transit of the moon extended in the horizon.

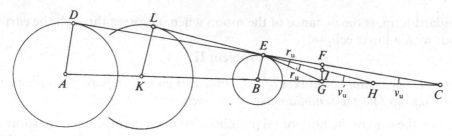

Figure 4

In the previous figure (4), let the sun approach the earth as far as K and let semidiameter KL be equal and parallel to AD. And let a new line be drawn touching both bodies at E and L, and let it be extended until it intersects the axis of the shadow of the earth in H so that EHB is the new semivertex. Accordingly, line FG will be intersected by line EH; let it intersect in I. Now because ECB is the original semivertex and EHB the new semivertex, the difference of the two is angle HEC, that is IEF. But IEF is the difference between the angles GEF of the original apparent semidiameter [of the shadow] and GEI of the new semidiameter in the same place of the transit of the moon GF. Therefore the differences are equal, q.e.d.

Now I say 'extended in the horizon' because if the moon, while being eclipsed, approaches the midheaven, it will then be closer to the observer than to the center of the earth by about one semidiameter of the earth; so in that event the theorem, no longer defined, would not be strictly true.

Comm. As the sun moves from A to K, v_u increases to v'_u and r_u decreases to r'_u. As is evident from Figure 4,

$$v'_u - v_u = HEC = r_u - r'_u \quad \text{or} \quad \Delta r_u = -\Delta v_u.$$

When the moon is near the midheaven, the intersection of the ecliptic and the meridian, it is closer to the observer at the surface of the earth than to the center of the earth, and the difference may reach as much as one semidiameter of the earth if the moon is in the zenith.

Theorem VI

If, from the difference of the apparent semidiameters of the sun due to its varying distances, you subtract the small difference of the horizontal parallaxes of the sun at the same distances, the remainder is the difference of the apparent semidiameters of the shadow, in the very same place of the transit of the moon, corresponding to the varying distances of the sun.

For by Theorem V the difference of the semivertices is equal to the difference of the semidiameters of the shadow of the earth in the same place of the transit

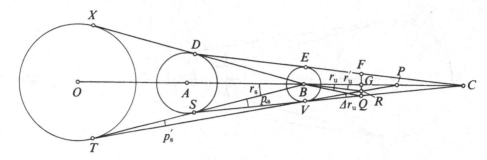

Figure 5

of the moon. But by the Corollary to Problem[28] I, the difference of the semivertices is less than the difference of the semidiameters of the sun by the small difference of the horizontal parallaxes of the sun. Therefore when this small difference is subtracted from the difference of the semidiameters of the sun, the remainder is the difference of the apparent semidiameters of the shadow, that is, of its angles.

Corollary. Since the small difference of the parallaxes of the sun due to the eccentricity of the sun is nearly insensible, we may use the difference of the semidiameters of the sun in place of the difference of particular semidiameters of the shadow in particular places of the transit of the moon; for as much as the diameter of the sun increases, so much approximately does the diameter of the shadow decrease.

Comm. By Th. V $\Delta r_u = -\Delta v_u$, and by Pr. I Cor. $\Delta r_s = \Delta(v_u + p_s)$ or $\Delta r_s - \Delta p_s = \Delta v_u$, and thus $\Delta r_s - \Delta p_s = -\Delta r_u$. Corollary. Since $\Delta p_s \approx 0$, we may assume $\Delta r_s \approx -\Delta r_u$.

Theorem VII

Assuming a fixed size of the angle of the apparent semidiameter of the sun, and supposing under the same angle varying distances of the sun from the earth, as much as the parallax of the sun decreases on the supposition of a greater distance, by just as much does the semidiameter of the shadow also decrease in any transit of the moon through it.

For in our figure (5), let *SBD* or *TBX* be the very same angle of the sun, and let it be supposed that the sun is either at a shorter distance from the earth through *BA* or at a longer distance through *BO*, and let the tangents *SVC*, *TVP* be drawn. Accordingly, in triangle *STV* the exterior angle *BSV* is equal to the sum of *STV* and *SVT*, that is, to *CVP*. But *BSV* is the parallax of the sun supposed closer, and *BTV* is the parallax of the sun supposed farther,

[28]   188:1 Theor:/Probl:.

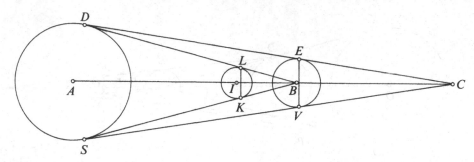

Figure 6

while the angle of its apparent size at center *B* remains the same. Finally, *PVC* is the difference of the semidiameters of the shadow in the same transit of the moon, which let be *FGQ*. Because by Theorem V *TV* produced is an interior line *VP* cutting *GQ* in *R*, and because *TPO* is the semivertex of the shadow due to the more distant sun, but *SCA* the semivertex due to the closer sun, therefore the smaller the parallax *VTB* is than *VSB*, the smaller also is angle *RBG*, the semidiameter of the shadow due to the more distant sun, than *QBG*, the same due to the closer sun, q.e.d.

Corollary. The smaller you assume the difference of the parallaxes of the sun, the smaller is the error if, in place of the difference of the semidiameters of the shadow in one fixed place of the transit of the moon, one make use of the difference of the apparent semidiameters of the sun due to the solar eccentricity.

Comm. Assuming that the sun may be at different distances with the same apparent diameter, there will be a change in the apparent diameter of the shadow. However, this change will be very small, equal only to the change of the solar parallax. In Figure 5,

$$\Delta r_u = QVR = TVS = p_s - p'_s \quad \text{or} \quad \Delta r_u = \Delta p_s.$$

Corollary. From Th. VI Cor., for small Δp_s, $\Delta r_s \approx -\Delta r_u$.

Definition I

We shall use the word 'illumination' technically for that entirely luminous part of the cone forming the shadow of the earth that is between the sun and the earth (just as the shadow of the earth is the remaining, dark part of the same cone on the farther side of the earth, terminating in the vertex). For if the entire moon is absent from that space, the entire hemisphere of the earth facing the sun enjoys the entire illumination of the sun. In the accompanying figure (6), let the centers of the sun and earth be *A*, *B*, and let two straight lines *SV* and

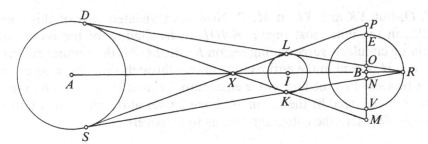

Figure 7

DE touch their bodies in the plane drawn through their centers, which lines continued will meet with each other and with the axis at *C*, representing the cone. Accordingly, the truncated part of this cone *SVED* will be called the 'illumination', just as the vertex of the cone *VCE* is the shadow.

Comm. By *lumen*, here translated 'illumination', Kepler means transmitted light – the *species* or image of light – or reflected light as distinguished from *lux*, light at a self-luminous source. The distinction is an old one.[29] Although Kepler may not follow it strictly, in the *APO* and the *EAC* he usually refers to the *lux solis* when its light is considered directly, to the *lumen solis* when its light is illuminating another body, and to the (reflected) *lumen lunae*. Here the illumination is the lighted truncated cone *SVED* in Figure 6 while the shadow is the darkened *VCE*. The circle *VE*, nearly a great circle, is the 'circle of illumination' of the earth (see Def. III) dividing the illumination from the shadow.

Definition II

The 'penumbra' of the moon is all the space in which some small part of the sun is covered by the moon.[30] In the accompanying figure (7), let the center of the moon be *I*, and let the shadow of the moon be formed by the lines *SK*, *DL* touching the bodies of the sun and the moon and meeting in *R*. Accordingly, in the conical space *KRL* not even a small part of the sun can be seen, and thus this is the cone of the shadow of the moon. Now let other lines be drawn through the axis *AR* of the shadow of the moon, touching the bodies of the sun and the moon at points on opposite sides, as *SXL*, *DXK* extended to *P*, *M*, and *XM*, *XP* will be a conical surface. And let all the lines be cut by the plane of a great circle of the earth *B*, to which the axis *AB* is perpendicular, *SR* and *DR*,

---

[29]   For example, see D. C. Lindberg, *Theories of Vision from al-Kindi to Kepler*, (University of Chicago Press, 1976), pp. 133–5 on the schoolman Buridan.

[30]   189: note 1 contains a cancelled first sentence that I have not translated (the text appears faulty) that defines the penumbra in terms of the cone formed by the rotation of a straight line touching opposite limbs of the sun and moon and intersecting the axis of the shadow of the moon between the sun and the moon.

in N, O, but XK and XL in M, P. Now the truncated part of this cone [is] $KMPL$, in the innermost region $KNOL$ is the shadow of the moon, in the remaining circularly surrounding region KNM, LOP is the penumbra, because in points M, P and in the entire circle drawn through them the entire sun can at last be seen. For the farther we enter into the interior from there, the more of the sun is covered by the moon until one arrives at the circle drawn through points N, O where the entire sun begins to be covered.

Comm. To the best of my knowledge, Kepler invented the term 'penumbra' in the *APO* VI,7 (*KGW* 2, 211) where he describes the penumbra of the earth. The penumbra of the moon is described in *EAC* VI,v,8 (*KGW* 7, 497) using a figure essentially the same as that in the *Sciametria*, but there Kepler mentions the essential fact that the diameter of the penumbra, although shown larger in the figure (7) for reason of clarity, is smaller than the diameter of the disc of the earth (see Def. III), that is, MP is really less than VE, a point surely omitted by oversight in the *Sciametria*. Strictly, the penumbra is a part of a truncated cone $KMPL$ surrounding the truncated shadow cone of the moon $KNOL$. Within $KNOL$ the moon may cover the entire sun in a total eclipse, while in the region $KMPL$ it may cover part of the sun in a partial eclipse.

Definition III

The 'disc of the earth' is the nearly great 'circle of illumination' of the earth, intersecting the axis of the shadow of the earth perpendicularly near the center of the earth. It is called a 'disc' because we imagine the illuminated surface of the earth to be projected into the plane of this circle.

Comm. In the *APO* VI,3 (*KGW* 2, 204ff.) Kepler described the 'circle of illumination' of the moon and of the earth. In Figure 6 the former is the circle, nearly a great circle, perpendicular to AB passing through the points of tangency K and L dividing the lighted and darkened parts of the moon. The circle of illumination of the earth is the nearly great circle through V and E dividing the lighted and darkened parts of the earth and the 'illumination' $SVED$ from the shadow VCE (see Def. I). The 'disc of the earth' is the plane of the circle through VE, and it is nearly a great circle since it is very close to the center of the earth B. Much the same description is given in *EAC* VI,v,8 (*KGW* 7, 497).

Theorem VIII

The angle of the semidiameter of the illumination in the place of the transit of the moon seen at the center of the earth is equal to the sum of the horizontal parallax of the moon and the semivertex of the shadow of the earth.

In the figure (8) let the place of the transit of the moon through the

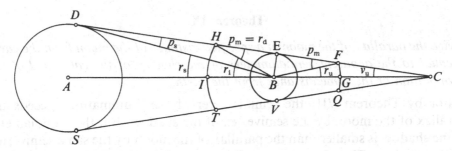

Figure 8

illumination be *HIT*, *HI* the semidiameter of the illumination, let *HB* be drawn, and *IBH* will be the angle at the center of the earth *B*. Now in triangle *HCB* exterior angle *HBI* is equal to the sum of the interior and opposite angles *BHC*, *BCH*. But *BHC*, that is *BHE*, is the horizontal parallax of point *H* or of the moon in it, and *BCH* is the semivertex of the shadow. Therefore, etc., q.e.d.

Comm. Here Kepler uses 'transit of the moon' (*transitus lunae*) to mean the distance of the moon when it crosses through the cone of illumination in a solar eclipse. The 'semidiameter of the illumination' r_i is the semidiameter of a circle formed by the intersection of the cone of illumination with a plane perpendicular to the axis of the cone at the distance of the transit of the moon. As is clear in Figure 8, in triangle *BCH* exterior angle $r_i = p_m + v_u$. In a cancelled addition to Th. XII (*KGW* 20.1, 193:3), Kepler gives the example $p_m = 1°$, $v_u = 0;14°$, and thus $r_i = 1;14°$. In his figure Kepler draws *HIT* and *QGF* lying on a circle with center *B*, although the arcs, $HIT = 2r_i \approx 2\frac{1}{2}°$ and $QGF = 2r_u \approx 1\frac{1}{2}°$, are rather small.

Problem III

Given the parallaxes of the luminaries and the apparent semidiameter of the sun, to find the semidiameter of the illumination.

To the horizontal parallax of the moon add the apparent semidiameter of the sun, from the sum subtract the parallax of the sun; or, what amounts to the same thing, add in one sum the parallax of the moon less the sun and the semidiameter of the sun; in both ways the result is the semidiameter of the illumination.[31]

Comm. What is meant is again the semidiameter of the illumination in the place of the transit of the moon. In Figure 8 in triangle *HDB* exterior angle $p_m = (r_i - r_s) + p_s$, and thus

$$r_i = (p_m + r_s) - p_s = (p_m - p_s) + r_s.$$

[31]   190:16 *om*. Vide probl: I F.

Theorem IX

Twice the parallax of the moon at a specified distance of the moon from the earth is equal to the apparent semidiameters of the shadow of the earth and of the illumination at the same distance from the earth.

For by Theorem VIII the semidiameter of the illumination exceeds the parallax of the moon by the semivertex of the shadow. But the semidiameter of the shadow is smaller than the parallax of the moon by the same semivertex of the shadow. Therefore the semidiameters of the illumination and of the shadow added together are equal to twice the parallax of the moon, by the excess compensating the equal defect. Therefore, etc.

Briefly from the figure (8):

> IBH exceeds BHE by angle BCH,
>
> GBF is exceeded by BFE by angle BCE [$\equiv BCH$];
>
> since BHE, BFE are equal,
>
> [therefore $2BHE = IBH + GBF$].

Comm. $r_i - p_m = v_u$ (Th. VIII); $p_m - r_u = v_u$ (Th. II); and thus

$$r_i - p_m = p_m - r_u \quad \text{or} \quad 2p_m = r_i + r_u.$$

Theorem X

The angle of the semidiameter of the moon seen from the center of the earth is equal to the sum of the semidiameters, of the sun seen from the center of the earth and of the shadow of the moon at the center of the earth as seen from the moon.

In the figure (9) let points L, B be joined; therefore in triangle LRB the exterior angle LBI is equal to the sum of the interior and opposite angles BLR, BRL. But LBI is the semidiameter of the moon seen from the center of the earth, and BLR, that is BLO, is the semidiameter of the shadow [of the moon] formed at the center of the earth B and seen from point L of the moon. Finally, BRL, that is ARD, is the semidiameter of the sun seen from point R, which is always very close to B the center of the earth. Therefore, etc. q.e.d.

Corollary. The sum of the semidiameters of the sun and the moon is equal to the sum of the diameter of the sun and the semidiameter of the shadow of the moon.

Comm. Kepler now takes r_s from R rather than B – since the apparent semidiameter of the sun is much the same from either point – forming triangle BLR in which exterior angle $r_m = r_s + r_{um}$, the last being the apparent radius of the moon's shadow, as seen from the moon, where it intersects the disc of the earth. By taking R at the point of the earth farthest from I, r_{um} is made as large

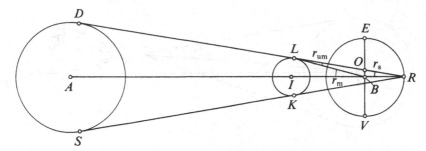

Figure 9

as possible, but if, as is often true, $r_m < r_s$, the complete shadow of the moon will not reach the center, or even the surface, of the earth. Corollary. It follows from the theorem that

$$r_s + r_m = r_s + (r_s + r_{um}) = 2r_s + r_{um}.$$

Problem IV

Given the apparent semidiameters of the luminaries, to find the semidiameter of the shadow of the moon.

When the semidiameter of the sun is subtracted from the semidiameter of the moon, if [the latter is] greater, the remainder is the semidiameter of the shadow of the moon as though seen from the moon.

Comm. From Th. X, $r_{um} = r_m - r_s$. This also appears in *TR* XXXI Precept 150 (*KGW* 10, 207). The rule is given in *EAC* VI,v,8, (*KGW* 7, 498) as $r_{um} \approx r_m - r_s$, the approximation due to measuring r_s from R rather than B, a rather fine point of $1/3469$ part of the distance of the sun. From $r_{m\,max} = 0;16,22°$ and $r_{s\,min} = 0;15°$, it follows that $r_{um\,max} = 0;1,22°$. Since the earth subtends an angle of $2;6,44°$ at the moon when at least distance, the maximum width of the path of totality in a solar eclipse is about $\frac{1}{46}$ the diameter of the earth. When $r_m = r_s, r_{um\,min} = 0°$. Kepler says in *EAC* that if the subtraction cannot be done, if r_m is not greater than r_s, there is not a complete shadow of the moon, which is the configuration for an annular eclipse in which the vertex of the moon's shadow does not reach the surface of the earth.

Theorem XI

The semidiameter of the penumbra including its shadow in the center is composed from the semidiameters of the sun and moon, and from as great a part of the horizontal parallax of the sun as the apparent diameter of the sun is a part of the parallax of the moon.

43

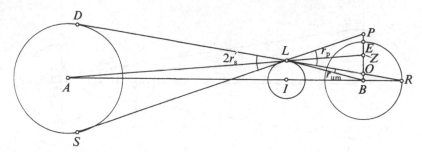

Figure 10

[First proof]

In the figure (10), let a straight line be drawn from the center of the sun A, through the limb of the moon L, to BP, which will be ALZ, and let points L, B be joined. Now, the proportions AZ/AL and AZ/ZL are the same as the proportions of the diameter of the sun observed from L to the diameter of the sun observed from Z,[32] near the center of the earth B, and to its diminution[33] in Z, which is true in such small angles not withstanding Proposition [8] of Euclid's *Optics*. But as AZ/AL, so is the parallax [of the moon] L to the parallax [of the sun] A by the same [argument]. Therefore, as the diameter of the sun observed from L is to its diminution as far as Z, so is the parallax [of the moon] L to the parallax [of the sun] A. And by permutation, as the diameter of the sun observed from L is to the parallax of the moon L, so is the diminution of diameter of the sun as far as Z to parallax of the sun A. And conversely, as the parallax of the moon L, is to the parallax of the sun A, so is the diameter of the sun seen from L to its diminution seen from Z. Therefore SLD, the diameter of the sun [seen] from L, is made up from the diameter of the sun as it appears in Z and from as great a part[34] of the parallax of the sun as the entire apparent diameter of the sun is a part of the parallax of the moon.

But OLP equal to SLD is one part of the semidiameter of the penumbra of the moon [as] seen from the moon. The other part is BLO, the angle of the semidiameter of the shadow of the moon as seen from L. Therefore BLP, the angle of the semidiameter of the penumbra seen from L, is made up from[35] two semidiameters of the sun, the part of the parallax of the sun mentioned, and the semidiameter of the shadow of the moon. But by Theorem X, the semidiameter

[32]   192:4 *om.* ejus augmentum.   [33]   192:4 diminitionem/diminutionem.

[34]   In place of the following, an addition has: 'of the diameter of the sun as the parallax of the sun is a part of the parallax of the moon'. The two are algebraically equivalent, i.e. $p_s(2r_s/p_m) = 2r_s(p_s/p_m)$.

[35]   In place of the following, an addition has: 'the entire diameter of the sun and as great a part of it as the parallax of the sun is part of the parallax of the moon, and from the semidiameter of the shadow of the moon'. Again, the two are equivalent.

of the sun and the semidiameter of the shadow of the moon together are equal to the semidiameter of the moon seen from the center of the earth. Therefore, the remaining semidiameter of the sun together with the part of the parallax [of the sun] and the semidiameter of the moon form the penumbra, q.e.d.[36]

Comm. This theorem is not entirely consistent, and seems to have given Kepler some trouble for he wrote two different versions. The first version is as follows. In Figure 10 the radius of the penumbra in the plane of the disc of the earth is *BP*, and its apparent radius as seen from the moon, including the shadow of the moon, is *BLP*, which may be divided into two parts $r_p = OLP$, the semidiameter of the penumbra itself, and $r_{um} = BLO$, the semidiameter of the shadow of the moon on the disc of the earth. Now, $r_p = OLP = SLD = 2r_s'$, the apparent diameter of the sun as seen from the moon. Drawing *AL* from the sun and extending it to meet *BP* at *Z*, and letting $2r_s$ be the apparent diameter of the sun seen from *Z* near the center of the earth,

$$\frac{2r_s'}{2r_s} = \frac{AZ}{AL} \quad \text{and} \quad \frac{2r_s'}{2r_s' - 2r_s} = \frac{AZ}{ZL} = \frac{p_m}{p_s},$$

since the angles are small and thus Euclid, *Optics* 8 – that the apparent sizes of equal and parallel magnitudes at unequal distances from the eye are not proportional to the distances – may be ignored, as Kepler remarks. Therefore

$$2r_s' = 2r_s + p_s \cdot \frac{2r_s'}{p_m} \approx 2r_s + p_s \cdot \frac{2r_s}{p_m},$$

since $2r_s/p_m \approx 2r_s'/p_m$. And since $r_p = 2r_s'$,

$$r_p = 2r_s + p_s \cdot \frac{2r_s}{p_m}.$$

By Th. X, the shadow of the moon $r_{um} = r_m - r_s$, and thus the entire penumbra including the shadow of the moon

$$r_p + r_{um} = r_s + r_m + p_s \cdot \frac{2r_s}{p_m}.$$

[Second proof]

In the figure (11), connecting *K* with *B* and *I* with *N*, because *MP* is the diameter of the penumbra by Definition II, therefore *MIB* or *MKB* is the angle of its apparent semidiameter as though observing from the moon.[37] But *MKB* has two parts, *MKN* and *NKB* or *NIB*, of which the latter is the semidiameter

---

[36] 192: note 3 is an addition of four lines that I do not understand and have not translated.

[37] The line *IN* and the angles *MIB* and *NIB*, although referred to, are not used in the demonstration, and have been omitted from Fig. 11.

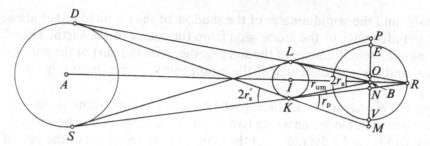

Figure 11

of the shadow of the moon formed upon the disc of the earth, but seen from the moon as in the previous [Theorem X], while the former, or angle MKN, is equal to angle SKD of the diameter of the sun seen in point K of the moon. And as DM/DK or SN/SK, so inversely is angle SKD to angle SBD of the diameter of the sun seen from center B. For in the case of such small angles, Proposition [8] of Euclid's *Optics* is no obstacle to us. But also as SN/KN, so inversely is the parallax of the moon to the parallax of the sun, and therefore as $SN/(SN-KN)$, that is SN/SK, so is the parallax of the moon to the difference of the parallax of the moon and the parallax of the sun, which difference[38] is the parallax of the moon less the sun. Therefore, as the parallax of the moon is to the parallax of the moon less the sun, so is SKD/SBD. And conversely, as the parallax of the moon less the sun is to the total parallax of the moon, so is SBD, the diameter of the sun seen at the center of the earth, to SKD or MKN, one of the parts of the penumbra.

But the excess of the total parallax of the moon over the parallax of the moon less the sun is about $\frac{1}{60}$; therefore the excess of MKN over SBD[39] is also about $\frac{1}{60}$ part, and consequently about $30''$ or half the parallax of the sun. And thus when both elements MKN and NKB are put together in one MKB, we shall have the diameter of the sun, the semidiameter of the shadow of the moon, and about half the parallax of the sun. But by the Corollary of Theorem X, the sum of the diameter of the sun and the semidiameter of the shadow of the moon is equal to the sum of the semidiameters of the sun and the moon. Therefore the semidiameter of the penumbra is equal to half the parallax of the sun and the sum of the semidiameters of the sun and moon.

Comm. The second proof is slightly different. The ratio $2r_s/p_m \approx 0;30°/1° = \frac{1}{2}$, and in this second proof it is shown that $r_p+r_{um} = r_s+r_m+\frac{1}{2}p_s$. In Figure 11 the apparent semidiameter of the penumbra as seen from the moon, including the shadow of the moon, is now angle MKB, which is divided into $r_p = MKN$, the semidiameter of the penumbra, and $r_{um} = NKB$, the semidiameter of the

---

[38]   193:28 diameter/differentia, *add.* est F.   [39]   193:33 ABD/SBD.

shadow of the moon on the disc of the earth. Now, $r_p = MKN = SKD = 2r'_s$, the apparent diameter of the sun seen from the moon. And where $SBD = 2r_s$, the apparent diameter of the sun seen from the earth,

$$\frac{2r'_s}{2r_s} = \frac{SKD}{SBD} = \frac{SN}{SK},$$

since the angles are small and again Euclid, *Optics* 8 may be neglected. Hence,

$$\frac{2r'_s}{2r_s} = \frac{SN}{SK} = \frac{d_s}{d_s - d_m} = \frac{p_m}{p_m - p_s} = \frac{r_p}{2r_s}.$$

But since $p_m \approx 1°$ and $p_s \approx 0;1°$, so $p_m - (p_m - p_s) = 0;1° = \frac{1}{60} p_m$. And since $2r_s \approx 0;30°$,

$$r_p - 2r_s = \frac{1}{60} r_p \approx \frac{1}{60} 2r_s \approx 0;0,30° = \frac{1}{2} p_s,$$

that is, $r_p = 2r_s + \frac{1}{2} p_s$. But by Th. X Cor. $r_s + r_m = 2r_s + r_{um}$. Hence

$$r_p = 2r_s + \frac{1}{2} p_s = r_s + r_m - r_{um} + \frac{1}{2} p_s,$$

and thus the semidiameter of the penumbra including the shadow of the moon

$$r_p + r_{um} = r_s + r_m + \frac{1}{2} p_s.$$

Problem V

Given the semidiameters of the luminaries [and the horizontal parallax of the sun], to determine the semidiameter of the penumbra of the moon.

Let the semidiameters of the luminaries and about half the horizontal parallax of the sun be added together in one sum, and in this way the result will be the semidiameter of the penumbra of the moon.[40]

Comm. By semidiameter of the penumbra is meant the penumbra including the shadow of the moon, that is, the sum of the semidiameters of the penumbra and the shadow of the moon $(r_p + r_{um})$. Hence from Th. XI, $(r_p + r_{um}) = r_m + r_s + \sim \frac{1}{2} p_s$. The same rule is given in *TR* XXXI Precept 150 (*KGW* 10, 207) and in *EAC* VI,v,8 (*KGW* 7, 498) with the remark that a demonstration is contained in the *Hipparchus* and the example: $r_s = 0;15°$, $r_m = 0;16,20°$, $\frac{1}{2} p_s = 0;0,30°$, and thus $r_p + r_{um} = 0;31,50°$.

Theorem XII

The horizontal parallax of the moon at any point of its distance from the earth is equal to the apparent semidiameter of the disc of the earth [seen from the moon].

---

[40]   194:9 luna/penumbrae lunae F.

For because we wish to find the proportion of the semidiameters of the disc of the earth and of the penumbra, the angles under which each appears must also be determined in the same place. But necessarily we determine the angles of the penumbra and of the shadow [of the moon] at the moon, from which, of course, the penumbra *MKN* (Figure 11) spreads out. Therefore the semidiameter of the disc of the earth should also subtend an angle at the moon. But the angle the semidiameter of the earth subtends at the moon is the angle of the horizontal parallax of the moon.

Comm. By the semidiameter of the disc of the earth r_d is meant the angle subtended at the moon by the radius of the earth, which is the lunar parallax p_m. In Figure 8 half the disc of the earth is BE, and thus with the moon at H, $r_d = p_m$. This rule appears in *TR* XXXI as Precept 149 (*KGW* 10, 207). In the preceding example in the comment to Pr. V, the semidiameter of the penumbra with the shadow of the moon $r_p + r_u = 0;31,50°$. Since here $r_d = p_m \approx 1;3°$, the diameter of the penumbra can cover about one-half the surface of the earth in a central solar eclipse. This also shows that the figure of the penumbra (Figures 7, 10, 11), in which it is larger than the earth, is not correct (although useful for clarity).

A cancelled Th. XIII, following from an incomplete, cancelled Th. XII (*KGW* 20.1, 190–91) gives the (incorrect) rule $r_d = (p_m - p_s) - \sim \frac{1}{2} p_s$, and in *EAC* VI,v,8 (*KGW* 7, 497) Kepler gives the equivalent $r_d = p_m - \frac{3}{2} p_s$ with the remark, 'I demonstrate in my *Hipparchus*'. Evidently, the correct Th. XII was reached after the publication of *EAC* VI in 1621, and before the publication of *TR* in 1627.

Problem VI

Given the horizontal parallaxes of the luminaries and their apparent semidiameters, to find the sum of the semidiameters of the disc of the earth and of the penumbra of the moon.

Add together in one sum the semidiameters of the luminaries, the horizontal parallax of the moon, and half the horizontal parallax of the sun, and there will result the sum of the semidiameters of the disc and of the penumbra (from Problem V and Theorem XII).

Comm. Again by semidiameter of the penumbra is meant $r_p + r_{um}$. Since from Pr. V,

$$(r_p + r_{um}) = r_m + r_s + \sim \tfrac{1}{2} p_s,$$

and from Th. XII $r_d = p_m$, so

$$r_d + (r_p + r_{um}) = r_m + r_s + p_m + \sim \tfrac{1}{2} p_s.$$

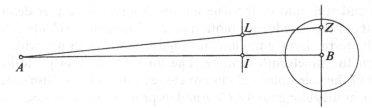

Figure 12

Theorem XIII

The lines of the motion in longitude of the moon from the sun, or in latitude from the ecliptic, are projected into the plane of the disc of the earth under angles at the moon having the proportion to the angles, under which they appear from the earth, that is the sum of the parallaxes of the moon and the sun to the parallax of the moon alone.

In the figure (12) let I be a point of the moon located under the ecliptic, IL an arc of latitude, let AL be drawn, and let it be extended to point Z in the plane of the disc. Now $AI/IL = AB/BZ$. But AB/AI is as the parallaxes of the moon and the sun to the parallax of the moon alone. Therefore, etc., q.e.d.

Comm. In Figure 12 the sun is at A and the earth at B with a line perpendicular to AB drawn in the disc of the earth. Suppose the moon to move through an arc IL of elongation from the sun or to be at a latitude IL. Draw AL and extend it to the disc of the earth at Z, and the projection of IL into the disc of the earth will be BZ. Since $AI/IL = AB/BZ$, so,

$$\frac{BZ}{IL} = \frac{AB}{AI} = \frac{d_s}{d_s - d_m} = \frac{p_m + p_s}{p_m}.$$

The purpose of this theorem is principally to project on to the earth the motion of the moon through the path of totality of a solar eclipse.

Problem VII

To augment the hourly motion [of the moon] correctly.

Let the parallax of the moon be divided by the parallax of the sun, let the quotient divide the arc of latitude or the hourly motion of the moon, and then let this quotient be added to the entire [hourly motion or arc of latitude].

Comm. The *horarius lunae* is the true hourly motion of the moon or at syzygy, as is the case here, the difference between the true hourly motions of the moon and sun, that is, the true hourly elongation. Kepler describes the components of the true hourly motion and its limits in *EAC* VI,v,7 (*KGW* 7, 493–4). The term 'horary motion' in English was still in use with the same meaning in the nineteenth century. The augmented hourly motion is the projection of the true hourly motion or elongation on to the disc of the earth. Let the true hourly elongation $LI = \eta$ and its projection on the disc of the earth $ZB = \eta'$. Then from Th. XIII,

$$\eta' = \eta \cdot \frac{p_m + p_s}{p_m} = \eta + \eta \cdot \frac{p_s}{p_m}.$$

Kepler's equivalent rule is

$$\eta' = \eta + \eta/(p_m/p_s)$$

and either formula applies also to projecting the moon's latitude β on to the disc of the earth. Since $p_s/p_m \approx \frac{1}{60}$, the increase is very small, about $0;0,30°$ per hour in η and at most about $0;1,30°$ in β in eclipses.

Theorem XIV

Given two unequal circles intersecting each other, if a straight line drawn through the centers of both and extended to the circumferences on both sides be cut by a straight line through the intersections of the circumferences, then if the center of the smaller circle is on the circumference of the larger: the square of the semidiameter of the smaller will be equal to the rectangle contained by the entire diameter of the larger and its shorter segment cut off by the line through the intersections.

Let the larger circle (Figure 13) be AKN, which the smaller circle, with center A located on the circumference of the larger, will intersect. Let the intersections be V, X, and let a straight line be drawn through them, intersecting the line drawn through the centers E, A and continued to the circumferences at B, Y. Let the intersection be D. I say that the square of the smaller semidiameter AB is equal to the rectangle contained by YA and AD.

For let center A be joined to intersection X. Now because $AB = AX$ and DX is perpendicular to AY, so $DA/AX = AX/AY$. And in continued proportions, the square of the mean is equal to the rectangle contained by the extremes. Therefore, $[AB^2 = AX^2 = DA \cdot AY]$.

Corollary. If I had said that the rectangle contained by the segments of the diameter of the smaller circle cut off by the circumference of the larger, namely, BA, AC, is equal to the rectangle contained by the shorter segment $[DA]$ of the diameter of the larger circle cut off by the line VX through the intersections and by either the sum or difference of the diameter AY of the larger circle and the

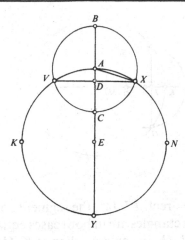

Figure 13

difference of the segments *BA, AC*, it will turn out the same. For rect. $BA \cdot AC$ = BA^2, and thus the difference of the segments *BA, AC* is nothing. But nothing subtracted from *AY* leaves *AY*, nothing added to *AY* makes *AY*.

Comm. It is to be shown that in Figure 13 $AB^2 = YA \cdot AD$. Draw chord $AX = AB$. Since $DA/AX = AX/AY$, so

$$AB^2 = AX^2 = DA \cdot AY.$$

Corollary. Since $BA = AC$, thus $BA \cdot AC = BA^2$, $BA - AC = 0$, and

$$BA \cdot AC = DA(AY \pm (BA - AC)).$$

Theorem XV

Other things remaining as before, if the center of the smaller circle is inside the circumference of the larger: the rectangle contained by the segments of the smaller diameter, which the larger circumference has divided, is equal to the rectangle contained by the shorter segment of the larger diameter, which the line through the intersections cuts off, and by the difference between the larger diameter and the difference of the said segments of the smaller diameter.

Let the center *A* of the smaller circle be placed inside the circumference of the larger (Figure 14), and let the line through the intersections be *FG*, cutting the line drawn through *E, A* in point *I*. In line *EA* extended, let points of the larger circumference be *N, K*, of the smaller *L, M*, and let *K, M* be interior. And from *M* let $MO = LK$ be extended towards center *A* so that $KO = KM - LK$ and $ON = NK - KO$. I say that rect. $LK \cdot KM =$ rect. $KI \cdot ON$.

In order that this may be demonstrated without great confusion, consider that *IF* is the common perpendicular from point *I* of the diameter to point *F* of both circumferences. Therefore, $IF^2 =$ rect. $KI \cdot IN$, the segments of the

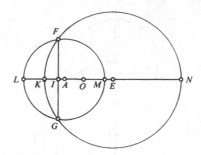

Figure 14

larger diameter, and $IF^2 = $ rect. $LI \cdot IM$, the segments of the smaller diameter. And consequently these rectangles are in both cases equal to each other [$KI \cdot IN = LI \cdot IM$]. However, $LI \cdot IM$ is greater than $LK \cdot IM$ by $KI \cdot IM$. Let the common $KI \cdot IM$ be subtracted [from $KI \cdot IN$ and $LI \cdot IM$], and accordingly $LK \cdot IM = KI \cdot MN$. But now $LK \cdot IM$ is less than the proposed $LK \cdot KM$ by $LK \cdot KI$. And likewise $KI \cdot MN$ is less than the proposed $KI \cdot ON$ by $KI \cdot OM$, that is, $KI \cdot LK$ by construction [since $OM = LK$]. The equals are therefore smaller by some equal than those that were proposed, each one answering to its own. Therefore the proposed are also equal to each other.

Comm. Referred to Figure 14, the theorem states that

$$LK \cdot KM = KI(KN - (KM - LK)).$$

From M let $MO = LK$ be extended towards A so that

$$KO = KM - LK, \quad ON = KN - KO,$$

and thus it may be shown that

$$LK \cdot KM = KI \cdot ON.$$

Now, in each circle

$$IF^2 = KI \cdot IN = LI \cdot IM.$$

But

$$LI \cdot IM - LK \cdot IM = KI \cdot IM,$$

and thus subtracting $KI \cdot IM$ from both sides of $KI \cdot IN = LI \cdot IM$, the remainders are $KI \cdot MN = LK \cdot IM$. And subtracting these from $LK \cdot KM$ and $KI \cdot ON$,

$$LK \cdot KM - LK \cdot IM = LK \cdot KI,$$

$$KI \cdot ON - KI \cdot MN = KI \cdot OM = KI \cdot LK,$$

since $OM = LK$ by construction. Thus $LK \cdot KM = KI \cdot ON$.

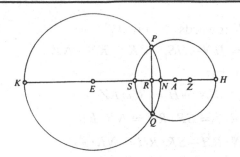

Figure 15

Theorem XVI

Other things again remaining the same as in Theorem XIV, if the center of the smaller circle is outside the circumference of the larger: the rectangle contained by the segments of the smaller diameter, which the larger circumference has divided, is equal to the rectangle contained by the shorter segment of the larger diameter, which the line through the intersections cuts off, and by the sum of the larger diameter and the difference of the said segments of the smaller.

Let the center A of the smaller circle be placed outside the circumference of the larger (Figure 15), and let the line through the intersections be PQ, cutting the line drawn through E, A in point R. And in this line drawn through the centers, let points of the larger circumference be N, K, as before, of the smaller H, S, but let N, S be interior. And from H let $HZ = SN$ be extended towards center A so that $ZN = HN - NS$ and $ZK = KN + NZ$. I say that rect. $HN \cdot NS$ = rect. $RN \cdot KZ$.

For again, as before, since the common perpendicular RP is a mean in both circles, it is demonstrated that rect.[41] $SR \cdot RH$ = rect. $NR \cdot RK$. But $SR \cdot RH$ is less than $SN \cdot RH$ by $NR \cdot RH$. Let the common $NR \cdot RH$ be added [to $SR \cdot RH$ and $NR \cdot RK$], and therefore $SN \cdot RH = NR(RK + RH) = NR \cdot KH$. But now $SN \cdot RH$ is greater than $SN \cdot NH$ referred to in the proposition by $SN \cdot NR$. And likewise $NR \cdot KH$ is greater than $NR \cdot KZ$ referred to in the proposition by $NR \cdot ZH$, that is $NR \cdot SN$ by construction [since $ZH = SN$]. The equals are therefore greater by some equal, each one answering to its own referred to in the proposition. Therefore those referred to are also equal to each other.[42]

Comm. Referred to Figure 15, the theorem states that

$$SN \cdot NH = RN(KN + (NH - NS)).$$

[41] 197:33 triangula/rectangula F.

[42] 197:43 An additional partial sentence reads: 'Therefore, since rect. $HN \cdot NS$ = rect. $RN \cdot KZ$, or to the two $RN \cdot NK$ and $RN \cdot NZ$ [i.e. $HN \cdot NS = RN \cdot KZ = (RN \cdot NK + RN \cdot NZ)$], but the same $HN \cdot NS = N\delta^2$ ...' In Kepler's figure $N\delta$, tangent to the larger circle at N, meets the smaller circle at δ, and thus $HN \cdot NS = N\delta^2$, but this is never applied.

From H extend $HZ = SN$ towards A so that

$$ZN = HN - NS, \quad ZK = KN + NZ,$$

and thus it may be shown that

$$SN \cdot NH = RN \cdot KZ.$$

Now, in each circle

$$RP^2 = SR \cdot RH = NR \cdot RK.$$

But

$$SN \cdot RH - SR \cdot RH = NR \cdot RH,$$

and thus adding $NR \cdot RH$ to both sides of $SR \cdot RH = NR \cdot RK$, the sums are $SN \cdot RH = KH \cdot NR$. And subtracting from these $SN \cdot NH$ and $RN \cdot KZ$,

$$SN \cdot RH - SN \cdot NH = SN \cdot NR,$$

$$NR \cdot KH - NR \cdot KZ = NR \cdot ZH = NR \cdot SN,$$

since $ZH = SN$ by construction. Therefore $SN \cdot NH = NR \cdot KZ$.

Problem VIII

Given the semidiameters of a luminary and of the shadow, and the length of the arc of either the lighted or darkened perimeter of the luminary, to find the segments of the diameter of either the luminary or the shadow produced by the line through the intersections, of which the shorter segment can be called the apotome or sagitta of the luminary or of the shadow.

For given the diameter of a circle and an arc, the sagitta of the arc is given in the same measure, which, multiplied by the remainder of the diameter of the luminary forms the square of half the line through the intersections. This square subtracted from the square of the semidiameter of the shadow leaves the square of the segment which is the remainder [of the semidiameter] as far as the center of the shadow after the sagitta is subtracted. Therefore the square root subtracted from the semidiameter of the shadow shows the sagitta of the shadow.

Another way: Given the semidiameter of the luminary and half the eclipsed or lighted arc, the sine of the arc is given in the same measure as both semidiameters [of the luminary and the shadow]. And given the sine in the measure of the semidiameter, the cosine is also given, which, subtracted from the semidiameter,[43] forms the sagitta.

Comm. In this problem the luminary may be either the moon or sun and the shadow either the shadow of the earth in a lunar eclipse or the darkened moon in a solar eclipse. However, Kepler seems to have in mind the configuration for a lunar eclipse, and Pr. XI is an application of PR. VIII to a lunar eclipse. Suppose now that Figure 16 shows an eclipse of the moon with lighted arc *FLG*

[43] 198:39 diametro/semidiametro F.

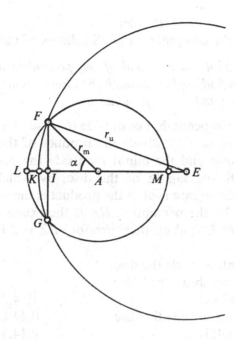

Figure 16

and darkened arc *FMG*. The 'apotome' (cut off segment) or 'sagitta' (arrow) is the versed since, i.e. vers $\alpha = 1 - \cos\alpha$. We wish to find the sagitta *LI* of the eclipsed luminary and *KI* of the shadow. Let the radius of the moon $r_m = AF$, the radius of the shadow $r_u = EF$, and half the lighted arc $LF = \alpha$. Kepler's steps are:

$$r_m \text{ vers } \alpha = LI, \quad \text{the sagitta of the luminary,}$$

$$LI \cdot IM = LI(2r_m - LI) = FI^2,$$

$$r_u^2 - FI^2 = IE^2,$$

$$r_u - IE = KI, \quad \text{the sagitta of the shadow.}$$

An alternate method is:

$$r_m \sin\alpha = FI, \quad r_m \cos\alpha = (r_m^2 - FI^2)^{\frac{1}{2}} = IA,$$

$$r_m - IA = LI, \quad \text{the sagitta of the luminary.}$$

The last two steps are as before. In Pr. XI Kepler has another method of finding *IE* and *IA* after finding *FI* that is suitable for use with a table of log cos α and is very accurate for small arcs. Consider *FIE* and *FIA* to be right spherical trangles. Then

$$IE = \cos^{-1}(\cos EF / \cos FI), \quad IA = \cos^{-1}(\cos AF / \cos FI).$$

Problem
Useful for the computation of all eclipses of the sun

Given the semidiameters of the disc and of the penumbra and the distances between their centers, to find the line through the intersections and the length of the arc of the disc intercepted by the penumbra.

First let the center of the penumbra be outside the disc. Accordingly, double the distance between the centers; divide the product of the minutes [of the penumbra] inside the disc and the minutes outside the disc by the double [distance]; there results the sagitta of the disc, by which multiply[44] the remainder of the disc; the square root of the product is the sine of half the arc of the disc intercepted by the penumbra. *NA* is the excess of the distance[45] between the centers over *EN*; altogether therefore $KZ = 2AE$.

Minutes[46] of the penumbra inside the disc	0;19,41	111458
Semidiameter of the penumbra 0;32,13		
[Diameter of the penumbra]	1; 4,26	
Minutes of the penumbra outside the disc	0;44,45	29325
Rectangle [0;19,41 · 0;44,45]	0;14,41	140783
Let the distance between the		
centers be 1;15;54, double	2;31,48	92819
[Sagitta]	0; 5,48	233602
Sine [!, in fact 0;5,48½]	0;18,40	116796
Semidiameter [of the disc]	1; 3,22	5471
Where the disc is 60, [the sine of the arc]	0;17,40	122267
Arc 17;7½°		

Again, let the center of the penumbra be inside the disc. Accordingly, add the distance between the centers to the semidiameter of the disc; from the sum subtract the amount of the semidiameter of the penumbra inside the disc, namely, as far as the center [of the penumbra]; divide the product of the minutes inside the disc and the minutes outside the disc by the remainder; there results the same sagitta [of the disc] as before. Note that *EK* or *EN* is equal to $EA + AK$. Therefore $ON = 2AE$ no less than in the previous case in which the center was outside [the disc].

Comm. This problem is an application of Th. XV and XVI, and refers to the configurations of Figures 14 and 15.[47] In the case of the center of the penumbra lying outside the disc of the earth, shown in Figure 15, *SH* is the part of the

---

[44]   198:6 duo/duc F.   [45]   198: note 1 distantia/distantiae

[46]   198:8 *om.* Sine.

[47]   It is out of place in the manuscript, between Th. XV and XVI, and may be intended to precede Pr. IX.

penumbra inside and NH the part outside. Likewise, if the center of the penumbra lies outside the disc, shown in Figure 14, LK is the part outside the disc and KM the part inside. Kepler's instructions for finding the sagitta RN or KI are equivalent to

$$RN = \frac{SN \cdot NH}{KZ} = \frac{SN \cdot NH}{2AE}, \quad KI = \frac{LK \cdot KM}{NO} = \frac{LK \cdot KM}{2AE}.$$

Note that in each case $NA = AZ$ and $KA = AO$, so that

$$KZ = 2(EN + NA) = 2AE, \quad NO = 2(EK - KA) = 2AE,$$

that is, twice the distance between the centers. Next, where the remainder of the disc $KR = r_d - RN$ and $IN = r_d - IN$, half the line through the intersections is

$$PR = (RN \cdot KR)^{\frac{1}{2}}, \quad FI = (KI \cdot IN)^{\frac{1}{2}},$$

and half the arc of the disc intercepted by the penumbra

$$\text{arc } PN = \sin^{-1}(PR/r_d), \quad \text{arc } FK = \sin^{-1}(FI/r_d).$$

Kepler's numerical example contains an error that is not consistent with these rules. The logarithms in the right column, which are not very accurate, are natural logarithms of the form $N = -10^5 \ln s$ where $s = (0-1;0)$, and are considered further in the commentary to Pr. XI below. For the case of the center of the penumbra outside the disc (Figure 15), we are given

$$r_p = 0;32,13°, \quad r_d = 1;3,22°, \quad AE = 1;15,54°, \quad SN = 0;19,41°.$$

Thus,

$$NH = 2r_p - SN - 0;44,45°, \quad SN \cdot NH = 0;14,41$$

and the sagitta

$$RN = (SN \cdot NH)/2AE = 0;5,48°.$$

So far so good, but now by error Kepler takes $0;5,48^{\frac{1}{2}} = 0;18,40$, from which he finds $\sin PN = 0;18,40/1;3,22 = 0;17,40$ and arc $PN = 17;7\frac{1}{2}°$. Correctly,

$$KR = r_d - RN = 0;57,34°, \quad PR = (RN \cdot KR)^{\frac{1}{2}} = 0;18,16°,$$

$$\sin PN = (PR/r_d) = 0;17,18, \quad \text{arc } PN = 16;45°.$$

Problem IX

Given the semidiameters of the disc of the earth and of the penumbra, and given the width of the crescent of the penumbra extending beyond the disc, to find the segments of the diameter of the penumbra produced by the line through the intersections.

Let the width of the crescent be multiplied by the remainder of the diameter of the penumbra, and (*a*) let the result be divided by the diameter of the disc reduced by twice the difference between the width of the crescent and a semidiameter of the penumbra larger [than the width of the crescent]; or (*b*) let it be divided by the diameter of the disc increased by twice the difference between the width of the crescent and a semidiameter of the penumbra smaller [than the width of the crescent]. If (a) the crescent is smaller in width than the semidiameter of the penumbra, subtract twice the deficit from the diameter of the disc; but if (b) it is larger, add twice the excess to the diameter of the disc; and in this way you will obtain the divisor. Or double the distance between the centers [of the disc and the penumbra], and you will obtain the divisor.

Comm. This is also an application of Th. XV and XVI. Refer first to Figure 14 for case (a), and let KN be the diameter of the disc of the earth and LM the diameter of the penumbra with LK the width of the crescent extending beyond the disc of the earth. For the width of the crescent less than the semidiameter of the penumbra, Kepler's rule (a) is equivalent to

$$KI = \frac{LK \cdot KM}{KN - 2(AL - LK)} = \frac{LK \cdot KM}{KN - 2KA} = \frac{LK \cdot KM}{NO} = \frac{LK \cdot KM}{2AE},$$

since as before $KA = AO$ and $NO = 2(EK - KA) = 2AE$, twice the distance between the centers. KI is the sagitta of the disc, and the segments of the diameter of the penumbra are then

$$LI = LK + KI, \quad IM = LM - LI.$$

Referring now to Figure 15 for case (b) of the crescent larger than the semidiameter of the penumbra, where SH is the diameter of the penumbra and NH the width of the crescent, rule (b) is equivalent to

$$RN = \frac{SN \cdot NH}{KN + 2(NH - AH)} = \frac{SN \cdot NH}{KN + 2NA} = \frac{SN \cdot NH}{KZ} = \frac{SN \cdot NH}{2AE},$$

since $NA = AZ$ and $KZ = 2(EN + NA) = 2AE$. Now RN is the sagitta of the disc, and the segments of the penumbra are

$$SR = SN - RN, \quad RH = SH - SR.$$

Problem X

Given the diameters of a luminary and of the shadow and the magnitude of an eclipse, to investigate the segments of the diameter of the shadow produced by the line through the intersections.

Multiply the width of the lighted part by the width of the darkened part, and then subtract the smaller from the larger. If the lighted part is greater, add the difference to the diameter of the shadow, but if the darkened part is greater, subtract the difference from the diameter of the shadow. And divide the product of the original multiplication by this sum or remainder; it will show the segment of the diameter of the shadow produced by the line through the intersections.

Comm. This is much the same as Pr. IX. Again it may be applied to either a lunar or solar eclipse, although Kepler is concerned primarily with the former. Assume first that the darkened part of the luminary is greater, that the magnitude of the eclipse is greater than one-half the eclipsed body. Then in Figure 14, *KM* is the magnitude of the eclipse, *LK* the lighted part of the luminary, and *KN* the diameter of the shadow (or the eclipsing body). Hence,

$$ KI = \frac{LK \cdot KM}{KN - (KM - LK)}, \quad NI = KN - KI. $$

Note that $KN - (KM - LK) = KN - 2KA = ON = 2AE$. Now assume that the lighted part is greater, that the eclipse is less than one-half, and in Figure 15 *SN* is the magnitude of the eclipse, *NH* the lighted part, and thus

$$ RN = \frac{SN \cdot NH}{KN + (NH - SN)}, \quad KR = KN - RN. $$

In this case $KN + (NH - SN) = KN + 2NA = KZ = 2AE$.

Problem XI

Given the size of the diameter of a luminary and also of the shadow, and given the eclipsed or lighted arc, to find the magnitude of the eclipse.

Determine the sagittae of both the luminary and the shadow, and as long as less than a semicircle of the luminary is in the shadow, add the sagitta of the luminary to the sagitta of the shadow to find the magnitude of the eclipse. But if the darkened perimeter is greater than a semicircle, then, in place of the sagitta of the luminary, the remainder of the diameter should be added, that is, always the part of the diameter in the shadow. (Or subtract the sagitta of the shadow from the sagitta of the luminary to find the remaining lighted part.) In this way the total number of minutes of the eclipse will be formed.

Example of the operation through logarithms:
Let the remaining lighted perimeter be 81°.

$$ \text{half} = 40;30° \quad \log \quad 43155 \quad \text{quadrantal sine} $$

Let the semidiameter of the

$$ \text{moon be } 0;17,1° \quad \log \quad \underline{126000} \quad \text{sexagesimal} $$
$$ \log \quad 169155 \quad \text{sexagesimal} $$

This indicates 0;11,2°, which is half the line through the intersections. Hence, the length of the sagitta by antilogarithms:

Let the semidiameter of the shadow be	0;48,36°	antilog 9.98
Half the line through the intersections	0;11, 2°	antilog 0.515
Therefore, the remainder of the semidiameter of the shadow	0;47,18°	9.465
[Semidiameter of the shadow	0;48,36°]	
Sagitta of the shadow	0; 1,18°	
Semidiameter of the moon	0;17, 1°	antilog 1.225
[Half the line through the intersections	0;11, 2°]	0.515
Remainder of the semidiameter	0;12,57°	0.710
[Semidiameter of the moon	0;17, 1°]	
Remainder of the diameter because more than a semicircle is in eclipse	0;29,58°	
[Sagitta of the shadow]	0; 1,18°	
Magnitude of the eclipse	0;31,16°	

Comm. The method of finding the sagitta of the luminary and the shadow (or eclipsing body) was given in Pr. VIII. If the eclipse is less than a semicircle of the luminary (Figure 15), the sagitta of the luminary is SR and the sagitta of the shadow RN. The magnitude of the eclipse in angular measure m is then $m = SR + RN$. If the eclipse is greater than a semicircle of the luminary (Figure 14), the sagitta of the luminary is LI and of the shadow KI, and the magnitude is found by adding to KI the remainder of the diameter of the luminary $IM = LM - LI$, that is $m = KI + (LM - LI)$. Alternatively, the lighted part LK may be found from $LK = LI - KI$, and then the magnitude is $m = LM - LK$.

The computation in the example is done by logarithms of the sort found in the *Rudolphine Tables*.[48] A marginal note to the first logarithm is: 'The quadrantal-sine logarithm (*sinosus quadrantalis logarithmus*) is where [the logarithm of] the total sine [i.e. sin 90°] is 0. The sexagesimal (*sexagenarius*) [logarithm] is where [the logarithm of] 60′ is 0.' Both kinds come from the table *Heptacosias logarithmorum logisticorum* (Seven hundred computational logarithms) in the *TR* (*KGW* 10, Tables, pp. 2–11) which gives a column of natural logarithms, $N = -10^5 \ln x$ to from eight to five places, that may be used with four columns of entry of which two are applied here. In the 'Sexagesimal minutes' (*Sexagesimus scrupula*, here *sexagenarius*), $x = s$ (0 − 1;0, $\Delta s = 0;0,5$). In the 'Arc of the quadrant' (*Arcus quadrantis*, here *sinosus quadrantalis*) $x = \alpha = \sin^{-1} s$ (0° − 90°, $\Delta \alpha = \Delta \sin^{-1} s$ to 0;0,1°). Hence the column of logarithms may be read as $N = -10^5 \ln s$ or as $N = -10^5 \ln \sin \alpha$.

---

[48] For a description of these logarithms, and Kepler's *Chilias Logarithmorum* of 1624 (*KGW* 9, 277ff.), see H. H. Goldstine, *A History of Numerical Analysis from the 16th through the 19th Century*. Studies in the History of Mathematics and Physical Sciences, 2 (New York: Springer-Verlag, 1977), pp. 41–5.

Finally, what Kepler calls 'antilogarithms' come from a table 'of more exact antilogarithms especially for eclipses' (*KGW* 10, Tables, p. 23) that gives the natural logarithms of the cosine, $N = -10^5 \ln \cos \alpha \, (0° - 1;40°, \Delta\alpha = 0;0,10°)$.

The example is for a lunar eclipse, shown in Figure 16, in which $AF = r_m = 0;17,1°$, $EF = r_u = 0;48,36°$, and half the lighted arc $LF = \alpha = 40;30°$. Hence,

$$FI = r_m \sin \alpha = 0;11,2°. \qquad \text{(Corr. 0;11,3)}$$

Regarding *FIE* as a right spherical triangle,

$$IE = \cos^{-1}(\cos EF / \cos FI) = 0;47,18°, \qquad \text{(Corr. 0;47,20)}$$

and the sagitta of the shadow

$$KI = r_u - IE = 0;1,18°.$$

Then, taking *FIA* as a right spherical triangle, the remainder of the semidiameter of the moon,

$$IA = \cos^{-1}(\cos AF / \cos FI) = 0;12,57°,$$

and since more than a semicircle is eclipsed, the remainder of the diameter

$$MI = IA + r_m = 0;29,58°.$$

Finally, adding to this the sagitta of the shadow, the magnitude of the eclipse is

$$m = MI + KI = 0;31,16°.$$

Problem XII

Given the diameter of the eclipsed luminary, the length of the arc of the lighted or darkened perimeter, and the magnitude of the eclipse, to determine the semidiameter of the shadow of the earth.

This problem is more ingenious than useful. For given the semidiameter of the luminary, as the total sine is to the sine of half the lighted arc (Figures 14 and 15), *LF* or *HP*, so this semidiameter will be to half the line through the intersections *FI* or *PR*. Then as the total sine is to the sagitta of the same arc, so is the semidiameter[49] of the luminary to the line *LI* or *HR* terminated at the line through the intersections. Subtract from this line the magnitude of the eclipse [!] expressed in the same measure as the semidiameter of the luminary, namely *LK* or *HN*, and the remainder will be the sagitta of the shadow *KI* or *NR*. But as the sagitta is to the previous half of the line through the intersections, namely to *FI* or *PR*, so is this line through the intersections to the remainder of the diameter of the shadow, namely, to *IN* or *RK*.

Comm. Since the configuration described is a lunar eclipse, and the object is

---

[49]   200:19 diameter/semidiameter F.

to find radius of the shadow of the earth, the luminary here can only be the moon. The instructions refer to the lettering of Figures 14 and 15 together. The line through the intersections,

$$FI = r_m \sin LF, \quad PR = r_m \sin HP,$$

and the sagitta of the luminary,

$$LI = r_m \text{ vers } LF, \quad HR = r_m \text{ vers } HP.$$

Subtracting the diameter of the luminary *less* the magnitude of the eclipse, $2r_m - m$ (Kepler mistakenly says to subtract the magnitude of the eclipse), the result is the sagitta of the shadow,

$$KI = LI - LK, \quad NR = HR - HN.$$

The remainder of the diameter of the shadow,

$$IN = FI^2/KI, \quad RK = PR^2/NR.$$

Theorem XVII

If two unequal circles intersect each other, a straight line be drawn through the centers, and a perpendicular erected on it from the center of the smaller; and from the center of the larger circle to the perpendicular a line be applied equal to the sum of both semidiameters: the square of this perpendicular will be equal to the rectangle contained by the segment of the diameter of the smaller circle [within the larger circle] and by the sum of this segment and twice the distance between the centers.

Let there be two circles (Figure 17), the larger *LEM* with center *P*, the smaller OFX[50] with center *A*, intersecting each other in *V*, *X*, and let a straight line *PA* be drawn through the centers, intersecting the circumferences, the larger in *E*, the smaller on the farther side of its center in *F* and on the closer side[51] of its center in *O*. And from *A* let *AB* be erected perpendicular to *PF*, but from *P* let a straight line be terminated in *BA* made up from $PL + LB = PE + AF$, which will be *PB*. I say that $BA^2 = \text{rect. } EO(EO + 2AP)$.

For let a circle with center *P* and radius *PB* be described which will intersect *PF* produced in *N*, *Q*. Now, since *QAB* is a right angle, $AB^2 = \text{rect. } QA \cdot AN$. But $QA = EO$ because *AE* is an element common to both and *AO* or *AF* and *EQ* or *LB* have been made equal. But $AN = EO + 2AO$.[52] For because $EO = AQ$ and $AQ + AP = PQ$, therefore $EO + AP = PQ$, that is *PN*. Therefore, let *AP* be added again to *PN*; the entire $AN = EO + 2AP$. [Hence $AB^2 = QA \cdot AN = EO(EO + 2AP)$.]

---

[50]   200:31 *OF/OFX* F.   [51]   200:33 eis/cis F.

[52]   201:2 composita/compositae F.

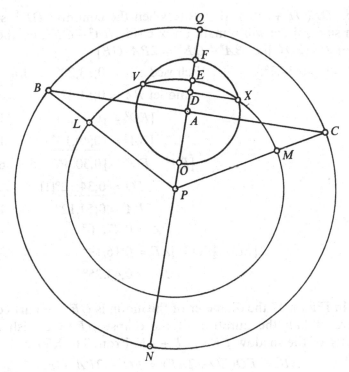

Figure 17

Comm. In Figure 17 it is to be shown that $AB^2 = EO(EO+2AP)$. Now $AB^2 = QA \cdot AN$. Since by construction $QE = AO$,

$$QA = QE + EA = EA + AO = EO,$$
$$AN = PN + AP = QP + AP = QA + 2AP = EO + 2AP,$$

and thus

$$AB^2 = QA \cdot AN = EO(EO + 2AP).$$

Problem XIII

Given the magnitude of an eclipse and the proportion of the semidiameter of the moon to the arc of half the duration [of the eclipse], to find the proportion of the same semidiameter of the moon to the semidiameter of the shadow and to the arc of latitude.

The proportion of [the diameter of the moon] *OF* is given (Figure 17), both to the eclipsed part *OE* and to the arc *AB* of half the duration. Now let OE^2 be subtracted from AB^2, let the remainder be transformed into a rectangle of which one side will equal $2EO$, the other side will be *PA* (to which let be added $AE = EO - AO$); or one side will equal *EO*, the other will be $2PA$. For because

63

BA^2 = rect. $OE(OE+2PA)$, therefore when the common OE^2 is subtracted from both sides, there will remain on one side BA^2-OE^2, on the other side $OE \cdot 2PA$ or $PA \cdot 2OE$ [i.e. $BA^2-OE^2 = 2PA \cdot OE$].

Let $FO = 0;34,2°$,

Eclipsed $EO = 0;33,20°$	log 58790
and let $BA = 0;54,22°$	9872
$[EO^2 =]0;18,31°$	117580
$[BA^2 =]0;49,15°$	19744
$[BA^2-EO^2 =]0;30,44°$	66900
$FO = \underline{0;34, 2°}[!]$	56711
$2PA = 0;54,12°$	10190

Therefore,

$$PA = 0;27, 6°$$
$$[EO-\tfrac{1}{2}FO =]AE = \underline{0;16,19°}$$
$$PE = 0;43,25°$$

Comm. In Figure 17 the diameter of the moon is OF, the part eclipsed OE, and the arc of half the duration of the eclipse AB. We wish to find the semidiameter of the shadow $PE = AP+AE$. From Th. XVII,

$$AB^2 = EO(EO+2AP) = EO^2+2PA \cdot OE.$$

Hence,

$$2PA = (AB^2-EO^2)/EO, \quad AE = EO-\tfrac{1}{2}FO, \quad PE = PA+AE.$$

Kepler again uses logarithms, in this case exclusively 'sexagesimal' logarithms $N = -10^5 \ln s$ (see *comm.* to Pr. XI), but not very accurately. We have

$$FO = 0;34,2°, \quad EO = 0;33,20°, \quad BA = 0;54,22°,$$

and thus

$$BA^2-EO^2 = 0;49,15-0;18,31 = 0;30,44.$$

So far so good, but now he mistakenly divides by FO instead of EO, and all that follows is defective. Correctly, using EO,

$$2PA = 0;30,44/0;33,20 = 0;55,19°.$$

Hence

$$PA = 0;27,40°, \quad AE = 0;16,19°, \quad PE = 0;43,59°.$$

Theorem XVIII

If a larger circle be tangent to two smaller circles equal to each other, one within, the other without, in the same region from the center of the larger; and the centers of the smaller circles be joined, both with the center of the larger, and also with each other by an extended straight line, and a perpendicular be drawn from the center of the larger to that line: the rectangle contained by the difference of the two lines that join the centers of the smaller circles with the center of the

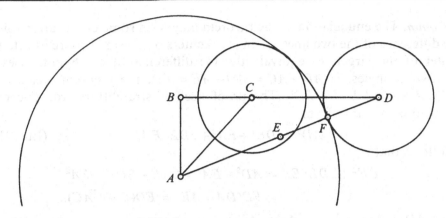

Figure 18

larger, and by the sum of both added to the square of the line between the center of the interior circle and the perpendicular, will be equal to the square of the line between the center of the exterior circle and the same perpendicular.[53]

Or: If one side of a right triangle be cut by a straight line from the opposite angle, forming a smaller triangle within the larger with the right angle in common: the rectangular parallelogram contained by the difference of the hypotenuses, and by the sum of both, together with the square of the side in the smaller triangle, will be equal to the square of the side in the larger triangle.

Let side *BD* of triangle *ABD* (Figure 18) be cut in *C* by straight line *AC* drawn from the opposite angle *A*, so that *BAC* will be a smaller triangle with a common right angle *B*, and in *AD* let *AE* be marked off equal to *AC*. I say that rect. $ED(DA + AC) + CB^2 = BD^2$.

For because *B* is a right angle,

$$CA^2 = EA^2 = AB^2 + BC^2.$$

But
$$AD^2 = DE^2 + EA^2 + 2DE \cdot EA.$$

Therefore, if for EA^2 you substitute $AB^2 + BC^2$,

$$AD^2 = DE^2 + 2DE \cdot EA + AB^2 + BC^2.$$

But
$$DE^2 + 2DE \cdot EA = ED(DA + AE) = ED(DA + AC).$$

Therefore,
$$AD^2 = ED(DA + AC) + AB^2 + BC^2.$$

Further
$$AD^2 = AB^2 + BD^2.[54]$$

Therefore,
$$ED(DA + AC) + AB^2 + BC^2 = AB^2 + BD^2.[55]$$

Subtract the common AB^2 and there will remain on one side $ED(DA + AC) + BC^2$, and on the other BD^2, both equal [i.e. $ED(DA + AC) + BC^2 = BD^2$], q.e.d.

[53]   202:8 particularem/perpendicularem F.   [54]   202:38 *om. AC* F.

[55]   202:40 quadrato *AB*/quadratis *AB, BC* F.

Comm. The enunciation of the theorem may need some clarification. Since the difference of the two lines joining the centers of the smaller circles with the center of the larger, or equivalently, the difference of the hypotenuses of the two triangles, is $AD-AC = AD-AE = ED$, the theorem means that $ED(DA+AC)+BC^2 = BD^2$. The proof itself is straightforward. The only point to note is that from

$$AD^2 = DE^2+EA^2+2DE\cdot EA, \qquad\qquad \text{(Euc. II,4)}$$

it follows that

$$DE^2+2DE\cdot EA = AD^2-EA^2 = (AE+ED)^2-EA^2$$
$$= ED(DA+AE) = ED(DA+AC),$$

since $AD = AE+ED$ and $AE = AC$ by construction.

Problem XIV

Given the diameter of the moon DE and the arc of immersion DC with the arc of half the delay CB in a total eclipse, to find the semidiameter of the shadow AF and the arc of latitude AB.

Construction (Figure 18). Let the arc of immersion DC be added to the arc of half the delay CB, and the result will be the arc of half the duration DB. Then let the remainder[56] $DB^2 - BC^2$ be transformed into a rectangle with a side equal to the diameter of the moon DE, and a side will be equal to $DA + AC = 2AF$, [twice] the semidiameter of the shadow [i.e. $DB^2 - BC^2 = DE(DA+AC) = DE\cdot 2AF$]. Next subtract from AF the semidiameter of the moon, $AF-FE = EA = CA$, and from $AC^2 - CB^2 = AB^2$, the square root of the remainder is the arc of latitude AB.

Let the diameter of the moon $DE = 0;31,44°$, the arc of immersion $DC = 0;36,30°$, the arc of half the delay, $CB = 0;39,46°$, and therefore the arc of half the duration, $[DC+CB =]DB = 1;16,16°$. But

$$DB^2 - CB^2 = 1;36,52-0;26,22 = 1;10,30.$$

Divide this by twice the diameter of the moon

$$(DB^2 - CB^2)/2DE = 1;10,30/1;3,28 = 1;6,39° = AF,$$

[the semidiameter of the shadow], from which subtract the semidiameter of the moon

$$AF-FE = 1;6,39° - 0;15,52° = 0;50,47° = AE = AC.$$

From the square subtract the square of half the delay

$$AC^2 - CB^2 = 0;42,58-0;26,22 = 0;16,36 = AB^2,$$

and the square root is the latitude $AB = 0;31,35°$.

[56] 203:6 *om.* cui addatur F.

Comm. Yet a fourth way – cf. Pr. II, XII, XIII – to find the semidiameter of the shadow in a lunar eclipse, but this time a total eclipse. In Figure 18, DC is the arc of immersion, CB the arc of totality or of half the delay, and $DB = DC + CB$ the arc of half the duration of the eclipse. Since $AE = AC$, were the center of the moon at E, it would be internally tangent to the shadow, and thus DE is equal to the diameter of the moon. From Th. XVIII, $DB^2 - BC^2 = DE(DA + AC)$. But

$$DA + AC = DA + AE = (DF + AF) + (AF - EF) = 2AF,$$

since $DF = EF$. Thus $DB^2 - BC^2 = DE \cdot 2AF$ and the semidiameter of the shadow $AF = (DB^2 - BC^2)/2DE$. Then, subtracting the semidiameter of the moon $AF - FE = AE = AC$, and $(AC^2 - CB^2)^{\frac{1}{2}} = AB$, the arc of latitude at the middle of the eclipse.

Kepler's example, somewhat rearranged in the translation, is clear enough although there are two errors. Correctly, $DB^2 = 1;36,57$ and $AB = 0;31,34$. However, these result from inaccuracies in his logarithmic solution that is placed in the margin.[57] We transcribe it here, adding in the first column the corresponding letters of Figure 18.

$\frac{1}{2}DB$	0;38, 8	log	45331
$(\frac{1}{2}DB)^2$	0;24,13		90662
$4 \cdot (\frac{1}{2}DB)^2 = DB^2$	1;36,52		
CB	0;39,46		41100
CB^2	0;26,22		82200
$\frac{1}{2}(DB^2 - CB^2) = c$	0;35,15		
DE	0;31,44		63677
$c - DE$	0; 3,31		283600
$(c - DE)/DE = AF - 1$	0; 6,39		219923
$AF - FE = AC$	0;50,47		16700
AC^2	0;42,58		33400
$AC^2 - CB^2 = AB^2$	0;16,36		128400
AB	0;31,35		64200

[57]   A last line of the problem, 203:37, 'For reason of certainty to work also computationally', may refer to this solution. On the logarithms, which are of the form $N = -10^5 \ln s$, see the comment to Pr. XI. Note that in the first three lines, because $s = (0 - 1;0)$, in order to find $DB^2 > 1;0$, Kepler uses $4 \cdot (\frac{1}{2}DB)^2 = DB^2$. For the same reason, instead of $(DB^2 - CB^2)/2DE = AF > 1$, he computes

$$[\tfrac{1}{2}(DB^2 - CB^2) - DE]/DE = AF - 1.$$

Problem XV

Given the semidiameter of the sun, given also the semidiameter of the shadow from the previous problem [XIV], that is, not by means of the parallax of the sun, and finally given the parallax of the moon, to determine the parallax of the sun.

Add the semidiameters of the sun and the shadow, from the sum subtract the parallax of the moon, the remainder is the parallax of the sun, for by Theorem III the sum of the semidiameters of the sun and shadow is equal to the sum of the parallaxes of the sun and moon.

Comm. Having independently found the semidiameters of the sun and shadow and the lunar parallax, it is now possible, in principle, to find the solar parallax. By Th. III, $p_s + p_m = r_s + r_u$, and thus $p_s = r_s + r_u - p_m$. It does not, however, appear that Kepler believed this procedure to be practically applicable.

Problem XVI

Other things remaining the same, but in place of the simple parallax of the moon, given the parallax of the moon less the sun from observation of an eclipse of the sun, to determine the same parallax of the sun.

Add the semidiameters of the sun and the shadow as before, from the sum subtract parallax of the moon less the sun, the remainder is twice[58] the parallax of the sun.

Comm. Here instead of the lunar parallax we take the parallax of the moon less the sun, $p_m - p_s$, and from Th. IV.

$$r_s + r_u - (p_m - p_s) = 2p_s.$$

Pr. XV and XVI are not extant in Kepler's manuscript, the text in *KGW* being taken from Frisch's edition. Since there appears to be no gap in the manuscript, one may wonder where Frisch found them (and hope he did not make them up).

Appendix: cancelled and omitted sections

For reasons of length we have not translated cancelled sections and have omitted three unnumbered problems at the end of the *Sciametria*. Here they are described briefly with page references from *KGW* 20.1. Cancelled theorems and problems are not the same as those with corresponding numbers in the main text.

Cancelled Theorem XII (190): $r_i + r_m = r_d + (r_p + r_{um})$. The proof is barely begun. The theorem is incorrect, and cannot be reconciled with Pr. III and the following Th. XI and new Th. XII or Pr. VI from which it would follow that

$$r_d + (r_p + r_{um}) = r_m + r_s + p_m + \tfrac{1}{2}p_s = r_i + r_m + \tfrac{3}{2}p_s.$$

[58] 204:8 parallaxeos/duplum parallaxeos F.

In fact Kepler wrote in the margin, 'Spurious. Three-halves (*sesqui*) of the parallax should be added.' A lengthy deleted section (192–3) marked 'Ad Th: XII Hipparchi' seems to be concerned with difficulties in the first proof of Th. XI. At the end of the first, aimless paragraph, Kepler writes *Nihil efficio* (I accomplish nothing), and then after two further short paragraphs he begins the second proof of the theorem.

Cancelled Theorem XIII (191): $r_d = (p_m - p_s) - \sim \frac{1}{2}p_s$. This may be proved from cancelled Th. XII, $r_d = r_m + r_i - (r_p + r_{um})$, and substitution from Pr. III and Th. XI, from which

$$r_d = r_m + (r_s + p_m - p_s) - (r_s + r_m + \tfrac{1}{2}p_s) = (p_m - p_s) - \tfrac{1}{2}p_s.$$

However, it appears from the proof that Th. XI had not yet been written, and in any case the result is contradicted by the new Th. XII, that $r_d = p_m$, and is incorrect. As noted in the commentary to Th. XII, in *EAC* VI,v,8 (*KGW* 7, 497) Kepler gives the equivalent (incorrect) rule $r_d = p_m - \frac{3}{2}p_s$ with the remark, 'I demonstrate in my *Hipparchus*'.

Cancelled Problem VII (194): Given p_m, p_s, and r_s, to find r_d. The solution appears to be

$$r_d = p_m - p_s \cdot \frac{p_m + p_s}{r_s}.$$

However, since $(p_m + p_s)/r_s \approx 4$, this would mean that $r_d \approx p_m - 4p_s$, contradicting both the cancelled Th. XIII and the new Th. XII.

Cancelled Problem VII [formerly VIII] (195): Given p_m, p_s, and r_s, to find $(r_p + r_{um})$. The solution is to find $r_d + (r_p + r_{um})$ by Pr. VI (text V), find r_d by (cancelled) Pr. VII (text VI), and subtract to find $(r_p + r_{um})$. This would give the odd result

$$(r_p + r_{um}) = r_m + r_s - 3\tfrac{1}{2}p_s,$$

contradicting Th. XI and Pr. V (although it is possible that by Pr. V and VI Kepler here means something not in the present manuscript).

All the preceding theorems and problems are faulty in one way or another, and must have been written and cancelled before the present versions of Th. XI and XII and Pr. V and VI. And the present version of Th. XII contains a note, 'Having examined Th. XI, the entire XII [i.e. the cancelled XII] will be deleted and the following [theorem] will be understood in this way.'

In the middle of Th. XV is a long deleted paragraph (196–7) that appears to be a superseded proof in which a diagram (not given) was marked by Greek letters. It was presumably replaced by the present second paragraph of the theorem.

The numbered theorems and problems are followed by two problems and one theorem in which Kepler does not so much explain a method as show an example. The first problem is to find the radius of the earth's shadow from two lunar eclipses, respectively, of large and small magnitude, north and south of the ecliptic, and near lunar apogee and perigee. The eclipses are those of 26 June and

20 December 1620, recomputed, I presume, with the lunar and eclipse sections of the *TR*, and Kepler finds 0;44,7°, and 0;49,27°, nearly the minimum and maximum radius and then the mean value 0;46,47°. The second problem is to find the eccentricity of the moon from the same eclipses, actually from the maximum and minimum radii of the shadow just found, and assuming the parallax and apparent diameter of the sun and the distance of the moon from the earth. The method is not good, and the result, 0.04775, is much too large, as is obvious when compared with the better values 0.04444 in the following chapter on the true eccentricity of the moon and 0.04362 in the *EAC* and the *TR*. The following theorem is to estimate the moon's mean argument of latitude from some number of solar eclipses (although lunar eclipses would do just as well). Kepler uses eclipses of 14/24 December 1601 and 2/12 October 1605, a period of 4 years less 73 days, or $(4-\frac{1}{5})$ years, in which the node must have regressed by the 73° the sun falls short of four revolutions. The period of the regression is therefore $(360/73)$ $(4-\frac{1}{5})$ years, or about 19 years, amounting to a little more than 0;3° per day added to the moon's motion in longitude to give its argument of latitude.

3 Descartes, Pappus' problem and the Cartesian Parabola: a conjecture

HENK J. M. BOS

Acknowledgement

Kirsti Andersen read and discussed several drafts of the present article; I am thankful for her advice and support.

Sometime in 1631 the Dutch mathematician and orientalist Jacob van Gool (or Golius) suggested to Descartes that he should try his new mathematical method in solving the problem, mentioned by Pappus, of the 'locus to three, four or more given lines'.[1] This is the problem that, through Descartes' treatment of it in his *Geometrie*[2] of 1637, has become famous as 'Pappus' problem'. Pappus' problem is a so called 'locus problem', that is, it requires that a curve be determined from a given property shared by all its points. In the case of Pappus' problem the given property can be varied in many ways; consequently there are various cases of Pappus' problem to be distinguished, each leading to particular classes of solution curves. In his *Geometrie* Descartes explained in general how the equations of the loci could be derived and he presented a thorough study of the case of 'three and four lines'; in that case the loci are conic sections. In addition he dealt with one special further case. Here the locus was a third-degree curve which, because of its special role in the *Geometrie*, later became known as the 'Cartesian Parabola'. Indeed the curve occurs in the *Geometrie* in four different settings; apart from being the locus in one case of Pappus' problem, it appears as the result of a particular procedure of curve tracing which I shall call the 'turning ruler and moving curve procedure', as an example in the application of Descartes' method for finding normals to curves, and as the curve used in a standard construction of roots of fifth- and sixth-degree equations.

---

[1] Cf. R. Descartes, *Oeuvres de Descartes*, ed. Ch. Adam and P. Tannery, 12 vols (nouvelle présentation) (Paris: Vrin, 1964–1974), vol. 2, p. 235 note.

[2] Descartes' *La Geometrie* constitutes one of the 'essais' in his *Discours de la Methode*, Leiden: 1637 (pp. 297–413). The text is in his *Oeuvres* 6, pp. 367–485. There is a facsimile edition with English translation: D. E. Smith and M. L. Latham eds., *The Geometry of René Descartes* (New York: Dover, 1954). In translating the *Geometrie*, I have in general followed Smith and Latham's translation, but occasionally I found it necessary to modify their version.

The separate occurrences of the same curve suggest connections. The case of the normal method needs no particular explanation – to illustrate his method Descartes naturally chose a curve already introduced earlier. But the connections between the other three occurrences are more enigmatic. I have been intrigued for a long time by the suggestive coincidences around the Cartesian Parabola, and the invitation to contribute to the 'Festschrift' for Tom Whiteside provided an occasion to work out some ideas about the curve, its tracing procedure and its role in Descartes' earliest studies on Pappus' problem, and to present them in honour of (and to the critical scrutiny of) this master of the restoration and recreation of seventeenth-century mathematical thinking.

The problem of Pappus is as follows:[3]

> **Problem** ('Pappus' Problem'). Given n straight lines L_i in the plane (see Figure 1), n angles θ_i and a line segment a. For any point P in the plane one defines oblique distances d_i to the lines L_i; these are the lengths of segments that are drawn from P towards L_i making with L_i the angle θ_i. It is required to find the locus of points P for which a certain ratio, involving the segments d_i and depending on the number of lines, is constant. The relevant ratios are:
>
> For 3 lines: $\qquad\qquad\qquad\qquad\qquad d_1^2 : d_2 d_3;$
> For 4 lines: $\qquad\qquad\qquad\qquad\qquad d_1 d_2 : d_3 d_4;$
> For 5 lines: $\qquad\qquad\qquad\qquad\qquad d_1 d_2 d_3 : a d_4 d_5;$
> For 6 lines: $\qquad\qquad\qquad\qquad\qquad d_1 d_2 d_3 : d_4 d_5 d_6;$
> In general for an even number $2k$ of lines: $\quad d_1 \ldots d_k : d_{k+1} \ldots d_{2k};$
> And for an uneven number $2k+1$: $\qquad d_1 \ldots d_{k+1} : a d_{k+2} \ldots d_{2k+1}.$

For brevity I shall refer to the curves that appear as loci for some instance of Pappus' problem as 'Pappus curves' or 'Pappus' loci', and to special instances of Pappus' problem as 'Pappus problems', if necessary indicating the number of given lines ('5-line Pappus problem', 'Pappus problem in k lines'). It should be noted that, unlike the other problems in an uneven number of lines, the 3-line problem is a special case of the problem in four lines, arising when two lines coincide. The 3-line problem which one would expect by analogy,

[3] Descartes formulated the problem in *Geometrie* pp. 304–7, Pappus mentions the problem in the *Collectio* VII–33–40 (Pappus Alexandrinus, *Collectionis quae supersunt*, ed. F. Hultsch, 3 vols. (Berlin: Weidman, 1876–1878) (reprint Amsterdam: Hakkert, 1965), vol. 2, pp. 676–81; Pappus Alexandrinus, *Book 7 of the Collection*, ed. tr. A. Jones, 2 vols. (New York etc.: Springer, 1986), vol. 1, pp. 118–23). I use modern notation in presenting the problem; Descartes (not to speak of Pappus) did not use indexed letters for bringing out the problem in its general form. Yet from their formulation it is clear that they meant the generality which modern notation makes expressible.

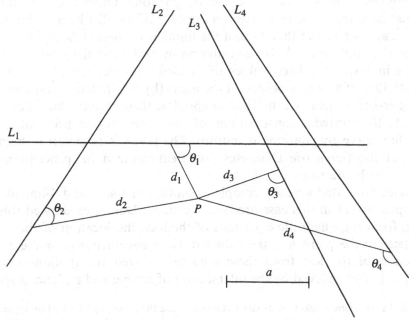

Figure 1

namely the one in which the proportion is $d_1 d_2 : a d_3$, occurs neither in Descartes' nor in Pappus' statement of the problem. However, I shall argue below that Descartes did indeed meet and solve that problem.

Descartes' solution of Pappus' problem as developed in the *Geometrie*[4] may be summarised as follows. He introduced a coordinate system with its origin at the intersection of L_1 and one of the other lines, its X-axis along L_1 and its ordinate angle equal to θ_1. With respect to that system $d_1 = y$. He showed that for any point P with coordinates x and y, each d_i could be written as

$$d_i = \alpha_i x + \beta_i y + \gamma_i,$$

where the α_i, β_i and γ_i were constants expressed in terms of the θ_i and the segments along the L_i between their points of mutual intersection; as the position of these lines in the plane was given, the α_i, β_i and γ_i were known. If the values of the angles θ_i are changed, the corresponding distances d_i are multiplied by some factor. Descartes was aware that in the special case that all the lines L_i are parallel, one has $d_1 = y$, $d_i = \beta_i y + \gamma_i$ for all i. In all cases the given constant proportions could be transformed into equations in the form:

$$y(\alpha_2 x + \beta_2 y + \gamma_2) \ldots = \delta(a)(\alpha_l x + \beta_l y + \gamma_l)(\alpha_{l+1} x + \beta_{l+1} y + \gamma_{l+1}) \ldots;$$

(where $l = k+1$ or $k+2$ depending on the number of given lines) the factor a in the right hand side occurred only if the number of lines was uneven and δ

4   Descartes, *Geometrie*, pp. 310–14, 323 sqq. and passim.

was the given constant value of the proportion. These are polynomial equations[5] in x and y (or in y alone in the case of parallel lines). The d_i are linear in x and/or y, and therefore, if the number of lines is $2k$ or $2k-1$, the degree of the equation is k. By considering in particular the degrees of the equations in y (i.e. for fixed values of x) Descartes derived results on the constructibility of separate points on the locus (by straightedge and compass or by higher-order means). I shall not recapitulate these results,[6] but I note that they relate to pointwise constructions of the locus, not to procedures for tracing that curve by continuous motion. The point is noteworthy because throughout the *Geometrie* Descartes put great value upon procedures for tracing curves by motion. —

Descartes presented a nearly complete solution of the 3- and 4-line locus.[7] The Pappus curves in this case are conic sections. Descartes showed this by deriving, from the quadratic equations of the loci, the location of their axes and vertices in the plane and the values of their parameters. For the actual construction of the loci from these data he referred to Apollonius' constructions. These[8] proceed by the intersection of a cone and a plane in space.

[5] If the problem is taken in its strict classical sense the d_i, as well as the line segment a and the ratio δ, should be interpreted as positive, whence the equation should be

$$|y(\alpha_2 x + \beta_2 y + \gamma_2) \cdots | = \delta | a(\alpha_i x + \beta_i y + \gamma_i)(\alpha_{i+1} x + \beta_{i+1} y + \gamma_{i+1}) \cdots |,$$

which is equivalent to

$$y(\alpha_2 x + \beta_2 y + \gamma_2) \cdots = \pm \delta a(\alpha_i x + \beta_i y + \gamma_i)(\alpha_{i+1} x + \beta_{i+1} y + \gamma_{i+1}) \cdots.$$

The solution of one Pappus problem, therefore, consists of two curves. For a given set of n straight lines the collection of Pappus loci with respect to these lines and arbitrary constant values for the proportion thus constitutes a one-parameter family of curves represented by the equation

$$y(\alpha_2 x + \beta_2 y + \gamma_2) \cdots = \delta a(\alpha_i x + \beta_i y + \gamma_i)(\alpha_{i+1} x + \beta_{i+1} y + \gamma_{i+1}) \cdots,$$

where δ now ranges over all (positive and negative) numbers. Usually Descartes started by considering one point on the locus and adjusting the coordinate system such that its x and y coordinates were positive. He then read off the values of the coefficients α_i, β_i, γ_i from the figure and tacitly assumed that the expressions thus gained applied generally. Moreover he usually took the constant δ to be 1. The effect of these choices was that in dealing with a Pappus problem he considered one solution curve only. Yet the figures he provided suggest that he was well aware of the other solutions and realized that an obvious adjustment of the equation would produce them.

[6] Cf. my 'On the representation of curves in Descartes' *Géométrie*', *Archive for History of Exact Sciences* 24 (1981):295–338, on p. 301.

[7] Descartes, *Geometrie*, pp. 324–34; as Descartes himself noted in a letter to Debeaune of 20 February 1639 (*Oeuvres*, vol. 2, p. 511), he had overlooked the case where the coefficient of y^2 in the equation of the curve is zero.

[8] Apollonius, *Conica* I, props. 52–60, cf. Apollonius, *Les coniques d'Apollonius de Perge*, ed. tr. P. Ver Eecke (Bruges: Desclée de Brouwer, 1923) (reprint Paris: Blanchard, 1963), pp. 97–115.

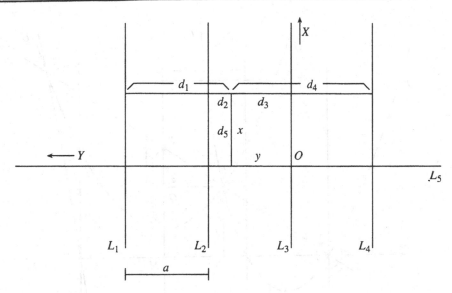

Figure 2

Thus, although in the case of the 3- or 4-line locus Descartes provided more than mere pointwise constructions, he did not offer a procedure for tracing the curves by continuous motion.

Besides explaining the general approach to Pappus' problem and solving it in detail for $n = 3$ or 4, Descartes discussed one special case. This is the problem 'in five lines'; its solution is the *Cartesian Parabola*. One of the intriguing aspects of this case is that it is the only instance of a Pappus problem in which Descartes provided a procedure for tracing the locus. Although usually referred to as 'the problem in five lines' it is in fact a very special case of that problem – Descartes claimed it was the simplest possible case – namely the case that four of the lines are parallel and equidistant while the fifth is perpendicular:

Problem.[9] Given (see Figure 2) four parallel, equidistant lines $L_1,...,L_4$ (distance a), and one line L_5 perpendicular to them. It is required to find the locus of points whose perpendicular distances d_i to L_i satisfy

$$ad_3 d_5 = d_1 d_2 d_4.$$

Note that Descartes chose a particular order of the d_i in the ratio; I shall return to the reasons of that choice below. Taking the Y-axis along L_5, with positive values marked to the left, and the X-axis along L_3, Descartes straightforwardly derived the equation of the curve:

$$axy = (2a-y)(a-y)(a+y) = y^3 - 2ay^2 - a^2y + 2a^3.$$

[9]   Descartes, *Geometrie*, pp. 335–8.

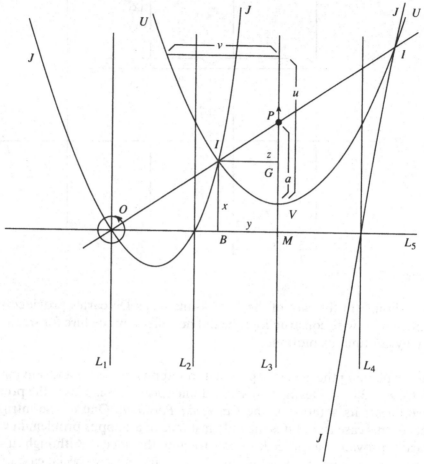

Figure 3

He then provided the following procedure to trace the required curve by motion:

Construction.[10] ('Cartesian Parabola' traced by the intersection of a turning straight line and a moving parabola.)
(1) Consider (see Figure 3) a parabola UVU with vertical axis along L_3 and *latus rectum* $= a$ (which means that its equation in rectangular coordinates u and v is $au = v^2$). The parabola can move up and down while keeping its axis along L_3. Moving with it is a point P on the axis inside the parabola with distance a to the vertex V. (2) Consider also a straight line OP which can turn around the intersection O of L_1 and L_5 while P moves along L_3.

10   Descartes, *Geometrie*, p. 337.

(3) During the combined motion the points I of intersection of the parabola and the straight line move over the plane; they trace a new curve $JOIJJIJ$; that curve is the required locus.

To prove that the curve thus generated was the required locus Descartes derived its equation with respect to the coordinate system as in Figure 3. With IG perpendicular to L_3, the following relations are directly deducible from the figure:

$$PG:IG = IB:OB$$

(by similar triangles),

$$GV:IG = IG:a$$

(because I is on the parabola) and

$$PG+GV = PV = a.$$

Now $IB = x$, $IG = y$ and $OB = 2a-y$, and the equation follows by easy calculation:

$$axy = y^3 - 2ay^2 - a^2y + 2a^3;$$

hence the generated curve and the 5-line locus are the same.

I shall call the tracing procedure used in the construction above the 'turning ruler and moving curve procedure'. Descartes had introduced it earlier in the *Geometrie*[11] as a general procedure for generating geometrical curves: if any curve or straight line is substituted for the parabola, a new curve arises. Descartes considered curves to be acceptable in geometry if they were traced by acceptable motions (I shall return to this condition below). Evidently he saw the turning ruler and moving curve procedure as a prototype of such curve tracing motions and hence as prime means to obtain geometrically acceptable curves. Descartes showed explicitly that a straight line generated a hyperbola (cf. below p. 88), he mentioned that a circle would produce a conchoid, and he announced that (as we have seen) a parabola generated a curve that would serve in the solution of the 5-line problem.

After showing that the Cartesian Parabola solved the special Pappus problem in five lines, Descartes returned twice to the curve. In explaining his method of finding normals to curves he applied it to three different curves, one of which was the Cartesian Parabola traced by the turning ruler and moving curve procedure. On that occasion[12] Descartes derived the equation of the curve for arbitrary values of the parameters that determine the tracing procedure, namely (see Figure 3) $OM = b$, $GV = c$ and the *latus rectum d* of the parabola. The equation is

$$y^3 - by^2 - cdy + bcd + dxy = 0.$$

Finally he used the curve in a standard construction for roots of fifth- and sixth-degree equations. That construction proceeds by tracing an appropriate

[11] Descartes, *Geometrie*, pp. 319–23.   [12] Descartes, *Geometrie*, p. 344.

Cartesian Parabola and intersecting it with an appropriate circle; the roots of the equation were found as the ordinates of the points of intersection. I shall discuss it in more detail below.

Descartes left the origin of the turning ruler and moving curve procedure totally unexplained. For the readers of the *Geometrie* the fact that the solution curve of the 5-line problem was precisely the one traced by a turning ruler and a moving parabola, although proved, remained a mere coincidence. Descartes was somewhat more explicit in his motivation for the choice of the Cartesian Parabola in the construction of roots of fifth- and sixth-degree equations. He wrote that it was the simplest curve that could be used for the purpose, that in the context of Pappus' problem it came immediately after the conics, and that the successive cases of Pappus' problem generated all the curves that are to be accepted in geometry in their proper order.[13] But there was no explanation why that would be sufficient reason for choosing the Cartesian Parabola, rather than any other third-degree curve, to serve in a standard construction of the roots of fifth- and sixth-degree equations. However, a further analysis of the solution of the 5-line locus suggests that the relation between the turning ruler and moving curve procedure and Pappus' problem in Descartes' geometry was much closer than a fortuitous coincidence. Indeed I shall argue that Descartes probably found the procedure while studying Pappus' problem and searching for a method to reduce higher order Pappus curves to lower order ones. I shall also show that this connection between the two themes clarifies various enigmatic aspects of Descartes' doctrine of geometry.

Consider, again, the 5-line problem (see Figure 4). From Descartes' construction of the locus, as well as from the proof that that construction is correct, it is clear that the segment *PG* is the crucial auxiliary element in the argument. Let *I* be a point on the locus and consider the line *OIP* with *P* on L_3. If we call $PG = z$, we have

$$d_5 : d_1 = z : d_3.$$

Moreover, the condition of the problem implies

$$a d_5 d_3 = d_1 d_2 d_4,$$

whence

$$d_5 : d_1 = d_2 d_4 : a d_3.$$

Combining these results yields

$$az = d_2 d_4.$$

Now *z* is equal to the distance *IH* of *I* to a horizontal line L_P through *P*. Calling this distance d_P, we have

$$a d_P = d_2 d_4,$$

[13] Descartes, *Geometrie*, p. 402.

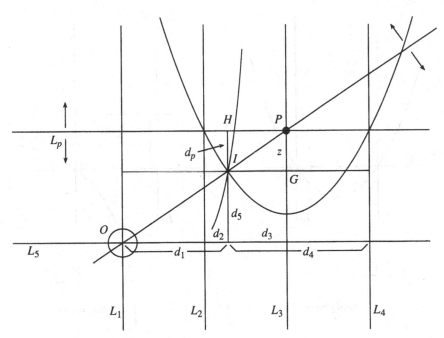

Figure 4

which means that the point I lies on a 3-line Pappus locus[14] with respect to the three lines L_2, L_4 and L_P. Such a 3-line locus is (as we will see) a parabola with axis along L_3. If now OP is conceived as a ruler turning around O and forcing P to move along L_3, the line L_P moves up or down and, because L_2 and L_4 are verticals, one may conceive the system of three lines L_P, L_2 and L_4 as moving up and down with P. Hence so does the 3-line locus: the parabola moves up and down together with L_P, and throughout the motion the point I on the Pappus curve is at the intersection of the ruler and the parabola. *This is precisely the combination of motions which Descartes described.*

As noted, the argument assumes the knowledge that the 3-line locus which occurs is a parabola. It seems highly likely that Descartes studied the relevant 3-line loci, either in their own right (because he may naturally have started with studying simple special cases) or once he had seen that he could reduce the 5-line locus to a 3-line one. The relevant 3-line loci are those with two parallel lines L_1 and L_2 and one perpendicular L_3. It would be easy for him (with or without analytic methods) to realize that there are essentially two types of such 3-line loci, namely (cf. Figure 5)

type 1: $d_1 d_2 = cd_3$
type 2: $d_1 d_3 = cd_2$.

[14]   In the sense of the analogon to the fifth-, seventh-, etc. line case, cf above p. 72.

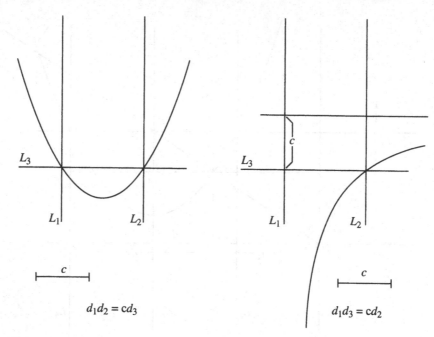

Figure 5

The first yields a parabola through the intersections of L_3 with L_1 and L_2; its axis is the vertical equidistant from L_1 and L_2. In the case arising as above from the 5-line locus we have $c = a$ and the distance of L_1 and L_2 is $2a$. The second type yields a rectangular hyperbola through the intersection of L_3 with L_2; L_1 is its vertical asymptote and its horizontal asymptote lies at distance c from the horizontal axis.

Thus Descartes' solution of the 5-line locus essentially consists in reducing it to a Pappus problem in a smaller number of lines, through the consideration of the proportionalities that arise when a line is drawn from a point on the locus to one of the intersections of the transversal with the other given lines. The turning ruler and moving curve procedure is the direct kinematic interpretation of that reduction.

I propose as a conjecture that Descartes found this reduction and its kinematic interpretation at an early stage of his study of Pappus' problem, probably soon after Golius had suggested the problem to him. Moreover I propose that this combination of an analytical method of reduction of a problem to one of a simpler type, and a kinematic method to generate intricate curves from simpler ones, was decisive in the formation of Descartes' ideas on geometry, on its proper demarcation and on its legitimate methods of construction.

In the remainder of this article I shall list a number of features of Descartes' work which lend support to my conjecture and I shall explore the clarification it offers in understanding Descartes' more general thoughts on geometry.

There is a certain self evidence to the procedure of reducing the 5-line locus to a 3-line one, which makes it plausible that Descartes found it easily. Indeed it is natural enough (cf. Figure 4) to draw and prolong OI, list the proportionalities it induces, and realize that one of these proportionalities, namely $d_5 : d_1 = z : d_3$, may be used to reduce the given ratio $ad_5 d_3 : d_1 d_2 d_4$ by eliminating d_5, d_1 and d_3. The resulting ratio $az : d_2 d_4$ suggests a Pappus problem in three lines and the final step is the insight that, at least in the particular case of the locus to four parallel lines and one transversal, the problem refers to a moving system of three lines and therefore yields a moving curve.

Our earliest source on Descartes' solution of Pappus' problem is the letter he wrote in January 1632 to Golius[15] as an additional commentary and summary of the solution he had sent earlier. He referred to an 'écrit' containing that solution; the écrit, unfortunately, is lost. It is generally assumed[16] that the solution which Descartes sent to Golius was basically the same as the one he published five years later in the *Geometrie*. The text of the letter is indeed compatible with such an assumption. But Descartes' words defy precise interpretation at several points. It is not necessary to discuss the letter here in detail for despite the interpretative difficulties it does suggest that Descartes had found the techniques explained above for deriving the equations of the Pappus curves. But his words of 1632 also suggest an interest in the generation of these curves by motion,[17] and thus the letter is compatible as well with the assumption that already during his first studies of Pappus' problem Descartes had found the reduction method and the turning ruler and moving curve procedure.

In a few sentences following his solution of the 5-line locus in the *Geometrie* Descartes made some remarks about possible generalizations. These remarks can be interpreted easily if we assume that they were inspired by the reduction of loci via the turning ruler and moving curve procedure. Descartes stated that in his approach to the 5-line locus the distances d_i need not be perpendicular, that the transversal need not be perpendicular either and that the parallels

---

[15] Descartes to Golius, January 1632, *Oeuvres*, vol. 1, pp. 232–6.

[16] E.g. G. Milhaud, *Descartes savant* (Paris: Felix Alcan, 1921), pp. 124–48, in particular p. 125.

[17] Cf. 'Datis quotcunque rectis lineis, puncta omnia ad illas iuxta tenorem quaestionis relata, contingent unam ex lineis quae describi possunt unico motu continuo, et omni ex parte determinato ab aliquot simplicibus relationibus' *Oeuvres*, vol. 1, p. 233.

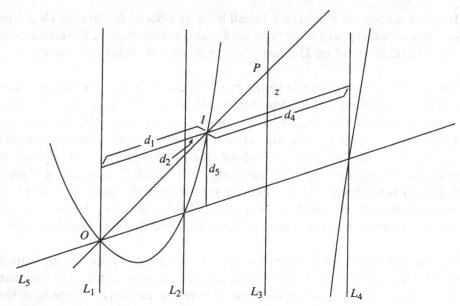

Figure 6

need not be equidistant.[18] Taking oblique rather than perpendicular distances affects the value of a in the equation $ad_5 d_3 = d_1 d_2 d_4$. In the derivation above (pp. 78–9 cf. Figure 4) a need not be equal to the distance of the parallels, nor, indeed need these parallels be equidistant. It is also easily seen (cf. Figure 6) that if the transversal is not perpendicular one can take the $d_1, ..., d_4$ parallel to it, by which the same proportionalities apply as above and analogously the problem is reduced to a 3-line locus with respect to two vertical lines and one oblique transversal. It is not difficult to derive (analogously to the argument on pp. 79–80 above, see Figure 7) that these 3-line loci are either parabolas with vertical axis and passing through the two points of intersection of L_P, L_2 and L_4, or hyperbolas passing through one of the points of intersection, having the other vertical as asymptote and their second asymptote parallel to the transversal.

Descartes also claimed that the procedure applied in some instances of the 5-line problem where the four lines are not parallel. That remark is more difficult to justify. He may have seen that in order to apply the procedure the lines L_2 and L_4 have to be parallel (otherwise the moving 3-line locus would change its shape), and L_1 and L_3 have to be parallel (otherwise the basic

18   'Or encore que les paralleles données AB, IH, ED, et GF ne fussent point esgalement distantes, et que GA ne les couppast point a angles droits, ny aussy les lignes tirées du point C vers elles, ce point C ne laisseroit pas de se trouver tousiours en une ligne courbe, qui seroit de cete mesme nature. Et il s'y peut aussy trouver quelquefois, encore qu'aucune des lignes données ne soient parallelles.' *Geometrie*, pp. 338–9.

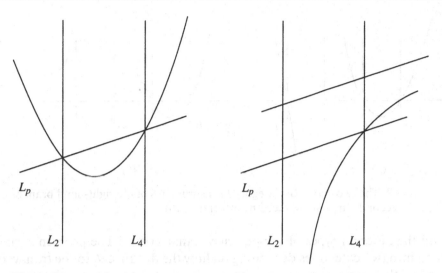

Figure 7

proportionality introduced by the ruler no longer applies), but the two pairs of parallels need not have the same direction. Yet Descartes claimed that none of the four lines need be parallel, and if interpreted along the same line as his earlier remarks, that claim is not tenable. The remark suggests that at an early stage he thought the procedure he had worked out for the configuration of parallel lines and one transversal could somehow be generalized to cover the case of arbitrary given lines.

I now turn to Descartes' assertion that the case of the 5-line locus he studied was the simplest case. The statement is usually interpreted to mean that the configuration of four equidistant parallels and one perpendicular transversal is the simplest of five lines (after the one of five equidistant parallel lines). This is how Newton interpreted Descartes' claim when he criticized it as one of the errors of the *Geometrie*.[19] That interpretation is incomplete because it leaves unexplained the fact that this simplest configuration of lines gives rise to several types of Pappus curves only one of which was singled out as the simplest by Descartes. I find, however, that Descartes' assertion is meaningful if we take into account how these various Pappus curves are traced by the turning ruler and moving curve procedure.

To explain this I shall first survey the different variants of Pappus' problem with respect to four parallel lines L_1, ..., L_4 and one perpendicular transversal

---

[19]   I. Newton, *The Mathematical Papers of Isaac Newton*, ed. D. T. Whiteside, 8 vols, (Cambridge University Press, 1967–1981), vol. 4, pp. 336–45. Newton argued that by analogy the parabola or the hyperbola, originating as locus with respect to two parallels and one perpendicular transversal, would be the simplest second-degree curve, simpler, therefore, than the circle, which he considered absurd.

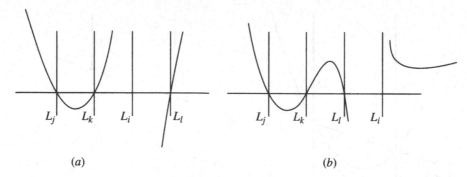

(a)   (b)

Figure 8. The two curves of Category I (the ordinates of the right-hand branch of the second curve are reduced by a factor 1/10).

L_5 and the different types of Pappus curves that arise.[20] The problems can be divided into two categories depending on how the distance d_5 to the transversal occurs in the given proportion. From now on I take the X-axis along L_5, denoting d_5 by y. The two categories can be subsumed in the two equations:

category I: $ayd_i = d_j d_k d_l$

category II: $ad_i d_j = yd_k d_l$.

Descartes' case belong to category I.

If we vary the order of the ds in the defining equations of the loci of category I we find that – disregarding symmetries[21] – there are essentially two such loci; they are drawn in Figure 8. One is the curve which Descartes proposed; the other is different in that the branch extending from $-\infty$ to $+\infty$ is not monotonic but has two local extremes.[22] In both cases a turning ruler and moving curve construction can be found by taking P on L_i and O on either L_j, L_k or L_l. For each of these choices the moving curve turns out to be a 3-line locus of type 1, that is, a parabola. Thus each locus of this category can be traced in three different ways by a turning ruler and moving parabola. The three such procedures for Descartes' curve are illustrated in Figure 9a, those for the other curve in Figure 9b; the first of those in Figure 9a is the one which Descartes actually gave in his *Geometrie*.

[20]   C. Rabuel undertook the same survey in his *Commentaires sur la Géométrie de M. Descartes* (Lyon 1730), pp. 254–82.

[21]   That is: the symmetry with respect to the X-axis and the symmetry with respect to a vertical (the latter is obtained by numbering the lines from right to left); thus $ayd_1 = d_2 d_3 d_4$, $-ayd_1 = d_2 d_3 d_4$ and $ayd_4 = d_1 d_2 d_3$ are considered the same.

[22]   In Newton's classification of cubics both curves belong to the class of 'tridents' or 'Cartesian Parabolas' since both have an equation of the form $xy = Ax^3 + Bx^2 + Cx + D$, cf. Newton, *Mathematical Papers*, vol. 7, pp. 630–31, Figure 76. In the present essay I shall use the term 'Cartesian Parabola' to refer exclusively to the type proposed by Descartes.

(a)

(b)

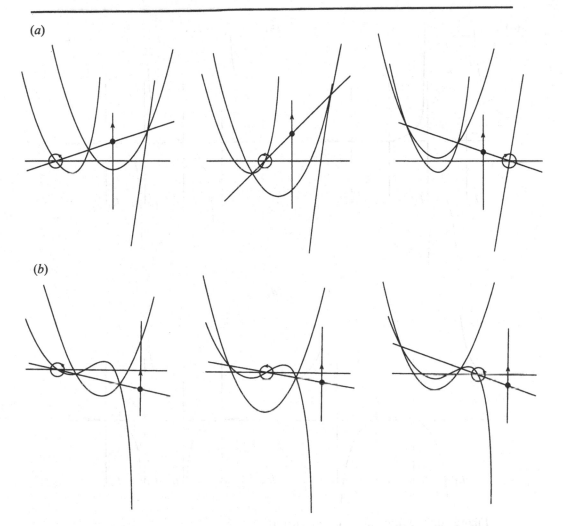

Figure 9. Ways of tracing the curves of category II by a turning ruler and a moving parabola.

For the loci of category II a similar analysis yields that there are (again disregarding symmetries) four different types of loci (see Figure 10).[23] For each

23   Figures 10a–d illustrate the four types of category II; IIa: $ad_3 d_4 = yd_1 d_2$; IIb: $ad_2 d_4 = yd_1 d_3$; IIc: $ad_2 d_3 = yd_1 d_4$; IId: $ad_1 d_4 = yd_2 d_3$. Curves of category II belong to what Newton called the 'hyperbolisms of the hyperbola'; he subdivided this class in four species, numbered 57–60. Species 59 and 60 are in fact special cases of 58 and 57 respectively, featuring extra symmetry (cf. Newton, *Mathematical Papers*, vol. 7, pp. 626–9, Figures 61–64). The occurrence of this symmetry in types IIa–d depends on the distances of the parallel lines; in particular the equidistant cases may feature the symmetry. The identification of the types is as follows: Type IIa: species 57; type IIb: species 58 or in special cases 59, but the equidistant case belongs to species 58 (no

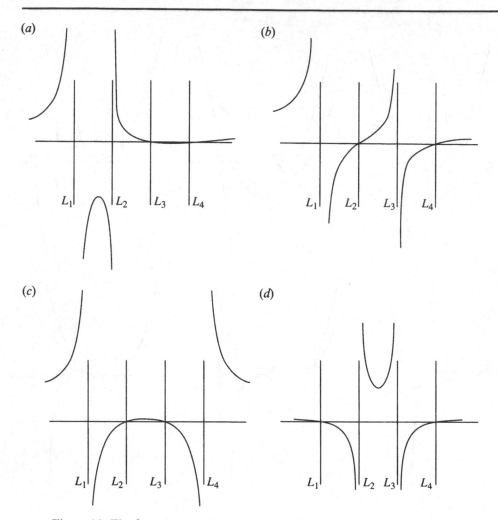

Figure 10. The four curves of category II.

of these types four different reductions to a moving 3-line locus can be found, namely by taking O on either L_i or L_j and P on either L_k or L_l. In each case the moving curve is of type 2, that is, a hyperbola. Figure 11 shows two of the four possibilities of tracing one of the loci of type II.

The survey of the 5-line Pappus curves, I find, does suggest a reason for Descartes' assertion that the case he actually presented was the simplest. Even if Descartes did not fully survey the possible cases, he probably considered some of them. The case he presented, being of category I, involves a parabola rather than a hyperbola, and the motion of a parabola in the direction of its axis seems a more natural and simple process than the motion of a hyperbola

symmetry); types IIc, d: species 57 or in special cases 60; the equidistant cases belong to species 60.

86

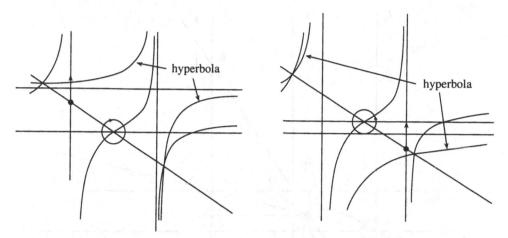

Figure 11. Two ways of tracing the curve of Figure 10b by a turning ruler and a moving hyperbola.

in the direction of one of its asymptotes. Furthermore, Descartes' case is the only one of the first category (cf. the six cases of Figure 9) in which the point P, connecting the motion of the ruler with that of the parabola, is on the axis of the parabola, thus lending a certain naturalness (equilibrium one might say) to the motion. In all other cases the position of P with respect to the curve has some arbitrariness.[24]

There is a further point which supports my conjecture about the origin of the turning ruler and moving curve procedure. If, as I suppose, the procedure of reducing a Pappus problem to one in a smaller number of lines was central to Descartes' early studies of Pappus' problem, then he probably also applied it to other configurations similar to the special one of five lines. We may assume in particular that he studied the case of four lines, three of which are parallel and one perpendicular. In that case (see Figure 12) the defining property of the locus is

$$yd_i = \delta d_j d_k.$$

Applying the same method as above, the point O can be chosen on either L_j or L_k. Assuming that it is on L_j, P must be on L_i. Setting $z = PG = IH$ and eliminating d_i and d_j leads to

$$z = \delta d_k.$$

This is, one might say, a 2-line locus; at any rate the locus is directly recognized

[24]   In his survey of the 5-line loci Rabuel (cf. note 20) gave tracing procedures involving a moving parabola for both curves of category I. They are in principle the same as the ones featured in Figure 9, though Rabuel added a complicated system of linked rulers in order to have the connecting point on the axis of the parabola. He did not suggest any fundamental role of these tracing procedures in the development of Descartes' geometrical ideas.

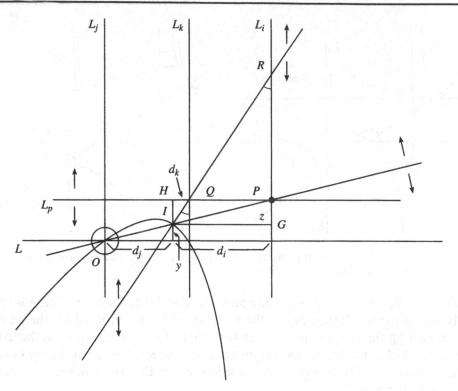

Figure 12

as a straight line intersecting L_k in Q, Q being at the same height as P. That straight line moves up and down with P, thus if R is its intersection with L_i, the distance RP is constant and so is the angle at R. The locus is therefore traced by a turning ruler and moving curve procedure with OP as ruler and the straight line RQ as moving curve. *This is precisely the first example of the turning ruler and moving curve procedure which Descartes discussed in the Geometrie.*[25] He used it there to explain the procedure and he proved that the resulting curve was a hyperbola. He did not mention any connection with Pappus' problem, but I find the coincidence striking enough to interpret it as positive evidence for my conjecture and to suppose, therefore, that Descartes' method to trace the hyperbola originated in the same way as the procedure for the Cartesian Parabola during his early studies of Pappus' problem.[26]

[25] Descartes, *Geometrie*, pp. 319–22.

[26] It should perhaps be added that the same approach is applicable in one of the two cases of the locus with respect to two parallel and one perpendicular line, namely the case with the equation $d_3 d_1 = c d_2$. One finds a turning ruler and moving curve procedure for the locus (a hyperbola in this case) where the moving curve is a straight line parallel to the transversal. In the other case, $c d_3 = d_1 d_2$, the locus is a parabola, but the approach turns out not to be applicable.

As mentioned above, Descartes wrote down a short and rather cryptic remark about the second category of the 5-line locus, the one given by

$$ad_i d_j = yd_k d_l.$$

He wrote:

> The required point lies on a curve of different nature, namely, a curve such that, all the ordinates to its axis being equal to the ordinates of a conic section, the segments of the axis between the vertex and the ordinates bear the same ratio to a certain given line that this line bears to the segments of the axis of the conic section having equal ordinates. I cannot say that this curve is less simple than the preceding; which nevertheless I believed should be taken as the first, since its description and calculation are somehow easier.[27]

From the analysis of the 5-line problem given above one would expect Descartes to have found a turning ruler and moving curve procedure for this case, with a moving hyperbola instead of the parabola. The passage from the *Geometrie*, however, does not suggest curve tracing or motion at all and cannot therefore be interpreted as referring to a turning ruler and moving curve procedure. The passage does imply reduction to a conic and Rabuel[28] suggested an explicit interpretation in that way, valid for the special symmetrical case (cf. Figure 10c)

$$ad_2 d_3 = yd_1 d_4.$$

One might surmise, then, that the underlying idea would be the reduction of the 5-line locus to one in fewer (three of four) lines. Such an interpretation is indeed possible for all loci of category II (including Rabuel's as a special case). A variant interpretation, where x rather than y is taken as 'ordinate', is also possible. For lack of space I shall not give details here, but the calculations involved do show that neither interpretation involves any immediate link with curve tracing by motion. Hence it seems improbable that Descartes found the procedure he referred to in the passage we are discussing while searching for solutions that can be traced.

Thus, although Descartes' passage on the second category of the 5-line locus does not exactly contradict my conjecture, it does not lend positive support either. I like to see it as an echo of some intermediate, primarily algebraic,

---

[27] 'Ce point cherché est en une ligne courbe d'une autre nature, a sçavoir en une qui est telle, que toutes les lignes droites appliquées par ordre a son diametre estant esgales a celles d'une section conique, les segmens de ce diametre, qui sont entre le sommet et ces lignes, ont mesme proportion a une certaine ligne donnée, que cete ligne donnée a aux segmens du diametre de la section conique, ausquels les pareilles lignes sont appliquées par ordre. Et ie ne sçaurois veritablement dire que cete ligne soit moins simple que la precedente, laquelle iay creu toutefois devoir prendre pour la premiere, a cause que la description, et le calcul en sont en quelque façon plus faciles.' *Geometrie*, p. 339.

[28] Rabuel, *Commentaires*, p. 271, cf. my 'Representation of curves', p. 316, note 21.

investigation of the problem, undertaken sometime after the completion of the 'écrit' and perhaps as a result of doubts about the generality of the earlier solution. The uncertainty, voiced in the final sentence of the passage about whether, after all, this curve or the other is the simplest, may be read as indication of such doubts.

The arguments above, I think, lend support to the conjecture that at some early stage of his dealings with Pappus' problem Descartes found the turning ruler and moving curve procedure as a kinematic interpretation of the argument by which certain special n-line Pappus problems could be reduced to $(n-2)$-line problems. Moreover, it seems probable that these insights decisively influenced his thinking about geometry. I find in particular that two features of Descartes' doctrine of geometry gain in understandability if we suppose that at some stage Descartes was struck by the special power of the turning ruler and moving curve procedure with respect to solving Pappus problems and tracing Pappus curves. These features are the demarcation of geometry and the doctrine of construction.

Descartes claimed repeatedly that curves should satisfy certain conditions in order to be acceptable in geometry. The basic condition, already referred to above, concerned the manner in which the curves were traced:

> provided they can be conceived of as described by a continuous motion or by several successive motions, each motion being completely determined by those which precede; for in this way an exact knowledge of their measure is always obtainable.[29]

Curves so traced were 'geometrical'; others were merely mechanical and did not properly belong to geometry.

Descartes also claimed that the class of acceptable, 'geometrical' curves consisted precisely of those curves that have an algebraic equation, and that any curve with an algebraic equation might occur as a Pappus curve with respect to some configuration of the given lines. Thus Descartes used three different criteria to demarcate the class of 'geometrical' curves: being traceable by acceptable motion, being a solution of a Pappus problem, and having an algebraic equation. I shall use the terms 'traceable curves', 'Pappus curves', and 'algebraic curves' to denote these classes respectively. Descartes was convinced that the three classes coincided and that they comprised the only legitimately geometrical curves. Let me now survey the arguments he adduced to sustain this conviction.[30]

[29] Descartes, *Geometrie*, p. 316: 'pourvû qu'on les puisse imaginer estre descrites par un mouvement continu, ou par plusieurs qui s'entresuivent et dont les derniers soient entierement reglés par ceux qui les precedent. car par ce moyen on peut tousiours avoir une connoissance exacte de leur mesure.'

[30] See my 'Representation of curves' for more detailed information.

Descartes' derivation of the equation of a Pappus curve, as explained above, showed that all Pappus curves were algebraic. His reason for believing that, conversely, all algebraic curves were Pappus curves was, apparently, a kind of dimension argument: he claimed that $2n$ lines could be placed in so many different ways in the plane that any combination of values of the coefficients could occur in the resulting nth-degree equation.[31] The argument is convincing at first sight (and correct for lower values of n) but, as Newton was the first to point out, it is not valid in general.[32]

The equivalence between Pappus curves or algebraic curves on the one hand and traceable curves on the other is a much more complicated matter. In the *Geometrie* Descartes derived the equations of two curves traced by the turning ruler and moving curve procedure, namely the hyperbola and the Cartesian Parabola. The derivation in the case of the hyperbola was meant to show that equations could be derived for all curves traced in a similar manner, and the example is indeed convincing in that respect. As to the converse statement, that all algebraic curves are traceable, the argument in the *Geometrie* is complex and ultimately inconclusive. Without going into details[33] it can be summarized as follows. By the doctrine of construction elaborated in the *Geometrie*, the roots of any polynomial equation can be geometrically constructed. Hence for any algebraic curve one can, given x, construct the corresponding y-values, and thereby the corresponding points on the curve. This procedure of pointwise construction can be done for arbitrary x-values. The arbitrariness (which does not apply for pointwise construction of non-algebraic curves like the spiral and the quadratrix) corresponds somehow to continuity of tracing[34] and Descartes concluded from this correspondence that algebraic curves can indeed be traced by continuous motion. The argument is not fully worked out and it remains unconvincing.

The question, then, is: whence Descartes' conviction, despite the weakness of his arguments, that algebraic curves are traceable and for that reason legitimately geometrical? I find that that conviction and Descartes' attitude to the related questions in the *Geometrie* become better understandable in the light of my conjecture. If Descartes' approach to Pappus' problem involved the reduction argument and the turning ruler and moving curve procedure for tracing successive solutions, he may well for some time have cherished the hope that he would be able to find explicit tracing procedures for all Pappus curves. Thereby all Pappus curves (and therefore all algebraic curves) would be traceable, whereby the round of arguments for the equivalence of the three

[31] The argument can be glimpsed in the 1632 letter to Golius, cf note 17; it is clearly stated in the *Geometrie*, e.g. p. 324.

[32] See the appendix in my 'Representation of curves', pp. 332–8.

[33] For a fuller discussion see my 'Representation of curves', pp. 323–5.

[34] This crucial step in the argument is on pp. 339–40 of the *Geometrie*.

classes would be smoothly closed. Understandably, then, he modelled his chief example of acceptable curve tracing on the turning ruler and moving curve procedure, stressing its power to generate new curves by repeated application. I would indeed claim that the 'several successive motions, each motion completely determined by those which precede' in Descartes' description of geometrically acceptable motion refer precisely to the iterative application of the turning ruler and moving curve procedure.[35]

Whether or not Descartes for some time believed, rather than merely hoped, that he could fully generalize the methods he had found for special cases of Pappus' problem must remain undecided. But at some stage he must have realized that he had no general method for finding explicit tracing procedures for arbitrary Pappus curves. After such a realization he would naturally concentrate on the algebraic aspects of the problem, which is what we see him doing in the *Geometrie*. However, he obviously retained the programmatic and structural ideas about geometry and he tried to save the arguments for them. The result was that in the *Geometrie* Pappus' problem was no longer explicitly linked to curve tracing and that Descartes generally based his arguments more upon the nature of the equations of the curves than upon their generation by motion. Still several elements of his original arguments about explicit curve tracing were retained in the *Geometrie* where in fact they no longer served much purpose.

I now turn to *construction*, a pivotal theme in the *Geometrie*. Indeed Descartes proposed a new and universal answer to the question how to construct in geometry, in particular in those cases where ruler and compass are not sufficient. His doctrine of construction[36] may be summarized as follows: Geometrical construction has to be performed (in generalization of the classical use of ruler and compass, i.e. of straight lines and circles) by the intersection of curves. These curves should be geometrically acceptable and as simple as possible. For Descartes that meant that they should be 'geometrical', hence algebraic, and of lowest possible degree. He showed that any geometrical construction problem could be translated in algebraic terms as the construction

---

[35] Incidentally we have here an answer to the question why Descartes could refer in the *Discours de la Methode* with great admiration to the 'long chains of reasoning' of mathematicians (*Oeuvres*, vol. 6, p. 19) and at the same time write a *Geometrie* in which axiomatics and formal deductive argument are almost entirely absent. I would maintain that for Descartes the successive generation of acceptable curves by the turning ruler and moving curve procedure is an example of a long chain of reasoning whereby, however many steps are involved, the certainty of the result, and hence the acceptability of the generated curve, remains unaffected.

[36] See my 'The structure of Descartes' *Geometrie*', in *Descartes: il Discorso sul Metodo e i Saggi: Atti del convegno per il 350o anniversario della pubblicazione del Discours de la Méthode e degli Essais* (ed. G. Belgioioso e.a., Florence 1990) pp. 349–69.

of the roots of some polynomial equation. He then sketched what he claimed to be a general doctrine for the construction of roots of algebraic equations. The constructions depended on the degrees of the equations. If that degree was 1 or 2 the roots should be constructed by the intersection of straight lines and circles. If the degree was 3 or 4, the construction should be performed by the intersection of a parabola and a circle. If it was 5 or 6 the construction was to be performed by the intersection of a circle with the Cartesian Parabola. Up to degree 6 Descartes worked this out in considerable detail explaining precisely how the equations should be reduced to standard forms and how subsequently the basic parameters of the two constructing curves (straight line and circle, parabola and circle, Cartesian Parabola and circle respectively) should be adjusted in order to yield the roots. For linear and quadratic equations these methods were known, in the other cases they were new and the construction for third- and fourth-degree equations constituted a decisive improvement with respect to the methods that were extant. Having come so far Descartes merely added

> It is only necessary to follow the same general method to construct all problems, more and more complex, ad infinitum.[37]

Descartes' construction of the roots of third- and fourth-degree equations has a certain elegance,[38] but the next case, featuring the Cartesian Parabola, is decidedly cumbersome. He prescribed the following

> **Construction**.[39] Given a sixth-degree equation of the form
> $$x^6 - px^5 + qx^4 - rx^3 + sx^2 - tx + v = 0,$$
> it is required to construct the roots.
> Construction. (1) Draw a Cartesian Parabola by the turning ruler and moving curve procedure, adjusting the parameters OM, PV and the *latus rectum n* of the parabola (see Figure 3) as:
>
> $$OM = \frac{p}{2}$$
>
> $$n = \sqrt{\frac{t}{\sqrt{v}} + q - \frac{p^2}{4}}$$
>
> $$PV = 2\frac{\sqrt{v}}{pn}.$$

[37] 'il ne faut que suivre la mesme voye pour construire tous ceux qui sont plus composés à l'infini', *Geometrie*, p. 413.

[38] Descartes, *Geometrie*, pp. 389–95; cf. my 'Arguments on motivation in the rise and decline of a mathematical theory; the "Construction of equations", 1637–*ca.* 1750', *Archive for History of Exact Sciences*, 30 (1984): 331–80, on pp. 337–42.

[39] Descartes, *Geometrie*, pp. 402–11; cf. Whiteside's note in Newton, *Mathematical Papers*, vol. 1, p. 495, note 15.

(2) Draw a circle with radius R and centre (x_M, y_M) (with respect to the origin M), adjusting these values as follows:

$$x_M = \frac{2\sqrt{v}}{pn} - \frac{t}{2n\sqrt{v}},$$

$$y_M = \frac{m}{n^2},$$

with

$$m = \frac{r}{2} + \sqrt{v} + \frac{pt}{4\sqrt{v}},$$

and

$$R^2 = \frac{t}{2n\sqrt{v}} - \frac{s + p\sqrt{v}}{n^2} + \frac{m^2}{n^4}.$$

(Note that each of these lengths can be constructed by ruler and compass from the given coefficients.) (3) The x-coordinates of the points of intersection of the two curves are the required roots of the equation.[40] (Descartes proves the correctness of this construction by a lengthy computation.)

Such was the canonical way of finding the roots of fifth- and sixth-degree equations as advocated by Descartes and indeed as accepted in principle by most mathematicians until well into the eighteenth century. I have written out the formulas in full to show how very complicated the construction was that Descartes saw as the crowning result of his investigations concerning geometrical construction.[41] Indeed, to allow myself one anachronistic remark, the construction procedure seems decidedly useless and odd. So why and how did Descartes devise it?

Descartes did not explain how he found the values of the parameters. It is likely that he did so by writing out the equations of the Cartesian Parabola and the circle with undetermined coefficients, eliminating y and putting the resulting sixth-degree polynomial equal to a multiple of $x^6 + px^5 + \cdots + v$. That method indeed leads, by straightforward if laborious calculation, to the values he gave. But he could only do so if he had already decided that the Cartesian Parabola, rather than any other cubic, should be used in the construction. As regards algebraic ease or simplicity, the choice of the Cartesian Parabola is not at all obvious. One would, for instance, rather think of

$$y = ax^3 + bx^2 + cx + d,$$

[40] The construction presupposes that the values of p, q, r, etc. are positive and such that the square roots occurring in the expressions above do exist. Descartes claimed, correctly, that any fifth- or sixth-degree equation can be transformed into an equation satisfying these requirements by means of a translation $x \rightarrow x + \alpha$ and if necessary a multiplication by x.

[41] Cf. *Geometrie*, p. 413 and *Oeuvres*, vol. 1, p. 492 (letter to Mersenne, January 1638).

and if one adopts this curve, the coefficients are found more easily and they are less involved than the ones for the Cartesian Parabola.[42] Clearly, then, Descartes was determined to use the Cartesian Parabola in the construction of the roots of fifth- and sixth-degree equations, and he was willing to accept considerable algebraic complication to reach that goal.

His preference for the curve related to its property of being a solution of the Pappus' problem in five lines. Descartes wrote at the beginning of the section on constructing roots of fifth- and sixth-degree equations:

> I therefore believe that I have accomplished the best that is possible when I have given a general rule for constructing problems by means of the curve described by the intersection of a parabola and a straight line, as previously explained; for I am convinced that there is nothing simpler in nature that will serve this purpose. You have seen, too, that this curve directly follows the conic sections in that question to which the ancients devoted so much attention, and whose solution presents in order all the curves that should be received in geometry.[43]

Thus also in the doctrine of construction we find Pappus' problem as the origin of a programmatic choice. The Cartesian Parabola should be used in the construction not merely because it belongs to the simplest possible configuration of five lines, but because it is the simplest within a procedure that successively generates all acceptable geometrical curves. Here too I find Descartes' preferences more understandable in the light of my conjecture; an earlier hope or conviction that the turning ruler and moving curve procedure would serve to generate all Pappus curves explains both the choice of the Cartesian Parabola in the doctrine of construction and the value which apparently Descartes attached to this choice.

Although, as I hope to have shown, my conjecture provides explanations or clarifications of several enigmatic features of the *Geometrie*, it by no means solves all riddles with which that book confronts the historian. Descartes' classification of curves, for instance, remains incoherent even if one supposes, as I do, that also in that respect his ideas were influenced by what he found during his early study of Pappus' problem.

Let me conclude by remarking that, whatever the explanatory force of my

[42] I omit the calculations for reasons of tedium and space.

[43] Descartes, *Geometrie*, p. 402: 'C'est pourquoy ie croyray faire en cecy tout le mieux qui se puisse, si ie donne une reigle generale pour les construire, en y employant la ligne courbe qui se descrit par l'intersection d'une Parabole et d'une ligne droite en la façon cy dessus expliqué. car i'ose assurer qu'il n'y en a point de plus simple en la nature, qui puisse servir a ce mesme effect; et vous avez vû comme elle suit immediatement les sections coniques, en cete question tant cherchée par les anciens, dont la solution enseigne par ordre toutes les lignes courbes qui doivent estre receuës en Geometrie.'

conjecture, it does highlight the role of Pappus' problem in the development of Descartes' geometrical thought. To the casual reader of the *Geometrie* that problem may seem merely a didactic device to explain Descartes' new methods, with the additional advantage of classical appeal and standing. It was much more; it was the problem that provided him, in 1631–32 with a new ordered vision of the realm of geometry and that shaped his convictions about the structure and the proper methods of geometry.

4 Honoré Fabry (1607–1688) als Mathematiker – Eine Reprise

EMIL A. FELLMANN

Abstract

The mathematical work of Honoré Fabry (1607–1688), a Jesuit philosopher, theologian and physicist, would probably have fallen into complete oblivion if Leibniz had not mentioned him in his review of Newton's *Tractatus duo*... (1704) in connection with the dispute over priority of the invention of the calculus.

Moritz Cantor (1829–1920), being severely prejudiced against Fabry, successfully blocked interest in his *mathematica* and it was not until the 1950s that a first thorough analysis was carried out.

Following Cavalieri's method of indivisibles which basically admits only a static interpretation of these elements in an Archimedean sense, Fabry considered this method's novel feature to be the 'fluent generation' of geometrical objects, the *fluxus* representing time. This rather dynamical concept of the generation of the continuum by the *fluxus* of an indivisible leads qualitatively to the calculus of fluxions of the Newtonian type.

In his rather didactic text *Synopsis geometrica* (1669), Fabry deduces geometry genetically according to this principle. His 'new method', connected with a 'new form', essentially consists of

(i)   the introduction of a notion of *fluxus* instead of indivisibles;

(ii) a division of the geometric figures known at that time into twelve classes;

(iii) a skilful use of common properties of certain figures belonging to the same class, a quality denoted 'homogeneity' by Fabry.

Using this method, Fabry managed to operate quite successfully in his research on infinitesimals. Leibniz characterized it, along with the work of Cavalieri, Gregorius a St Vincentio and Pascal, as important for the discovery of the *Calculus*.

In his *Opusculum geometricum* (1659), which must be viewed in the context of Pascal's famous challenge of 1658, Fabry presents a wealth of significant single achievements: an ingenious quadrature of the cycloid, various quadratures and cubatures which amount to special cases of $\int x^n \sin x \, dx$, $\int \sin^p x \, dx$, and $\int\int \arcsin x \, dx \, dy$, as well as determinations of the centre of mass for segments of sine areas and cycloids, including the solids of revolution around their axes and tangents of the vertex. Three typical examples will be demonstrated here.

The epilogue contains a short biography of Honoré Fabry as well as a very concise evaluation of his – not always fortunate – achievements as a physicist. It closes with a remark about the accuracy of Leibniz's historical judgement, specifically in the case of the pugnacious mathematician Honoré Fabry.

Prolog

Die Geschichte der Wissenschaft geht sehr oft – wie das 'Schicksal' selbst – merkwürdig verschlungene Wege, deren Nachzeichnung dem Historiker nicht immer leicht möglich ist. Dies gilt in besonders hohem Masse, wenn es sich um die Bewertung der Leistung und Wirkung eines Forschers handelt, der zweifelsfrei nicht zu den 'Sternen erster Grösse' zählt, wie das etwa bei Honoré Fabry[1] als Mathematiker der Fall ist. Zu seiner Zeit war dieser streitbare Jesuit zunächst als Physiker, Theologe und Philosoph, später dann in seiner Eigenschaft als Grossinquisitor am Poenitentiarkollegium in Rom unter Papst Innozenz IX eine sehr bekannte und einflussreiche Persönlichkeit; er stand mit vielen Koryphäen der Wissenschaft in Briefwechsel – bekanntlich auch mit Gassendi, Huygens, Leibniz und Tschirnhaus, welch Letzteren er auch persönlich kannte.[2] Zu einer direkten Bekanntschaft mit Leibniz kam es allerdings nicht, denn der *uomo universale* besuchte Rom erst im Jahr nach Fabrys Tod. Dennoch ist es Leibniz zu verdanken, dass Fabry auch als Mathematiker nicht ganz der Vergessenheit anheimgefallen ist, auch wenn der entscheidende Umstand durch ein *omen dubiosum* belastet sein mag: ich denke an die berühmt-berüchtigte Stelle in Leibniz' Rezension von Newtons *Tractatus duo* ..., die den Prioritätsstreit um die Erfindung des Infinitesimalkalküls so sehr angeheizt hat und deren verschiedene Aspekte, Auslegungen und Geschichte D. T. Whiteside im achten Band seines monu-

---

[1]   Eine Kurzvita findet sich – ausser hier im 'Epilog' – im *Dictionary of Scientific Biography*. vol. IV, und eine ausführlichere, die mathematischen Arbeiten Fabrys jedoch nicht berücksichtigende Biographie stammt von P. de Vrégille, 'Un enfant de Bugey – Le Père Honoré Fabri 1607–1688', *Bulletin de la Société Gorini*, vol. 3 (1906):5–15. Eine vollständige Bibliographie von Fabrys Werken bietet: C. Sommervogel, *Bibliothèque de la Compagnie de Jésus*, vol. III (Bruxelles–Paris, 1892), pp. 511–21; vol. IX, pp. 309–10.

[2]   G. W. Leibniz, *Sämtliche Schriften und Briefe*, (Akademie-Verlag Berlin, 1923–), Dritte Reihe, zweiter Band: *Mathematischer, naturwissenschaftlicher und technischer Briefwechsel* (1987), ed. H.-J. Hess, cf. Index.

mentalen Lebenswerks[3] miteinbezogen hat. Wie man weiss, verstieg sich Leibniz in seiner Besprechung zu einer – mindestens höchst unglücklich formulierten – zweideutig auslegbaren Analogie, die von Newton (und den Newtonianern) als unfair empfunden werden konnte, auch wenn dies nicht in Leibniz' Absicht gelegen haben mag. Diese ominöse Stelle möge hier einmal mehr wiedergegeben werden – jedoch nicht, um den kalten Braten erneut aufzuwärmen, sondern weil sie in unserem Kontext von Bedeutung ist:

> Pro differentiis igitur Leibnitianis D. Newtonus adhibet semperque adhibuit *fluxiones, quae sint quam proxime ut fluentium augmenta aequalibus temporis particulis quam minimis genita*; iisque tum in suis *Principiis Naturae Mathematicis*, tum in aliis postea editis eleganter est usus, quemadmodum et Honoratus Fabrius in sua *Synopsi Geometrica* motuum progressus Cavallerianae Methodo substituit.[4]

In der Uebersetzung von D. T. Whiteside:

> In place of Leibnitian differences, accordingly, Mr. Newton employs and has ever employed 'fluxions', which are [defined] to be 'very closely near as the augments of the fluents begotten in the very smallest equal particles of time'; and has elegantly used them in his [*Philosophiae*] *Natur[alis]* *Principia Mathematica* and also in other subsequent publications, in the manner that Honoré Fabry, too, in his *Synopsis Geometrica* substituted advances of motion for the Cavalerian method.[5]

Nicht nur Newton und seine Freunde legten diese Worte von Leibniz als 'booby trap' aus, sondern auch einige Kontinentale wie etwa der Leibniz-biograph Guhrauer,[6] doch vor allem Moritz Cantor, dessen mathematische Geschichtsschreibung seit der Jahrhundertwende bis nach dem Zweiten Weltkrieg für den deutschen Sprachraum als repräsentativ galt. Nachdem Cantor die einschlägigen Stellen zitiert, gemäss welchen Leibniz bei der Lektüre von Huygens' *Horologium oscillatorium* seine Unwissenheit in mathematischen Belangen eingesehen hatte und dann diese Lücke haupt-sächlich mit dem Studium der Werke von Cavalieri, Gregorius a St Vincentio, Pascal und Fabry geschlossen habe, kommentiert Cantor (im Kontext mit dem Prioritätsstreit) das oben angeführte Zitat folgendermassen:[7]

> Im Ganzen war also der Ton der Besprechung ein sehr wohlwollender..., wenn nicht ein Satz in ihr vorgekommen wäre, dessen schriller Misston durchgehört werden musste... Newton wird mit Fabri verglichen, der den

---

[3]   Isaac Newton, *The Mathematical Papers of Isaac Newton*, ed. D. T. Whiteside, 8 vols. (Cambridge University Press, 1967–81), vol. 8. cf. Index.

[4]   *Acta Eruditorum* (Januar 1705): 30–6, p. 35.   [5]   Cf. Anm. 3: pp. 26–7.

[6]   G. E. Guhrauer, *Gottfried Wilhelm Freiherr v. Leibniz – Eine Biographie*, vol. 1 (Breslau, 1846), p. 311.

[7]   M. Cantor, *Vorlesungen über Geschichte der Mathematik*, vol. 3 (Leipzig, 1901²), pp. 293–4.

Fortschritt der Bewegungen an Stelle der Methode Cavalieris setzte. Fabri kannte Cavalieris Schriften, kannte sein Verfahren und veränderte es in nicht der Rede werten Nebenumständen. Er hat sich damit nur selbst geschadet. Seine *Synopsis geometrica* gehört zu den wenigst bekannten Schriften der damaligen Zeit und würde ohne die Erwähnung in dem Satze, von dem wir gerade reden, wohl ganz vergessen sein. Und mit diesem Fabri wird Newton verglichen, wird mit ihm durch den Vergleich auf eine Linie gestellt!

Nun – nicht alles in diesem Kommentar ist falsch. Falsch ist sicher die Plagiatsbezichtigung Fabrys an Cavalieri, und dies hätte Cantor vermeiden können, indem er selbst einen Blick in die *Synopsis geometrica*[8] von 1669, oder noch besser in Fabrys *Opusculum geometricum*[9] von 1659, das auch in der *Synopsis* abgedruckt war, geworfen hätte. Dort erklärt Fabry nämlich in aller Deutlichkeit sein Abhängigkeit von seinem 'hochverehrten Lehrer' und markiert recht genau die Verzweigungspunkte, an welchen er von Cavalieris Methode abweicht respektive über sie hinausgeht. Noch schlimmer hingegen erscheint mir die Tatsache, dass Cantors harsches (Vor)Urteil bezüglich des Mathematikers Honoré Fabry die Meinung der Mathematikhistoriker für das nächste halbe Jahrhundert geprägt hat. Eine Ehrenrettung sowohl für Leibniz als auch für Fabry versuchten zwar in der Zeit der Weimarer Republik Dietrich Mahnke[10] und Ferdinand Lindemann,[11] doch erfolgte eine genauere Analyse der Fabryschen *mathematica* erst in den Fünfzigerjahren[12] – leider zu spät, um in der heute massgeblichen deutschsprachigen *Geschichte der Mathematik* von Joseph Ehrenfried Hofmann (1900–1973) noch berücksichtigt werden zu können.[13]

Der Mathematiker

Bekanntlich hat B. Cavalieri (1598–1647) mit der Einführung des Indivisibels als derjenigen scholastischen *forma* (*substantialis*), welche aus den Punkten als

[8]   H. Fabry, *Synopsis geometrica cui accessere tria opuscula…*(Lyon, 1669).

[9]   Antimus Farbius [= H. Fabry], *Opusculum geometricum de linea sinuum et cycloide* (Rom, 1659) (im folgenden abgekürzt mit *O.G.*).

[10]   D. Mahnke, 'Neue Einblicke in die Entdeckungsgeschichte der höheren Analysis', *Abhandlungen der Preussischen Akademie der Wissenschaften 1925. Physikalisch-mathematische Klasse* Nr. 1, (Berlin, 1926).

[11]   F. Lindemann, 'Fabri, Barrow und Leibniz', *Sitzungsberichte der Bayrischen Akademie der Wissenschaften. Mathematisch-naturwissenschaftliche Abteilung* (München 1928), pp. 273–84.

[12]   E. A. Fellmann, 'Die mathematischen Werke von Honoratus Fabry (1607–1688)', *Physis* 1 (1959):6–25, 73–102. – Im Text abgekürzt mit: 'Fellmann, *Fabry*'.

[13]   J. E. Hofmann, *Geschichte der Mathematik*. Sammlung Göschen 226 (Erster Teil), 875 (Zweiter Teil), 882 (Dritter Teil) (W. de Gruyter & Co., Berlin 1953–57).

den materialen Elementen das Kontinuum prägt, zu den ersten Schritten zum Newton-Leibnizschen Calculus beigetragen. Cavalieris bekannte Methode lässt jedoch eine primär nur statische Interpretation dieser Indivisibilien im archimedischen Sinne zu,[14] während die mehr dynamische Auffassung einer Erzeugung des Kontinuums aus dem *fluxus* eines Indivisibels zur späteren Fluxionsrechnung führt. Honoré Fabry, auf den sich Leibniz in seiner Jugendschrift *Physica nova* (1670) bei der funktionellen Umdeutung des Indivisibilienbegriffes beruft, hat das Neue in der Methode Cavalieris in der 'fliessenden Erzeugung' der geometrischen Gebilde erblickt, indem der *fluxus* die Zeit veranschaulichen soll.

In seiner *Synopsis geometrica* von 1669 wird die Geometrie gemäss dieser Maxime genetisch deduziert. Die Linie entsteht *per fluxum puncti*, die Fläche *per fluxum lineae* und der 'reine intelligible Raum' *per fluxum superficiei*, womit der Begriff der Bewegung bewusst als Prinzip in die Geometrie eingeführt wird. Die 'neue Methode', verbunden mit einer 'neuen Form', um die es Fabry geht, besteht im wesentlichen

(a) in der Einführung des Fluxusbegriffes anstelle der Indivisibilien,

(b) in einer Einteilung der um die Mitte des 17. Jahrhunderts bekannten geometrischen Figuren in zwölf Klassen und

(c) in der geschickten Verwendung der gewissen Figuren gemeinsamen Eigenschaften gemäss den Definitionen dieser Klassen, der 'Homogenität'.

Der Begriff dieser Klassen möge kurz erörtert werden. Es gibt (nach Fabry) ebensoviele verschiedene Klassen, als es verschiedene 'Aspekte der Erzeugungsweise' (*geneseos rationes*) gibt. Diese und deren gegenseitige Beziehungen, die 'Homogenität', nennt Fabry *principia directa et intrinseca*. Die *erste Klasse* enthält alle jene Gebilde, die durch Fluxus eines konstanten Elementes entstehen wie Parallelogramme, Prismen etc. Die *zweite Klasse* umfasst alle jene Figuren, deren Elemente *s* sich proportional einer linearen Funktion der Höhe ändern, also etwa

$$s :: h^1.$$

Hierzu gehören nicht bloss Flächengebilde einfacher Natur, sondern auch Körper wie Zylinder und Kreiskegel, deren Axialschnitte Rechtecke oder

---

[14] In B. Cavalieris *Geometria indivisibilibus ... promota* (Bologna 1635), taucht der Begriff *fluxus* ganze dreimal auf (wenn ich richtig gezählt habe), hingegen wird er in den *Exercitationes geometricae sex* (Bologna 1647), bereits als methodischer Ansatz praeformiert. In diesem Kontext wird auch (erstmals?) der Terminus *fluxio* eingeführt: '...*Tertius et ultimus modus est, quam appello fluxionem, scilicet dum modum partes moventur ad congruentiam* ...', (p. 217). – Cavalieri betreffend cf. K. Andersen, 'Cavalieri's method of indivisibles', *Arc. Hist. Ex. Sc.*, vol. 31, Nr. 4, 1985, pp. 291–367.

Dreiecke sind. Die *dritte Klasse* entsteht, wenn die Elemente einer Figur proportional dem Quadrat der Höhe wachsen, also wenn

$$s :: h^2.$$

Weil dies dem Charakter der Pyramide entspricht ($Q_1 : Q_2 = h_1^2 : h_2^2$), nennt Fabry die dritte Klasse auch 'die Klasse der Pyramiden'. In die *vierte Klasse* gehören alle jene Gebilde, deren Elemente sich proportional höheren ganzzahligen Potenzen der Höhe ändern, also

$$s :: h^n \quad (n = 3, 4, \ldots).$$

Es kommt natürlich häufig vor, dass eine geometrische Figur je nach Wahl ihrer Elemente *s* gleichzeitig verschiedenen Klassen angehört. So umfasst etwa die *fünfte* diejenigen Gebilde, deren Elemente sich proportional der Schnitt-fläche einer Halbkugel ändern, weshalb die Pyramiden nicht nur zur dritten, sondern auch zur fünften Klasse gehören. Aufgrund dieser Eigenschaft nennt nun Fabry Pyramide und Halbkugel *homogen*. Geometrische Figuren sind *homogen*, wenn sie in irgendeiner Beziehung derselben Klasse angehören. Die *sechste Klasse* wird von all jenen Figuren gebildet, welche die 'Kom-plemente' der Figuren der vierten Klasse sind, d.h. deren Elemente sich derart ändern, dass sie der Differenz der Elemente der Figuren der ersten und der vierten Klasse entsprechen. Die *siebte Klasse* enthält nebst den Figuren der sechsten auch die gewöhnlichen Parabeln, insofern diese nach zur Scheitel-tangente parallelen Linienelementen betrachtet werden. Alle parabolischen Trilinea, nach achsenparallelen Elementen aufgelöst, bilden den Inhalt der *achten Klasse*. Die *neunte*, die 'Klasse des Quadranten', enthält alle Figuren, deren Elemente sich proportional zum Sinus ändern, während die *zehnte*, die 'Klasse der Halbkugelflächen', alle jene Figuren beinhaltet, deren Elemente sich proportional den Parallelkreisen einer Kugelfläche ändern. Die *elfte* ist die 'Klasse der Zykloide' und die *zwölfte* diejenige der Hyperbeln, in welche auch alle Gebilde vom Charakter der Archimedischen Spirale gehören wie die Konoide.

Mit diesem – zugegebenermassen etwas dürftigen – 'Homogenitätsprinzip' operiert Fabry bei seinen infinitesimalmathematischen Untersuchungen nicht ohne Erfolg; Leibniz bezeichnete diese – neben den Arbeiten von Cavalieri, Gregorius a St Vincentio und Pascal – als wichtig für die Entdeckung seines *Calculus* (1675). Wärend in der *Synopsis geometrica* der Akzent auf dem Didaktischen liegt, bringt Fabry im *Opusculum geometricum* eine Fülle von Einzelleistungen, die im Kontext mit Pascals berühmt gewordener – und auch ominös belasteter – Herausforderung vom Jahre 1658 gesehen werden müssen.[15] Wir finden bei Fabry unter anderem neben einer geistreichen

15 B. Pascal, *Oeuvres de Blaise Pascal*..., L. Brunschvicg, P. Boutroux, F. Gazier (eds.), 14 vols. (Paris 1908–1914), speziell Bd.VIII und IX; *Oeuvres complètes de Blaise Pascal*, J. Chevalier (ed.), (Paris 1954), 'Pléiade', pp. 173–343.

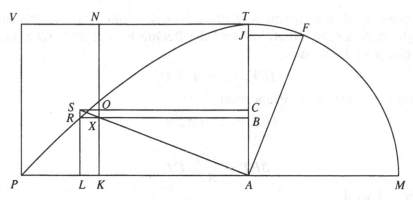

Figur 1

Zykloidenquadratur, von welcher Leibniz so begeistert war, verschiedene Quadraturen und Kubaturen, die (analytisch ausgedrückt) auf Spezialfälle von $\int x^n \sin x \, dx$, $\int \sin^p x \, dx$ und $\iint \arcsin x \, dx \, dy$ hinauslaufen sowie Schwerpunktsbestimmungen an Sinusflächen- und Zykloidensegmenten samt deren Rotationsgebilden um beide Achsen. Als *pars pro toto* mögen hier drei Beispiele vorgeführt werden. (Dabei ist zu beachten, dass wir es dem Leser leicht machen: Fabry bedient sich in seinem 'Literalkalkül' ausschliesslich der synthetisch-geometrischen Darstellung und operiert natürlich auch nicht mit (Newtonschen) Differentialinkrementen.)

Beispiel 1 (Sinusflächensegment)

Den Schlüssel zur Lösung einer Gruppe der 'Pascalschen Probleme' fand Fabry in der geometrischen Bewältigung des bestimmten Integrals $\int_0^z \sin x \, dx = 1 - \cos x$, das bereits von Kepler ausgewertet worden war, um für die Abweichung der Planetenbahnen von der Kreisbahn eine physikalische Ursache ermitteln zu können.[16] Fabry nennt nun (*O.G.* Prop. VI, Tab. 3, Fig. 11, hier Fig. 1) die Sinussegmentfläche *PRL* ein *segmentum rectum*, und das *segmentum versum BRT*, das dem bestimmten Integral $\int_0^z \arcsin z \, dz$ entspricht, berechnet er folgendermassen: Es sei *APT* die Sinusfigur, *AMT* der Kreisquadrant, $AK = AM = 1$, $PA = VT = \pi/2$, $TF = PL = x$ und $LR = JF = \sin x$. Nun gilt wegen der Homogenität (Figuren 9. Klasse) die Proportion

$$APT : LPR = AT : JT, \tag{1}$$

weil sich (nach Pappus) Kugelkalotten zueinander verhalten wie ihre Höhen.

16 Cf. S. Günther, 'Ueber eine merkwürdige Beziehung zwischen Pappus und Kepler', *Bibliotheca Mathematica* 2(1888):81–7; G. Eneström, 'Sur un théorème de Kepler équivalent à l'intégration d'une fonction trigonométrique', *Bibliotheca Mathematica* 3(1889):65–6.

Demgemäss wird das Segment $LPR = 1 - \cos x$, und da die Sinusfigur inhaltsgleich dem Quadrat AN ist und die Rechtecke AR und AO einander flächengleich sind, gilt auch

$$APT : AO = AT : AC$$

und nach Erweiterung der Proportion

$$APT : BRT = AT : CJ$$

oder

$$BRT = \frac{APT}{AT} \cdot CJ.$$

Für $AM = 1$ wird

$$BRT = CJ, \tag{2}$$

wobei zu beachten ist, dass es sich dabei um eine Flächenbeziehung handelt: CJ ist mit dem Faktor 1 zu versehen, sodass die rechte Seite von (2) zu einem Rechteck wird. Nach kurzer Rechnung[17] findet man mittels Streckenanalyse in Fig. 1

$$CJ = \cos x - (\pi/2 - x) \sin x, \tag{3}$$

welch hübsches Resultat Fabry einzig der Beziehung $\int_0^x \sin x \, dx = 1 - \cos x$ verdankt, die bereits in (1) enthalten ist. Tatsächlich ergibt sich auch das Segment der Arcus-sinus-Fläche als

$$BRT = APT - \text{Rechteck } AR - LPR = 1 - (\pi/2 - x) \sin x - \int_0^x \sin x \, dx$$

$$= \cos x - (\pi/2 - x) \sin x,$$

und dies entspricht der Auswertung des bestimmten Integrals

$$\int_0^z \arcsin z \, dz = z \cdot \arcsin z + \sqrt{1 - z^2} - 1.$$

Beispiel 2 (Zykloidensegment)

Von hier aus gelangt Fabry mühelos zur Quadratur eines Segmentes der Zykloidenfläche KBN (Fig. 2). Est ist nämlich

$$KBN = KBL + LBN. \tag{4}$$

Stellt GHJ die Sinusfläche und die Kurve BNQ die Halbzykloide dar, so ist jede Abszisse $LN = EF$, d.h. die Figur LBN ('Zipfelmütze') ist dem Segment EJF homogen, mithin flächengleich, da die beiden Figuren gleiche Grundlinien und Höhen haben. Dieses Segment EFJ ist jedoch gemäss (3) berechnet und braucht bloss dem Kreissegment KBL addiert zu werden, damit die Quadratur

---

[17] Cf. Fellmann, *Fabry*, pp. 76–7.

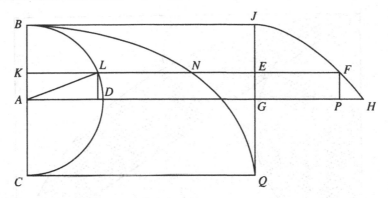

Figur 2

vollendet wird. Setzt man $AB = AC = 1$, $LD = HP = x$, so ist $AK = PF = \sin x$, $KL = \cos x$, $BL = \pi/2 - x$, und das Kreissegment KBL ergibt sich als

$$KBL = \tfrac{1}{2}(\pi/2 - x - \sin x \cos x). \tag{5}$$

Addiert man nun die Gleichungen (3) und (5), so findet man das gesuchte Zykloidensegment[18] als

$$KBN = (\pi/2 - x)(\tfrac{1}{2} - \sin x) + (1 - \tfrac{1}{2} \sin x) \cos x,$$

was für $x = -\pi/2$ auch tatsächlich die bekannte Fläche der Halbzykloide $F = 3\pi/2$ liefert.

Mit diesen Quadraturen und dem Begriff des statischen Momentes (den übrigens Leibniz von Fabry kennengelernt hat[19]) ist auch der Weg zu den oben genannten Kubaturen und Schwerpunktsbestimmungen geebnet, und es ist nur noch eine Sache der Virtuosität in der Handhabung der Methode, ein Raumgebilde, dessen Volumen durch das Doppelintegral

$$\int_0^1 \int_0^{\sqrt{1-x^2}} \arccos x \, \mathrm{d}x \, \mathrm{d}y$$

dargestellt wird, der geometrischen Erforschung zu unterwerfen, wie Fabry an diesem Beispiel überzeugend nachzuweisen imstande ist.

Beispiel 3 (eine damals neue Zykloidenquadratur)

Um die Mitte der Dreissigerjahre des 17. Jahrhunderts gelang einigen Mathematikern der strenge Nachweis, dass die von einer (gemeinen) Zyloide – der 'Schicksalskurve des 17. Jahrhunderts' – umspannte Fläche genau gleich dem Dreifachen des erzeugenden Kreises ist. Die Prioritätskontroverse zwischen Torricelli und Roberval resp. deren Anhängern ist Gegenstand jeder

18 *loc. cit.* pp. 90–1.

19 J. E. Hofmann, *Leibniz in Paris 1672–1676* (Cambridge University Press, 1974), p. 51.

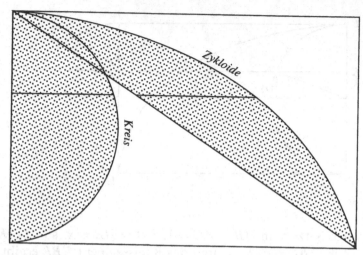

Figur 3

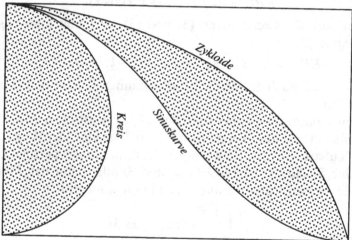

Figur 4

einschlägigen 'Geschichte der Mathematik', sodass – auch im Hinblick auf die ansehnliche Menge an Sekundärliteratur[20] – hier nicht darauf eingegangen wird. Hingegen soll eine weitgehend vergessene oder kaum zur Kenntnis genommene Zykloidenquadratur gezeigt werden, welche Fabrys ureigenste Erfindung[21] ist und die ihrer Originalität und Schönheit halber hier nicht unterschlagen werden soll. Natürlich kannte Fabry die diesbezüglichen Resultate Torricellis (Fig. 3) und Robervals (Fig. 4), welche an dieser Stelle bloss figurativ angedeutet seien (die punktierten Flächen sind inhaltsgleich),

[20]   K. O. May, *Bibliography and Research Manual of the History of Mathematics* (Toronto and Buffalo 1973), p. 425.

[21]   *O.G.*, Prop. XXIII und XXIV; Fellmann, *Fabry*, pp. 87–90.

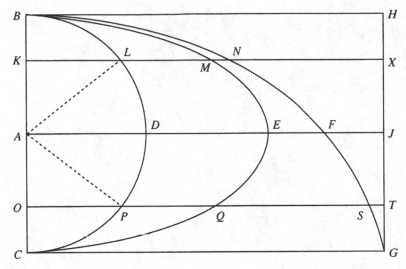

Figur 5

doch bereicherte er den Problemkomplex mit einer wahrhaft geistreichen Idee (Fig. 5).[22] Zwischen dem Halbkreis *BDC* und der Halbzykloide *BNFSG* wird eine Halbellipse *BEC* mit der grossen Halbachse *AE = 2AD* eingeschoben. Nun ziehe man zwei beliebige parallele Transversalen *KX* und *OT* derart, dass *CO = BK = HX = GT*. Aufgrund der genetischen Eigenschaft der Zykloide, dass die Abszisse eines jeden ihrer Punkte gleich der Summe des abgewälzten Kreisbogens und dessen Sinus ist, lässt sich zeigen, dass stets gilt[23]

$$NX = QS. \tag{6}$$

Das ist der Kern von Fabrys origineller Quadratur, denn aus (6) folgt sofort, dass die von der Ellipse und der Zykloide begrenzte Figur *BECGFB* dem Trilineum *BFGHB* homogen ist, woraus *per fluxum* die Flächengleichheit dieser Trilinea resultiert. (Man beachte den Spezialfall, dass *F* der Halbierungspunkt von *EJ* ist.) Der Rest ist eine Kopfrechnung: Die Fläche der Halbellipse ist gleich der ganzen Kreisfläche, die Rechtecksfläche gleich deren Doppeltem und die von der Zykloide halbierte Differenzfläche aus Rechteck und Halbellipse ist ebenfalls dem ganzen Kreis inhaltsgleich, mithin hat die Halbzykloide den anderthalbfachen Inhalt des erzeugenden Kreises.

[22]   Fabrys Originalfigur wurde der Transparenz halber etwas vereinfacht; die Bezeichnungen wurden beibehalten.

[23]   Der Beweis von Gleichung (6) kann leicht folgendermassen geführt werden: Mit $AB = r = 1$, $\overset{\frown}{BL} = x$ und $BH = \pi$ wird $KL = \sin x$, $LN = x$ und somit $NX = \pi - x - \sin x$. Aus $OS = 2\sin x + QS = OP + \overset{\frown}{BDP} = \sin x + (\pi - x)$ ergibt sich $QS = \pi - x - \sin x$, mithin $NX = QS$.

Figur 6. Honoré Fabry, SJ, im Alter von 75 Jahren. Oelbild in der Generalskurie des Archivum Romanum SJ des Vatikans. Ich danke den Patres für die seinerzeitige Ueberlassung des Negativs und die Publikationserlaubnis.

Epilog

(A) Vita

Honoré Fabry war ein origineller, aber wilder Geselle.[24] Er entstammte einer Richterfamilie von Valromey und wurde am 5. April 1607 in Virieu-le-Grand in der Dauphiné geboren. Nach seiner Erziehung im Institut zu Belley trat er am 10. Oktober 1626 in das Noviziat der Jesuiten in Avignon ein, wo er bis 1628 verblieb. Im Herbst dieses Jahres begab er sich nach Lyon ins Collège de la Trinité,[25] wo er seinen Kursus in scholastischer Philosophie unter Claude Boniel[26] absolvierte. Nach einem zweijährigen Lehraufenthalt im Collège zu Roanne kehrte Fabry 1632 nach Lyon zurück, um dort seinen Kursus in Theologie zu beginnen, den er – nach seiner Ordination zum Priester (1635) – im Jahre 1636 beendete. Noch im selben Jahr wird er zum Professor für Logik am Collège in Arles ernannt, wo er während zweier Jahre Vorlesungen über Philosophie mit Einschluss der Physik hielt. In diese Zeit fällt Fabrys – von Harvey unabhängige – Entdeckung des grossen Blutkreislaufs, worüber er öffentlich lehrte.[27]

Neben seinem neuen Amt als Präfekt am Collège in Aix-en-Provence leitet Fabry 1638–39 eine Art Zirkel, der ihm unter anderem die Bekanntschaft wie auch eine lang andauernde Korrespondenz mit Gassendi einbringt. Anschliessend wird er zur Absolvierung seines dritten Probationsjahres unter P. Barnaud nach Lyon zurückgerufen und avanciert 1640 zum Professor für Logik und Mathematik sowie zum Dekan am Collège de la Trinité. Während der folgenden sechs Jahre lehrt Fabry Metaphysik, Astronomie, Mathematik und Physik. Diese Periode war wohl die glänzendste und fruchtbarste seines Lebens; mehrere später gedruckte Bücher sind Ausarbeitungen der in dieser Periode gehaltenen Vorlesungen. Fabry eröffnet die Reihe der zahlreichen berühmt gewordenen Professoren, die das Collège de la Trinité hervorgebracht hat: zu seinen Schülern gehören Pierre Mousnier, der später viele Vorlesungen seines Meisters ediert hat, der durch seine Freundschaft mit Newton bekannte Mathematiker François Regnaud,[28] G. D. Cassini und P. de la Hire. Im

---

[24] Zwei seiner theologisch-philosophischen Schriften mussten von ihm selbst als Chef der Inquisition auf den *Index librorum prohibitorum* (1672 und 1673) gesetzt werden, und ein Brief Fabrys über den 'Frieden von Clemens IX' wurde am 26. März 1669 auf Parlamentsbeschluss in Paris verbrannt. – Cf. Sommervogel, I, Spalten 516 und 518.

[25] Das Collège de la Trinité wurde 1565 von den Schöffen von Lyon den Jesuitenpatres übergeben, jedoch infolge des Attentats des Jesuitenzöglings Jean Chatel auf Henry IV 1594 geschlossen. Nach ihrer Vertreibung aus Frankreich kehrten die Jesuiten 1604 zurück und gründeten das Collège neu; es wurde jedoch erst 1660 vollendet.

[26] Cf. Sommervogel, I.   [27] Cf. *Journal des Sçavans* (1666), pp. 395–400.

[28] Auch Reynaud, Regnauld, Raynaud geschrieben.

gleichen Kreise finden wir ferner C. F. M. de Châles[29] sowie den Astronomen und Mathematiker Bertet. Zwischen all diesen Gelehrten und den beiden Huygens (Vater Constantijn und Sohn Christiaan), Descartes, Leibniz, Mersenne und anderen entwickelte sich eine aktive Korrespondenz, an welcher Fabry beachtlichen Anteil hatte. Gegenstand dieser lebhaften Aktivität sind fast alle damals akuten Fragen der Wissenschaft: Heliozentrismus, Saturn-ringe, Gezeitentheorie, Magnetismus, Optik, Kinematik und Dynamik. In der Mathematik stehen infinitesimale Methoden und das Kontinuumproblem im Vordergrund.

Die positiv-kritische Aufnahme gewisser kartesianischer Konzeptionen[30] verwickelte Fabry in eine heftige Kontroverse mit seinen Vorgesetzten, die schliesslich zu seiner Relegation in Lyon und zur 'Strafversetzung' nach Rom führte, wo er am 12. September 1646 eintraf. Obwohl sein Aufenthalt nur provisorisch sein sollte, wurde Fabry noch im gleichen Jahr dem Poeni-tentiarkollegium (Inquisition) angegliedert, in welchem er dann 34 Jahre lang – zuletzt als Grossinquisitor – amten sollte. Trotz des bedeutenden Engagements in der Kirchenpolitik und der Theologie – Fabry galt als erster Kenner des Jansenismus – blieb ihm offenbar genügend Zeit für seine breitangelegten wissenschaftlichen Forschungen.

In Rom lernt Fabry den Mathematiker und späteren Kardinal Michelangelo Ricci kennen, der auch mit E. Torricelli befreundet war. Ricci empfiehlt Fabry dem Grossherzog Leopold II von Medici, und dieser ernennt den schon einflussreichen Jesuiten zum korrespondierenden Mitglied der (liberalen) *Accademia del Cimento*. Im Jahre 1660 eröffnet Fabry mit einer anonymen Schrift[31] die Kontroverse mit Huygens über die Saturnringe,[32] die erst nach fünf Jahren und mit grossem Aufwand seitens Leopolds II zu Gunsten von Huygens gegen Fabry entschieden wurde. Doch letzterer erwies sich als fairer Gegner: er entschuldigte sich öffentlich und bekannte sich zur Auffassung von Huygens (ihm blieb allerdings nach dem Machtspruch des Grossherzogs und dem schlagenden Ausgang des teleskopischen Experiments auch nichts anderes übrig).

In der *Brevis annotatio*[31] findet sich nun eine Bemerkung etwa folgenden Inhalts: 'Solange man keinen strengen Beweis für die Bewegung der Erde hat, ist die Kirche zuständig zu entscheiden. Ist aber der Beweis einmal gefunden,

---

[29] Auch Dechales, Dechasle geschrieben.

[30] Cf. Morhoff, *Polyhistoria litterarum*, Bd. 2, p. 115; A. Baillet, *La vie de Monsieur Des-Cartes* (Paris 1691–93), vol. 2, p. 299; René Fülöp-Miller, *Macht und Geheimnis der Jesuiten* (Berlin 1929), p. 160f.

[31] *Eustachii de Divinis Septempedani brevis annotatio in systema saturnium Christiani Hugenii* (Rom, 1660).

[32] Ch. Huygens, *Systema Saturnium...*, Den Haag 1659; *Brevis assertio systematis Saturnii...* (Den Haag, 1660).

dann soll es keine Schwierigkeit geben zu erklären, dass die betreffenden Bibelstellen in einem mehr symbolischen Sinne ausgelegt werden müssen'. Diese Sentenz hätte vielleicht später unter Papst Clemens IX, auf den Fabry starken Einfluss hatte, durchgehen können, doch unter Alexander VII brachte sie Fabry – als Mitglied des heiligen Offiziums! – 50 Tage Gefängnis ein; Fabrys Freilassung wurde nur durch die Intervention des Grossherzogs Leopold II erwirkt. Doch diese bittere Erfahrung hinderte den streitbaren Jesuiten nicht daran, seinen *Dialogi physici* von 1665 ein in gleichem Geiste abgefasstes Kapitel mit dem Titel *De motu terrae* einzufügen, obwohl Alexander VII noch immer auf dem päpstlichen Stuhl sass. In dasselbe Jahr 1665 fällt auch Fabrys Entdeckung des grossen Andromedanebels, den er zunächst als einen neuen Kometen auffasste.

1668 trat Fabry einen etwa einjährigen 'Erholungsurlaub' in die alte Heimat an, um den Druck verschiedener seiner Werke zu veranlassen und zu überwachen. Noch bis 1680 wirkte er im hl. Offizium in Rom und verbrachte die letzten acht Jahre seines Lebens abseits der Stadt auf dem Lande, wo er neben einer Geschichte der Inquisition eine stattliche Anzahl von Abhandlungen historischen, apologetischen und moralischen Inhalts verfasste.[33] Fabry verstarb am 8. März 1688 im Alter von 81 Jahren.

(B) Würdigung

Im Gebiet der Physik hatte Fabry eine weniger glückliche Hand als in der Mathematik. Immerhin müssen folgende Leistungen als beachtenswert angesehen werden: die fortwährende Verwendung des Begriffs des statischen Momentes; ein Erklärungsversuch der Gezeitenerscheinungen durch die Wirkung des Mondes (wenn auch über den Luftdruck als Medium); eine auf dem Prinzip der Streuung beruhende Erklärung der Bläue des Himmels sowie Untersuchungen über die Kapillarität. Gänzlich verfehlt ist sein Erklärungsversuch der Kohäsion, während das 'Loch-Nadel-Experiment' richtig gedeutet wird im Sinne Grimaldis, von welchem Fabry höchstwahrscheinlich abhängig ist. Isaac Newton erwähnt in seiner zweiten Abhandlung über das Licht und die Farben, er habe erstmals von Grimaldis Experimenten durch die Vermittlung eines gewissen italienischen Autors ('some Italian author') vernommen,[34] den er in seinem Brief an Oldenburg vom 7. Dezember 1675 als Fabry identifiziert.[35]

---

[33] Die Manuskripte der letzten Schaffensperiode Fabrys liegen – weitgehend unveröffentlicht – in der Bibliothèque de la Ville de Lyon.

[34] Gemeint ist: H. Fabry, *Dialogi physici, quorum primus est de lumine,* ...(Lyon, 1669).

[35] H. W. Turnbull (ed.), *The Correspondence of Isaac Newton*, I (Cambridge University Press, 1959), p. 384.

Eine summarische Würdigung Fabrys als Mathematiker findet sich in Fellmann, *Fabry*.[12] Zusammenfassend lässt sich sagen, dass Fabrys Beiträge zur Infinitesimalmathematik des 17. Jahrhunderts sowohl in inhaltlicher als auch formaler Hinsicht nicht unwesentlich sind und dass Leibniz – entgegen M. Cantors Auffassung, welche lange Zeit das allgemeine Urteil über die mathematischen Arbeiten des Jesuiten geprägt hat – in seiner eingangs erwähnten Besprechung des Newtonschen *Opus* keinesfalls Newton beleidigen wollte noch konnte, als er dessen Fluxionsrechnung mit der Systematisierung der Cavalierischen Indivisbilienmethode durch Fabry verglichen hatte. Im Gegenteil hat der grosse Synoptiker Leibniz sehr genau die *ideengeschichtlichen Zusammenhänge* fixiert, als er in Fabry einen – zwar bloss geometrischen – Vollzug des Ueberganges vom statischen Indivisibilienbegriff zum dynamischen Fluxionsbegriff erkannte. Wenn der Prioritätsstreit zwischen Leibniz und Newton tatsächlich wegen dieser als Beleidigung empfundenen Leibnizschen Rezension ausgebrochen sein sollte, so wäre dies nur ein Symptom für die Treffsicherheit historischer Urteile des Philosophen.[36]

[36] Dass man in England Fabrys *Mathematica* gekannt, oder mindestens in Besitz hatte, geht allein schon aus der Tatsache hervor, dass John Collins ein Exemplar der *Synopsis geometrica* besass. Er hat seinerzeit dieses Exemplar der 'Mathematical School in Christ's Hospitall' vermacht; später kam es durch Kauf in schweizerischen Privatbesitz.

— II —

Newton's manuscripts

5 Sotheby's, Keynes and Yahuda – the 1936 sale of Newton's manuscripts

P. E. SPARGO

Acknowledgement

In the General introduction to Volume I of his monumental *Mathematical Papers of Isaac Newton*, Whiteside describes in meticulous detail the history of Newton's papers after his death in 1727. However, in this fine study the treatment of the early history of Newton's papers is very much fuller than their history in the twentieth century and in fact ends with only a brief reference to the great Sotheby sale of Newton's papers in July 1936 – a situation which is perfectly understandable in view of the fact that Whiteside's primary concern was of course Newton's mathematical papers, the bibliographic history of which had in effect ended in the late nineteenth century with their donation to the University Library, Cambridge, by the fifth Earl of Portsmouth.

With respect to those numerous manuscripts classified as non-scientific, however, their history in the twentieth century, and particularly their sale at Sotheby's in 1936, is of overwhelming importance, and in order to complete Whiteside's history of the Newton papers this historic sale, as well as the subsequent tragic dispersal of Newton's manuscripts, needs to be described. This essay is therefore written to complement Whiteside's pioneer work in this field.

Tom Whiteside and I met for the first time in April 1970 shortly after I arrived in Cambridge on sabbatical leave. Since his first graphic piece of advice to me on that occasion ('Don't waste your time here going to seminars on Newton – get into the University Library and study the manuscripts') I have on innumerable occasions benefited from his unstinting professional generosity, kindness and encouragement, as well as the warm hospitality of both Ruth and himself. He has assisted and encouraged me more than he can ever know and it is with respect, admiration and warm affection that I dedicate this small Newtonian contribution to Tom Whiteside, our master and guide in all things Newtonian.

When Newton died intestate in March 1727 his papers, together with his other possessions, passed into the hands of his niece Catherine Barton who ten years previously had married John Conduitt, Newton's successor at the Mint. In 1740 their only child, also Catherine, married the Hon. John Wallop, later to become Viscount Lymington when his father was created first Earl of Portsmouth. Newton's papers, together with the material Conduitt had collected for a projected life of Newton, thus entered the possession of the Portsmouth family, where they were to remain intact for more than 150 years. During this time the papers were only seriously explored on two occasions: by Bishop Samuel Horsley in preparation for his five-volume edition of the works of Newton (1779–85), and again by Sir David Brewster when writing his two-volume biography of Newton, published in 1855.

In 1872 the fifth Earl of Portsmouth entrusted all the Newton papers in his possession to the University of Cambridge, where they were arranged and catalogued by a committee consisting of H. R. Luard, Sir George Gabriel Stokes, John Couch Adams and G. D. Liveing. On completion of the committee's work in 1888 that portion of the papers regarded as being scientific was presented by the Earl of Portsmouth to the University. Thus those papers dealing with mathematics and chemistry, together with some correspondence, books and miscellaneous papers, as well as the papers dealing with Newton's relations with Flamsteed, were deposited in the University Library, Cambridge, where they have remained ever since.

The remainder of the collection was returned to Lord Portsmouth at his family seat, Hurstbourne Park, in Hampshire. Although the house and most of its contents were destroyed by a terrible fire in January 1891, the Newton papers survived virtually unscathed. They remained at Hurstbourne until 1936 when, in July of that year, they were offered for sale at Sotheby's, the London auction house. This very substantial quantity of material, several million words in extent, consisted principally of Newton's alchemical, theological, historical and Mint papers, as well as some important correspondence. It also contained the material which John Conduitt had gathered so hopefully two centuries earlier for his projected biography of Newton, as well as Godfrey Kneller's magnificent portrait of Newton.

Sometime around March 1936 a metal trunk was delivered to 34/35 New Bond Street, the premises of Sotheby and Co., the well known London auctioneers.[1] The trunk contained the Newton manuscripts, which were to be sold at auction by order of Viscount Lymington, son of the eighth Earl of Portsmouth and Trustee for the family. They had presumably come directly to Sotheby's from Hurstbourne Park. Under heavy financial pressure as a result of death duties and other family expenses – such as, it was rumoured, an impending divorce – the Portsmouth family had decided to sell the Newton manuscripts as well as Hurstbourne Park itself, which was put on the market in April 1936.

On examination at Sotheby's the manuscripts were found to be in considerable disorder and the heavy task of sorting and organising them, as well as preparing a detailed sale catalogue, was commenced at once by John Cameron Taylor, the Chief Cataloguer in the Book and Manuscripts

---

[1] From 1819 to 1917 Sotheby's was located at 13 Wellington Street. After the final sale in these premises on 25 May 1917, the company moved to 34 & 35 New Bond Street, where it is still located. (In certain matters of detail in this section I have been able to draw on Frank Herrmann's fine history, *Sotheby's – Portrait of an Auction House*, (London: Chatto and Windus, 1980), which I gratefully acknowledge.)

Department at Sotheby's.[2] The sale had been set down for mid-July and it soon became a race against time to have the catalogue ready by the required date. John Taylor came under increasing pressure from the management of Sotheby's, who were becoming more and more agitated by the delay in the completion of the copy for the printer. In reply Taylor stressed time and again not only the amount of work involved in the preparation of the catalogue but also the great historic significance of the documents he was cataloguing.[3] However, in the end the deadline was met and on Monday 22 June 1936 the first public announcement of the sale appeared in *The Daily Telegraph*.[4] The following day *The Times* devoted the whole of its column 'The Sale Room' to a discussion of the forthcoming sale as well as carrying a small Sotheby advertisement of the auction.[5]

During the next three weeks a few further small advertisements appeared in *The Times*, *The Morning Post* and *The Times Literary Supplement*.[6] Rather unexpectedly the most impressive publicity came through *The Illustrated London News*, which on 4 July carried photographs and a brief description of

[2] John Taylor is an exceptionally interesting figure. His father, a warehouseman by day and a student by night, mastered Latin, Greek and Hebrew and eventually became a Baptist minister. After John had been forced to leave school at the age of fourteen through the family's straitened circumstances, he worked in an industrial laboratory in Leeds, matriculated by evening study and then moved to London, where he worked in Hodgson's, the London book dealers. Although passionately interested in chemistry, he decided to first take a degree in classics at London University. When the First World War broke out Taylor enlisted and after being injured in the trenches in 1915 he was transferred to the Ministry of Munitions, where he took charge of a nitrating chamber used in the production of the enormous quantities of nitroglycerine needed for the war. He ended the war in the chemical research laboratories, where there and later he carried out distinguished research work. After the war Taylor spent some time as a cataloguer for Hodgson's, but soon joined Sotheby's as a full-time cataloguer in the books and manuscripts department. There his first-hand experience as a scientist stood him in good stead as he catalogued the scientific and medical books handled by Sotheby's. During the Second World War Taylor served in Military Intelligence where he once again rendered distinguished service, particularly in the development of the micro-dot for compressing large quantities of information into a small area. He returned to Sotheby's at the end of the war and continued with the firm until his retirement in 1968. He died in 1979. May I record my delight – and the great privilege – of meeting John Taylor in the company of the late John Carter and Lord John Kerr over a protracted and memorable lunch at the Westbury Hotel in London in September 1972.   [3]   Frank Herrmann, *Sotheby's*, p. 291.

[4]   *The Daily Telegraph*, London; 22 June 1936, p. 5: 'Isaac Newton as alchemist. His theories on transmutation. Papers to be sold.'

[5]   *The Times*, London; 23 June 1936, p. 18: 'The Newton papers'.

[6]   *The Times*, London; 23 June 1936, p. 30; 30 June 1936, p. 30; 7 July 1936, p. 28; 14 July 1936, p. 30; *The Morning Post*, London; 29 June, 6 July and 13 July, all p. 20; *The Times Literary Supplement*, London; 4 July 1936, p. 568.

two of Newton's notebooks together with a few details of the sale itself.[7] There appears to have been no other pre-sale publicity of any sort in Britain – a far cry from the intense interest which a sale of such monumental importance would arouse today. It is interesting, however, to note that the two longest and most enthusiastic pieces of pre-sale publicity occurred not in Britain but in the United States. A copy of the sale catalogue had reached that country just in time for *The New York Herald Tribune* of Sunday 12 July – the day before the sale opened – to devote the whole of its 'Notes for Bibliophiles' column to a lengthy description of the sale. Similarly, *The New York Times Book Review* of 12 July devoted three columns to a full and enthusiastic description by Philip Brooks of the following day's sale: 'This will be one of the most phenomenal one-man auctions ever held'. It described the catalogue as a 'handsome and elaborate affair' and its contents as 'an educational feast'; and finally, 'In offering the present collection with its millions of words, mostly un-published...the auctioneers are justified in claiming that its importance in the sale room "can show few parallels".'[8]

In every way the publicity surrounding the Newton sale was heavily over-shadowed by that accorded the impending sale at Christie's of Mr Henry Oppenheimer's celebrated collection, which included a number of drawings by the greatest of the Old Masters.[9] This remarkable sale opened amidst intense publicity on Friday 10 July, only three days before the Newton sale, and all eyes had been focused on it for some weeks before. From the point of view of a counter-attraction the Newton sale could hardly have been held at a worse time.

If it is noteworthy how little comment the announcement of the Newton sale caused amongst the public at large it is even more remarkable – at least by today's standards – how little it caused in the scholarly world, or at least any that surfaced in the newspapers of the period. There were no solemn editorials or indignant letters to the editor regretting the fact that a substantial proportion of the papers of England's greatest scientific genius were about to be broken up and dispersed at auction, with many important manuscripts doubtless destined to leave the country. Even the University of Cambridge and Trinity College, Cambridge, seemed strangely unmoved by the impending dispersal of a major portion of the manuscripts of their greatest son.

[7] *The Illustrated London News*, London; 4 July 1936, p. 18.

[8] *The New York Herald Tribune Books*, 12 July 1936; Section VII, p. 18: 'Newton's manuscripts'; *The New York Times*, 12 July 1936; Section VI, p. 21: 'The Newton sale'.

[9] For information relating to the Oppenheimer sale, see *The Times*, London, 11 July 1936, p. 9; 14 July 1936, p. 14; 15 July, p. 11; 18 July, p. 8; 22 July 1936, p. 13; 30 July 1936, p. 11.

Perhaps attention was also diverted from the Newton sale and the subsequent dispersal of the manuscripts by the serious, and deteriorating, international situation. First there were the continuing problems in Palestine, then a British mandate; on the very morning on which the sale opened it was announced in *The Times* that a further three battalions of British troops were to be sent to that troubled land. Together with this there was the persistent tension between Germany and Austria. Finally, also on the morning of the opening of the sale there was the ominous announcement that in Spain the seizure of the radio station in Valencia had been followed by rioting. A few days later insurrection broke out in numerous garrisons in Spanish Morocco – the Spanish Civil War had started.

The catalogue itself was a masterly piece of work, being remarkably accurate and reliable – especially when one considers the speed with which it was produced and the extraordinarily diverse nature of the Newton manuscripts. It was scholarly in its treatment of the subject without being pedantic and was full without being over-lengthy. Its value as a research tool has long been recognised and it has been continually sought after, referred to and quoted by Newton scholars since the 1936 sale. John Taylor produced many fine sale catalogues in his years at Sotheby's, but none finer than the Newton catalogue.

Two versions of the sale catalogue were produced. The illustrated version, with its dark green cover, contained sixteen full-page plates and sold for 7s. 6d. The other, with its cream cover, contained no illustrations, and its selling price is not recorded.[10] Both versions, 144 pages in length, were printed by H. Davy of 92 Fleet Street, London. There is no record of how many copies of these catalogues were printed but it is unlikely to have been more than a few hundred. As a result both have become items of the greatest rarity and have for many years been virtually unobtainable. As usual with Sotheby auctions the manuscripts were on view for three days prior to the sale. However, we have no record of how many people examined them in this period.

The sale opened at 'one o'clock precisely', to quote the contemporary advertisements, on Monday 13 July 1936 at Sotheby's premises in New Bond Street. It extended over two days, recommencing at one o'clock on the Tuesday. Of the 332 lots offered for sale, the first 174 were sold on the first day and the remainder on the second. It was, according to the accounts of some eyewitnesses, not particularly well attended, only about fifty people being present. John Taylor himself only dropped in for a few minutes 'to see how

[10] Frank Herrmann, *Sotheby's*, p. 218: 'In the Twenties and Thirties all these catalogues could be obtained for an *annual* subscription of only 2 pounds, with 5s. extra for those relating to autograph material.'

things were going'.[11] The bidding was restrained and business was conducted in an atmosphere of calmness and decorum. Apart from the usual London dealers, or their representatives, the sale was also attended by a number of important private buyers. The first was John Maynard Keynes, a Fellow of King's College, Cambridge, and already well known in academic and economic circles as the author of the *General Theory of Employment, Interest and Money* which had been published in February 1936. Keynes, although he was to play an enormously important role in the subsequent history of the Newton manuscripts, only attended the sale as a result of the urging of his brother Geoffrey and was in fact brought to the sale by him.[12] Others attending included the French autograph dealer Emmanuel Fabius, who travelled from Paris especially for the sale,[13] and Viscount Lymington.

Before the sale opened Sotheby's had received a substantial number of bids from prospective buyers who were not able to be present at the sale and who had not requested any of the booksellers who would be attending the sale to act directly on their behalf. These bids left 'on the house' were carefully entered, in the Sotheby price code, in the marked-up copy of the sale catalogue used by the auctioneer at the sale.[14] If, for any particular lot, the highest bid received from those present at the sale did not match the bid left on the house, then the latter would be accepted by the auctioneer as the successful bid. If the bids received from those present at the sale were higher, then the bid left on the house would of course be unsuccessful.

The marked-up catalogue shows that seventy-four bids were received from sixteen prospective purchasers, ranging from Roy Sowers, a bookseller located in Los Gatos, California, who left twenty-six bids on the house, to a Miss J. R. Garrood, who left only one. Only nine of these bids were successful: three from 'Hadley' (A. de Coppet), two from Roy Sowers, and one each from Professor John Read, of the University of St Andrews, 'Ludford' (still unidentified), C. W. Turner and 'G. Gregory Bookstore'.

In all these cases the purchaser obtained the desired lot at a price substantially lower than that to which he had informed Sotheby's he had been prepared to go – perhaps evidence of an intense desire to acquire one or two highly prized items. It is curious that the reverse is also true, for the unsuccessful bids left on the house were almost invariably substantially lower

---

[11] From a private discussion with John Taylor in 1972. Of particular interest in this connection, is Frank Herrmann's comment in *Sotheby's*, p. 199 of 'the iron rule of those days [*ca.* 1920] that the book cataloguers could not attend sales'.

[12] Private communication, Sir Geoffrey Keynes to P. E. Spargo, 22 January 1974.

[13] Private communication, M. Emmanuel Fabius to P. E. Spargo, 17 May 1974: 'Yes, I went especially to London to participate in person in the sales of 13th and 14th July 1936.'

[14] This copy of the sale catalogue is still preserved amongst Sotheby's records.

than the actual prices fetched for those items – perhaps some purchasers were just trying their luck.

Two further points are worthy of note here. The first is that John Maynard Keynes left a bid on the house – but only one. This is not only evidence of the fact that he was aware beforehand of the sale, but powerful corroboration of his brother Geoffrey's clear memory many years later that John Maynard was at first only mildly interested in the sale, had to be encouraged by Geoffrey to attend and only became excited as the first day of the sale proceeded. The second is that all but one of Roy Sower's twenty-six bids left on the house were for alchemical manuscripts. Either Sowers was particularly interested in handling this type of material or, more probably, he was bidding on the instructions of a (still unidentified) American client who was especially keen to acquire a substantial portion of Newton's alchemical manuscripts. If the latter is the case then his client must have been disappointed indeed for only two of the twenty-five bids left on the house for alchemical manuscripts were successful.

In all there were thirty-seven purchasers, of whom only nine bought ten or more lots.[15] At the other end of the spectrum fourteen purchasers bought only one or two lots. By far the most prolific purchaser was the London firm of Maggs Brothers, which acquired eighty-nine lots. This was followed by J. M. Keynes, who purchased thirty-nine lots; the Cambridge firm of Heffer & Son twenty-four; the New York Dealer Gabriel Wells twenty-three; the anonymous 'Ulysses' sixteen; Emmanuel Fabius thirteen; the London firm Francis Edwards twelve and Viscount Lymington, who purchased under the *nom de vente* of 'Maxwell', ten. Thus nine buyers purchased 247 lots, or almost precisely three-quarters of those which were on sale. Although the great majority of the purchasers were British, a number of American firms such as Rosenbach, Scribners and Gabriel Wells were also represented. Only two continental purchasers can be identified with certainty – Emmanuel Fabius of Paris and Dr Erik Waller of Uppsala, Sweden, who apparently did not attend in person but purchased under the *nom de vente* of 'Gyles'. Several purchasers, such as 'May' and 'Manning', have still not been identified and some of the lots they purchased remain unlocated to this day.

Two personal anecdotes about the sale are worth recording. Sir Geoffrey Keynes notes that in connection with the purchases which he and his brother John Maynard made at the sale, 'We had an agreement that he would not bid for Boyle lots (126, 156), if I did not compete elsewhere & so I got mine.'[16] Also, Viscount Lymington, writing to Keynes some two months after

---

[15]   Shortly after the sale Sotheby's, following its custom, published a list, 'for four shillings', detailing the purchaser of each lot at the sale, as well as the price paid.

[16]   Private communication, Sir Geoffrey Keynes to P. E. Spargo, 22 January 1974.

the sale, recalled that he had overlooked Lot 295 'until just after the hammer fell' and that he therefore did not deserve it 'owing to my carelessness'.[17]

The lack of excitement which greeted the announcement of the sale was, if anything, exceeded only by that which the sale itself caused, either in the academic community or amongst the public at large. Apart from the few letters to *The Times* mentioned below only one mild expression of regret or disappointment at this appalling – perhaps unique – example of manuscript dispersal appeared either in the editorial or the correspondence columns of British newspapers. This was an anonymous report in *The Times Literary Supplement* on the Saturday following the sale, in which the following comment appeared:

> One cannot help regretting that it was not possible to buy the entire collection for an English library. Someone, surely, if properly approached, would have been prepared to find, say, £10,000 to acquire a valuable body of papers such as this, including some three million words in Newton's own hand. It is said, however, that the only practical suggestion for a sale came from abroad.[18]

There are no doubt manifold reasons for this remarkable lack of interest both in the sale itself and in the subsequent fate of the manuscripts. It is, however, perhaps more credible when seen in the context of the time. The world was still emerging from the Great Depression, many of the effects of which were only to disappear with the outbreak of the Second World War some three years later. Thus the quantity of ready cash available for what was no doubt considered by most people to be the purchase of luxuries was very much more limited than it is today. Also, inflation in the 1930s was relatively low and the concept of purchasing manuscripts and other works of art not for their intrinsic value or attractiveness but rather as hedges against inflation was pursued with none of the passion which it evokes today, when scholarly collectors form only one group amongst all manuscript collectors. Finally, there is the fact that *objets d'art* still continued to hold the centre of the stage with collectors; books and manuscripts – and particularly the latter – lagged well behind.

There was also an extraordinarily powerful counter-attraction to the Newton sale for, as mentioned above, at Christie's, the other great suite of London sale rooms, the sale of Mr Henry Oppenheimer's world-renowned collection of engravings, etchings and water-colours had opened on Friday 10 July, i.e. only three days before the Newton sale, and continued well after it. This monumental fine art sale, which occupied Christie's rooms for fully

---

[17] King's College Library, Cambridge, Keynes Collection: Viscount Lymington to J. M. Keynes, 8 September 1936.

[18] *The Times Literary Supplement*, London, 18 July 1936, p. 604.

fifteen days and raised the enormous sum of £141,748, was one of the great art sales of the inter-war years, containing as it did drawings by artists such as Rembrandt and Rubens.[19] It is therefore hardly surprising that an auction containing such glittering prizes should draw attention away from the Newton sale.

The prices realised at the Newton sale were, even by the standards of the time, modest; by the standards of today they are little short of unbelievable. The whole sale comprised manuscripts containing some three million words, the great majority in Newton's own hand. This great treasure house of primary source material in the history of science, containing as it did all Newton's papers on alchemy, chemistry and Mint affairs, all of the valuable material collected by John Conduitt for his proposed life of Newton, numbers of important unpublished letters to and from Newton, early personal notebooks, very substantial quantities of manuscripts dealing with Newton's work in chronology, theology and the development of the calculus, plus two magnificent portraits (and a death mask!) sold for only £9,030 10s. 0d.[20] Even allowing for the fact that the purchasing power of the pound sterling was much greater in 1936 than it is today, this is still an incredibly low sum. Today a page or two of manuscripts in Newton's hand – on even relatively unimportant subjects – command a price far greater than a hundred pages did in 1936. Even Keynes was taken aback at the low prices realised at the sale for, writing to Emmanuel Fabius some two months afterwards he agrees with him that the Newton manuscripts 'went under the hammer more cheaply than I had expected'.[21] Unfortunately, we have no record of Viscount Lymington's reaction to the prices realised at the sale.[22]

The sale evoked remarkably little general newspaper publicity. Small reports appeared in at least three papers: *The Morning Post*, *The Manchester Guardian* and *The Times Literary Supplement*.[23] *The Times* itself totally ignored the sale: Isaac Newton's manuscripts had been completely overshadowed by Mr Henry

[19] For details of the Oppenheimer sale see note 9.

[20] From this sum must also be subtracted the $12\frac{1}{2}$% commission charged by Sotheby's for handling the sale. '$12\frac{1}{2}$% commission is charged for offering the following: Autograph letters, books, manuscripts...; $7\frac{1}{2}$% is charged for paintings, silver, etc.' – From a prospectus issued by Sotheby's in the early 1930s advertising their services. See Frank Herrmann. *Sotheby's*, p. 204.

[21] King's College Library, Cambridge: Keynes Collection. Carbon copy of a letter of J. M. Keynes to Emmanuel Fabius, 17 September 1936.

[22] It is particularly disappointing that the autobiography of the 9th Earl of Portsmouth, *A Knot of Roots* (London: Bles, 1965), makes no mention of the Newton sale, or of the Earl's reaction to it.

[23] *The Morning Post*, London, 15 July 1936, p. 5. (Five centimetres of column as against the Oppenheimer sale's thirty five!); *The Manchester Guardian*, Manchester, 15 July 1936, p. 10; *The Times Literary Supplement*, London; 18 July 1936, p. 604.

Oppenheimer's *objets d'art*. However, a week later the sale did finally enter the columns of *The Times* in the shape of a news item announcing Lord Wakefield's purchase of Newton's Mint papers from the New York dealer, Mr Gabriel Wells, who had purchased them at the sale. Gabriel Wells had generously agreed to re-sell them to Lord Wakefield at cost provided the latter presented them to the nation. Lord Wakefield at once presented the papers to the Royal Mint.[24]

In retrospect it is interesting to conjecture whether Lord Wakefield, that generous patron of the arts, would not have purchased the whole of the Newton papers for presentation to the nation if he had been approached appropriately and in time. As an art collector, however, no doubt his attention was also distracted by the Oppenheimer sale.

Lord Wakefield's gift was the cause of an interesting exchange of letters in the correspondence columns of *The Times* during the next few weeks. This was between William Shaw, the economic historian, who some forty years before had published some of Newton's Mint reports, and Sir Robert Johnson, Director of the Royal Mint. Shaw entertained serious doubts as to the accessibility to scholars of Newton's Mint papers if they were to be stored at the Royal Mint; Sir Robert Johnson's light-hearted replies did little to allay his fears.[25] *The Times* also reported with approval Sir Robert Hadfield's donation to the Royal Society of various letters purchased at the sale, as well as the purchase by the National Gallery of the Kneller portrait of Newton (Lot 330).[26]

As far as the public was concerned the dust had settled and the sale was now over. Nothing, however, could have been further from the truth, for now began a low profile, but nonetheless intense, competition for the acquisition of the manuscripts dispersed at the sale. The two principal *dramatis personae* in this minor drama were J. M. Keynes and A. S. Yahuda. United in their passion for collecting books and manuscripts, Keynes and Yahuda shared little else in common.

John Maynard Keynes was in numerous respects the typical British scholarly man-of-affairs. Born in 1893 in Cambridge, where his father was the University Registrary, and educated at Eton and Cambridge, Keynes joined the British civil service on leaving university. After a spell in the India Office in

---

[24] *The Times*, London, 22 July 1936, p. 12: 'Gift of Lord Wakefield'. See also *The Times Literary Supplement*, 25 July 1936, p. 619 and *The New York Times*, New York, 22 July 1936, p. 5: 'British Mint gets relics. New Yorker aids in gifts of papers of Sir Isaac Newton'.

[25] *The Times*, London, 27 July 1936, p. 13; 7 August 1936, p. 11; 12 August 1936, p. 15; 15 August 1936, p. 6 and 19 August 1936, p. 6.

[26] *The Times*, London, 23 July 1936, pp. 14 and 17.

1913 he moved to the Treasury. At the Peace Conference in Versailles in 1919 he was the chief representative of the British Treasury but disagreed so strongly with what he believed to be the unnecessary and unwise harshness of the reparations imposed on Germany that he resigned in protest. He justified his position in his influential *The Economic Consequences of the Peace*. Keynes rose rapidly in reputation as a brilliant, creative economist and his *New Theory of Money and Trade* is still widely regarded as one of the most influential books of our era. For many years a Fellow of King's College, Cambridge, Keynes was both an enthusiastic patron of the arts and also an energetic, knowledgeable and shrewd collector of books and manuscripts. During the course of his life he assembled a magnificent library, rich in both books and manuscripts which, on his death in 1946, he bequeathed to his old college in Cambridge.[27]

Abraham Shalom Ezekiel Yahuda was in many ways the typical oriental scholar. Born in Jerusalem in 1877 of a cultured Baghdad family his precocity was such that he published his first book, *Kadmoniyyot ha-Aravim* ('Arabs' Antiquities') at the age of fifteen. After studying Semitics at Heidelberg and Strasburg, and lecturing at the Berlin Hochschule (Lehranstalt) fuer die Wissenschaft des Judentums, in 1915 he was appointed to a special chair at the University of Madrid. In 1942 he was appointed professor at the New School for Social Research in New York. An extraordinarily competent linguist he published extensively on theological matters as well as on Jewish–Arab relations. During his life Yahuda travelled widely and acquired an extensive and valuable collection of books and manuscripts. He died in New Haven, Connecticut, in 1951, but such was the high regard in which he was held that a year after his death his body was moved to Israel where he was given a state burial in Har-Menuhoth, Jerusalem, resting place of Israel's leaders.

In 1953 Yahuda's widow, who had come into possession of his library at the time of his death, announced her intention of donating his library, which of course included his Newton manuscripts, to the Jewish National and University Library in Jerusalem. Unfortunately, however, Mrs Yahuda's will made no provision for carrying out the donation, with the result that one of the trustees, Yahuda's nephew, Benjamin Yahuda, claimed that the library should remain in the estate and not be treated separately. This resulted in a series of

[27] See A. N. L. Munby, 'The Keynes Collection of the works of Sir Isaac Newton at King's College, Cambridge', *Notes and Records of the Royal Society of London*, Volume 10, 1950–51, pp. 40–50. See also 'Lord Keynes and his books' by A. N. L. Munby, *The Times Literary Supplement*, 19 October 1946, p. 512 and Milo Keynes (Editor) *Essays on John Maynard Keynes* (Cambridge University Press, 1975), Chapter 28 'The Book Collector', by A. N. L. Munby (pp. 290–8).

acrimonious court cases in the United States and Israel, delaying until 1966 the arrival of the library in Jerusalem.[28]

If Keynes attended the Sotheby sale with some reluctance his actions immediately afterwards certainly compensated for his early lack of enthusiasm. Although, as we have noted above, he started off at the sale by buying slowly, his interest had quickened to the point where he bought relatively heavily on the second day. Now that the sale was over he set to with energy and determination systematically to acquire substantial numbers of the lots which he had missed, particularly on the first day of the sale. He especially sought those lots which had been bought by British dealers and which were therefore potentially available for rapid re-purchase – an action which was made much easier than might have been expected by the fact that Keynes was already a well-known and valued customer of the dealers in question.

Keynes's quest had clearly become not only important to him, but also urgent for on Monday 20 July 1936, less than a week after the sale was over, he wrote to Messrs Maggs Brothers, the London book dealers, to enquire whether lots 107, 253 and 300, all of which had been purchased by Maggs at the sale, were available for purchase. Unfortunately the original of this letter has been lost but Maggs's reply still survives.[29] This, and dozens of other similar letters to and from Keynes after the sale, the carbon copies of which he carefully filed, are preserved amongst Keynes's papers in the library of King's College, Cambridge.

The letter to Maggs Brothers was the first of literally dozens of enquiries and transactions entered into by Keynes during the next few months. On many occasions he wrote as many as three or four letters per day to booksellers enquiring about the location and possible purchase of Newton manuscripts and on 3 August 1936, he actually wrote six. On 4 August, only three weeks after the sale, Keynes informed Maggs of the plan that had now finally taken shape in his mind: 'I have now decided to form a very substantial collection of

---

[28] For information on the life and work of Yahuda, see his obituary notice in *The New York Times*, 14 August 1951, as well as the entry under his name in the *Encylopaedia Judaica* and in *Who Was Who, 1951–1960*. A long report of an interview published in *The Jewish Chronicle* (London) for 9 May 1919, contains much valuable information concerning his early life while details of the court battle over Yahuda's books and manuscripts are recorded in *The New Haven Register* (Connecticut) for 27 July and 21 August 1958, 9 June 1959, 11 April 1961 and 13 September 1966. It is interesting to note that all reports describe Yahuda's library as containing valuable books and manuscripts dealing with 'Hebrew and Semitic culture and history'; the Newton MSS are never mentioned. I am very much indebted to Helen Martin, of West Grove, Pennsylvania, for her invaluable assistance in tracing details of Yahuda's life and the fate of his library.

[29] King's College Library, Cambridge: Keynes Collection (hereinafter referred to as 'Keynes Collection'); Maggs Brothers to J. M. Keynes, 22 July 1936.

these papers with the idea of keeping them permanently in Cambridge and with this object in mind I have again been going through the catalogue carefully!' In the same letter Keynes also revealed that he was especially interested in one type of manuscript, 'I am particularly anxious to have these Lots 215 and 217 since they would complete the collection for biographical purposes which I am endeavouring to accumulate. Apart from these lots I have now acquired the whole of Section 5 at the sale.'[30]

Keynes's enthusiasm for his Newton collection continued to grow steadily and a little over a week later, in a particularly revealing letter dated 13 August, he confessed to Maggs that 'I am afraid that my appetite grows on what it feeds on and there are still more items I should be interested to have, namely Lots 17, 37, 44, 62, 91, 92, 94, 97, 142, and 158. There are also four more which I should like to see on approval, namely 48, 63, 70, and 74'.[31] This letter is especially noteworthy in that it reveals Keynes's marked change in attitude towards the alchemical manuscripts, which had been catalogued as Lots 1–121, and in which he had shown comparatively little interest at the sale itself. He was now, however, buying them urgently and on 8 September 1936, writing to the book dealer Gabriel Wells, he informed him that 'I only gradually came to the decision to make my collection comprehensive. It is also only gradually that I came to the view that the papers concerning alchemy were really interesting'.[32]

The following day, writing to Viscount Lymington, Keynes with justifiable satisfaction summed up his progress to date and his plans for the future of his Newton collection: 'I have now purchased between one-third and one-half of all the lots you disposed of, and, from the point of view of interest, I think they comprise considerably more than half. My idea is to keep them together and give or bequeath them for permanent retention at Cambridge. I hope you will sympathise with the idea of keeping as much as possible together and intact'.[33] Lymington replied 'I am very glad indeed to hear that you have been successful in purchasing so many of the lots which we disposed of, and more than glad not only that you collected them, but that sooner or later they shall remain permanently in Cambridge'.[34] Keynes's feelings of pride in his role in the 'saving' of the Newton manuscripts were to surface from time to time in later years. Thus, writing to the Master of Trinity College, Cambridge, J. J. Thomson, in April 1939, Keynes informed him that 'When the Portsmouth Collection of Newton MSS was sold some two years ago, I acquired the major

[30]   Keynes Collection; J. M. Keynes to Maggs Brothers, 4 August 1936.
[31]   Keynes Collection; J. M. Keynes to Maggs Brothers, 13 August 1936.
[32]   Keynes Collection; J. M. Keynes to Gabriel Wells, 8 September 1936.
[33]   Keynes Collection; J. M. Keynes to Viscount Lymington, 9 September 1936.
[34]   Keynes Collection; Viscount Lymington to J. M. Keynes, 11 September 1936.

portion of them so as to prevent the great loss of their value which would have resulted from their widespread dispersal'.[35]

In his attempt to reassemble as many Newton manuscripts as possible after the sale Keynes dealt principally with the well-known London dealers of the time: Maggs Brothers, Pickering & Chatto, Bernard Quaritch, and Tregaskis, as well as the New York-based Gabriel Wells and Heffers in Cambridge. However, his correspondence shows sporadic contact with many others besides. In general, lots were offered to Keynes with a mark-up of some 30% to 80% on the sale price. The notable exceptions were Gabriel Wells, who always marked up by only 10%, and Maggs Brothers, who almost always marked up the sale prices by 20%. An incident worthy of placing on record took place a week after the sale when, on 22 July, Gabriel Wells offered Lot 240 to Keynes at £42, against the sale price of £40. Keynes, however, found this 5% markup unacceptable and insisted that Wells take 10% instead. Wells agreed and the lot passed to Keynes on 3 August for £44.[36] Small wonder that Keynes's relationships with booksellers were so good and that he was so well served by them.

Although Keynes was undoubtedly very anxious to acquire as many Newton manuscripts as possible he was certainly no indiscriminate buyer; he examined each lot carefully and did not hesitate to return to a dealer a lot sent on approval but which he believed to be over-priced. Thus against one of the items on Maggs's invoice of 17 August 1936 there appears in Keynes's hand the cryptic words 'Too expensive'.[37] Similarly, he returned Lot 41 to Stonehill, who had offered it to Keynes at £10 – double the sale price.[38]

Before we leave Keynes's post-sale activities two other incidents relating to his Newton manuscripts deserve mention. The first concerns access to Keynes's Newton papers. Although usually generous in the assistance he was prepared to give scholars, Keynes nevertheless applied some restrictions when it came to his Newton papers. Thus, on 7 April 1943, replying to Griffith Davies, Assistant Secretary of the Royal Society, who had written to Keynes requesting access to his Newton papers by D. M. Matheson, Secretary of the National Trust, Keynes replied that 'I am not very much inclined [to put the Newton papers at his disposal] unless he really is a specialist in those subjects'.[39] There

35 Keynes Collection; J. M. Keynes to J. J. Thomson, 22 April 1939.
36 Keynes Collection; Gabriel Wells to J. M. Keynes, 22 July 1936; J. M. Keynes to Gabriel Wells, 3 August 1936.
37 Keynes Collection; invoice of Maggs Brothers to J. M. Keynes, 17 August 1936.
38 Keynes Collection; Stonehill to J. M. Keynes, 7 September 1936. On 18 September 1936 Stonehill was able to assure A. S. Yahuda that this lot was still available. The latter presumably also returned it, for it is now in the Babson College Library, Babson Park, Massachusetts.
39 Keynes Collection; J. M. Keynes to Griffith Davies, 7 April 1943.

the matter seems to have ended. The second concerns the storage of Keynes's Newton manuscripts during the Second World War, when their destruction by bombing was believed to be a very real possibility. When E. Ashworth Underwood, the historian of medicine, sought access to Keynes's Newton papers in 1943, he was informed by him that 'As regards my Newton Collection, this, I am sorry to say, is sunk in cellars for the time being and is not available'.[40]

We turn now to the role of A. S. Yahuda in the history of the Newton manuscripts after the sale. There is no evidence that Yahuda himself attended the 1936 sale or that he arranged for a London dealer to bid on his behalf. However, there is no doubt of the fact that Yahuda had very soon become aware of the sale, for on 1 August 1936, a little more than a fortnight after the auction, Yahuda wrote to Gabriel Wells enquiring about five lots from the sale which he was interested in purchasing. Two of these had already been sold (to Keynes – although of course Yahuda did not know this), but he was able to purchase Lot 245 for £45 (the sale price) plus 15%.[41] Some ten days later Yahuda again contacted Gabriel Wells, enquiring about the purchase of a further four lots.[42] Once again he was only successful in acquiring one (Lot 230). These first two enquiries were the first of many to various booksellers concerning the purchase of Newton manuscripts. In some cases he was successful; in many others he was disappointed. The reason for his failure was usually that Keynes had been there before him, having started his round of intense post-sale purchasing a few weeks before Yahuda.

It is almost certain that at the time of the sale Keynes's and Yahuda's paths had not crossed and that they were therefore unaware of each other's existence. However, it was inevitably only a matter of time before they discovered that they were both competing in the same field. The contact seems to have been brought about by one of the booksellers involved in handling the Newton manuscripts, for on 1 August 1936 Yahuda's enquiry of the same day about the possible purchase of Lot 235 was answered by Gabriel Wells, 'I believe Lot 235 I let Mr Keynes have'.[43] Some six weeks now elapsed before contact was established between the two collectors when, on 15 September,

[40]   Keynes Collection; J. M. Keynes to E. Ashworth Underwood, 9 June 1943. A similar letter to D. M. Matheson, Secretary of the National Trust, dated 16 April 1943 describes his Newton manuscripts as being 'sunk I am afraid, out of sight in a cellar and not very easy to get at.' (I have found, incidentally, no evidence to support the rumour that Keynes spent his time in taxis during the war reading his Newton manuscripts!).

[41]   Yahuda Papers, Jewish National and University Library, Jerusalem (hereinafter referred to as 'Yahuda Papers'); A. S. Yahuda to Gabriel Wells, 1 August 1936.

[42]   Yahuda Papers; A. S. Yahuda to Gabriel Wells, 12 August 1936.

[43]   Yahuda Papers; Gabriel Wells to A. S. Yahuda, 1 August 1936.

Yahuda wrote directly to Keynes. He enquired whether Keynes possessed Lots 229, 235, 242 and 263, all dealing with Newton's theology, and went on to explain that he was particularly anxious to secure Lot 263, a major eighty-four page manuscript dealing with Solomon's Temple, as he had bought other lots dealing with this topic and he believed that these were probably part of Lot 263. Significantly, he made the following suggestions:

> I thought that you might be persuaded either to sell that lot or to exchange it against other Newton lots which I have in my possession.[44]

Keynes seems to have been less than enthusiastic about entering into correspondence with Yahuda for, contrary to his normal habit of answering letters from booksellers by return of post, he waited nearly seven weeks before informing Yahuda that he did have three of the lots in question. At the same time he enquired of Yahuda concerning four lots (50, 72, 87 and 225) in which he was interested.[45] A week later Yahuda confirmed that he had the four lots about which Keynes had enquired and on 1 December he posted them to him.[46] Although we have no further correspondence about any arrangement entered into by Keynes and Yahuda regarding these four lots, the fact that two of them are now in the Keynes Collection in the library of King's College, Cambridge, and at least two manuscripts purchased by Keynes after the sale (Lots 236 and 258) are now in the Jewish National and University Library in Jerusalem (bequeathed by Yahuda) is a strong indication that a swap did in fact take place. This is confirmed by the following note in Yahuda's hand on a letter of Keynes's dated 1 December 1936:[47]

44   Keynes Collection; A. S. Yahuda to J. M. Keynes, 15 September 1936.
45   Keynes Collection; J. M. Keynes to A. S. Yahuda, 4 November 1936.
46   Keynes Collection; A. S. Yahuda to J. M. Keynes, 11 November and 1 December 1936.
47   Yahuda Papers; The note is in pencil in Yahuda's hand on Keynes's letter of 1 December 1936 to Yahuda.

exchanged with Keynes:

$$
\begin{array}{rr}
\text{Lots } 33 - & 58 \\
72 - & \underline{46} \\
& 104 \\
15\% & \underline{16} \\
& \underline{120}
\end{array}
$$

$$
\begin{array}{rr}
\text{for lots } 236 - & 10 \\
258 - & 44 \\
263 - & 38 \\
\end{array}
$$

$92 + 20\%\ 110 + 10 \text{ for } 236 = 120$

I interpret this to mean:

To J. M. Keynes from A. S. Yahuda

Lot number	Price at sale (£)
33	58
72	46
	Total 104
Plus 15%	16
Total	120

To A. S. Yahuda from J. M. Keynes

Lot number	Price at sale (£)
236	10
258	44
263	38
	Total 92
Plus 20%	18
236	10
Total	120

Thus Yahuda was assuring himself of the equivalent value of the lots which had been exchanged. Why Lot 236 is included twice in the one list is not obvious – perhaps it was simply an error.

At this point the correspondence ceases. It resumes again in April 1938 when Keynes enquires of Yahuda:

> Could you tell me, by the way, what happened about that possible exchange which you were going to try for when we last met? You will remember that you took away one of my MSS and were going to let me have another in exchange.[48]

Although Yahuda replies that after a careful examination he is satisfied that 'we have fully settled our exchange business',[49] Keynes is far from satisfied:

> The Newton MS which I thought you had was one purchased by me which you believed to be a part of a MS of which you had purchased the other portion. Probably this description will recall it to your memory. You took this away with you and were going to let me have in exchange, if you could get it, an item which I wanted in the hands of Ulysses with whom you were going to bargain.[50]

In a further letter, Yahuda again repeats his belief that the exchanges had been settled. However, if Keynes can supply him with the number of the missing lot this will help.[51] Keynes, after a delay of several months, informs Yahuda that he has 'not been in a position to look things up', but again reminds Yahuda of the nature of the missing lot.[52] Finally, on 6 August, Yahuda, about to leave for the Continent for several weeks, promises to go through his Newton manuscripts on his return.[53] There the correspondence ends and as far as can be ascertained the two men never corresponded again.

In the light of the firmness of Keynes's conviction that Yahuda still had one of his Newton manuscripts perhaps his claim should be examined with some care. One way in which this can be done is to scrutinise Keynes's annotated copy of the list of sale purchasers and prices, which is fortunately still in existence.[54] On this list Keynes has circled all those lots which he had either purchased at the sale or which he acquired afterwards. A careful check reveals that, apart from the lots which we know he exchanged with Yahuda, only one which he circled on his list is not now in the Keynes Collection in the Library

48   Keynes Collection; J. M. Keynes to A. S. Yahuda, 3 April 1938.
49   Keynes Collection; A. S. Yahuda to J. M. Keynes, 24 April 1938.
50   Keynes Collection; J. M. Keynes to A. S. Yahuda, 25 April 1938.
51   Keynes Collection; A. S. Yahuda to J. M. Keynes, 28 May 1938.
52   Keynes Collection; J. M. Keynes to A. S. Yahuda, 4 August 1938.
53   Keynes Collection; A. S. Yahuda to J. M. Keynes, 6 August 1938.
54   This list forms part of the Keynes Collection.

of King's College, Cambridge. This is Lot 40 ('Out of La Lumiere Sortant des Tenebres'), a four folio page manuscript which was purchased by Keynes at the sale for £7 10s. 0d. It is now in the Jewish National and University Library in Jerusalem, to which it was bequeathed by Yahuda on his death. Is this the 'missing' manuscript to which Keynes referred? One cannot be sure, but it certainly may well be.

Yahuda was remarkably successful in his campaign to assemble a sizeable collection of Newton manuscripts. Whereas Keynes had purchased thirty-seven lots at the sale, Yahuda had, as far as we know, purchased none. However, by October 1938, according to a list in his hand in Jerusalem, he had acquired thirty-nine lots, of which many were substantial lots of major importance. These were distributed as follows: alchemy, nine; personal papers, two; chronology, three; theology, twenty-three; miscellaneous, two. The section on theology was particularly impressive, comprising a major proportion of Newton's important writings in this field. Several of the alchemical manuscripts were also of considerable importance. Perhaps Yahuda's enthusiasm for acquiring Newton manuscripts is most graphically expressed in Gabriel Wells's statement to Keynes, 'Professor Yahuda has kept after me all along'.[55]

For his part Keynes, as we have seen, not only purchased fairly heavily at the sale, acquiring thirty-nine lots, but he also commenced his post-sale purchases earlier than Yahuda and with a net cast more widely amongst booksellers with whom he was generally well known and on good terms. It was therefore, inevitable that he would succeed in assembling a larger collection of Newton papers than Yahuda. This indeed he did for on his death in April 1946 the wonderful collection of books and manuscripts which he bequeathed to King's College contained no fewer than 130 lots from the 1936 sale.

Both Keynes and Yahuda pursued their quest of acquiring numerous manuscripts offered at the 1936 sale with remarkable energy and tenacity. What were their reasons for doing so? Of course, one level of explanation is simply that both were committed and enthusiastic collectors of books and manuscripts – as well as being sufficiently wealthy to indulge their interests with relatively little financial constraint. But there must, however, have been other, deeper, reasons. In the case of Keynes we have already noted that on 4 August – a scant three weeks after the sale – he had already confided to Maggs his explicit desire 'to form a very substantial collection of these papers with the idea of keeping them permanently in Cambridge'.[30] In the case of Yahuda, however, his papers appear to contain no such explicit statement of purpose and one has no option but to speculate. Throughout his life Yahuda was an

[55] Keynes Collection; Gabriel Wells to J. M. Keynes, August 1936.

ardent supporter of Jewry and the Jewish religion. The great majority of the manuscripts purchased by him at the sale were theological (manuscripts which Einstein described as providing insight into Newton's *geistige Werkstatt*, his spiritual workshop'[56]) and dealt heavily with Mosaic religion and Old Testament Prophecy. At a time when Nazi anti-semitism was rampant through much of Europe, with Jewry facing the greatest threat in its history, it would have been entirely natural that Yahuda would have wished to collect as many as possible of the theological writings of an acknowledged intellectual giant such as Newton, for perhaps in these papers he hoped to find support for the great vision of the 'Restoration of the Jews'. If this is indeed the case how appropriate that Yahuda's widow should have bequeathed his papers to the Jewish National and University Library in Jerusalem, where they now rest.[57]

Comparisons, however, are neither appropriate nor profitable. Instead tribute should be paid to two remarkable men both of whom shared a vision of assembling for eventual donation to open libraries as many Newton manuscripts as possible. The result was the creation of the two largest collections of Newton manuscripts outside the Portsmouth Collection in the University Library, Cambridge. All of us owe them an overwhelming debt of gratitude.

[56]   Yahuda Papers; Albert Einstein to S. A. Yahuda, September 1940. (Quoted in Frank E. Manuel, *The Religion of Isaac Newton* (Oxford University Press, 1974), p. 27.)

[57]   I acknowledge with appreciation the contribution of Karin Figala to some of the ideas expressed in this paragraph.

6 *De Scriptoribus Chemicis*: sources for the establishment of Isaac Newton's (al)chemical library

KARIN FIGALA, JOHN HARRISON† AND
ULRICH PETZOLD

Acknowledgements

First of all, we wish to thank D. T. Whiteside for his kind and valuable help in dating the Newton manuscript to be discussed in this paper. One of the authors particularly (K F) is deeply obliged to him, since she has made demands both on his scholarship and on his friendly advice in the course of all her own Newtonian research in the last decades. We are indebted to Alan E. Shapiro for his communications in the field of watermark dating of Newtonian (al)chemical manuscripts and for his – and Peter E. Spargo's – stimulating discussion of an early version of this paper. Last, but not least, we thank Margaret Kimball, the Stanford Archives and Manuscripts librarian, for kindly providing us with copies of the basic material for our study, and Richard Lorch for his efforts in translating parts of this paper.

A study of *De Scriptoribus Chemicis* was planned a long time ago when John Harrison – some years after the publication of his invaluable study, *The Library of Isaac Newton* (1978) – asked K F for her opinion on this manuscript, its possible source and significance. The contribution in honour of D. T. Whiteside, their common friend, seemed to both an ideal opportunity for realizing this long delayed project. Unfortunately, the development of this study took a tragic turn with the sudden and unexpected death of John Harrison. There was no opportunity to discuss the final version of this paper which had been substantially changed, when in the course of transcribing our basic manuscript UP identified further fragments of a related nature and suggested their incorporation together with a discussion of chronological aspects within the growth of Newton's library. Thus we had to decide to rework the draft left by Harrison to avoid reiterations and/or contradictions. For reasons of conclusiveness and consistency we have been unable to restrict such interventions to a few essential points. Harrison's text is now incorporated mainly into the initial section of this study, and into the introduction to our annotated transcription of Newton's *De Scriptoribus Chemicis* manuscript. We were, of course, able to base the introduction on Harrison's preliminary work but we also include our own further identifications of works traceable to collected editions or compendiums in Newton's possession. We refrained from designating the portions of any contributor and request that this study be read as our common effort. We hope that we have concluded our joint investigations in accordance with John Harrison's wishes and would like it to stand in part as a tribute to his own path-breaking work on the library of Isaac Newton.

(K F and U P)

Introduction: Newton's manuscript *De Scriptoribus Chemicis* and its source

Thanks largely to the work of J. Harrison,[1] the contents of Newton's library as it appeared to his executors is now very precisely known to us. Thus we know what chemical and alchemical books – which concern us here – Newton once had possessed, except, perhaps, for volumes that had been lost during his lifetime or afterwards. Of course, we do not know a priori when or in what order this (al)chemical library grew. But a temporal reconstruction of his library seems to be rather important in so far as it would give us valuable insights into the development of Newton's interest in alchemy beyond the knowledge we may derive from the study of his manuscripts in this domain. It could, in principle, also help us to date parts of these manuscripts by using titles of books much as paleontologists use 'key fossils', i.e. to judge the date of origin of manuscript records by the presence or absence of quotations of verifiable books.

Unfortunately, the whole corpus of Newton's alchemical papers contain very little information from which we could deduce directly when any volume was acquired. On the other hand a whole series of hand-written documents and notes can be listed offering indirect evidence about Newton's ownership as well as about his quest for relevant literature. Thus our presentation and short evaluation of such material may serve as a first step towards more detailed studies concerning the formation of Newton's (al)chemical library.

The primary source for this study was a manuscript, headed 'De Scriptoribus Chemicis'[2] and comprising five closely written pages of text which had been drawn up in Newton's hand on his typical quarto-sheets. It was formerly preserved with the bulk of Newton's alchemical papers[3] and is now located in the Stanford University Libraries – Department of Special

---

[1] J. Harrison, *The Library of Isaac Newton* (Cambridge University Press, 1978); for all references to books from Newton's library this work is quoted in the following as *HL*, succeeded by Harrison's item number.

[2] See Harrison, *Library*, pp. 8–9, where he makes mention of the *De Scriptoribus Chemicis* manuscript. KF was able to identify its source only after Harrison had published his *opus magnum*. An additional book-list, drawn up by Newton after 1697, is described *ibid.*, p. 9. From dash-markings mentioned by Harrison and discrepancies between the content of this list and Newton's final library, we may conclude a purpose quite similar to the material to be discussed here. However, these books concern different subjects, 'mainly classical literature, history, and mythology'.

[3] See section I, lot 6 in *Catalogue of the Newton Papers sold by order of the Viscount Lymington*...[London: Sotheby, 1936]; for a slightly modified version of the Sotheby descriptions see 'Appendix A: The alchemical papers of Sir Isaac Newton', in B. J. T. Dobbs, *The Foundation of Newton's Alchemy or, The Hunting of the Greene Lyon* (Cambridge University Press, 1975), pp. 235–48. See also the introduction to our Appendix ('Transcription').

Collections, Manuscripts Division: Newton Collection (M132), Container 2, Folder 4.[4] Contrary to its comprehensive title this manuscript is not a commentary on (al)chemical writers and their work but a mere listing of eighty-four authors and titles of books, arranged in alphabetical order and supplied with successive page references; the majority of the works are in Latin, with the exception of a few in French. Furthermore, the majority of entries refer to printed books, again with the exception of a few titles 'in MS'. Unlike Newton's common application of references to and quotations from printed sources, these entries are strictly arranged bibliographically, i.e. almost all include author, title, place and date of publication, and frequently size. As a conspicuous peculiarity of this listing we find a dash at the beginning of fourteen of the works or, heading two other entries, a cross and once Newton's characteristic ☞ sign, while the remainder has no sort of introductory symbol. The whole set of notes is completed by an additional short listing of another six works, but now in English and without further bibliographical details. Newton had written this portion at the bottom of his last page, immediately following the final entry of his main list but clearly separated by indention and a short line.[5]

The manuscript gives the firm impression that the main sequence, i.e. the first eighty-four items, was drawn up in a single operation: the handwriting, its regularity, and the evenness of presentation point strongly to this conclusion. Newton's short supplementary listing and one further insertion may seem to disturb this view, but he obviously added material to his notes while progressing with his completion of the list. The later entries, together with the said insertion that explicitly bears '1692' as a date of publication, were written with a sharper, final quill pen, and the internal evidence of the actual years of their publication and their quite different sources combine to make it plain that they were later supplements to the main body of titles. Professor B. J. T. Dobbs gave special attention to these supplements and dated the Stanford list to Newton's 'late period by the handwriting'.[6] Our own estimation of a much earlier dating of the main portion of this manuscript, i.e. to the early 1670s, was quite recently supported by the judgement of D. T. Whiteside. From his wide experience with the characteristics of Newton's handwriting, he places the date of origin into 'the late 1660s even before April 1669 when he [Newton]

---

4   See also *Stanford University Libraries: Department of Special Collections – Manuscripts Division Register*, proc. J. Halpern (type-script, 1981); quoted as Stanford M132.

5   See our Appendix ('Transcription'). All subsequent references to any title from Newton's *De Scriptoribus Chemicis* list (abbreviated *DSC*, followed by item numbers) are by item numbers; thus items [1]–[84] constitute the main list, items [Add 1]–[Add 6] the short supplementary listing; cf. also the introductory remarks to this transcript.

6   Dobbs, *Foundations*, p. 112; for Newton's late insertion see *DSC* [24].

bought the *Theatrum chemicum*'.[7] Further, (internal) arguments for this early dating will be discussed below.

The latest internal date, apart from the confounding '1692' insertion, occurs within Newton's own plain reference to his original source: '[...] promittebat Borellus A. C. 1654',[8] a reference that can be verified undoubtedly by collation. Among the books on (al)chemical matters which Newton owned he had a copy of Pierre Borel's *Bibliotheca chimica. Seu Catalogus librorum philosophorum hermeticorum. In quo quatuor millia circiter authorum chimicorum, vel de transmutatione metallorum, re minerali, & arcanis, tam manuscriptorum, quam in lucem editorum, cum eorum editionibus, vsque ad annum 1653. continentur. Cum eiusdem Bibliothecæ Appendice & Corollario. ...* (Paris, 1654, in 12°).[9] This work, consisting of an introductory twelve pages, followed by 276 pages of the main text in alphabetical order of authors and titles (including the 'Appendix' on twenty-four pages, the 'Corollarium' on twelve pages, and an 'Epistola chimica'), 'was the first of its kind', as Ferguson acknowledged,[10] and was regarded by Bolton as 'the first extensive catalogue of chemical books'.[11] It was from Borel's book that Newton chose to copy out the entries which comprise the Stanford manuscript except for the final six entries (one of these was not mentioned by Borel and five were published after 1654) and the insertion of 1692 recording an English translation published at London.

There is little doubt that Newton looked upon Borel's *Bibliotheca* as his

---

[7] Private communication to J. Harrison (1989). On Newton's purchase of the *Theatrum chemicum* in April 1669 see note 58.

[8] See *DSC* [43]; see also [1]: '[...] Borellus vidit [...]'.

[9] *HL 246*; all references to the *Bibliotheca* in the present study are to this 1654 edition.

[10] J. Ferguson, *Bibliotheca chemica*, 2 vols. (Glasgow: Maclehose, 1906; repr. London: Verschoyle, 1954), vol. 1, p. 116. Borel was born at Castres in or about 1620, studied medicine at Montpellier University and, after obtaining his doctor's degree at the University of Cahors in 1643, returned to practise medicine in his native town. In 1653 he went to Paris where he actively pursued studies of natural history, chemistry, optics, astronomy, antiquities, philology, and bibliography. He published several books on such subjects and also left a considerable list of projected works, none of which, however, was issued until his death in 1671 or afterwards. A reprint of his *Bibliotheca* – differing just in setting and pagination – was published at Heidelberg in 1656, which would suggest that the book was favourably received. For more details (and corrections) see J. Chabbert, 'Pierre Borel (1620?–1671)', *Revue d'Histoire des Sciences* 21(1968): 303–43; cf. also *The Correspondence of Henry Oldenburg*, eds. A. R. & M. Boas Hall, 13 vols. (Madison: University of Wisconsin Press; London: Mansell; London/Philadelphia: Taylor and Francis, 1965–86), esp. vol. 1, pp. 320–2, no. 119.

[11] H. C. Bolton, *Select Bibliography of Chemistry*, 3 vols. (Washington: Smithsonian Institution, 1893–1901), vol. 1, p. 7. An edition of the *Bibliotheca* of 1676 mentioned by Bolton cannot be confirmed.

prime guide on bibliographical material relating to matters chemical and alchemical. The present location of his copy of this work of reference is unfortunately unknown nor was its condition ever described in detail. Further indications of Newton's confidence in Borel as a reliable authority, however, are provided by scattered manuscript notes in some surviving books from his own library. For example, in his copy of *Tractatus aliquot chemici singulares summum philosophorum arcanum continentes...*(Geismar, 1647),[12] at the end of the book's preface which is signed 'L.C.' Newton added 'id est Lud. Combachius. Vide Bibl. Chem. p. 64'. Similarly on the title-page of his *Enarratio methodica trium Gebri medicinarum, in quibus continetur vera lapidis confectio...*([London], 1678),[13] Newton wrote 'Vide Borelli Bibliothecam Chemicam p. 20' where the same title is included within the entry for a manuscript in the possession of 'D. de Loberie, Paris'. It is perhaps more significant to see how Newton appears to have deciphered the code employed by Michael Maier at the foot of p. 160 of his *Themis aurea; hoc est, De legibus Fraternitatis R. C. tractatus...*(Frankfurt, 1618).[14] The last seven lines of Maier's text are printed in an apparently meaningless jumble of letters. Newton's annotations returned the lines to understandable Latin, and underneath he added the key to the permutation of five letters which enabled him to break the code. In this case he failed to acknowledge that Borel had already supplied the solution. In his 'Epistola chimica' addressed to Thomas de Riolet, Borel had discussed 'Ænigmatis Maieriani, in Themide aurea, Referatio', and on the following page offered his 'Explicatio'.[15]

Borel's compilation was not the only bibliographical tool on (al)chemical matters which Newton ever had at hand. At least during the very late 1680s he came into the possession of a small volume, entitled *A Catalogue of chymicall books. In three parts. In the first and second parts are contained such chymicall books as have been written originally, or translated into English: with a large account of their titles, several editions and volumes....* (London, 1675),[16] which had been collected and published by the London bookseller William Cooper. This booklet was a most useful complement to the *Bibliotheca chimica*, since Cooper had restricted himself to English editions, and afforded valuable insights into the development of contemporary (al)chemical publishing up to

---

[12]   *HL 1623.*   [13]   *HL 554.*   [14]   *HL 1049*, and Harrison, *Library*, p. 21.

[15]   Borel, *Bibliotheca*, pp. 275 seq.

[16]   *DSC* [Add 1]. On Cooper see Ferguson, *Bibliotheca chemica*, vol. 1, p. 135, and S. J. Linden, *William Cooper's Catalogue of Chymicall Books 1673–88: A verified edition.* Garland reference library of the humanities, 670 (New York: Garland, 1987); also L. Rostenberg, 'Restoration scientific publishing: [4.] William Cooper, specialist in alchemy', *The AB Bookman's Yearbook* 1980/1: 77–106, on 102 seq., and her recent *The Library of Robert Hooke. The Scientific Book Trade of Restoration England* (Santa Monica, Calif.: Madoc, 1989); see note 44.

Newton's days. Significantly, Newton purchased a late reissue of the original *Catalogue*, enlarged by an inserted 'Continuation and appendix...to...1688'. Knowledge of this additional work of reference, however, does not contribute substantially to the knowledge of Newton's bibliographical pursuit for his own copy bears no conspicuous signs of use.[17] In the case of Borel's *Bibliotheca*, there are no recorded signs at all. Thus any study of Newton's continuing concern for printed sources on alchemy has to be based exclusively on manuscript records.

De Scriptoribus Chemicis and related manuscripts: new approaches to Newton's (al)chemical library

At about the time that Newton left Cambridge – the scene of his extensive studies in alchemy, theoretical and practical – he wrote the most important of the manuscripts documenting the formation of his private library on the subject. An inventory headed *Lib. Chem.*[18] records the holdings of the collection about 1696–97, shortly after he settled in London. From this we know that at this time Newton possessed 112 (113) titles in 139 volumes of an almost exclusively alchemical nature.[19] Another datable document, a book-seller's delivery note 'Books for M[r] Newton', gives information about a further twelve titles of alchemical and related literature that he acquired

[17] The 'Continuation' (1688) was not considered by Harrison; cf. the collation of Newton's copy [Babson 403] in *A Descriptive Catalogue of the Grace K. Babson Collection of the Works of Sir Isaac Newton*...(New York: Reichner, 1950), pp. 183–4, and Linden, *Cooper's Catalogue*, pp. xli–xliii (who only knows of two copies of the enlarged 1688 reissue – but not of this one). The Babson *Catalogue* refers to 'one entry...marked, apparently in Newton's writing'.

[18] Babson Institute Library, Babson Park, Mass.: Babson [418]; see *Catalogue of the Babson Collection*, p. 192 (with reproduction of f. 1r facing p. 177). The catalogue entry ('A List of 119 Alchemical Books, or author's names, arranged under nationalities, forming part of Newton's Library, with his press-marks') is misleading as the compilers mixed up the descriptions of two different manuscripts forming together lot 3 of the Sotheby sale in 1936. Babson [418] is only part 2 of the original Sotheby item whereas part 1 ('A list of 119 [*recte*: 121] Writers on Alchemy, arranged under nationalities') is now preserved at St Andrews University; cf. note 24. In 1982 J. Harrison prepared a preliminary study of *Lib. Chem.* (and sent his manuscript to KF for cross-checking). In view of the importance of the manuscript in the establishment of Newton's (al)chemical library, the present authors hope to edit Harrison's introductory essay together with his scholarly work of identifying Newton's short-titles in this list in a forthcoming study. None of the following references (in the following abbreviated *LC*) to books appearing in Newton's *Lib. Chem.* manuscript of 1696–97 could have been made without Harrison's preliminary work.

[19] See Harrison, *Library*, pp. 8–9, and 41–2. The count of titles and volumes has been slightly revised compared to Harrison's.

around 1702, in this case predominantly in French.[20] Together, *Lib. Chem.* and this post-1700 bill may be considered as an important reference for the following reflections for they tell us the titles of those books that Newton demonstrably possessed and give us a *terminus ante quem* – or even an actual date – for their acquisition. In the course of our discussion, however, we hope to show that there is more evidence for the acquisition of alchemical books by Newton about or after 1700 than these documents alone, and the consideration of years of publication of other books in his library, would suggest.

There is another group of records and notes of quite a different kind. Of these 'bibliographical papers' the earliest and most important is the one reproduced and commented upon here, the *De Scriptoribus Chemicis*. Further notes based on the same source, Borel's *Bibliotheca chimica*, have come to light from our casual inspection of Newton's alchemical papers, but they differ from the first list both in fullness of detail and in dating. These are, briefly, a revised version of *De Scriptoribus Chemicis*, and a concise draft of extra items. Also to be considered here are the above-mentioned additions and insertions to these originally homogeneous excerpts – additions that must be dated individually and which have been taken from more recent sources. The importance of this group of manuscripts as a whole lies in their mainly being lists of desiderata. It is true, that the first of these excerpts, the *De Scriptoribus Chemicis* list, also contains some references to extant manuscripts (owned by French or English collectors) and a few pseudo-biographical notes on chemical writers. At first sight this may lead one to believe that it is not simply a list of titles that Newton wished to incorporate into his (al)chemical library. However, internal evidence as well as a comparison with the later excerpts from the *Bibliotheca* make it quite plain that this was its main purpose. As his main inventory of desiderata, Newton's repeated transcriptions from the *Bibliotheca* would gain particular importance for by comparing the various lists – so far as they are datable by handwriting or internal evidence – at least an approximate period can be deduced for the accession of books in his library.

A short *excursus*, presenting Newton's different attempts to compose

[20]   Bodleian Library, Oxford: MS New College 361/II (Ekins Papers), f. 78r–78v; see Harrison, *Library*, p. 9. R. S. Westfall, 'Alchemy in Newton's library', *Ambix* 31(1984): 97–101, discussed the 'Ekins list' rather as documenting Newton's interest in the French language than in alchemy at the time after 1700; see also R. S. Westfall, *Never at Rest. A Biography of Isaac Newton* (Cambridge University Press, 1980), p. 531, and also his 'Newton and alchemy', in *Occult and Scientific Mentalities in the Renaissance*, ed. B. Vickers (Cambridge University Press, 1984), pp. 315–35, on p. 332 seq. On the controversy in general and the 'Ekins list' in particular see also K. Figala & U. Petzold, 'Alchemy in the Newtonian circle: personal acquaintances and the problem of the late phase of Isaac Newton's alchemy', in J. V. Field and Frank A. J. L. James, eds., *Renaissance and Revolutions: Humanists, Scholars, Craftsmen and Natural Philosophers in Early Modern Europe.*

authors' lists or establish principles of classification within the vast amount of alchemical writers, may further elucidate the peculiarities of what we call 'bibliographical records'.[21] The heading of *De Scriptoribus Chemicis* suggests the close relationship, if not a direct correlation, to another Newton manuscript, entitled 'Of Chemicall Authors & their writings'.[22] Similarly to *De Scriptoribus Chemicis* its 'English counterpart' is a comprehensive listing of 120 authors' names (including anonymous tracts) in alphabetical order, each entry summarizing details on their lives and/or significance of works. But unlike the former, this manuscript bears no references to available editions of any of the works specified. Instead it bears unmistakable quotations of its sources throughout: *Of Chemicall Authors* was compiled from the 'historical' representation of alchemy as Newton found it in the work of the German Michael Maier, especially in his *Symbola aureæ mensæ duodecim nationum...* (Frankfurt, 1617).[23] This manuscript from the early 1670s was followed by a series of classifying listings which again in essence were based on the work of Maier, but had been expanded by Newton to some more contemporary names. Thus he drafted a table of authors' names arranged by nationalities[24] and compiled a bundle of chronological listings arranged by approximate dates (of origin or publication of major works) which comprise elaborated sets (e.g. 'Authores antiquissimi' or 'vetustissimi' to medieval and fifteenth- to seventeenth-century authors) as well as preliminary drafts, sets without subdivision and sets of selected authors ('Authores magis utiles' or 'optimi').[25] The whole of these rather semi- (or bio-) bibliographical records gives the impression of being the nucleus of Newton's own proposed chronology of alchemical writers or, in analogy to his unique key word

---

[21] The authors owe the insertion of this paragraph – contrary to our initial tendency towards disregarding any consideration of such 'non-bibliographical' materials – to the valuable suggestions of P. E. Spargo who also made his own copies of two documents (the following Babson and St Andrews MSS) available to us. A detailed discussion of the significance of Newton's classified authors' lists would need a separate study.   [22] Babson [419]; dated to 1670–75 by handwriting.

[23] See e.g. *ibid.*, f. 1r: 'Symbola aureæ mensæ Majeri lib 6.', etc.; on Newton's holdings of Maier's works see note 84. Other sources are 'Mairi [*sic*] Emblem 7' (f. 2v; cf. note 85), 'Bernh: Trevisan[us]' (f. 1r, etc.; text publ. in *Theatrum chemicum*, vol. I, and *Tractatus aliquot chemici...*[1647, cf. note 12]), 'Flammellus [...] Hieroglyph [...] Cap 5' (f. 4r; most probably MS excerpt; cf. note 90).

[24] St Andrews University, Chemical Department, Special Collections: MS Newton 3 (no heading).

[25] King's College, Cambridge (Keynes Collection; in the following generally quoted as Keynes MSS): Keynes MS 13*A* (early part: mid- or late 1680s)/*B* (middle part: early 1690s)/*C* (late part: after 1700); part *B* contains an extra Borel excerpt (see note 36). On Keynes MS 13 in general see also Dobbs, *Foundations*, p. 174.

compilation *Index chemicus*,[26] of an 'Index auctorum chemicorum'. Thus the main characteristics of the *Of Chemicall Authors* group of manuscripts are their common source – Maier's *Symbola* – their commented listing of writers and works and the absence of bibliographical data. In contrast, the *De Scriptoribus Chemicis* group is based on a common source of completely different nature – Borel's *Bibliotheca* – and merely enumerates authors, titles of their works, and data specifying particular editions. It is the last-named class of authors' lists that we understand by Newton's 'bibliographical records'.

Of course we dare not deny any interdependencies between both the *Of Chemicall Authors* and the *De Scriptoribus Chemicis* groups. There are some few demonstrable links[27] as well as a remarkable correspondence within the choice of authors and titles. But in the context of compiling information on (al)chemical primary sources *Of Chemicall Authors* might be regarded as a possible recommended-reading guide. *De Scriptoribus Chemicis*, however, is undoubtedly aimed at definitely available prints of separate or collected text editions. Newton's reasons for compiling the list, the use for which it was intended, and his motivation for singling out these particular items may be revealed by the following considerations.

First, there are the above-mentioned marginal signs to indicate single titles or publications. These markings do not relate in every instance to Newton's own library; they do not relate in any instance to that of Trinity College where the holdings of (al)chemical books were very modest during his long residence there.[28] Some of these particular works are repeated, in another form, in as many as three subsequent lists within the *De Scriptoribus Chemicis* group, but may be shown to have been in Newton's final library. Others, marked as well as unmarked editions, appear only in the first list, but were incorporated into his library, too. Yet others recur once or twice, without having ever been in Newton's possession. This procedure of selecting, further selecting by marking, and repeating within a subsequent choice suits perfectly an arrangement of

26 See R. S. Westfall, 'Isaac Newton's Index chemicus', *Ambix* 22 (1975): 174–85.

27 See Keynes MS 13, f. 3v: 'Democritus de Arte sacra cum Synesij et Pelagij Comment. Coloniæ 1574 in 16.'; this is the only entry within authors' lists that gives bibliographical data copied by Newton from Borel or from his own previous versions of *De Scriptoribus Chemicis*; see *DSC* [47]/[81*a*] and [51]. Some textual equivalents in Babson [419] and *DSC* items originate from correspondences of Maier's *Symbola* and Borel's *Bibliotheca*; a few pseudo-biographical *DSC* items, too, may be considered as such links.

28 P. Gaskell, *Trinity College Library: the first 150 years* (Cambridge University Press, 1980), pp. 241–8, where under the Catalogue headed 'Medici' there are eight works which could be considered (al)chemical; see also Trinity College Library: Add.MS a.101 and Add.MS a.107.

desiderata lists 'in progress'. The interpretation is confirmed by the comparison of these listings with their 'result', Newton's final library. Thus the category of repeated and/or marked items comprises foremost hardly obtainable sixteenth-century issues; the holdings of such volumes in Newton's library are rather fragmentary. The category of non-recurring entries comprises a great number of works that Newton possessed at least in translation or by collected editions; all the contemporary – and therefore available – English language titles, published in the second half of the seventeenth century and recorded by Newton in addition to his Borel excerpts, came into his library without exception.

Further evidence comes from an amazing feature of *De Scriptoribus Chemicis*: the total absence of such prominent and – especially for Newton's own attitude towards alchemy in his early period – extremely influential authors as Michael Maier and Michael Sendivogius. Borel had given detailed catalogue descriptions of their work and, in turn, Newton included both names into his authors' lists of the *Of Chemicall Authors* group.[29] If he did not do so in *De Scriptoribus Chemicis*, we might conclude that Newton merely considered works which had not been accessible to him at that time, that is, which had been desiderata. Thus even the omission of a title or author may give us valuable information concerning books that Newton had known – or even possessed – at the very beginning of his (al)chemical career, i.e. before writing his bibliographical records. Finally, there is a clear indication of the purpose of these records when, in his first list, Newton inserted (and crossed out again) a remark about a work that according to Borel's misleading entry appeared in three (instead of two) parts or *decades*: 'Decade tertio careo'.[30]

As stated above, Newton's earliest excerpts from Borel's *Bibliotheca* (*De Scriptoribus Chemicis*, henceforth cited as *Stanford A*) can be dated from the handwriting to the early 1670s or perhaps even to the late 1660s. This hypothesis finds some internal support in that *Stanford A* would well have been written before the acquisition of the *Theatrum chemicum* in 1669. For otherwise the entries with the addition 'in Th. Ch.'[31] at the beginning of the list

[29] See e.g. Babson [419], f. 2v: 'Michael Maierus [...] scripsit Symbola Aureæ mensæ duodecim nationum, Hieroglyphica, Emblemata nova (sive Atlantam [*sic*] fugientem) AD 1616.': *ibid.* f. 3r: 'Anonymus Sarmata (D.L.G.A. [i.e. 'Divi Leschi genus amo' an anagram used by Sendivogius]) [...] anno 1616 vivus tractatus scripsit [...]'; St Andrews Univ. MS 3: under 'Germani' and 'Sarmatæ' respectively; Keynes MS 13: *passim*. See notes 84 etc.

[30] Borel, *Bibliotheca*, pp. 111–12: 'Harmonia imperscrutabilis [*sic*] Chimico Philosophica, seu [...] Decades 3. [...]'; *ibid.*, p. 196: 'Rhenani Decades 2. [...]'; for Newton's remarks see *DSC* [71].

[31] *DSC* [2] and [3]. First reference to the *Theatrum* seems to have been made by Newton in [3] only, keeping strictly to the original; repeated reference, now in both items, obviously has been added in a second step.

would be unnecessary. Especially striking are those cases in which Newton copied bibliographical details from Borel and, by looking ahead in the *Bibliotheca*, added a remark 'Extat et in Th[eatr]. Ch.',[32] when the same title recurs elsewhere and more comprehensively. The sparse references to the *Theatrum* scarcely contradict the early dating, since it is to be assumed that Newton planned to buy this most important of all collections of alchemical texts long before he actually acquired it and therefore disregarded more detailed quotations. Although we dare not decide whether Newton made his first and most extensive excerpts from Borel before or after the *Theatrum* came into his possession, it is reasonable to assume that *Stanford A* was written at a time when Newton was not fully acquainted with the contents of these six bulky volumes.

Some fifteen or twenty years later he made a second extract from Borel (cited here as *Stanford B*).[33] Although only one and a half pages long and much shorter than the earlier *Stanford A*, it contains a large number of common items. The difference lies principally in eliminating the heading, in taking some additional authors into account, in a more precise quotation of titles, and particularly in the omission of all descriptive remarks concerning the contents or composition of individual works. *Stanford B* is now well-nigh unreadable, since the ink has soaked through the paper. It could therefore have been written about 1680–81 (the so-called 'bad ink period':[34] when Newton frequently seems to have used ink of poor quality, discolouring the writing paper), although by the handwriting it is more likely from some time in the late 1680s. This manuscript is now with a bundle of notes and extracts enumerating, *inter alia*, printed works of Raymond Lull.[35] Since Newton's second excerpt from Borel also contains editions of Lull's works, emphasized by underlining, both series of records seem to be related to one another. The handwriting of the Lull records, however, suggests that they were written in the early or mid-1690s, that is, still later than the excerpts from Borel in *Stanford B*.

The latest document found thus far that is based on Borel's *Bibliotheca*, is a twelve-line insertion (referred to as *Keynes 13*), beginning 'Desiderantur Henrici Kunrath [...]', in one of Newton's above-mentioned drafts for a preliminary chronology of alchemical authors.[36] *Keynes 13* must have been

---

[32]   *DSC* [18] and [22], especially Newton's addition of page numbers where Borel first made mention of the *Theatrum* as including these tracts, too.

[33]   Stanford M132, container 2, folder 3, f. 1r–1v (our pagination).

[34]   See Dobbs, *Foundations*, p. 256, on the dating of 'bad ink' manuscripts and its restrictions.

[35]   Stanford M132, container 2, folder 3, ff. 2r–2v, 4r–5v, 7v (our pagination; for details cf. note 74).

[36]   Keynes MS 13*B* (early 1690s), f. 3r; the latest internal date within the main (non-bibliographical) body is '1689' (f. 3v).

drawn up about 1690 or in the first half of the 1690s because one of the books cited there appears in Newton's *Lib. Chem.* as having been in his library in 1696–97.[37] This concise listing has the additional peculiarity that it is not in alphabetical order.[38] It is further remarkable that Newton wrote the entire *Keynes 13* note at one time, although it cannot be traced back to Borel exclusively. This time he must have had at least one further source at hand, for the final two entries, both in French, did not originate in the *Bibliotheca*. Newton's last item, however, can be associated with one of the twelve (al)chemical volumes acquired in 1702 with 'Books for Mr Newton'.[39]

Because of such non-Borel additions, *Keynes 13* belongs to a final group of bibliographical memoranda from Newton's hand, i.e. notes of mixed origin. As stated above, the early *Stanford A* list contains insertions and appended matter added at some later time. One such supplement to the extract from Borel, clearly separated from the foregoing and written at one time, comprises six titles.[40] All of these are in English, published between 1652 and 1675 (except for one different work that appeared only in 1690), and had almost certainly come to Newton's notice from advertisements or publishers' announcements. Five of the additional titles appear in virtually identical words in catalogues ('Books sold by...') that the bookseller and publisher William Cooper bound in with two of his collected editions, entitled: *Aurifontina chymica*...(London, 1680) and *Collectanea chymica*...(London, 1684).[41] The one exception is the last of Newton's six supplementary entries, 'The Chymical weddin [*sic*] translated by Mr F.' that first appeared in 1690 as *The Hermetick romance: or The chymical wedding...translated by E. Foxcroft*. This must have been long extant in manuscript, as the said Ezechiel Foxcroft died in 1674/5.[42] It is

---

[37]   *DSC* [17]; see note 70.

[38]   The concordance of the first six items with the page numbers in Borel's *Bibliotheca* is as follows: p. 130 (*DSC* [75a]), p. 10 ([13]), p. 50 ([37]), p. 96 ([66]), p. 226 ([83a]), p. 171 ([17]/[81a]).

[39]   *DSC* [Add a]; no source established. 'Le Cabinet Chymiques [*sic*]' can only be tentatively identified, perhaps as *Bibliothèque des Philosophes (chymiques)*...(see note 59); 'La Tourbe François [...]' is listed in MS New College 361/II as: 'Philosophie naturelle de Trevisan'.   [40]   *DSC* [Add 1] to [Add 6].

[41]   *HL 103, 410* (also in *LC*); see also Linden, *Cooper's Catalogue*, pp. 119 seq. The titles copied by Newton (*DSC* [Add 1] to [Add 5]) are specified by Cooper as follows: 'A Catalogue of Chymical Books in Three Parts.', 'Five Treatises of the Philosophers Stone.', '*Starkey*'s Pyrotechny.', '— his Liquor Alchahest.', '[*Boyle*] — his Tracts of the Growth of Metals in their Ore.' (see *Aurifontina*, unpag., following p. 272).

[42]   *DSC* [Add 6] (also in *LC*); translation of J. V. Andreae's *Chymische Hochzeit: Christiani Rosenkreütz*...(Strasburg, 1616). For sources giving the year of Foxcroft's death see K. Figala, 'Newton as alchemist', *History of Science* 15(1977): 102–37, on p. 139 n. 3 (please note that on p. 103 'Old Style' should read 'New Style' in both cases).

possible that this translation of one of the most influential treatises from the borderland of alchemy and Rosicrucianism had been advertised sometime in the 1680s; for when drawing up notes that are certainly to be dated after 1690, Newton used the correct title 'The Hermetic Romance'.[43]

The actual year of publication (1690) of this outstanding book is accordingly of little help in dating the set of non-Borel additions to *Stanford A*, unless we suppose that Newton bought his own copy of the *Aurifontina* as a possible source for his bibliographical completions as long as ten years after its appearance. However, this does not seem plausible, since it is known that he rated Cooper's publications very highly and thus should have acquired the book immediately on publication. And by 1690 Newton's second biblio-graphical tool, the 1688 issue of William Cooper's *Catalogue*, would have been his preferable source concerning new publications.[44] It is therefore more likely that the six supplementary titles were added to *De Scriptoribus Chemicis* shortly after 1680. Hence it follows that the entirety of the non-Borel items was written in two steps, since *Stanford A* obtained its final form only about ten years later when Newton amended one of his entries from about 1670 by adding bibliographical data of an English translation: he became aware that a short tract by Arisleus had been published as a tailpiece to *Penotus παλιμβιος: or the alchemists enchiridion in two parts.... Together with a small treatise...by that very ancient Philosopher, Arislaus, concerning the Philo-sopher's Stone...* (London, 1692). What is more, Newton seems to have

[43] See Keynes MS 13A (mid- or late 1680s), f. 1v: 'The Chymical Wedding'; but *LC* (Babson [418], 1696–97): '[2.6.] 8 The hermetic Romance'; Keynes MS 13C (after 1700), f. 4r, 'Authores optimi': 'The Hermetic Romance'. On the other hand, in his *Index chemicus* (Keynes MS 30) Newton quoted the 'Chymical Wedding' obviously from the printed book, as it is shown by his page references, e.g. on f. 61r (key word 'Metalla'); on the dating of the 'Index' corpus see Westfall, 'Newton's Index chemicus'. We have not been able to prove any printed advertisement before the publication of the book. In the London term catalogues it was only announced for the Trinity term, 1690, as 'The Horinctick [*sic*] Romance, or the Chymical Wedding [...]'; see *The Term Catalogues 1668–1709 A.D....*, ed. E. Arber, 3 vols. (London: privately printed, 1903–6), vol. 2, p. 325.

[44] Alchemical books in Newton's library, published by Cooper: *DSC* [Add 1] (*Catalogue of chymicall books*; also in *LC*); *DSC* [Add 4] (George Starkey; not in *LC*); *HL 1478, 1407, 554* (Eirenaeus Philalethes; also in *LC*); *HL 659* (Geber; also in *LC* but not assignable with certainty); *HL 103, 410* (cf. note 41); others, e.g. *HL 513* (Edmund Dickinson, published at Oxford; also in *LC*) had been sold by Cooper *et al.* Just two alchemical titles published (exclusively) by Cooper cannot be proved to be in Newton's library: *The philosophical epitaph of W. C. Esquire....* (London, 1673), and Kenelm Digby's posthumous *A choice collection of rare secrets...* (London, 1983, reissue of a 1682 edition). On Cooper as publisher of the works of Eirenaeus Philalethes see Linden, *Cooper's Catalogue*, pp. 149–56, and B. J. T. Dobbs, 'Newton's copy of "Secrets reveal'd" and the regimen of the work', *Ambix* 26(1979): 145–69.

bought the book – there is evidence that he owned a copy at least about 1696–97.[45] Finally, an interesting sidelight on Newton's continuing concern for alchemy, and on the way he used *De Scriptoribus Chemicis* throughout a long period – from the early 1670s to the early 1690s – may be seen from those later entries. But at about his move to London in 1696 these interdependent bibliographical records seem to have served their purpose for the time, since Newton wrote down the title (and price!) of another alchemical work only elsewhere.[46]

We may now summarize the chronology of *De Scriptoribus Chemicis* and related memoranda. Around 1670, about the time when he purchased the *Theatrum chemicum* or a little later, Newton made his first extractions from Borel's *Bibliotheca chimica*, the result being *Stanford A* in its original form. About 1680 he added six non-Borel titles, all of them in English and published after 1650. In the late 1680s he repeated the first step in a more decisive manner and produced *Stanford B*. About 1690 he once more collected chosen titles from Borel in *Keynes 13*, which he supplemented by two French works published after 1670. Finally, Newton returned to *Stanford A* by adding a last insertion in (or after) 1692.

This reconstruction is confirmed by a close analysis of the idiosyncrasies in both content and form of the individual manuscripts or fragments. *Stanford A* may be broadly characterized as a first draft, or perhaps a collection of working notes, whereas *Stanford B* and *Keynes 13* may be regarded rather as fair copies or summaries of well-considered items selected for a particular purpose. In *Stanford A* Newton tried to find and collect as detailed bibliographical information as possible, but in drawing up his later lists he knew very precisely what he was looking for. Further distinctive features are revealed by the maintainance or disappearance of single works through the

---

45 *DSC* [24]; not *HL*; but see *LC*: '[2.6.] 24 The Alchemists Enchiridion'. Unfortunately we had no opportunity to discuss this point with John Harrison. In his unpublished study of the *Lib. Chem.* list, however, Harrison anticipated the identification of the said book with this item from *Lib. Chem.*

46 Yale University Library, New Haven: Mellon MS 78.4, f. 4v: 'Centrum naturæ concentratum. Or yᵉ salt of Nature regenerated. [... (London, 1696)]'; see *Alchemy and the Occult. A Catalogue...of the Collection of Paul and Mary Mellon...*, eds. I. MacPhail *et al.*, 4 vols. (New Haven: Yale University Library, 1968–77), vol. 4, pp. 473–6 (including facsimile). Since this book is recorded in *LC* ('[2.7.] 17 Salt of nature regen'), Newton must have bought his copy (*HL 25*) immediately after its publication; another fragmentary title that follows the note on *Centrum naturæ* in the Mellon MS can be identified with one of Newton's non-alchemical books (*HL 302*). This short memorandum is obviously not related to the Borel excerpts, although we are not able to specify Newton's source. Most probably, it had been copied casually from a bookseller's advertisement; the quotation is also different from that given in the *Term Catalogues*, vol. 2, p. 569 (Hilary term, 1695/6).

various stages, i.e. that some were repeated while others were omitted and replaced by new titles. The few additional (non-alphabetical) items in *Keynes 13* seem to have been arranged by Newton according to their importance (or rather in some accidental sequence) and may have functioned as a supplement to the reworked version of *Stanford B.*

To understand Newton's procedure in his first perusal of the *Bibliotheca* and thus the structure of *Stanford A*, we must enter into a short *excursus* on Borel's bibliography. Its title-page, cited in detail above, suggests the enormous number of 4,000 authors. In fact this is approximately the total number of entries, in many of which numerous, though evidently legendary, 'authores chimici' are introduced; furthermore, a large number of manuscripts are cited in addition to printed works; finally, individual items from collected editions are entered separately. The work is made intractable above all by Borel's repetition of numerous treatises, sometimes under different titles, sometimes under the authors' or the editors' names, and sometimes under the title of a collection or compendium. Therefore, the actual number of printed volumes or editions is substantially reduced, and the reader often has enormous difficulties in tracing them (to say nothing of the problems caused by printing errors and contradictory references to the years or places of publication). As a result, Borel's *Bibliotheca* has only limited use as a working basis for bibliographical investigations, or is, in the judgement of the historian of chemistry and bibliographer Bolton, 'unsatisfactory from the standpoint of modern [*sic*] bibliography'.[47] The only possibility of reliably surveying this extensive compilation seems to be Newton's method: to draw up detailed excerpts instead of marking the corresponding entries in the bibliography itself.

Stanford A shows that the choice of information was made from an apparently careful survey of the whole book. Newton worked through the *Bibliotheca* page by page, starting at p. 2 and ending at p. 264: he would copy an entry (often in abbreviated form) at first appearance, then add cross-references to repetitions on following pages or copy additional information found in repeated entries,[48] and later note the number of the page from which

[47] Bolton, *Bibliography*, p. 7.; cf. also his forerunner J. Chr. Wiegleb who criticized both reiterations and frequent absence of bibliographical data in his *Geschichte des Wachsthums und der Erfindungen in der Chemie, in der neuern Zeit*, 2 vols. (Berlin & Stettin, 1790–91), vol. 1, p. 7: 'Es würde solches noch schätzbarer seyn, wenn bey allen Ort und Jahr der Ausgabe angeführt worden wäre, wie es nur bey wenigen geschehen ist, und nicht so oft ein und dasselbe Buch auch noch unter andern Titeln wieder vorkäme.'

[48] See Newton's cross-references e.g. in *DSC* [16]: Newton copied the title of a collected edition from *Bibliotheca*, p. 14, and added – with his own numbering – the contents listed *ibid*.; he referred to an alternative quotation of the title on p. 19, and to a further tract, contained in this collection, on p. 23 – but without copying the information; he

he had copied an entry.[49] With this procedure Newton was not able to maintain a strict alphabetical order, for Borel initially listed the majority of anonymous treatises and collected editions under 'A' ('anonymi') or under I ('incerti') before repeating them under their titles. Here we have one explanation of disproportionate accumulations among the initial entries in *Stanford A* compared with the total of items in relation to their alphabetical distribution.[50] Independently of such particular characteristics of his source, Newton changed his style in the course of working through the *Bibliotheca*. He copied much more material at the beginning of the task than later, concentrating gradually on essentials, i.e. on single printed works and the relevant bibliographical details: references to the *Theatrum* are dropped (with the exception of indispensable cross-references); less, or no attention, is given to manuscripts,[51] as well as to non-bibliographical quotations.[52] Extraneous information of the latter kind is completely neglected in the later *Stanford B* list, in which Newton confined himself to the 'hard facts' that he copied from his source without remarkable accumulations (as noticed in the first pages of *Stanford A*). Thus *Stanford B*, along with *Keynes 13*, seems to be a condensed version of *Stanford A*, in which all excess cross-references and all data not absolutely essential for bibliographical purposes have been jettisoned. Also, mistakes have been rectified and variant titles reduced to their bibliographically correct form[53] – corrections facilitated by Newton's deepening knowledge of alchemical literature acquired during the 1670s and 1680s.

became aware of the identity of *DSC* [16] with his previous item [12] from p. 8 – and thus inserted there an 'infra'; at last he interlined two other tracts belonging to the same collection, now found on pp. 85 and 107 respectively. Still in *Stanford B* Newton copied the title of this compendium no less than three times in different readings linked up with a common marginal 'A' (see *DSC* [12], [16], [22a]). For other examples of Newton's way of accumulating information by cross-references see *DSC* [9], [18], [22], [26]. Compiling items [41]/[42] and [48]/[51], Newton had changed his style: now he thoroughly compared several pages before he went on with his next item.

49   In *DSC* [1] to [23] Newton used page numbers exclusively as page references; from [24] to the end (although not constantly) he added page numbers pertinent to his entries.

50   The first seventy-five items are copied from *Bibliotheca*, pp. 1–129 (or sections 'A' to 'I'), the last nine (!) items only from pp. 129–272 (or sections 'K' to 'Z', including 'Appendix' and 'Corollarium').

51   *DSC* [1], [7], [13], [15], [36], [48], [50], [55] refer to manuscripts mostly in private possession, [43] to a manuscript tract announced for publication.

52   *DSC* [29], [46], [52], [57], [72].

53   E.g. *DSC* [63]; nevertheless Newton did not become aware that *Farrago philosophorum* was identical to one treatise contained in *Theatrum chemicum*. In another puzzling case, *DSC* [18], Newton possibly overlooked his own reference to the *Theatrum* in *Stanford A* when he repeated this item in *Stanford B* – or was in search for a different treatise with the same title.

The growth of Newton's (al)chemical library: an attempt at chronology

This formal development from *Stanford A* to *Stanford B* and *Keynes 13* supports our initial hypothesis that Newton's excerpts from Borel must be interpreted as desiderata for his own (al)chemical library. More striking evidence, however, can be found by comparing the contents of these documents in their chronological order. A tabular and classified summary of the items selected by Newton (ignoring his non-Borel supplements) may illustrate the displacement of contents (see Table 1). The most obvious discrepancy is the abandonment of works on metallurgy[54] in the later lists. But despite such differences one should notice the extensive agreement of the two *Stanford* lists, since the later one contains only four additional authors or titles (there are two more in *Keynes 13*) beyond those in the earliest list.[55] All the others are repetitions of previously copied items; the bulk of those titles taken over from *Stanford A* to *B* (and partly to *Keynes 13*) are from important collected editions, which in *Stanford A* are throughout marked '–' or 'x' by Newton.[56]

The correspondence of the different Borel excerpts, as well as Newton's markings and a comparison of transcribed and, particularly, marked titles and editions with his own book inventory (*Lib. Chem.*) of 1696–97, supports an interpretation as outlined previously: When Newton compiled his first excerpts about 1670 he was drawing up a list of desiderata and marked the most essential items. In the late 1680s he made a new survey omitting all titles of works that he had acquired in the interim. Items repeated in *Stanford B* were thus still desiderata at that time. On the other hand, new items or authors may point to a change in Newton's attitude towards the literature of alchemy. The same applies to the items listed in *Keynes 13* about 1690. Apart from all Borel-based material, however, Newton's supplemented items are to be examined carefully. Since his second bibliographical work of reference (the above-mentioned 1688 reissue of Cooper's *Catalogue of chymicall books*) did not have any of the defects of Borel's labyrinthic compilation, Newton did not have to consider contemporary English titles while reworking his early manuscript desiderata list. This may explain the absence of any non-Borel titles in *Stanford*

---

[54] *DSC* [5], [8], [38], [54], [56], [60], [61]/[62], [77].

[55] See *DSC* [9a] (anonymous), [75a] (Heinrich Khunradt), [81a] (Michael Potier), [83a] (Gottfried Smoll[ius], Jean Collesson, and the fictitious Salomon Trissmosin).

[56] For '–'-markings see *DSC* [9], [10], [11], [16], [17], [27], [38], [39]/[40], [42], [51], [75], [76], [83]; 'x'-markings only in [12] and [14]. For the last-mentioned we have found no conclusive explanation, since one of the marked titles belongs to an important compendium (with reference to its more detailed repetition in [16]), but the second one does not seem to have been of any significance in the development of alchemical literature. In *Stanford B* Newton used similar markings only twice, in both cases accentuating titles he had taken over from *Stanford A* (see *DSC* [33], [83]).

Table 1.

	Stanford A (c. 1670)	Stanford B (late 1690s)	Keynes 13 (c. 1690)
Ancient and medieval authors (incl. MSS/non-bibliographical quotations)	38 (10/5)	8 (2/–)	3 (1/–)
Collected edns and anonymous works	20	13	2
16th/17th-cent. authors	17	11	1
Works on mineralogy, metallurgy, etc.	8	–	–
Summarized entries			
'De mineralibus'	1	–	–
'De Quintessentia'	–	1	–

B.[57] Generally, *Lib. Chem.* (1696–97), the bill 'Books for Mr Newton' (*c.* 1702) and the contents of his final library permit us to judge, whether, and when at the latest, Newton did purchase any of the books he searched for.

For two at least of the most important printed collections or compendiums of alchemical treatises the exact time of acquisition prior to 1696–97 is well documented: Newton recorded the purchase in April 1669 of the six-volume *Theatrum chemicum*...(Strasburg, 1659–61);[58] from his correspondence with Nicolas Fatio de Duillier it is clear that in the spring of 1692/3 Newton bought the two-volume French *Bibliothèque des Philosophes* (*chimiques*)...(Paris, 1672–78) from his young intimate.[59] As far as it is now possible to verify, there

[57] From comparing the supplements to *Stanford A* (*DSC* [Add 1] to [Add 6]) with *Lib. Chem.* it becomes evident that both works by George Starkey ([Add 3]/[Add 4]) would have been desiderata until after 1696–97: these are the only additional titles not traceable in Newton's *Lib. Chem.* inventory.

[58] See Harrison, *Library*, pp. 7–8, on Newton's early notebooks and recorded purchases of books including the *Theatrum chemicum* (*HL 1608*).

[59] *HL 221* (vol. 1 only); see Isaac Newton, *The Correspondence of Isaac Newton*, ed. H. W. Turnbull *et al.*, 7 vols. (Cambridge University Press, 1959–77), vol. 3, pp. 245, 260–3, nos. 404, 408–11 (February/March 1692/3; 2 vols.). This is the 'Bibliotheca Gallica' or 'ye ffrench Bibliotheque' that Newton referred to in Keynes MSS 45 and 28 respectively. On Keynes MS 28 see recently B. J. T. Dobbs, 'Newton's "Commentary" on the "Emerald Tablet" of Hermes Trismegistus: its scientific and theological significance', in *Hermeticism and the Renaissance: Intellectual History and the Occult in Early Modern Europe*, eds. I. Merkel & A. G. Debus (Washington, DC: Folger Shakespeare Library, 1988), pp. 182–91, who dates the manuscript 'in the early 1690s' (p. 183); since Newton's reference seems to be a later addition to his Hermes

are but few handwritten notes in Newton's surviving books on alchemy and chemistry that give details of purchase. One example, however, is a remark that he inserted in a copy of the anonymous *Sanguis naturæ...* (London, 1696): '[bought] at Sowles a Quaker Widdow'.[60] Considering Newton's desiderata records we may now estimate dates or periods and/or chronological order of further book purchases that attended Newton's (al)chemical studies.

If our interpretation is correct, only two of the volumes assembling basic alchemical texts came into Newton's possession during the 1670s and 80s: *Artis auriferæ, quam chemiam vocant, volumina...* (Basle, 1610 – first published 1572), and *Philosophiæ chymicæ IV. vetustissima scripta...* (Frankfurt, 1605), both of which are in *Stanford A* alone.[61] Further evidence that Newton acquired *Ars aurifera* in the early 1670s comes from one of his manuscripts from about 1675, which contains reading notes (*inter alia* 'Ex Turba') that are unmistakably copied from this edition.[62] Consequently the writings ascribed to Aristotle, Avicenna and Morienus do not recur in *Stanford B*, for these had entered Newton's library with the *Ars aurifera*.[63] Possibly during this same

text we may place the dating of the part in question of Keynes 28 more precisely to 1691 or 1692, also considering that the *Bibliothèque* must have been at Newton's hand before he bought it definitely.

[60] *HL 1446*; see also J. Neu, 'Isaac Newton's library: ten books at Wisconsin', *U. W. Library News* 14(4)(1970): 1–10, on p. 10; another copy: *HL 1445*. Newton bought neither copy immediately after they had been published, since this title does not appear in *LC* (1696–97).

[61] *DSC* [27], [39]/[40] (also in *LC*); on the latter volume see also Harrison, *Library*, p. 20.

[62] Keynes MS 25, f. 4r–4v: 'Ex Turba [...] (p. 7) Aliter legitur in p. 8 & in Theatr.'; page references agree with the text variant as printed in *Artis auriferæ...*, vol. I; a further quotation from the same paragraph ('Zimon in Turba') can be found in Mellon MS 79, f. 1v. Newton's remark in *Of Chemicall Authors* (Babson [419], early 1670s) on a work by 'Maria the Hebrew' ('is now extant') proves his familiarity with the same volume. There are some uncertainties attached to the purchase of *Ars aurifera*, since Newton owned two copies, both in the same edition of 1610 but one incomplete (only vols. I/II, *HL 91*). This copy bears short MS notes in Newton's hand (on p. 52 of vol. II) added to the same treatise which he also had – beyond his notes out of the 'Turba philosophorum' – extracted in Keynes MS 25, ff. 3v–4r: 'Ex Epistola Com. Trevisani ad Thom. Bonon.'; further notes not in his hand suggest that Newton bought this particular copy along with a copy of *Musæum hermeticum*, i.e. even before writing *Stanford A*, if we suppose an early date of acquisition of the *Musæum* (see notes 80 and 88). Thus Newton should have acquired his two-volume copy first, but replaced it by a complete three-volume edition (*HL 90*) at least during the early 1690s when he was occupied with the works of Raymond Lull (see note 74). This, too, was the copy that Newton entered into his *Lib. Chem.* inventory; see *LC*: '[2.4.] 16 Artis aurif. vol. 3.' On the dating of Keynes MS 25 (1670–75) see Dobbs, *Foundations*, p. 132.

[63] See *DSC* [25], [31], [78]. For a detailed table of contents of the various editions of *Ars aurifera* see Ferguson, *Bibliotheca chemica*, vol. 1, pp. 51–2, and *Alchemy and the Occult*, vol. 1, pp. 133–6. Of course, to most of the ancient (and medieval) authors'

period Newton purchased a remarkable composite volume made up of *Ars chemica*...(Strasburg, 1566, including the 'Septem capitula' by Hermes), John Garland's *Compendium alchemiæ*...(Basle, 1560), a treatise by the Paracelsian Gerard Dorn and, finally, Bernhardus Trevirensis' περὶ χημείας...(Strasburg, 1567),[64] since neither Hermes' nor Garland's names appear in *Stanford B* again.[65] Additional important works acquired during the 1670s or 1680s include Georg Agricola's *De re metallica*...(Basle, 1621)[66] and Albertus Magnus' *De rebus metallicis*...(Strasburg, 1541 – appended to Raymond Lull's *De secretis naturæ, sive quinta essentia*...).[67] On the other hand, the case of Martin Ruland's well-known *Lexicon alchemiæ*...([Frankfurt], 1612) is perplexing: although Newton definitely owned a copy,[68] the book is listed in *Stanford A* alone but was not entered into his *Lib. Chem.* inventory. This particular example may illustrate both the limits of our interpretation and the uncertainty of sources based on manuscript drafts or preliminary notes.

The next stage of Newton's search for alchemical texts comprises the period between the writing of *Stanford B* (the late 1680s) and Newton's move to London (1696). In these years he was able to obtain at least one volume of

names a 'Pseudo-' should be prefixed, especially to the Lullian corpus treated below; we use the attributed names without regard to modern distinctions.

[64]   *HL 85, 649, 536, 168*, bound in one volume; see *LC*: '[2.7.] 4 Hermetis capitula septem. Tab Smaragd cum commentario Hortulani. Consiliū conjugij [= *HL 85*]. Iterum Tab. Smaragd. cum commentario Hortulani. Arnoldus in Hortulanū. Garlandus de præp. Elix. Garlandus de mineral [= *HL 649*]. Lapis metaphys per Gerardum Dorn [= *HL 536*]. Trevisani Epistolæ et liber [= *HL 168*].' Newton's *Lib. Chem.* entry clearly shows that his copies of these originally individual tracts had been rebound as a single volume (see also note 80).   [65]   *DSC* [73], [69].

[66]   *DSC* [5] (not in *LC*; probably not classified 'chemical' by Newton in the 1690s). Newton's interest in the work of the famous German metallurgist may explain his following item [6]. But the ascription of an alchemical title to Georg Agricola was based on a misprint in Borel, falsifying both author's name and year of publication. The actual book by Daniel Agricola 'does not treat of Alchemy at all' (Ferguson, *Bibliotheca chemica*, vol. 1, p. 8). On Newton's study of Georg Agricola see K. Figala, 'Gedanken zu Isaac Newtons Studium von Agricolas Werken', in *Arithmos-Arrythmos: Skizzen aus der Wissenschaftsgeschichte. Festschrift für J. O. Fleckenstein*, eds. K. Figala & E. H. Berninger. Wissenschaftsgeschichte: Beiträge aus dem Forschungsinstitut des Deutschen Museums für die Geschichte der Naturwissenschaften und der Technik (Munich: Minerva, 1979), pp. 73–103.

[67]   *DSC* [8]; see also note 74.

[68]   *DSC* [83]. A second work by Ruland, *Progymnasmata alchemiæ, sive Problemata chymica*...(Frankfurt, 1607; also part of [83], 'Problemata [...]'), was repeated in *Stanford B* and marked 'x'. In contrast to the *Lexicon* this book was listed in *Lib. Chem.*; see *LC*: '[2.4.] 18 Rulandi progymasmata [*sic*] Alchymiæ, Marsilius Ficinus & Cosmopolita.' According to Harrison (unpublished) Newton's perplexing entry is based on marginal annotations in his copy (but in an alien hand) quoting Ficino and Cosmopolita (i.e. M. Sendivogius).

Alchemiæ, quam vocant artisque metallicæ doctrina...(Basle, 1572),[69] and – in the early 1690s – the *Opuscula quædam chemica*...(Frankfurt, 1614) which was the only one of the collected editions to be repeated once again in *Keynes 13*.[70] Roger Bacon's *De arte chymiæ scripta*...(Frankfurt, 1603) also found its way into Newton's library during these years.[71]

The latter volume is one of the few marked titles in *Stanford B* (unlike *Stanford A*), which suggests that Friar Bacon's significance in Newton's alchemical thought increased in the late 1680s. The same applies to the work ascribed to Raymond Lull. A comparison of *Stanford A* with *B* shows that only in these years did Newton take particular notice of the Lullian *corpus*: while *Stanford A* has a single entry, *Liber, qui codicillus, seu vade mecum inscribitur*...(Cologne, 1563), *Stanford B* lists no less than ten issues summarized as 'Lullij opera edita quædam'.[72] Underlinings made by Newton probably mean that these editions had been in his possession at the end of the 1680s, i.e. three (or four) titles plus the above-mentioned *Codicillus*.[73] More detailed information can be drawn from another manuscript sheet, now preserved with *Stanford B*. Here Newton tried to compile a detailed inventory of the works of Raymond Lull comprising six volumes in his possession and a summary of seven 'desideranda'.[74] To judge from the handwriting, the Lull

69   *DSC* [9]; see *LC*: '[2.7.] 5 Alchymiæ quam vocant &c Continens Tauladan, speculum Alchymiæ Baconis, Correctorium Richardi, Rosarius minor, Albertus magn. de Alchymia.' Newton's detailed description allows us to conclude that he merely owned vol. I of this two-volume edition; see *Alchemy and the Occult*, vol. 1, pp. 126–8 for more details on the contents. This volume corresponds to the 'not identified' *HL 1624* (identification confirmed by Harrison, private communication, 1989).

70   *DSC* [17]; see *LC*: '[2.8.] 28 Opuscula Riplei, Canones 10 incerti authoris', followed by '29 Speculum Alchymiæ [...] & Etschenreuteri Epistola.'; thus Newton's copy obviously was bound in two volumes.   71   *DSC* [33] (also in *LC*).

72   *DSC* [76], [76a].

73   *HL 1000* (= 'Basle, 1572' in *DSC* [76a]), *HL 996* or *998* or both (= 'Cologne, 1567' *ibid.*), *HL 995* (= 'Strasburg, 1541' *ibid.*; see also *DSC* [8]), *HL 997* (see *DSC* [76]).

74   Stanford M132, container 2, folder 3, f. 5r–5v (our pagination, part of a bundle of title listings summarizing the Lullian corpus): Raymundi Lullij opera', followed by tables of contents of altogether nine volumes, 'In Vol. 3 Theat. Chem.', 'In Vol. 4 Theat. Chem.', 'In Vol. 3 Artis Auriferæ', 'Lullij opera quædam Basileæ 1572 [= *HL 1000*]', 'Coloniæ 1563 [= *HL 997*]', 'Coloniæ 1567 [= *HL 996*]', 'Argentorati 1541 [= *HL 995*]', 'Coloniæ 1⟨6⟩567 [= *HL 998*]', 'Argentorati 1616 [= *HL 1001*]'; at the bottom of f. 5v: 'Desiderantur opera impressa [...]' (see *DSC* [76a]). Newton assigned his own numbering to every item of the contents obviously corresponding to its importance, thus: '1 Testamentum antiquius continens Theoricam & Practicam' (in *Theatrum*, vol. IV) to '27 De Conservatione vitæ' (in *HL 1001*). Interestingly, one volume (corresponding to *HL 995*) is described as being composed of '6 De secretis naturæ sive Quintessentia lib. 2. Deest tertius. 25 Albertus de mineralibus et rebus metallicis. 26 Encelius de Lapidibus & Gemmis.'; the same sequence was drawn up by Newton in a rough draft (now Babson [416]*B*:3). Since the original impression, published in

records were written in the early or mid-1690s, and thus may well be compared with the *Lib. Chem.* of 1696–97, which catalogues seven volumes as 'Lullys works'. None of the 'desideranda' found their way into Newton's library, but in view of the discrepancy in the number of volumes[75] we suspect that Newton integrated his copy of the anonymous *Ars magica...* (Frankfurt, 1631) into his series of Lull editions – the very same that he had marked most peculiarly ('☞') in *Stanford A*.[76] In summary, we may conclude that at that stage of his alchemical interests Newton endeavoured to obtain as complete a collection as possible of the published works of Raymond Lull, though his efforts met with only limited success.

Unattained desiderata, i.e. titles that cannot be proved to have been in Newton's *Lib. Chem.* inventory or in his final library, include some of the most important publications in the field of alchemy, such as *Veræ alchemiæ artisque metallicæ...doctrina...* (Basle, 1561 – a voluminous former edition of the above-mentioned *Alchemiæ, quam vocant artisque metallicæ doctrina*) that Newton entered in some length both in *Stanford A and B*.[77] Nor did he acquire the earliest collected edition of all, entitled *In hoc volumine de alchemia continentur hæc...* (Nuremberg, 1541),[78] as well as the first edition of Petrus Bonus' *Pretiosa margarita novella...* (Venice, 1546)[79] which still appears in

1541, did not contain the latter work, Newton's copy may have been a composite volume bound with Christoph Entzelt's *De re metallica...libri III...* (Frankfurt, 1551, reissued 1557); see *DSC* [56] (not *HL*). There was no opportunity to discuss this particular case with J. Harrison.

[75] See *LC*: '[2.6.] 25, 26, 27, 28, 29, 30, 31 Lullys works.' Harrison (unpublished) refrained from determining which of the total of eight Lull volumes in Newton's final library may have been included in the *Lib. Chem.* list. Tentatively, we may assume that Newton did not classify volumes as '(al)chemical' that treat with Lull's 'Ars magna' or 'universalis' (*HL 994*, and *HL 999*, bound with *32*). Thus six alchemical out of Newton's final Lull volumes contrast with the seven listed in *Lib. Chem.*, i.e. the identity of one volume is questionable.

[76] *DSC* [19]. We have not been able to find any other tolerably conclusive explanation for Newton's conspicuous marking. But this presumption is further supported by a marginal reference 'Ars magica [...]', written by Newton in one of his Lull volumes (*HL 1000*), although the cross-reference otherwise might be aimed to Lull's 'Magia naturalis, seu Compendium artis magicæ', as printed in *HL 998*. The *Ars magica* cannot be identified with any other entry in Newton's *Lib. Chem.* inventory; on the other hand, this work is closely related to *Trinum magicum...* (Frankfurt, 1609, continued 1611), edited by Caesar Longinus, a later reissue of which (1673) Newton owned (*HL 981*; not in *LC*). [77] *DSC* [12], [16].

[78] *DSC* [11]. For another doubtful compendium allegedly published in the same year see *DSC* [75].

[79] *DSC* [37]. The quotation chosen by Borel corresponds to the last lines of the internal heading of Bonus' 'Pretiosa margarita novella', as printed in Janus Lacinius' edition of the same title (Venice, 1546, and 1557, on f. 1). Even Ferguson (*Bibliotheca chemica*, vol. 1, pp. 115–6) causes some confusion concerning the differing titles, but in fact the

Keynes 13. That Newton failed to acquire these books is by no means surprising, if we note their years of publication. Most of them had been out of print for more than a hundred years when he started his hunt. Success in these cases depended on second-hand copies, for example, his copy of Lull's *Codicillus* and the afore-mentioned composite volume.[80] Perhaps Newton's residence at Cambridge was a handicap to purchasing such old works. In fact, in the late 1690s at London Newton must have been more successful when he found at least an incomplete copy of one of his long-time desiderata, the *De alchimia opuscula complura*...([Frankfurt], 1550).[81]

This survey, which has been mainly restricted to the most significant classical alchemical texts, clearly shows that Newton regarded these works as extraordinarily important. Indeed, they embrace almost the entire corpus of alchemical literature available until the three-volume first issue of the *Theatrum chemicum* appeared in 1602. Even after its publication, these old compendiums were by no means made obsolete, for the publisher of the *Theatrum*, Lazarus Zetzner, was little concerned to put out a systematic edition of classical texts, but rather he reprinted issues that had appeared up to the dates of publication of the particular *Theatrum* volumes. In this way he composed his first three volumes (1602), and again the fourth (first published 1613, with a reissue of the former volumes). The fifth, published in 1622 by Zetzner's heirs, contains a selection of older texts and the sixth and last, issued with the final edition (1659–61), translations of previous French and German publications. These facts may sufficiently explain Newton's insistence on obtaining the pre-*Theatrum* compendiums, for only after their acquisition would the whole

'Secretum omnium secretorum ac Dei donum' is not a part of but identical with the *Margarita* (1546). On the other hand, Newton, too, seems to have been on a wrong track, for he already had a corrected version of this treatise (published 1572, and reprinted in *Theatrum chemicum*, vol. V) at his disposal. Thus we may conclude that he was – of course in vain – looking for an additional work by Bonus well into the 1690s.

80   See also Harrison, *Library*, pp. 60, 77–8 (on second-hand purchases), 24–5 (on books annotated in alien hands, i.e. by previous owners). Unfortunately Harrison gives no details of the 'ten alchemical' books bearing notes not in Newton's hand. Some of the more important examples are his copies of Lull's *Codicillus* (*DSC* [76]), and – above all – Newton's two-volume copy of the *Ars aurifera* (cf. note 62) and his *Musæum hermeticum* of 1625 (cf. note 88). The latter two volumes both bear annotations in the same alien hand and thus may have been acquired by Newton on the same occasion. Another indication of second-hand purchases are composite volumes in his library, exemplified by the sixteenth-century volume mentioned in note 64, since Newton usually desisted from rebinding his alchemical books (see e.g. the split volume mentioned in note 70).

81   *DSC* [10]; this prominent early impression can not be traced in *Lib. Chem.* (1696–97). For additional supposed late purchases see, e.g., the works of G. Starkey (cf. note 57) and a work by M. Potier (*DSC* [83a], not *LC*).

range of the older alchemy have been available to him.[82] This tendency in Newton's selection of literature is emphasized by his taking into account works that are listed by Borel as being in manuscript only. Most of Newton's items of this group in fact were never – or only much later and in mutilated versions – printed, especially those still noted in *Stanford B* and *Keynes 13*.[83] Thus the presence of this set of manuscripts among the bulk of printed books by no means contradicts the nature of a desiderata list with respect to the establishment of an (al)chemical library; rather we must assess these items as essential supplements to the whole spectrum of printed literature.

As an argument supporting our interpretation of Newton's *De Scriptoribus Chemicis* selections and related documents as his main desiderata inventories, we put forward the neglect of items which actually were essentials for records in alchemical bibliography: e.g. the writings of Michael Maier and Michael Sendivogius. Both feature prominently among Newton's alchemical investigations, and long transcriptions from their works and other notes on them are to be seen among his papers. Furthermore, in his own final library Newton had nine works by Maier and three individual editions of Sendivogius' tracts.[84] His (al)chemical papers, however, reveal that some of the works from these quite 'modern' alchemists must have been in his possession very early, i.e. before the origin of *Stanford A* about 1670. In any case extracts from Maier's *Symbola* in Newton's hand – besides those discussed in the context of *Of Chemicall Authors* – may be dated to the second half of the 1660s.[85] Extensive

---

[82] A more detailed study on the composition of the *Theatrum* is still wanting, but this is not the place for discussions at length. For some examples of previous editions of tracts that had been reprinted in the *Theatrum* without alterations (i.e. including even original prefaces and dedications) see *DSC* [2], [18], [21], [22], [39]/[40] (tracts I and IV only), [63], [74], and Newton's misleading reference in [9]; *DSC* [41] and [83*a*] (Collesson) are examples of translations comprised in vol. VI of the final *Theatrum* edition. See also *Alchemy and the Occult*, vol. 2, pp. 358–94, and Ferguson, *Bibliotheca chemica*, vol. 2, pp. 436–40 (who speaks erroneously of its first appearance in four volumes); for a survey of the important early collections see *ibid*., and, more recently and detailed, R. Halleux, *Les Textes Alchimiques*. Typologie des sources du moyen âge occidental, 32 (Tournhout-Belgium: Brepols, 1979), chap. 5.

[83] See *DSC* [1], [13], [15], [55]. A short tract by Alphidius [13] was only published in German with Benedictus Figulus' *Thesaurinella olympica...* (Frankfurt, 1608), a tract ascribed to St Dunstan [55] in Lancelot Colson's *Philosophia maturata: an exact piece of philosophy...* (London, 1668). The latter one had been in Newton's possession (*HL 419*, also in *LC*: two copies). It may have been purchased before 1680, since this work was named in Cooper's advertisements but not copied by Newton for his supplements to *Stanford A* (see note 41). On Abraham the Jew [1] cf. also note 91.

[84] *HL 1044–52* (Maier); *HL 445, 1192, 1485* (Sendivogius; individual editions only).

[85] *HL 1048* (also in *LC*; early extracts in Keynes MS 29 (c. 1668–69), in a MS now at St Andrews University (c. 1669), in Babson [419] (early 1670s; cf. note 23); see Dobbs, *Foundations*, p. 131, Westfall, *Never at Rest*, pp. 291–2, and K. Figala, '"Die exakte

quotations from the same work, followed by extracts from four more of Maier's books, can be traced without exception to volumes that had been in Newton's possession; the same applies to references (dated to about 1670–72) to another work by the same author. All together, trying to date these important and comprehensive records correctly, would seem to lead to inconsistencies when dating by conventional methods, i.e. those based on handwriting. At present Newton's main Maier manuscript is commonly dated to his late period, but it can be shown that it is definitely quoted in one of his manuscripts from the 'middle early' period. Therefore utmost caution is advised concerning any conclusive statement about the access of Maier's works into Newton's library.[86] As for Sendivogius (or Sędziwój), some of Newton's notes on his writings can be dated to the years 1668 or 1669. Since Newton's early quotations from the anonymously published *Novum lumen chymicum* are joined to those from a tract by Jean d'Espagnet they imply the knowledge – if not the acquisition – of their common publication in *Bibliotheca chemica contracta*... (Geneva, 1653).[87]

Alchemie von Isaac Newton": Seine "gesetzmäßige" Interpretation – dargestellt am Beispiel einiger ihn beeinflussender Autoren', *Verhandlungen der Naturforschenden Gesellschaft in Basel* 94(1984): 157–228, on pp. 190 seq.

[86] Keynes MS 32 (about 1690?): 'Symbola aureæ mensæ duodecim nationum [...]', ff. 1r seq. (= *HL 1048*, also in *LC*); 'Lusus scrius [...] Dat. 1616. [...]', ff. 11r seq. (not *HL*?; see *LC*: '[2.4.] ⟨7⟩8 Maier lusus serius'; *HL 1044* is an English translation only); 'Mich. Majeri Atalanta fugiens, [...] Edit. 1618.', ff. 13r seq. (not *HL*, but see *LC*: '2.4.1 Maier Embl.'; a second entry, '[2.4.] 2 Maier Embl.', is *HL 1045*); 'Michaelis Maieri Viatorium [...]', ff. 25r seq. (= *HL 1052*, also in *LC*); 'Septimana Philosophica [...]', ff. 33r–44v (= *HL 1046*, also in *LC*). This eighty-eight page manuscript, partly paginated by Newton himself, can be identified by collation with the otherwise unknown 'Extract. Maier.' (or 'Maier. Extract.'), quoted in Newton's 'Notanda chymica' (Smithsonian Institution Libraries, Washington, DC: Dibner MS 77; formerly Burndy # 14), and as well in his *Index chemicus* (see Westfall, 'Newton's Index chemicus', p. 184). However, supported by the handwriting, the 'middle early' Dibner MS 77 has been dated to about 1670 or after, whereas Keynes MS 32 has been dated to the 1690s (we shared the assessment of Westfall, *Never at Rest*, p. 292, n. 37). This dating flagrantly contradicts these internal interdependencies. In addition, Dibner MS 77 also bears reference to 'Maier Hier.' (or 'Hieroglyph.'); this may be identified as Maier's *Arcana arcanissima*...([London], 1614, reissued 1625 as *De hieroglyphicis Ægyptiorum*...; not *HL*, but see *LC*: '[2.4.] 5 Maier Hieroglyph'). For an extensive bibliography of Maier's works see K. Figala & U. Neumann (forthcoming in *Archives Internationales d'Histoire des Sciences*).

[87] Babson [925]*A* (c. 1668): 'Loca difficilia in ⟨N⟩ovo Lumine Chymico exp⟨l⟩icata.', followed by 'Hermetick Secrets'; Keynes MS 19 (c. 1668–69): 'Collectiones ex Novo Lumine Chymico [...]' (with 'Collectionum Explicationes.'), followed by 'Arcanum Hermeticæ Philosophiæ Opus.' (with 'Explicationes.'), and 'Dialogus Mercurij Naturæ & Alchymistæ'); Jewish National and University Library, Jerusalem (JNUL): MSS Var. 259–4/5 (early 1670s): 'Novum Lumen chymicum. Sendivogij [...]', followed by 'Enchiridion Physicæ Joh. Spagneti. [...]', and 'Ab eodem Arcanum

In a similar way other prominent titles which are quoted in neither of Newton's desiderata lists can be traced in his alchemical reading notes and transcriptions, and for that reason probably came into his library before 1670. They should include George Ripley's *Opera omnia chemici*...(Kassel, 1649) and the *Tractatus aliquot chemica*...(Geismar, 1647), both edited by Ludwig Combach, as well as the *Musæum hermeticum*...(Frankfurt, 1625) and *Harmoniæ inperscrutabilis...Decas* [I–II]...(Frankfurt, 1625).[88] Apparently, Basilius Valentinus' *Currus triumphalis antimonii*...(Toulouse, 1646), edited by Pierre Jean Fabre, belongs to this group, too. But we cannot be sure in the latter case, since this edition was the source for a set of Newton's rather early reading notes, though the volume still appears in *Stanford A*. We may thus have to post-date excerpts made by Newton from this book from about 1668 to after 1670, or assume that he made his notes out of a loaned copy and then decided to buy his own.[89]

Among the early subjects of Newton's alchemical studies was the famous work of Nicolas Flamel, which, according to Professor Dobbs, was known to him through its English translation, *Nicholas Flammel, his exposition of the hieroglyphicall figures*...(London, 1624).[90] Moreover, his familiarity with this title is clearly shown by the spelling 'Flammel' (instead of Borel's 'Flamel') throughout his bibliographical records, notwithstanding that he himself did

Hermeticæ Philosophiæ Opus. [...]'; to this also may belong JNUL MS Var. 259–7 (part A): 'Ex Augurelli Chrysop⟨œ⟩a'. The *Bibliotheca chemica contracta* is the only source to present these tracts (Sendivogius/d'Espagnet/Augurello) jointly. For maintaining these early datings we have to suggest that Newton, at least for some time, had a copy of the 1653 edition at hand, although he is only known to have owned a reissue of 1673 (*HL 220*, also in *LC*).

[88] *HL 1405–6, 1623, 1130* (1625 edition; for the enlarged 1677(78) edition see *HL 1131*), *740* (all titles also in *LC*). For extracts from Combach's Ripley edition see Keynes MS 17 (early 1670s); for Combach's *Tractatus aliquot* see Newton's note in his copy (*HL 1623*). A further argument for Newton's early knowledge about this editor's activities may be seen in his explicitly mentioning the name in connection with Christophorus Parisiensis (see *DSC* [42], marked). Early reference to *Musæum hermeticum* of 1625, or rather to a tract contained therein ('Gloria mundi'), is made in JNUL MS Var. 260 (*c.* 1675), ff. 2r–5v *passim*. On the other hand, but still in agreement with the dating of MS Var. 260, there is some evidence for the acquisition of this volume in the early 1670s, i.e. after Newton had drawn up *Stanford A* (see notes 62 and 80); for *Harmoniæ... Decas* see Newton's remark in *DSC* [71].

[89] See *DSC* [59]; not *HL*, but probably in *LC*: '[2.4.] 20 Currus triumph. B. V.' Newton's final library only included editions with the commentary by Theodor Kerckring in Latin and English (*HL 882, 129*). Transcripts in Keynes MS 64 (c. 1668): '[...] Currus triumphalis Antimonij.', followed by 'Roger Baconis Oleum Stibij.' must have been made from Fabre's edition of 1646, because this one also comprises – *inter alia* – Bacon's tract.

[90] See Dobbs, *Foundations*, p. 130, on Newton's source for his Artefius and Pontanus transcripts in Keynes MS 14 (early 1670s). Early extracts from Flamel's treatise itself

not own the book. Newton may have borrowed a copy in the very beginning of his (al)chemical career and subsequently, between about 1670 (as shown in *Stanford A*) and through at least the 1690s, tried to find the whole *corpus* ascribed to (and connected with) this French alchemist.[91] George Starkey's *Pyrotechny asserted...* (London, 1654) is even more illustrative of the significance of Newton's bibliographical lists. A group of his earliest chemical, but not yet alchemical, notes prove that Newton consulted this work as a highly esteemed source-book at the beginning of his research on this subject. Astonishingly, however, Newton added its title as late as the 1680s when he supplemented his *De Scriptoribus Chemicis* with the help of William Cooper's book advertisements. There is every indication that Newton incorporated his own copy into his library only during a later period of his (al)chemical career.[92]

Conclusion

In a sense we have now come full circle. Detailed studies of the role and development of alchemy in Newton's thought have shown that he started with what may be called 'applied chemistry'.[93] Corresponding to this interest is the series of metallurgical and mineralogical works in his first list of desiderata of books on chemistry and alchemy in about 1670. Alternatively, the inner development of Newton's approach to alchemy is evident even from the same strictly bibliographical records: the book lists and fragments of the late 1680s and early 1690s comprise titles of unambiguously alchemical orientation. Moreover, by comparing his early knowledge and use of the literature of alchemy with his late desiderata, we are able to see a shift from 'modern' to classical authors and texts. It seems likely that the crucial impetus for this change was Maier's 'historical' outline of alchemy, elaborated in his *Symbola*. Accordingly, most of the ancient and medieval authors listed by Newton in *Stanford A* can be traced to that eminent work; though it must be noted that

> (including pen drawings by Newton) are in JNUL MS Var. 259–3. This MS seems to have been Newton's main source for later quotations from Flamel's work, since further extracts 'Out of Flammel' in Keynes MS 25 (1670–72) are based on the JNUL MS, as can be verified by collation; cf. Newton's reference in Keynes MS 25, f. 1r: 'Flam. p. 3 c 3.'; also reference in Babson [419] (cf. note 23).

[91] *DSC* [66], [67]. See also *DSC* [1], which is closely connected (and was believed as the main key) to the work of Flamel, and – probably – *DSC* [43] as a supplementary tract on the subject.

[92] *DSC* [Add 3] (not in *LC*); on Newton's chemical studies and his experimental notebook see M. Boas & A. R. Hall, 'Newton's chemical experiments', *Archives Internationales d'Histoire des Sciences* 11(1948): 113–58, also Dobbs, *Foundations*, pp. 87–8, and Westfall, *Never at Rest*, pp. 281 seq.

[93] See B. J. T. Dobbs, 'Conceptual problems in Newton's early chemistry: a preliminary study', in *Religion, Science and Worldview. Essays in honor of R. S. Westfall*, eds. M. J. Osler & P. L. Lawrence (Cambridge University Press, 1985), pp. 3–32.

Borel, too, made the widest use of the *Symbola* – and of a similar work by Giovanni Battista Nazari[94] – as sources for his entries on ancient and legendary 'authores chimici'.

However, beyond such a 'shift' one may also state 'continuity' by studying these bibliographical records. It seems likely that we have to expand the period of Newton's essential (al)chemical book purchases – the growth of his (al)chemical library – definitely to his London years. It is true that his move to the capital represents a significant hiatus: thus we may interpret the break of collecting bibliographical data and the cataloguing in about 1696–97. But Newton's bibliophile activities did not halt: this is the interpretation that we must make of the late acquisition of books which previously emerged within his long-time compilation of desiderata lists.

That Newton wrote, in *Stanford B* and *Keynes 13*, a second and obviously reworked bibliographical memorandum, beginning in the late 1680s (perhaps after having completed and published his *Principia* in 1687) suggests that his interest in alchemy was renewed or became considerably stronger at precisely this period. We may connect this interest with a prolonged search for the whole corpus of the alchemical tradition. From the studies of R. S. Westfall, especially, it is well known that Newton's enormous knowledge in this field found its most impressive manifestation in his voluminous manuscript *Index chemicus*.[95] Newton's famous remark to Hooke in 1675 that 'If I have seen further it is by standing on ye sholders of Giants',[96] is certainly true of his attitude towards the domain of alchemy. The bibliographical records, presented in this paper, namely *Stanford A* (*De Scriptoribus Chemicis*), *Stanford B* and *Keynes 13*, may be considered as important documents illustrating Newton's path towards this knowledge: his unbroken quest for (al)chemical books and the endeavour to establish an exquisite (al)chemical library, comprising the extant sources to reveal the secrets of the *prisca sapientia* until his own time.

We may conclude our survey with some few words on the possible relevance of such bibliographical material for further studies of Newton's (al)chemical papers. One of the main difficulties in this field is the lack of reliable methods

---

[94] See *DSC* [79]; this, too, was one of Newton's long-time desiderata.

[95] See Westfall, 'Newton's Index chemicus', *Never at Rest*, 'Newton and alchemy', *passim*; see also Dobbs, *Foundations*, p. 88, where she states that 'Newton went on...to probe the whole vast literature of the older alchemy as it has never been probed before or since.'

[96] Newton, *Correspondence*, vol. 1, pp. 416–17, no. 154; see also R. K. Merton, *On the Shoulders of Giants. A Shandean Postscript* (New York: The Free Press, 1965), whose fascinating *tour de force* through the history of books, tracts, text tradition, misleading quotations, etc. will meet with the widest approval of any historian of alchemy and early chemistry.

for dating the bulk of transcriptions, reading notes, drafts, etc. Certainly, there is an outstanding authority in dating these manuscripts from Newton's handwriting only – D. T. Whiteside. Furthermore, with the research on watermarks now done by A. Shapiro, it is hoped that one can obtain valuable help in determining cases which are undecidable based on handwriting alone, or even cases which are contradicted by internal evidence. On the other hand, the well-known date of acquisition of Newton's own copy of the *Theatrum chemicum* (in 1669) is widely used by almost all scholars on that subject in dating Newton's early (al)chemical manuscripts: the presence or absence of quotations from or references to the *Theatrum* may allow us to decide whether manuscripts had been written before or after April 1669. Thus a more precise knowledge of the dates of acquisition of other books or compendiums may give us a number of further 'key fossils' to date Newton's later (al)chemical papers. A first step, and thus a *terminus post quem*, is made by studying Newton's desiderata lists. Manuscripts, well-dated from handwriting, watermarks or internal evidence, may help us to verify both the dating of the bibliographical records, presented in this paper, as well as a *terminus ante quem* for acquisitions of important alchemical books. If it were possible to fix such dates of acquisition to a well-defined interval of time, these 'key fossils' would become a valuable supplement to other methods of dating manuscripts. But this demands further – intensive – study and identification of the vast number of Newton's textual references in his (al)chemical papers, which fortunately are, mostly, quotations from volumes he once had incorporated into his own library.

Appendix: Newton's *De Scriptoribus Chemicis*, an annotated transcription

In advance of their two-day sale of 332 lots of Newton papers and related material on 13 and 14 July 1936, Sotheby & Co. issued a detailed catalogue of the items on offer.[97] The compiler, John Taylor, listed them with the usual care and expertise shown by the London firm. The document being considered in this article, now in Stanford University Libraries, Department of Special Collections, Manuscripts Division, formed Lot 6 of Section I ('Manuscripts on alchemy. Almost all in the handwriting of Sir Isaac Newton') and was described as: 'Alchemical writers. 'De Scriptoribus Chemicis' [a Bibliography containing titles and particulars of over 80 Printed Books and several MSS.] 5 pp. Autograph. *sm.* 4to.' It was acquired by W. Heffer & Sons Ltd of Cambridge with a bid of £11–0–0. The subsequent journey of the manuscript from Heffers to Stanford was completed through the good offices of Frederick Edward Brasch, bibliographer, consultant, and generous benefactor to Stanford. The *De*

---

[97]   *Catalogue of the Newton Papers*; its title continues:... *Sold by order of the Viscount Lymington to whom they have descended from Catherine Conduitt, Viscountess Lymington, Great-niece of Sir Isaac Newton. Which will be sold by auction by Messrs. Sotheby and Co....*; see also P. E. Spargo (this volume).

Scriptoribus Chemicis was presented by him to the Stanford collection along with other books and manuscripts concerned with the history of science.[98]

When the manuscript was bought by Heffers in 1936 it found itself in Cambridge for a third time. It had been originally compiled there during Newton's time at Trinity College, then about 200 years later it was returned to Newton's university along with a vast quantity of the rest of his papers which were to undergo a thorough examination and classification over the period 1872–88. Two nineteenth-century Cambridge mathematicians had visited Hurstbourne Park, the seat of the fifth Earl of Portsmouth, in order to look through the entire collection of Newton material and report back to the university authorities. When Adams and Stokes quickly realized that their task was likely to be a formidable and lengthy one, Lord Portsmouth gave his permission for the entire corpus to be transferred back to Cambridge for a careful sorting and study.[99] The detailed survey took sixteen years, and its completion was marked in 1888 by the publication of a catalogue[100] in which the papers were divided into fifteen individual categories. Section I (mathematics) was undertaken by Professors Adams and Stokes. For the rest of the material, a vast amount covering chemistry, chronology, history, miscellaneous (mainly theological) papers, letters, etc., assistance was provided by G. D. Liveing, the University Professor of Chemistry, and by H. R. Luard, the University Registrary who dealt with the personal and non-scientific papers. The preface to the *Catalogue*, dated 26 May 1888, is signed H. R. Luard, G. G. Stokes, J. C. Adams, G. D. Liveing.

It was therefore quite obviously the last-named who assessed and sorted the chemical and alchemical items, and it is reasonable to assume that it was he who decided which papers were worth including as part of the 'scientific portion' to be presented to the university by Lord Portsmouth, and which should be returned to him at Hurstbourne Park. Though the classification of the chemical papers has been recognized as 'extremely competent',[101] it would seem that Newton's powerful and life-long fascination with alchemy struck no sympathetic chord within Professor Liveing, especially when we attribute to him the verdict given on p. xix of the 1888 *Catalogue* that 'Newton's manuscripts on Alchemy are of very little interest in themselves'. He was more impressed by the notes Newton had made concerning his own chemical experiments over his earlier

[98] A recent study of Brasch is found in H. Lowood, *Frederick E. Brasch and the History of Science* (Stanford University Libraries, 1987).

[99] For a full account of their visit, the subsequent transfer of the collection to Cambridge for sorting and close examination, and the eventual outcome after 'sixteen long years', see D. T. Whiteside, 'General Introduction', in *The Mathematical Papers of Isaac Newton*, ed. D. T. Whiteside, 8 vols. (Cambridge University Press, 1967–81), vol. 1, pp. xxx–xxxiii.

[100] *A Catalogue of the Portsmouth Collection of Books and Papers written by or belonging to Sir Isaac Newton, the Scientific Portion of which has been presented by the Earl of Portsmouth to the University of Cambridge* (Cambridge, 1888).

[101] See Whiteside, 'General Introduction', p. xxxiii.

years, so that these notes and a manuscript notebook were retained in Cambridge, evidently qualifying as 'scientific' and so being placed in a superior category to alchemical writings. Section II of the *Catalogue*, headed chemistry, was set out in six sub-divisions, the first of which was described as 'five parcels containing transcripts from various alchemical authors in Newton's handwriting, with notes and abstracts', with their contents catalogued to show 32, 35, 36, 12 and 31 documents respectively. The Stanford manuscript was item 14, in parcel 3, listed as 'Notes de scriptoribus chemicis' – somewhat briefer than the later *Sotheby Catalogue* entry.

The transcription of the Stanford manuscript (= *Stanford A*; Stanford University Libraries – Department of Special Collections, Manuscripts Division: Newton Collection (M132), Container 2, Folder 4; published by permission) reproduces Newton's original version as closely as possible, but a number of editorial insertions have been made in order to refer directly to each item and to page numbers in Borel's *Bibliotheca chimica*. All extra material of this nature is given in square brackets [], including capital letters to show the alphabetical division within Borel and (//) indicating change of pages within *Stanford A*. Except for the enclitic *que*, Newton's contractions have not been expanded. Subdivision into items was established according to Newton's indention of lines. All variant sections are given in angle brackets ⟨ ⟩, i.e. Newton's deletions, slips of the pen and corrections; interlineations are marked ↑↓. Newton's own 'editorial' additions, i.e. page references and material copied from following and/or repeated entries in Borel, are indicated by ⟨italic sections⟩, and his later supplements, not based on Borel by ⟪italic sections⟫.

Further information is given in the notes in part B at the end of the transcription referring to item numbers: notations of Newton's alternative spelling (restricted to major deviations or faults); corrections to Borel's entries and comments on actual editions or reprints in *Theatrum chemicum*, vol. I–VI (= *TC*); reference to John Harrison's *Library of Isaac Newton* (= *HL*) for volumes in Newton's possession. Newton's later excerpts from Borel (Stanford M132, container 2, folder 3, ff. 1r–1v = *Stanford B*; Keynes MS 13, f. 3r = *Keynes 13*) are incorporated into our annotations by reference only (for items repeated from *Stanford A*, without considering variants of quotation) or by detailed citation (for items occurring in the later excerpts only or repeated out of the original order; in these cases a cross-reference indicates the location within the actual order of quotation). Our numbering of Newton's items is as follows: [1] to [84] = set of excerpts copied from Borel; [Add 1] to [Add 6] = set of supplements added to Borel excerpts; item numbers indicated 'a' = additional items in *Stanford B*/*Keynes 13*, incorporated into the annotations following the transcript item numbers respectively. Reference to 'Borel' is made to his *Bibliotheca* edition of 1654 exclusively.

(A) Transcription of Stanford M132: Container 2, Folder 4: *De Scriptoribus Chemicis*

[*Item*:]   [*Borel*:]

De Scriptoribus Chemicis.

[*f. 1r*]

⟨B̶o̶r̶e̶l̶l̶u̶s̶ ̶v̶i̶d̶i̶t̶ ̶A̶⟩   [p. 2]

[1]

Abrahamum Judæum ↑MS↓ *Borellus vidit* cum fig.   [*A*]
et explicatione quadam idque Lutetiæ in 4.

[2]   Ægidius yᵉ [*sic*] ⟨v̶⟩Vadis. Dialog. int. naturam &   [p. 3]
fil. artis
        ⟨*in Th. Ch.*⟩

[3]   Ejusdem tab. diversorum metallorum in Th. Ch.   [p. 3]
        ⟨*in Th. Ch.*⟩ [*sic*]

[4]   Idem reperitur cum tractatu Penoti abditarum   [p. 3]
rerum Chemicarum Francofurti edit. 1595. 8.

[5]   Georgius Agricola de re metallica lib. 12 Basil. fol.   [p. 4]
16⟨2⟩1

[6]   Georgius Agricola Philopistius Germanus, Lapid   [p. 4]
Philosoph. Coloniæ 1521 [*sic*]

[7]   Alanus de Insulis MS in Angl.   [p. 5]

[8]   Albertus M. de rebus metallicis & minera⟨li⟩bus   [p. 5]
1. 5 1568 in 1 [*sic*] Colo. & 1541 a Gualt. Riff,
Argentorati

[9]   – Alchimiæ quam vocant Artisque metallicæ quam   [p. 7]
vocant doctrina certusque modus scriptis tum novis
tum veteribus tum veteribus [*sic*], duobus vol.
comprehensus Basil. per Pet. Pernam 1572 in primo
sunt 8 ⟨i̶n̶⟩tractatus in alio 13.
        ⟨*in quibus Margarita pretiosa (p. 50, 140*⟩

[10]   – De Alchimia opuscula quam plura veterum   [p. 7]
Philosophorum. In hoc libro sunt 9 tractatus
Francofurti 1550. 4.

[11]   – De Alchimia volumen Norimbergæ 1541. 4.   [p. 8]

[12]   × Alchemiæ veræ scriptores veteres Basil. 1561. fol.   [p. 8]
        ⟨*infra*⟩

[13]   Alphidij Secreta artis Chem. Lutetiæ apud D. de   [p. 10]
Lob⟨b⟩erie.

[14]   × Gaspar Amthor Chrysoscopion seu Aurilogium,   [p. 11]
Jenæ 1632. 4

[15]   Anaxagoræ conversiones MS in Angl.   [p. 11]

[16] – Elixiriorum varia compositio & modus cum   [p. 14]
      veteribus Alchemiæ Scriptoribus in fol. Basil. 1561.
      vizt in eo sunt 1 Epist vetus de Metallorum materia
      & artis imitatione. 2. De Lapidis Philosophici
      formatione Epilogus. ⟨ib.⟩⟨3⟩ Arcanum ut ex
      Saturno facias aurum perfectum ⟨4⟩ Caput de sale
      Alcali ib & in Theat. Ch. 5 An lapis valeat contra
      pestem. 6 Hist de argento in aurum verso. 7
      Tractatus de Marchasita ⟨–[?]⟩ ex qua fit
      Elixir.↑⟨8⟩ *Emanuelis liber 12 aquarum. Gratoroldi*
      [*sic*] *Defensio Alchemiæ*↓
        Vide etiam p. 19. 1. 19 & 23 1. 9. & p. 8 ⟨*& p. 85.*
      1. 20. p 107.⟩

[17] – Canones decem. Francof. 1614. 8 cum quibusdam   [p. 17]
      alijs opusculis Chemicis ⟨, *viz Heliæ speculo Alchemiæ*⟩
        ⟨*p. 113, 17, 171.*⟩

[18] Tractatus de secretissimo antiquorum Philosoph.   [p. 17]
      arcano L⟨–[?]⟩ipsiæ 1610↑*11*↓↓*12*↑. 8.
        pag 12, 17, 26, 221. Extat et in Th. Ch. [//]

[*f. 1v*]

[19] ☞Ars magica seu naturalis et artificiosa stupendos   [p. 17]
      effectus & secreta detegens. Francof. 1631 in 12.

[20] Introductio in vitalem Philosoph. Francof.   [p. 17]
      1623↑*7*↓. 4
        p. 1⟨*7*⟩*, 1*⟨*8*⟩*, 125.*

[21] Epistola cujusdam patris ad filium Lugd.   [p. 18]
      Bat 1601. 8

[22] Tractatus 7 a Justo a Balbian correcti et emendati,   [p. 18]
      ubi de Lap. Philos. 2 De minera Philosophi⟨c⟩a. 3.
      Compendium utile. 4 Tractatus parvus de mercurio
      Philos.
        p. 18, 40. Extat in Theatr Ch.

[23] Aqua aurea, Balsamum & Oleum nigrum cum   [p. 19]
      Alkemiæ [*sic*] scriptoribus 2 vol. 8.

[24] Arislai de lap. philos. in formam Dialogi opus cum   [p. 27]
      Denario Med. Penoti, Bernæ 1608. 8.
        p. 27. ⟪*Et Lond. 1692. 8.*⟫

[25] Aristotelis liber de practica lapidis in 4.   [p. 28]
        p 28.

[26] Arnaldi opera in fol. Lu⟨d⟩gd. 1520 & 1585 &   [p. 29]
      Francof. 1603 in 8.
        p. 31. 229.

[27]   – Ars Aurifera vol. 3 authores 48 antiquissimos   [p. 32]
        continens Basiliæ [sic] 1572 ⟨&⟩ in 8 & Franof. [sic]
        apud Aubrios.

[28]   ⟨A[?]⟩   [p. 32]
        Artefius et Flammel [sic]. Paris in 8 1609 & 4

[29]   Artefius de vita proroganda, qui se anno ætatis   [p. 33]
        1025 hunc librum scripsisse ait. Ex Naudæo.

[30]   A⟨ꜰ⟩vicennæ porta elementorum Basil. 8. 1572.   [p. 35]

[31]   Avicenna de tincturis metallorum Francof. 1530. 4.   [p. 35]

[32]   Augurellus carmine ⟨G⟩allico cum Hermetis 7 cap.   [p. 36]
        8 Pari⟨s⟩. 1626.

[33]   Rogerij Baconis ⟨s̶c̶r̶i̶p̶t̶a̶⟩ opuscula Francof. 1603   [p. 39]
        in 12   [B]

[34]   Rog. Bacon De l' admirable p⟨ui⟩ssance de l' Art   [p. 39]
        et de la Nature, [?] ou est traité de la pierre
        philosophale. a Lyon 1557. 8.

[35]   Jani Bacceri Thesaurus Chemicus experimentorum   [p. 40]
        certissimorum fide Justi Reinecceri Lipsiæ in 8 1609
        & Francofurti 1620 in 12.

[36]   Blemidas περὶ χρυσοποιίας antiquu⟨s⟩ liber Græcus   [p. 48]
        M.S. in Regia Bibl. Paris.

[37]   Petri ⟨♭⟩Boni [fferariensis] de secreto omnium   [p. 50]
        secretorum, [?] Dei dono, lib. in 8. Venetijs 1546.

[38]   – Cesalpinus de metallicis lib. 3 Norimberg. 1602 in 4. [p. 58]
        Idem de lapidibus.   [C]
            p. 58. [//]

[f. 2r]

[39]⟨–⟩Chymicæ Philosophiæ sex vetustissima scripta   [p. 60]
        Fran⟨c⟩of. in 8. 1605 apud Jo. Berne⟨t[?]⟩ [sic].
            p. 60.

[40]   – Philosophiæ Chemicæ ↑(sex)↓ vetustissima scripta   [p. 60]
        ex Arabico ser⟨m⟩one Latine facta Francof. 1605.
        in 8.

[41]   Christoph. Parisiensis⟨–[?]⟩ Elucidarium, ⟨&⟩   [p. 61]
        cujus pars 2ᵈᵃ est Arbor philosophalis. *La Medicine*
        [sic] *de troisieme ordre. Alfabet Apertorie [sic] de la*
        pratique envoy⟨e⟩ a s⟨on⟩ fils. La Somme. La
        Sommete en 4 parties (forte Summa minor) La
        Harpe. La Viole⟨t⟩e (alias Cithara sive
        Viole⟨t⟩ta⟨.⟩) Medulla artis. Particularia quædam.
        De Lapide Vegetabili
            pag. 61 et 182. ⟨Ab Italico Idiomate Gallica dat.
        15⟨84⟩⟩

[42]   – *Christophori Parisiensis libellus Chemicus a*
       Quercetano memoratus & a Combachio publicatus.
         pag. 181.

[43]   Jacques Cœur. ejus ⟨M. S. de⟩ Hieroglyphicorum   [p. 63]
       explicationem ⟨brevi promi⟩ & practicam quandam
       brevi in lucem mittendam *promittebat Borellus A.C.*
       1654.
         p. 63.

[44]   Oswaldi Crollij Basilica Chemica Francof.   [p. 68]
       1608⟨&⟩1609 ⟨et⟩ in 8 & 1647 in 4 & Genevæ in 8.
       Jdē Gallice. Idem cum comment Jo Hartmann⟨i⟩.

[45]   Crollius redivivus ab Anonymo de lap. philos. 1635   [p. 68]
       in 4 Francof. ⟨G⟩ermanicè.

[46]   Dardanus vel Dardanis, Antiquissimus   [p. 72]
       philosophus^↑Chemicus↓ seu Græcus seu Persa   [D]
       cujus opera Democritus a sepulchro ejus excepit &
       in Græciam transtulit et in ea commentaria scripsit.
         p. 72.

[47]   Democritus Abderita, Philosophus Græcus de arte   [p. 75]
       sacra seu Chemica cum Synesij et Pelagij comment.
       Interprete Dominico Piz⟨i⟩mentio Coloniæ apud
       Birkmannum 1574 in 16.

[48]   Democritus græce scripsit de Chemia ex Gesnero   [p. 75]
       qui in Bibliotheca sua ait, extare adhuc illius scripta
       Chemica apud Joh. Deé [*sic*] M.S. græce, sub hoc
       titulo περὶ χημείας id est yᵉ [*sic*] arte sacra cum
       commentarijs Synesij et Stephani.
         p. 75.

[49]   Arthur⟨i⟩ Dee ffasciculus Chemicus Paris 1631 in   [p. 74]
       12

[50]   *Prædictus* [*sic*] Democriti Physicorum et   [p. 75]
       mysticorum liber cum comment Synesij et
       Stephan⟨i⟩ MS apud D. Elichmannum Med. Lugd.
       Bat.

[51]   – *Democriti liber de arte sacra seu de rebus naturalibus*
       et mysticis cum comment. Synesij et Pelagij extant

[*f. 2v*] *latine cum Ant. Misaldi centu⟨–rijs⟩ [//]rijs 9*
           memorabilium Coloniæ, 1574 in 16 ap.
           Birkma⟨nn⟩.
             p. 75, 183, 217 [sic].

[52]   Sanctus Dominicus Alberti Magni magister   [p. 79]
         p. 79.

[53]	Cornellius Dre⟨p⟩belius seu Dreppels Belga de natura Elementorum liber cum Epist de mobilis perpetu⟨æ⟩i inventione e Belgico idiom. in Lat. versa a Petro Laurembergio Hamburg. 1621 in 8. & Francof. 1628. Reperiu⟨ntur⟩ et ejus opera in 12 cum Cosmopolita et ⟨a⟩Augurello, Genevæ impressa. *p. 82.*	[p. 81]
[54]	Lud. Dolce de Gemmis, Italice 1566 in 8 ⟨¥⟩Venetijs.	[p. 82]
[55]	Dun⟨s⟩tani Arch. Cant. Liber secretus de lapide majori M.S. in Anglia apud D. Mayernum Turquetum Medicum. *p. 83.*	[p. 83]
[56]	Christophorus Encel⟨iu⟩s de re metallica⟨–[?]⟩ 1557 in 8 Francof.	[p. 85] [*E*]
[57]	Ephrem [*sic*] Syrus *scripsit* cont Alchemiam vulgarem. *p. 86.*	[p. 86]
[58]	Epistolarum Philosophicarum Medicarum & Chemicarum volumen Francofurti. 1598 in fol.	[p. 87]
[59]	⟨P.J.⟩ Fabri annot *in 13 lib. chem* Tolosæ 1646 in 8.	[p. 91] [*F*]
[60]	G. Fabritius [*sic*] de ⟨m⟩etallicis Tigu⟨r⟩i. 1565. in 8.	[p. 92]
[61]	G. Fallopij lib. 3 de secretis in 4.	[p. 92]
[62]	Idem de mineralibus et metallis &c.	[p. 92]
[63]	*Joh. Chrysippi Faniani* farrago Philosophorum, hoc est varij processus et sententiæ h̄o̅r perveniendi ad lap. phil. Amberg. 1611 in 8. [*sic*]	[p. 93]
[64]	Fernelius lib. de abditis rerum ↑causis↓ cap. 18 1. 2 de lap. philos. optime disserit & se caluisse eum testatur processumque ejus docet. (*p. 93*)	[p. 93]
[65]	Flamelli Hierogl. *cum Artephio* ⟨e⟩*t Synesio in* [*sic*] *Lat. per* P. Arnaud. 1612 in 4. *p. 95.*	[p. 95]

[66] Le Desir desire *or* [*sic*] Tresor de Philosophie de N.   [p. 96]
Flammel [*sic*] dit autrem⟨e⟩nt, Le liure des 6
paroles in 8. avec divers autres Traitez ↑Propria
Flamelli manu. exaratus *putatur et*↓ ⟨E⟩xtat apud
D. Franciscum de Ger⟨z⟩an, dominum de
Sou⟨cy⟩ qui etiam aliam Flamelli Scheat [*sic*] habet
Paris.
     p 96.

[67] Le grand Esclairisment [*sic*] de Nich. Flammel [*sic*]   [p. 96]
in 8. Portio est libri, Christophori Parisiensis sub
hoc titulo editi.

[68] Joh. Nicholaus [*sic*] Furichius de lap. philos. seu   [p. 100]
Chyseidos [*sic*] lib. 4 cum ejusdem annot. 1622 in 8
et 1631 in 4 Argentorati.

[69] Garlandi *opera* cum Ventura extant in 8. Basil.   [p. 101]
1⟨5⟩60. *vizt* Dictionarium Chem. Compendium   [*G*]
Alchymiæ. De tincturis. In Arnaldum. De salibus.

[70] Tractatus absolutissimus ubi agitur de Gemmis,   [p. 109]
[*f. 3r*] de Alchimisticis [//] de Thesauris &c Francof. 1627
in 4 apud Emmelium.
     ⟨~~Harmoniæ Decas tertia F~~⟩

[71] Harmoniæ Chem. ⟨~~Decas tertia~~⟩ Decades   [p. 111]
↑2⟨~~vel~~⟩↓ ⟨3⟩ Francofurt. in 8. 162⟨5⟩   [*H*]
     ⟨~~Decade tertia eJaree.~~⟩
     p. 112, 196

[72] ⟨H⟩eliodorus↑Græcus↓ libum [*sic*] composuit   [p. 113]
metro Jambico de auri factura quem Theodosio
Imperatori dicavit, ex Nicephoro Callisto, Georgio
Cedreno & Gratarolo.
     p. 113.

[73] Hermetis cap. 7 Lipsiæ 1600 in 8 & in Th. Ch. cum   [p. 115]
Scholijs & Gallicè per Gabr. Joly 1626 a Paris *cum*
Au⟨g⟩u⟨r⟩ello et Epistola Trevisani ad Tho.
Boloniensem [*sic*].

[74] Philosophi cujusdam antiquissimi libellus de   [p. 123]
Alchimia Argentorati 1566 in 8.   [*I*]

[75] – Authores varij Chemici Venetijs 1541 in 4 et in 8.   [p. 124]

[76] – Lullij Codicillus seu ⟨C⟩antilena ad Regem Angl.   [p. 142]
Colon. 1576 in 16. et 1563 in 8.   [*L*]

[77]   De mineralibus scripser⟨e[?]⟩ Albertus, Agricola,   [p. 161]
       Cæsalpinus, Aldrovandus, Cæsius, *Guil. Fabritius*   [*M*]
       [*sic*], *Aubertus, Quercetanus, Encelius* ⟨&c. p. 159.⟩
       Gesnerus, Vincentius, Isaacus Hollandus, Guidius,
       Mylius.
          p 159, 161.

[78]   Morienus Hanoviæ ⟨1565⟩ in 4 & 1593 in 8. &   [p. 163]
       Parisijs 1559, 1⟨564⟩ & 1574 in 4 & MS in Bibl.
       Bo⟨d⟩leiana. ⟨*Castrensis Præfat in Morien.*⟩

[79]   Nazari concordanza di Philosoph⟨i⟩ 1599. in 4 in   [p. 166]
       Brescia Item della transmutatione metallica.   [*N*]

[80]   Nollij [*sic*] *opera.* [?]   [p. 167]
          p 167

[81]   Samuelis Nortoni Angli *opera varia*   [p. 168]
          p 168.

[82]   Joh. Rhenani aureus tractatus ⟨F⟩ libris tribus &c   [p. 196]
       Francofurt 1623 in 4   [*R*]

[83]   – Mart. Rullandi Lexicon Chemiæ, et Problemata   [p. 203]
       Chemica cum tractatu de L. P. Francofurti in 8.
       1607

[84]   Interpretation des secrets Hebrieus [*sic*] Chaldees & [p. 264]
       Rabi⟨n⟩s du Prince Dorcas Philosophe Ethiopien   [*Corol-*
       pour augmente⟨[?]r⟩ ⟨–[?]⟩ 1′ or & ⟨–[?]⟩ l'argent   *larium*]
       a dix p⟨er⟩our cent profit [*sic*] chaque semaine a
       Paris chez Pierre Ramier 1622 in 8 Par de Merac.

[Add 1]   ⟪*Catalogue of Chem. Books in 3 parts*⟫   [*not in Borel*]
[Add 2]   ⟪*Five Treatises*⟫
[Add 3]   ⟪*Starkeys Pyrotechny.*⟫
[Add 4]   ⟪*––– his liquor Alkahest*⟫
[Add 5]   ⟪⟨*The*[?]⟩ *Boyle*⟨*s*⟩↑*Tracts*↓ *of the growth of*
          Metals in their ore⟫
[Add 6]   ⟪*The Chemical weddin* [*sic*] *transla*⟨*d*⟩*ted by M^r*
          F.⟫

(B) Annotations to *De Scriptoribus Chemicis*

[1]   Also *Stanford B.*
[2]   *y^e*] de (Borel); *in Th. Ch.*] (not in Borel; see annot. to [3]/[4]).

[3] *in Th. Ch. in Th. Ch.]* in Th. Ch. (Borel; applying to [2]/[3]); see annot. to [4].

[4] Repr. in *TC* II; containing [2]/[3].

[5] = *HL 20.*

[6] *Georgius]* G. (Borel; *recte*: D[aniel].); *1521]* 1531 (Borel: *recte*: 1631); also *Stanford B* ('1531').

[7] Not identified; a short alchemical tract, 'Dicta Alani' (originally in German, Latin transl. publ. 1599 [*HL 717*], repr. in *TC* III) has been falsely ascribed to Alanus de Insulis (Alain de Lille).

[8] *1568]* (Borel; *recte*: 1569); *in 1]* (illegibly printed in Borel; *recte*: in 12). W. H. Ryff's edn of 1541 was publ. along with *Raimundi Lullii de secretis naturæ...libri II*; = *HL 995* (see also [76a] and text, note 74).

[9] *tum veteribus tum veteribus]* tum veteribus (Borel); = *HL 1624* (vol 1 only; see text, note 69); also *Stanford B*. This is an abridged and altered edn of G. Gratarolo's collection [12]/[16], also containing [23]. Newton's ref. to 'Margarita pretiosa' is misleading as this treatise (first publ. 1546; see [37]) was ed. in 1572 separately by M. Toxites and repr. in *TC* V.

[9a] Add. item in *Stanford B*:

> 'A Revelation of yᵉ most secret spirit declaring the most secrets of Alchymy Lond. 1532. 8.' (Borel, p. 7; *recte*: 1623).

[10] = *HL 493* (Cambridge University Library, Keynes Collection; vol. 1 only); also *Stanford B*. See [25]/[31].

[11] Also *Stanford B*. See [75].

[12] Also *Stanford B* (marked 'A'). Identical with [16].

[13] Also *Keynes 13.*

[15] Also *Stanford B.*

[16] *Gratoroldi]* Grataroli (Borel, p. 107); also *Stanford B* (marked 'A') as: 'Veræ Alchemiæ artisque metallicæ, citra ænigmate doctrina certusque modus scriptis tum novis tum veteribus nunc primum & fideliter majori ex parte editis comprehensus fol. Basil. 1561.' (Borel, p. 19; this is the correct title of [12]/[16]; cf. [22a]). See also [9] and, on the composition of this item, text, note 48.

[17] = *HL 1208*; also *Stanford B* as: 'Opuscula diversorum autorum Chemica partim e veteribus MSS eruta partim restituta Francofurti 1614 in 8.'; also *Keynes 13* (Borel, p. 171; correct title of [17]; cf. [81a]). See also [28].

[18] *1610/1611/1612]* (Borel, pp. 17/12/26, 221; *recte*: [1611]); also *Stanford B* ('1612', without ref. to *Theatrum chemicum*). The year of publ. was originally enciphered by a chronogram; repr. in *TC* V.

[19] = *HL 86.*

[20] *1623/1627]* (Borel, pp. 17, 25/18; *recte*: 1623); also *Stanford B* ('1623'; see [75a]). Ed. by – and usually ascribed to – J. E. Burggrav.

[21] Ed. by N. Barnaud; repr. in *TC* III.

[22] Publ. Leyden, 1599; repr. in *TC* III.

[22a] Add. item in *Stanford B*:

'Veræ Alchimiæ [...]' (Borel, p. 19; see [16]).

[23] *Alkemiæ]* Alchemiæ (Borel). Anonymous tract contained in [9].

[24] English transl. ('Lond. 1692.') not *HL*, but in Newton's possession; see text, note 45.

[25] Publ. with [10], repr. in [27].

[26] Also *Stanford B*. The edns. '1520' and '[Basileæ] 1585' both are Arnald's *Opera omnia*; the edn of 1603 is entitled: *Omnia, quæ extant, opera chymica*, ed. by H. Megiser. Its second part, *Speculum alchymiæ*, was repr. in *TC* IV, the tracts composing its first part had been published with [27], and also in Philip Ulstad's *Cælum philosophorum*..., Lyon, 1572 (also [Strasburg], 1630; *HL 1651, 1652*).

[27] *Basiliæ]* Basileæ (Borel); *Franof.]* Francof. (Borel); = *HL 90, 91*: Trinity College, Cambridge (third edn, Basel, 1610, vols. 1–3), and Royal Society Library (*ibid.*, vols. 1–2 only). The edns mentioned by Borel ('1572' and 'Frankofurti [*recte*: Basileæ, 1593, printed for De Marne & Aubry, Francfurt]') had been publ. in two vols. only, the 1572 edn with a slightly differing title: *Auriferæ artis...antiquissimi authores, sive Turba philosophorum*, instead of: *Artis auriferæ...volumina duo, quæ continent*...See also [25], [26], [31], [34], [78].

[28] *Artefius]* Artephius (Borel); *Flammel]* Flamel (Borel). For the edn in '4to' see [65]; the '1609' edn is the Latin *Artefii Clavis*... only, repr. in [17] and in *TC* IV.

[30] Publ. as: *Artis chemicæ, principes, Avicenna atque Geber*...(1572); correct title likewise in Borel, p. 35. Only edn!

[31] *1530]* (Borel; *recte*: 1550 probably). Publ. with [10], repr. in [27].

[32] Identical with '1626' edn in [73].

[33] = *HL 110*: Trinity College, Cambridge; also *Stanford B* (marked '*').

[34] Latin versions repr. in [27] and, with annotations by J. Dee, in *TC* V.

[35] *fide Justi]* (Borel; *recte*: Fidejusti). Ed. by Baccerus, but comp. by F. Reinneccerus. A '1620' edn cannot be proved but in the same year a reimpr. in 24mo of Roger Bacon's *De arte chymiæ scripta* was publ. at Frankfurt, entitled: *Thesaurus chemicus*...

[37] *, Dei dono]* Dei, Dono (Borel); also *Stanford B*; also *Keynes 13*. This is identical with *Pretiosa margarita novella*..., ed. by J. Lacinius (1546); see text, note 79, and [9]. Newton copied the surname, further emphasized by use of his significant [] brackets, from Borel's first entry on P. Bonus.

[39] *sex]* (Borel, *recte*: IV); *Bernet]* Berner (Borel). Identical with [40].

[40] *sex]* (see [39]); = *HL 1301*: Trinity College, Cambridge. Identical with [39], but likewise listed twice by Borel; tracts I & IV repr. in *TC* V.

[41] *Medicine]* Medecine (Borel); *Apertorie]* Apertoire (Borel); *1584]* 584 (Borel, p. 181, misprinted). Contraction (and translation) by Newton of Borel's extensive entries in Latin and French, omitting the following indications: the French versions were known in MSS only, the 'Elucidarius' [*sic*] was publ. in German (identical with [42]).

[42] Also *Stanford B*; cf. [81*a*]. Ref. to the German 'Elucidarius Christophori Parisiensis', as publ. in *Vier auserlesene Teutsche chemische Büchlein...*, Kassel, 1649, which must not be confused with L. Combach's *Tractatus aliquot...* (see text, note 12). A Latin transl. of this tract was publ. in *TC* VI.

[43] No impr. identified.

[44] = *HL 462+743*: Trinity College, Cambridge (edn Frankfurt, 1647, with J. Hartmann's *Praxis chymiatrica*).

[47] Also *Stanford B*; cf. [81*a*]; also in Keynes MS 13, f. 3v (see text, note 27). Identical with [51].

[48] *Deé]* Dée (Borel); *y^e]* de (Borel).

[49] = *HL 497*. Out of turn and obviously inserted by Newton while copying the entries on Democritus (see also his ref. 'prædictus' in [50]).

[51] *217] recte*: 215. More detailed repetition of [47], entered by Borel under 'Pelagius' and 'Synesius' respectively.

[53] Also *Stanford B*. The Geneva edn 'cum Cosmopolita et Augurello' is that of Sendivogius' *Novum lumen chymicum...* (1628); a reissue in 1639 (*HL 1192*: British Library) had been publ. without Drebbel's tracts. A French transl. of these was publ. with *Divers traitez...* (1672; see [Add *a*]).

[54] Ref. to C. Leonardi's *Speculum lapidum...* transl. by Dolce.

[55] Also *Stanford B*; see also text, note 83.

[56] Also publ. 'Francof[urti]', 1551. Not *HL*, but probably bound with *HL 995*; see [8], [76*a*], and text, note 74.

[57] *Ephrem]* Ephræn (Borel).

[58] Ed. by L. Scholtz.

[59] Latin version of Basilius Valentinus' *Currus triumphalis antimonii...* (explicitly mentioned by Borel but abbreviated by Newton), augmented by several other tracts, ed. by P. J. Fabre. Not *HL*, but probably in Newton's possession; see text, note 89.

[60] *Fabritius]* Fabricius (Borel). Publ. with C. Gesner's *De omnium rerum fossilium genere... libri aliquot...* (1565).

[61] Numerous edns, but almost all in 8vo.

[62] Ref. to Falloppio's *De medicatis aquis atque fossilibus tractatus...*, Venice, 1564 (also 1569), in 4to probably.

[63] Also *Stanford B* (without Faniani's name). Misleading ascription by Newton; in Borel the *Farrago* (ed. by A. Brentz) follows directly upon the Faniani entries; repr. in *TC* IV, as: 'Variæ philosophorum sententiæ...'.

[64] Also *Stanford B*. First publ. in 1548 at Paris; went through numerous edns, also with Fernel's *Universa medicina*.

[65] *in Lat. per]* traduit de Latin de (Borel; misleading transl. by Newton); = *HL 1309, 1310*: Trinity College, Cambridge (both copies third enlarged edn, Paris, 1682). Identical with edn in '4[^{to}]' in [28], although this is not a French transl. of Artephius' *Clavis* (1609, see [28]) but of his 'Liber secretus'; Flamel's and Artephius' tracts repr. in *Bibliothèque des philosophes* (*chymiques*)..., Paris, 1672–78 (*HL 221*: Trinity College, Cambridge [vol. 1 only, but originally both vols. in Newton's possession]; see text, note 59); an English transl. was publ. in W. Salmon's *Medicina practica*..., London, 1692 (*HL 1439* [edn 1707]).

[66] *or]* ou (Borel); *Flammel]* Flamel (Borel); *Scheat]* Scheád (Borel); also *Keynes 13*. Publ. with the French transl. of Sendivogius' works: *Cosmopolite, ou Nouvelle lumière de la phisique naturelle*...(edns Paris, 1618 and 1628/29), but not included in the edn in Newton's possession (*HL 445* [dern. éd., Paris, 1691]).

[67] *Esclairisment]* Esclaircissement (Borel); *Flammel]* Flamel (Borel); also *Keynes 13*. Publ. Paris, 1628.

[68] *Nicholaus]* Nicolaus (Borel); *Chyseidos]* Chryseidos (Borel).

[69] = *HL 649*: Trinity College, Cambridge. This '1560' edn is Garland's *Compendium alchimiæ* only, including his named tracts; it was reissued with L. Ventura's *De ratione conficiendi lapidis philosophici*..., Basel, 1571.

[70] Also *Stanford B*. Misprinted in Borel; *recte*: to be read as continuation of previous line, 'Ioh. Guidius de Mineralibus [Tracatatus absolutissimus...]'. On Newton's cancellation see [71].

[71] *Decades 2/3]* 3 (Borel, pp. 112/196; *recte*: 2); = *HL 740*: Trinity College, Cambridge (2 parts).

[72] *libum]* librum (Borel).

[73] *1600]* (Borel; *recte*: 1610); *Boloniensem]* de Bologne (Borel; also 'de Bononia', p. 49). Joly's edn was publ. as: *Trois anciens Traictez de la Philosophie naturelle*...; see [32]. The 'Septem tractatus seu capitula' had been publ. previously in *Ars chemica*..., Strasburg, 1566 (*HL 84, 85*: Trinity

College, Cambridge), a French transl. was extant in *Bibliothèque*...(see annot. to [65]), and an English version in *Medicina practica* (see annot. to [65]); Augurello's poem in Latin was repr. in [12]/[16], and in *TC* III, a slightly different version publ. with Sendivogius' *Novum lumen chymicum*...(edns Geneva, 1628, 1639; see annot. to [53]), and in *Bibliotheca chemica contracta*..., Geneva, 1653 (*HL 220*: Trinity College, Cambridge [edn Geneva, 1673]); the Latin text of the 'Epistola' by Bernhardus Trevisanus (i.e. Trevirensis) was repr. in [27], an English transl. publ. in *Aurifontina chymica*..., London, 1680 (*HL 103*: Trinity College, Cambridge).

[74] Also *Stanford B*. This is the treatise ascribed to Zadith (Senior), publ. Strasburg, *s.d.* (probably simultaneously with *Ars chemica*; see annot. to [73]), repr. in [39]/[40], and in *TC* V.

[75] Also *Stanford B*. Not identified, but probably misleading ref. to [11]: = Nuremberg, 1541, 4to (or to [8]: = Strasburg, 1541, 8vo), or, more probably, to an edn of the works of Geber, augmented by other tracts of several authors: = Venice, 1542 [*sic*], 8vo.

[75a] Add. item in *Stanford B*: see [20] (Borel, p. 125).

Add. item in *Keynes 13* (from Borel's division 'K'; no items copied in the *Stanford* lists):

'Desiderantur Henrici Kunrath Amphitheatrum æternæ sapientiæ Hamburgi 1611 in fol. Item Magnesia Catholica Philosophorum. Item de Athanore, igne Sophorum, Apocalypsi &c vel Ignis Magorum' (Borel, p. 130; *recte*: Hanoviæ [Hanau], 1609 [1602]).

[76] *1576]* (Borel; *recte*: 1572, or misprint of 1567; but in that year only other works by Lull had been publ. at Cologne); = *HL 997*: Trinity College, Cambridge (edn 1563; see Borel p. 267: 'Corollarium').

[76a] Add. item in *Stanford B*:

'Lullij opera quædam edita Norimberg. 1546↑& 1625↓, Basileæ↑1561 &↓1572 [= *HL 1000*: Trinity College Cambridge], Coloniæ↑1567↓ [= *HL 996, 998*] & 1592. Argentinæ, 1597 [Borel; *recte*: 1598 probably], Argentorati 1541 [= *HL 995*], Parisijs 1627. Lugd. Bat. 1502 [Borel; *recte*: 1602].' (Borel, pp. 141–7; underlining by Newton).

A similar listing was set up by Newton at the end of an inventory of Lull's works extant in his library (see text, note 74): 'Desiderantur opera impressa Norimbergæ 1546, 1625 Argentorati ⟨1541⟩ Basileæ 1561, 1600 Coloniæ 1592, 1576 Parisijs 1627'. – From these the '1502' (i.e. 1602) edn may ref. to *Secreta alchimiæ magnalia D. Thomæ Aquinatis*..., ed. by D. Brouchuisius (incl. Lull's 'Clavicula & Apertorium'), repr. in *TC* III; for the cancelled '1541' edn see [8]. All other issues (not in Newton's library) are described in detail by Borel (pp. 141ff, 267).

[77] *Fabritius]* Fabricius (Borel). Contraction/selection of two entries in Borel, pp. 159, 161, beginning 'De Metallis scripserunt [...]', and 'De Mineralibus, innumeri scripserunt, quorum præcipui sunt [...]' respectively; see also [8] ('Albertus'), [5] ('Agricola'), [38] (C[a]esalpinus'), [60] ('Fabritius'), [56] ('Encelius'), [70] (i.e. 'Guidius'). See also *HL 331* ('Cæsius'), *539* ('Quercetanus'), *787* ('Hollandus'), *1136* ('Mylius').

[78] The tract by Morienus was also publ. in [27], a French transl. in *Bibliothèque*... (see annot. to [65]).

[79] Also *Stanford B*. 'Della transmutatione metallica' is the main title; 'Concordanza' is an appendix, comprising four works by Arnald of Villanova.

[80] *Nollij]* Nolij (Borel). Contraction of a detailed listing of titles in Borel.

[81] = *HL 1180–87*: Trinity College, Cambridge (eight tracts bound in 1 vol.; publ. Frankfurt, 1630).

[81a] Add. items in *Stanford B*:

'Opuscula [...]' (Borel, p. 171; see [17] also *Keynes 13*); – 'Christophori Parisiensis [...]' (Borel, p. 181; see [42]) – 'Democritus [...]' (Borel, p. 183; see [47]/[51]) – 'Poterij Veredarius Hermeticus secreta secretissima detegens Francof. in 8 1622 [= *HL 1341*]. Ejusdem Philosophia pura materiam et processum ↑totum↓ apertissime detegens Francof. in 8 [publ. 1619]. Ejusdem Compendium Philosophicum in Trevisanū, Basil Valentinū &c materiam & processum demonstrans. in 12 1610.' (Borel, p. 189) – 'De Quintessentia [Quinta essentia (Borel)] scripserunt Vittestein, Lullius, Arnaldus, Drebel, Ripleus, Ulstadius, Rupesissa [Rupescissa (Borel)], Savonarola, Carerius, Guntherius [Guntherus (Borel)], Gerhardus &c' (Borel, p. 193; copied in full).

The 'Compendium' of M. Potier was listed on the title-page of *Elucidatio secretorum*... [Frankfurt, 1610, second edn; in German], but obviously not publ. For works on the 'Quinta essentia' see also [76]/[76a] ('Lullius'; spurious version of the work of 'Rupescissa'), [26] ('Arnaldus'), [53] ('Drebel'), [9] (vol. II incl. a tract by 'Savonarola'); see also *HL 1405–1406* ('Ripleus'), *1651–1652* ('Ulstadius'; see annot. to [26]), *667* ('Gerhardus').

[82] Also *Stanford B*. Repr. in his *Opera chymiatrica* (*HL 1397*: Trinity College, Cambridge [edn Frankfurt, 1668]).

[83] = *HL 1426; 1427*: Trinity College, Cambridge; Stanford University Library; also *Stanford B* ('Problemata' only, marked 'x').

[83a] Add. item in *Stanford B*:

'Smolij [Smollij (Borel)] Manuale venerandæ antiquitatis mysterium in 4 & in 24' (Borel, p. 212; publ. Hamburg, 1609, 1610).

Add. item in *Keynes 13* (from Borel's division 'V'; no items copied in the *Stanford* lists):

'Vellus aureum Trismosini.' (Borel, p. 226; publ. Rorschach, 1598(–99), add. vols. Basle, 1604; in German).

Add. item in *Stanford B*:

'L'Idee parfaite de la Philosophie Hermetique ou l' abrige [*sic*: abregé (Borel)] de la Theorie & Practique [*sic*: pratique (Borel)] de la Pierre des Philosophes. Par M. I. Colesson [Collesson (Borel)] a Paris 1631 in 8.' (Borel, p. 240: 'Appendix'; Latin transl. publ. in *TC* VI).

[84]   *Hebrieus]* Hebrieux (Borel); *profit]* de profit (Borel).

[*Additional*]   (Late entries not based on Borel, completing *Stanford A*)

[Add 1]   Publ. London, 1675; = *HL 442*: Babson College, Mass. (edn [1688]; reissue of 1675 edn with continuation; see text, notes 16 and 17).

[Add 2]   Publ. London, 1652; = *HL 621*.

[Add 3]   Publ. London, 1658; = *HL 1553*.

[Add 4]   Publ. London, 1675; = *HL 961* (edn 1684).

[Add 5]   Publ. London, 1674; = *HL 274*

[Add 6]   Publ. London, 1690 as: *The hermetick romance, or ...*; = *HL 1422*; see text, note 42).

[Add *a*]   Add. items in *Keynes 13*, not based on Borel:

'Le Cabinet Chimiques [*sic*]' (not identified; probably identical with *Bibliothèque* ... [publ. 1672–78]; see annot. to [65]) – 'La Tourbe Françoise la Parolle delasseé, Drebellius, L'ancienne Guerre des Chevaliers, composent ensemble une volume imprime a Paris chez d'Houry' (publ. as: *Divers traitez de la philosophie naturelle* ..., Paris, 1672; = *HCL 531*: Trinity College, Cambridge; see [53]).

7 Beyond the dating game: watermark clusters and the composition of Newton's *Opticks*

ALAN E. SHAPIRO

Acknowledgements

David McKitterick, who was at Cambridge University Library when I began this project and is now Librarian of Trinity College, Cambridge, encouraged me to carry it out and provided valuable guidance in what was an entirely new area to me. Research on watermarks puts unusual demands upon libraries and requires much assistance. The staff at Cambridge University Library was extremely accommodating, especially Godfrey Waller of the Manuscripts Room and the Photography Department. Peter Jones, Librarian, and Michael Halls, Modern Archivist, of King's College Library in Cambridge were also encouraging and very cooperative. Timothy Hobbs of Trinity College Library, and Mr Hawkins of the Public Record Office in Kew, where the Mint papers are located, should also be singled out for their assistance. I am grateful for the cooperation from the staffs of the Jewish National and University Library, Jerusalem, the Royal Society of London, the Bodleian Library, Oxford, the Beinicke Library at Yale University and Corpus Christi College Archive in Oxford. Without the detailed help provided by Betty Jo Dobbs and Karin Figala I could not have studied the alchemical papers. Simon Schaffer aided me with the theological papers. Dennis White, Department of Mathematics, University of Minnesota, provided me with much needed help in developing the probability arguments. Generous support from the John Simon Guggenheim Foundation, the National Science Foundation, and the Rowland Foundation allowed me to carry out this research.

D. T. Whiteside is renowned for his ability to date Newton's manuscripts by their handwriting alone. In the introduction to his magisterial edition of Newton's *Mathematical Papers* he affirmed that, 'we came ultimately to trust our ability to fix the date of composition of an unexamined portion of autograph manuscript accurately by sight to within half a dozen years (and sometimes even more narrowly still)'.[1] When I was transcribing the manuscripts for my own smaller edition of Newton's *Optical Papers*, I was confronted with the task of identifying, ordering, and dating the papers. I did not, however, develop a similar confidence in my ability to date the papers so narrowly by the handwriting throughout all of the more than sixty years from which they survive. Moreover, having observed the vigorous debates amongst

---

[1] D. T. Whiteside, ed., *The Mathematical Papers of Isaac Newton*, 8 vols. (Cambridge University Press, 1967–81), vol. 1, p. xi; henceforth cited as Newton, *Math Papers*.

eminent and experienced Newtonian scholars over the dates of particular documents, I sought a more objective means of dating.[2] Watermarks offered a possible solution to this problem of precise, objective dating. Accordingly, I set out to examine as many dated or readily datable manuscripts as possible amongst the thousands upon thousands that survive. Alas, it turned out that watermarks cannot provide the magic key that unlocks the date of Newton's papers to two or three years, as I had hoped.

To present a broad summary of my investigation, Newton used paper with the same watermark anywhere from a few months to fifteen or more years, but generally for two to three years. Since most of Newton's papers are undated, if we add two to three years on either side, we can reasonably safely assign an interval of eight or nine years to a typical watermark. The dates of his scientific papers can for the most part be independently assigned to a narrower interval than that, so that it might seem that watermarks are not particularly helpful. Nonetheless, it turns out that they are still a very useful tool to add to the historian's repertoire. When used cautiously they are, I believe, generally more reliable and definitely more objective than dating a document by its style or by the purported stage of intellectual development that it reflects. Since Newton's theological and alchemical studies were largely carried out in private and unpublished in his lifetime, much uncertainty – sometimes greater than that of the period of a watermark – often surrounds the date of these writings. We must also bear in mind that the historian generally has other information available besides handwriting and watermarks. Internal references such as to a published book or a public event, or external ones, such as in later recollections or other documents, can also assign a period during which it was composed. To be sure, all the evidence must agree, but when the evidence is slight, the conformation – as occurs in the great majority of cases – by the independent evidence of watermarks adds significantly to the confidence in the

---

[2] The manuscript 'De aere et aethere', which is discussed in note 19, has been dated to the beginning of the 1670s by Whiteside (personal communication 31 July 1979) and to the end of that decade by Westfall, with the Halls opting for an intermediate date; University Library, Cambridge [henceforth ULC], Add. MS 3970, ff. 652/3 (a solidus in folio numbers indicates a bifolium). The essay 'De gravitatione et aequipondio fluidorum' has been placed in the late 1660s and early 1670s by most scholars; A. Rupert Hall and Marie Boas Hall, eds., *Unpublished Scientific Papers of Isaac Newton: A Selection from the Portsmouth Collection in the University Library, Cambridge* (Cambridge University Press, 1962), pp. 89–156. Dobbs and others, however, have argued from handwriting and intellectual development for a date about fifteen years later, or immediately before the *Principia*; see Robert Palter, 'Saving Newton's text: documents, readers, and the ways of the world', *Studies in History and Philosophy of Science*, 18(1987):385–439; and also A. R. Hall's paper in this volume. Since this is in a bound notebook, it cannot be dated by watermarks, as I explain below.

assigned date. When, however, watermarks conflict with the other evidence and suggest redating, they are still more significant.

Besides their common use for absolute dating, a set of related papers or documents can be arranged and grouped by clusters of the same watermark. By a 'cluster' I understand simply successive sheets with the same watermark; and by 'clusters of the same watermark' or a 'set of clusters' I mean two or more clusters, or sets of successive sheets, with the same watermark that either are separated in the same manuscript by sheets with other watermarks or, less commonly, are in different manuscripts. In my own research on Newton's optics I have found cluster analysis to be, by far, the most useful application. In a long work like the *Opticks* (1704), which was written over the course of about fifteen years, the sets of clusters formed by a half-dozen watermarks can be used to follow and place in chronological sequence the composition of the various sections as they were completed, added, and replaced. The parts or clusters with the same watermark prove to have been written in the same time interval. By the same principle of clustering, as I call this result, if three *related* alchemical manuscripts, for example, have the same watermark, it is then probable that they were written in the same period. In dealing with watermarks it is always important to recognize that one is dealing with probabilities, and that, as in so much historical research, one must weigh all the evidence and use common sense and historical judgement.

In the first part of this paper I will briefly describe the formation of watermarks in the process of paper making and how one goes about studying them. In the next part I will provide some examples of the use of watermarks in dating (and redating) some of Newton's papers. In the following and longest part I will present a reconstruction of the composition of the *Opticks* and show the significant role that clusters of watermarks can play in such an analysis. I will conclude by justifying the use of the principle of clustering by a probabilistic argument that is based upon the nature of paper marketing in Newton's day and his habits in purchasing paper. The emphasis throughout will be on the optical papers, since they motivated me to undertake this study, and I know them best. This is not offered as a definitive investigation of the watermarks in Newton's papers. Rather it is a preliminary analysis that demonstrates that watermarks, and especially the principle of clustering, provide a valid historical tool for studying Newton's manuscripts that should be applicable to similar collections. While I judge the investigation preliminary, I have no intention of carrying it further. The amount of effort required for additional refinement does not seem commensurate to the information it would yield to the historian of science.

In order to understand how watermarks are actually used to study and date paper, we must first understand how they are formed in the manufacturing process.[3] A rectangular wire sieve fixed on a wooden frame (a mould) was dipped into a vat of pulp by the vatman, who dexterously shook the mould so that the fibers were uniformly distributed while the water was draining out. He then passed the mould to a co-worker, the coucher, who gently turned it over so that the wet paper could be pressed onto a piece of felt. The coucher returned this mould to the vatman and placed a new felt on the paper. Meanwhile the vatman was making another sheet of paper with the second mould of a pair, which he then passed to the coucher; back and forth, back and forth, several thousand times a day. When the pile of paper and felt was sufficiently high, it was put into a press to squeeze out the excess water; the paper was then hung out to dry. Since the same wooden rim (a deckle) had to be placed securely on each mould by the vatman in order to contain the pulp each time it was dipped, the two moulds were necessarily identical in size and formed a set. The sieve consisted of a grid of wires spaced about 25 mm apart parallel to the shorter side of the frame (chains) and about 1 mm apart parallel to the longer side. To identify the size, quality, or manufacturer of the paper watermarks – pictures or letters – were fashioned from wire and sewn to the grid. Because of the way in which the pulp is drained through, the grid and watermark are impressed into the paper, and they are clearly visible when it is held up to the light. The watermark on writing paper was placed so that when a sheet was folded in half to create a bifolium, it was centered on one of the halves. After about 1675 another design (the countermark) was generally placed on the other half of the sheet. Indeed, this practice became so widespread that after about 1685 virtually all the paper bought by Newton had a countermark. Another identifying mark (what I call an 'attendant' mark) was sometimes placed elsewhere on the half sheet with the watermark. On Newton's paper these were generally letters beneath the watermark; but on one common paper a letter appears in the lower corner; on another a small oval appears on the fold in the center of the sheet; and on still others it is integrated with the watermark itself. I have developed a compact notation to incorporate all this information: Letters preceded by a plus sign indicate the attendant mark, and those preceded by a solidus the countermark. Thus horn + HG/LL denotes a horn watermark with the letters HG below it and LL centered on the other half. When necessary I explain the watermark in more detail.

A consequence of the manufacturing process and the use of two moulds is that the same batch of paper will always contain two nearly identical water-

---

[3] On paper-making and watermarks see Philip Gaskell, *A New Introduction to Bibliography* (Oxford: Clarendon Press, 1979), pp. 57–77; and E. G. Loeber, *Paper Mould and Mouldmaker* (Amsterdam: The Paper Publications Society, 1982). Allan Stevenson, *The Problem of the Missale Speciale* (London: The Bibliographical Society, 1967) also contains a wealth of information on watermarks.

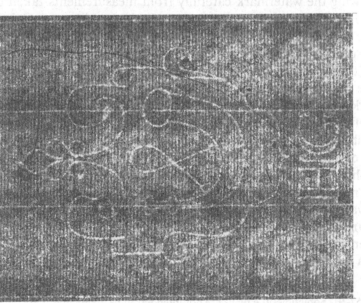

Figure 1. Beta radiographs of the twin watermarks horn + HG on ULC, Add. MS 3970, ff. 118 (on the right) and 121 (on the left), which are from Book II, Propositions 8 and 9 of the *Opticks*. Although they appear identical to the casual glance, the most notable difference between the two is that the watermark on the right is about 5 mm taller than its twin with the position of many other lines differing proportionally. The countermark is LL. (By permission of the Syndics of Cambridge University Library.)

marks, which are called twins. The mould-makers generally attempted to make the watermarks on the two moulds identical; but since they are handmade, the two always differ in some features that are detectable to the careful observer (Figure 1).[4] For example, one might be slightly larger than the other, as in the illustrated example; or one might just touch a chain, while the other misses it somewhat. Indeed, the chains are not identically spaced, and these alone can often serve to identify the twins. From studies of watermarks, it turns out that the situation is still more complex and that the watermarks may be quadruplets, sextuplets, and so on. This occurred when more than one set of moulds with the same watermark was ordered at the same time by a mill with more than one vat.

Casual description of watermarks, as I rapidly learned, is utterly inadequate to the task. One cannot simply record the type of watermark – arms of Amsterdam, horn, foolscap – and any attendant or counter marks, for there are too many watermarks of the same generic description. The most efficient way of proceeding is to make tracings of the watermarks (Figure 2), which is only possible when the papers are unbound, as is the case for most of Newton's manuscripts. The tracing can then be placed over other papers to determine if the watermarks are the same. To make tracings one needs, besides tracing paper and a pencil, a light box and a transparent sheet of plastic to protect the manuscript. For bound papers one must adopt the more time consuming procedure of drawing the watermark carefully from measurements taken off the manuscript. Just as Renaissance anatomists had to rediscover Galen's maxim that anatomy must begin with the skeleton, for 'as poles to tents and walls to houses, so are bones to living creatures, for other features naturally take form from them and change with them', so I had to be taught the significance of the chain lines and always to begin by drawing them.[5] With very similar watermarks, the chain lines and the watermarks' position relative to them virtually define the two. In making a tracing or drawing one obviously tries to choose a sheet with a clearly visible and well-defined watermark, but if a sheet is densely covered with writing or printing, discolored, or stained, then a beta radiograph can be ordered. The advantage of this method, which is available at most major research libraries, is that it provides an accurate image of the lines and watermark without any trace of the writing or printing on the sheet. Beta rays are absorbed only by the paper and not at all by the ink. The parts of the paper where the wires have left their impression are thinner and absorb fewer beta rays than elsewhere, so that these parts appear darker on the

[4] See Allan H. Stevenson, 'Watermarks are twins', *Studies in Bibliography*, 4(1951–2):57–91. Sometimes the twins are not intended to be identical but are mirror images or reversed.

[5] Galen, *On Anatomical Procedures*, trans., Charles Singer (London: Oxford University Press for Wellcome Historical Medical Museum, 1956), p. 2.

Figure 2. A tracing of the watermark English royal arms + 𝒜𝒥 on ULC, Add. MS 3970, f. 316, which is a sheet removed from Book III, Part I of the *Opticks*. The countermark is GLC. The original pencil tracing has been recopied in ink for reproduction here.

film. When a positive print is made, the lines and watermarks appear as bright lines, as in Figure 1. The only significant drawback of this method is its expense, which is somewhat more than a black and white photograph.

The problems of classifying and identifying watermarks are similar to those confronting the biologist. Watermarks fall into certain broad families, such as horns, pots, and various coats of arms, and there can be thousands of each one extending over centuries. Most of these are readily distinguishable from one another, but those fabricated by the same mould-maker in the same period can be quite similar in appearance. The addition of attendant and counter marks makes identification much simpler, for the arms of London with a countermark IV will form a different genus from that with EB. But even those with the countermark EB consist of more than one species in Newton's papers; for example, in one the E and B fall between a pair of chainlines, and in the other a chainline falls between the two letters. Similarly, the style and size of the letters in the attendant and countermarks may differ. A far more subtle problem arises from the existence of twins (and other siblings). Properly speaking each one is a different watermark, and – once one is aware of their existence – there is no difficulty in keeping track of their appearance. Since, however, each lot of paper contains a mixture of twins made simultaneously with a pair of moulds, they can be considered the same paper. By extension – and now loosely speaking – I sometimes consider the twins to be the same watermark for the purposes of dating and clustering. It is essential in studying watermarks to identify twins and other siblings, for otherwise you will judge what is in fact the same paper to be different papers.[6] Catalogues of watermarks

[6]   See Stevenson, *Missale speciale*, pp. 59–70, for a discussion of the meaning of the terms the 'same paper' and 'same watermark', and more generally for a brilliant example of the use of watermarks in the study of the production of a book.

Jerzy Zathey, the editor of Nicholas Copernicus, *Complete Works*, Vol. 1. *The Manuscript of Nicholas Copernicus' 'On the Revolutions'. Facsimile* (London/Warsaw/Cracow: Macmillan and Polish Scientific Publishers, 1972), did not avoid the various pitfalls encountered by the unwary in studying watermarks. In his introduction he identifies four different watermarks (designated as *C*, *D*, *E*, *F*) in the manuscript, and then further identifies three 'variants' of the first two (*C1a*, *C1b*, *C2* and *D1*, *D2a*, *D2b*) that he attributes to wear that occurred during the course of production of the paper; pp. 3–4, Plates I–III. These two sets of 'variants' in fact each consists of three different watermarks that bear only a family resemblance to one another, a serpent and a hand, respectively. Moreover, he fails to recognize the existence of twins, so that there are actually twelve different watermarks rather than the two (with two additional 'variants' of each) that he identifies. Since the standard catalogue of watermarks from this period similarly omits twins, the likelihood of matching and dating the watermarks is significantly diminished; see C. M. Briquet, *Les filigranes. Dictionnaire historique des marques du papier dès leur apparition vers* 1282 *jusqu'en* 1660. A facsimile of the 1907 edition, with supplementary material contributed by a number of other scholars, ed. Allan H. Stevenson, 4 vols. (Amsterdam: Paper Publications Society, 1968). Finally, the drawings of the watermarks in Copernicus' *Works* are not tracings but an artist's renderings of some ideal type, as can be seen by comparing them with the accompanying photographs. The drawings cannot therefore be used in any attempt to date the watermarks. Beta radiographs, which were developed by a Russian, D. P. Erastov, and first published in 1960, should have been included; see J. S. G. Simmons, 'The Leningrad method of watermark reproduction', *The Book Collector*, 10(1961):329–30 and plates. Although Noel M. Swerdlow warns that translators of the introduction, Edward Rosen and Erna Hilfstein, abbreviated and garbled the original Polish, the plates alone show that the editor must bear some responsibility for

for the seventeenth and eighteenth centuries such as those by Heawood and Churchill are of little value for the purpose of dating.[7] Their greatest weaknesses are that they contain only a very small proportion of watermarks manufactured in this period and do not represent twins. Heawood contains 4,078 watermarks and Churchill 578. Out of the more than 130 watermarks (excluding twins) that I have identified in Newton's papers, none were found in those compendia, though four were sufficiently close to suggest that they are twins. Moreover, I would conjecture that even if one found a perfect match it could be safely trusted only to an interval of twenty years or so, that is, ten years on either side of the given date, or about twice what I found in Newton's papers.

Personal habits in the use of paper vary greatly, and there is little economic incentive to use writing paper as efficiently and quickly as there is in book production. Consequently, I proceeded by making my own catalogue of watermarks in Newton's papers and their dates of use. The most reliable way to do this was first to utilize dated papers and correspondence and drafts of them, and then manuscripts and drafts of works whose dates are otherwise known by such evidence as the date of publication or references by Newton or others in correspondence. The editors of Leibniz' papers, I learned after I had completed this study, proceeded in a similar way and created an internal catalogue of watermarks appearing in his papers.[8]

The pattern of Newton's use of paper not unexpectedly appears to have changed in the course of over sixty years, but some (perhaps most) of this variance is undoubtedly an artifact of preservation. As is invariably the case, a greater proportion of papers survive from later than earlier years. With the passage of time early drafts and notes are discarded as they are judged to no longer be of use, while others are lost, destroyed, or damaged. I. B. Cohen has provided a nice example of this process: virtually no papers related to the composition of the first edition of the *Principia* (1687) survive. This is in striking contrast to the well over 1,000 surviving manuscript pages for improvements after the first edition was printed and to an equal number of pages related to the calculus priority dispute in the eighteenth century.[9] The

this travesty; see Swerdlow, 'The holograph of *De revolutionibus* and the chronology of its composition', *Journal for the History of Astronomy*, 5(1974):186–98.

7 Edward Heawood, *Watermarks Mainly of the 17th and 18th Centuries* (Hilversum: The Paper Publications Society, 1950); and W. A. Churchill, *Watermarks in Paper in Holland, England, France, Etc, in the XVII and XVIII Centuries and their Interconnection* (Amsterdam: Menno Hertzberger, 1935; rpt. 1967).

8 Personal communication, Domenico Bertoloni Meli to Shapiro, 12 December 1988.

9 I. Bernard Cohen, *Introduction to Newton's 'Principia'* (Cambridge: Harvard University Press, 1971), pp. 78–9.

pattern is further distorted by the fact that the surviving correspondence, the most consistent source of dated papers, is not uniformly distributed over the period. For example, the 1680s are especially poorly represented with only twenty-two letters with watermarks surviving the decade from 1679 to 1688, while seventy-three remain from 1689 to 1698. This problem is exacerbated for the early correspondence, for much of it is in private possession and inaccessible for the purpose of examining watermarks. In these early years Newton frequently used bound notebooks whose watermarks could not be included in this study. These were bought already bound from a stationer, and their paper bears no relation to any of his other writing paper. For these reasons then, watermarks cannot be used to date writings before 1672, the period when Newton was in the 'prime of his invention'. The year 1672, however, offers a bumper crop of letters and datable papers because of the controversies following the publication of the 'New theory about light and colors' in February 1672 and his further optical research. Through the mid 1670s there is sufficient material to use watermarks to assist in dating, but they again become inadequate until the mid 1680s, when Newton composed the *Principia*. From this period until the end of his life – another forty years – watermarks can be used with the most confidence. Besides correspondence (and drafts and copies), there is other dated material, such as a series of chemical experiments from December 1678 to April 1695 (ULC, Add. MS 3973), and Treasury documents. Many other papers can be dated by their relation to published material, such as the *Principia*, *Opticks*, and *Commercium epistolicum* (1712).

Through the 1670s Newton seems to have used a greater variety of paper for shorter intervals. Before the 1680s I have found only one watermark that was used longer than twenty months. Two were used for six months in 1672 and 1678; and all others that I have traced appear on only one or no dated sheets.[10] This no doubt reflects the relatively few surviving manuscripts and the still less dated and accessible ones from this early period. Yet, it does fit the pattern that one might expect from a young man, who would buy small quantities of paper as the need arose. As he grew impatient with the frequent interruptions of small tasks and felt more established, he would purchase more paper less frequently. After 1680 the same watermarks appear on far more sheets and for longer stretches. In all periods Newton used paper with more than one

---

[10]   Five watermarks appear on more than two dated manuscripts in the 1670s; the one anomalous watermark, crowned horn + $\mathscr{A}\mathscr{I}$, that was used for fifteen to twenty years is discussed below (see Figure 3). By a conservative estimate I have found over forty-five watermarks from the mid 1660s, through the 1670s that are either on only one dated paper or undated ones. Since I recognized early on that there are so few dated and accessible papers before 1672, I did not bother making tracings or a careful inventory of most of these watermarks.

watermark, but in the later period (especially after about 1695) more papers are used at the same time. Large quantities of readily available paper at the Mint probably accounts for much of the changing pattern of usage. It is difficult, if not impossible, to deduce precisely how much of Newton's use of paper – that is, the pattern of watermarks and periods of use – depends on personal habit and how much on paper marketing. If a watermark first shows up in 1685 and then reappears two and six years later, are we to attribute it to a stack of paper that Newton placed under other paper or stored elsewhere, or rather to the periods when it was marketed, or to both?

It is essential to recognize that independent of personal habits the mixed pattern of watermarks, or different papers, that always occurs during any extended period is inherent in the production and marketing of paper at this time. A pair of moulds when used daily would last about one year, and if laid aside to produce paper of different sizes and qualities, as was commonly the case, about three to ten years.[11] Four years, however, would be a reasonable estimate for the lifetime of a typical pair of moulds. A skilled vatman and coucher would produce about 5,000 sheets of writing paper a day of which about 4,000 were usable.[12] Thus during their lifetime a pair of moulds would turn out well over a million sheets of paper with two nearly identical watermarks. Paper was then a very valuable commodity, and all the evidence indicates that it moved rapidly through the system from maker to printer. The same paper, that is, with the same twin watermarks, could be distributed for up to ten years, though usually for much less than that. It is also quite likely that paper would linger longer, suppose another year, in a stationer's shop than at a printer's, who tended to use paper nearly as soon as it was received. Stationers did not sell paper from only one maker, since they did not purchase it directly from the manufacturers. Virtually all the paper used in England at this time came from northern France and Holland. The paper in a given region was bought by wholesalers who then sold it to English importers, who in turn sold it to stationers and printers. Thus the production of different makers (with different watermarks) would be distributed simultaneously. If we assume the distribution network was reasonably stable, we should expect paper from the

[11] Loeber, *Paper Mould*, p. 2. Stevenson in Briquet, *Les filigranes*, p. *32, estimates the life of a pair of moulds in continuous use to be one or two years. Gaskell cites an English mill in the 1780s where a pair of moulds on average lasted a little over seven months. 'Watermarks', he states, 'had an even shorter life, and might drop off the mould in about six months, when they would be repaired or replaced, or simply left off altogether' (*Introduction to Bibliography*, p. 63). This somewhat shorter estimate of the lifetimes for continuous use does not alter the nature of my analysis. A typical mill in Newton's day a century earlier had only one vat, whereas this mill had six.

[12] Gaskell, *Introduction to Bibliography*, pp. 59–60; Stevenson, *Missale speciale*, p. 282; Rupert C. Jarvis, 'The paper-makers and the excise in the eighteenth century', *The Library*, ser. 5, 15(1959): 100–116, on p. 107.

same maker with the same watermark to turn up periodically at the same stationer over as long as ten or eleven years, but typically for half that interval. The stationer would also sell paper with other watermarks throughout this period.[13] The production and marketing process explains why during the course of, say, five years a mixture – and not simply a succession – of different papers always occurs. The user's preferences – such as how often and in what quantities he buys paper, and how long he lays it aside – is superimposed over the pattern inherent in the marketing process itself.

The writing paper bought by Newton was generally pot size (named after a watermark) or about 40.5 cm wide by 31 cm high ($16 \times 12\frac{1}{4}$ in), though he also used foolscap, which is about 1.5 to 2.5 cm ($\frac{1}{2}$ to 1 in) larger in each dimension. The sheets were sold folded in half (to form a bifolium of two leaves), and they could be purchased singly, in quires (24 or 25 sheets), or in reams (20 quires of 480 or 500 sheets). I am sure that Newton never bought a ream of paper at one time, for even the manuscript of the *Principia* consists of only 230 sheets, of which there are 106 sheets of the most common watermark. When writing a major work like the *Principia* or *Opticks* Newton purchased between two and five quires at a time. Otherwise I suspect his individual purchases consisted of one or two quires. This is a reasonable quantity, since a quire has about fifty leaves, which is the same as the standard size writing tablet sold now in the United States. For the most part Newton used the paper as sold, folded with two leaves about 20.3 cm wide (8 in), but for his chemical and alchemical writings he generally folded the sheet once again and slit one fold to create a booklet with four leaves.

When I began this study, I considered the possibility that Newton had used different paper for his correspondence, mathematics and physics, theology, and alchemy and chemistry. I now strongly suspect that for the latter he did utilize a separate stack of paper, which he kept in his laboratory. Besides the different format for this paper, one of the two longest runs for a particular paper involves alchemy–chemistry. The watermark horn/IR appears in correspondence from August 1692 to June 1695.[14] It also appears on a number of papers, including revisions for the *Principia* and alchemical writings, that are assigned to the early 1690s by all scholars. There is, however, a series of

[13] An inventory of paper offered for sale by two London merchants in 1674 shows the variety of paper available at any one time. They offered a total of sixty-eight different papers of which about twenty-five were the size used by Newton, seventeen pot and eight foolscap. See R. W. Chapman, 'An inventory of paper, 1674', *The Library*, ser. 4, 7(1927):402–8.

[14] Newton to Locke, 2 August 1692, Bodleian Library, MS Locke c. 16, ff. 151/2; and Newton to Flamsteed, 29 June 1695, Corpus Christi College (Oxford) Archives [henceforth CCC], MS 361, no. 28. See note 26 where this watermark is shown to occur still later, or some time after 3 March 1695/6.

Figure 3. A beta radiograph of the watermark crowned horn + *AI* on ULC, Add. MS 3979.3, f. 1, a draft of Newton's letter to Flamsteed, 16 April 1681. (By permission of the Syndics of Cambridge University Library.)

chemical experiments dated 'Aug. 1682' with the same watermark (ULC, Add. MS 3973.4, ff. 13–16). Dobbs, on the basis of handwriting and contents, had attributed another alchemical manuscript to the early 1680s (Keynes 28; Sotheby 31), which is almost certainly on the same paper.[15] Three half sheets (ff. 5–10) have the countermark IR alone, and though it is difficult to identify paper without its watermark, I am pretty sure it is the same paper. Since only four sheets of this paper seem to have been used before the early 1690s, it is tempting to believe that after a few sheets were used in 1682, the stack of paper was buried in his laboratory and finally uncovered a decade later. Only one other paper appears to have been used over a longer time. Though its watermark, crowned horn + *AI* (Figure 3) is on letters to Flamsteed in February and April 1681 (CCC, MS 361, nos. 2, 4), it also appears on other

[15]   'Keynes' indicates a manuscript number in the Keynes collection in King's College Library. 'Sotheby' indicates the lot number in the auction catalogue of the sale of Newton manuscripts in 1936, *Catalogue of the Newton Papers Sold by Order of the Viscount Lymington* (London: Sotheby and Co., 1936); see Peter Spargo's contribution in this volume.

Figure 4. A beta radiograph of the watermark arms of Amsterdam on ULC, Add. MS 3970, f. 582, which is from Wickins's transcription of the 'Hypothesis' of 1675. The countermark is AI between a pair of chain lines. (By permission of the Syndics of Cambridge University Library.)

documents that can reasonably confidently be dated to a decade on either side of 1681. This paper is used in the 'Geometria,' which Whiteside assigns to 1693, and the 'Fundamentum Opticae', the first Latin draft of the *Opticks* composed in the late 1680s; on the early side of 1681, it is used in some calculations for geometrical optics, dated to about 1670 by Whiteside.[16] The attributed dates may possibly be shifted closer to 1681 by two or three years on each end, but the fifteen years or so that this paper was used would still be the longest interval, and make this watermark virtually useless for the purposes of dating.

In using watermarks to attempt to date manuscripts, it must always be borne in mind that it is simply one more piece of evidence, and in itself not decisive. Indeed, when a new Newton manuscript is encountered, one always has other

[16] For the 'Geometria' see, for example, ULC, Add. MS 3963, ff. 15/16, 33–9; and Newton *Math Papers*, 7:352. For the geometrical optics, see ULC, Add. MS 3970, ff. 633/4, 636/7; and Newton, *Math Papers*, 3:534–41.

guidance, such as handwriting and content. And just as a watermark can suggest a redating of a document, so the other evidence can suggest that the period for a given watermark be extended. A more interesting possibility, which we will consider, is when the watermarks themselves suggest that their periods must be revised. If a document consists of a continuous text written at one time without insertions and has two watermarks whose dated intervals do not overlap, then the dates of the two must be extended. After I had carried out my initial study of watermarks and realized that they were insufficiently precise for dating scientific, especially optical, papers, I believed that they might still be useful for alchemical and theological papers whose dates are generally far more uncertain. Subsequently I contacted the leading scholars in these areas to enter a cooperative arrangement in which they provided me with their proposed dating of those papers so that I could test the reliability of watermarks and in turn possibly aid them in dating troublesome documents and confirming the rest. It is reassuring that the vast majority of their datings were in agreement with those by watermarks, when two or three years are added to the interval of use as determined by dated documents. Let me give a few examples where the watermarks suggest redating. An alchemical manuscript in the Mellon Collection at Yale University (Mellon MS 79, Sotheby 36) is assigned to 'about 1700' in the published catalogue. Betty Jo Dobbs's attribution to the mid 1670s in her handlist of alchemical manuscripts is confirmed by the watermark, arms of Amsterdam with countermark *A I* (Figure 4).[17] It is the same as that used by Newton for the 'Hypothesis', which he sent to the Royal Society on 7 December 1675, and for a draft of his letter to Hooke two weeks later.[18] Incidentally, this watermark turns out to be the same as on Newton's essay 'De aere et aethere', whose date has been much debated. The Halls, who first published this manuscript, argue on the basis of the development of Newton's thought that it was written between 1673 and 1675, or before the 'Hypothesis'. Westfall, on the other hand, has attributed it to 1679, also arguing from the development of Newton's ideas.[19] I had

---

[17]   Lawrence C. Witten II and Richard Pachella, eds., *Alchemy and the Occult. A Catalogue of Books and Manuscripts from the Collection of Paul and Mary Mellon Given to Yale University Library*, 4 vols. (New Haven: Yale University Library, 1977) vol. 4, p. 479. The editors do indicate the watermark, but by failing to note the countermark they make that information not only useless but misleading.

[18]   For the 'Hypothesis' see ULC, Add. MS 3970, ff. 544–7, which is part of Newton's autograph copy (the rest is on a paper with a fleur de lys), and ff. 573–82, which is a copy in John Wickins' hand. Newton's letter to Hooke on 21 December 1675 is in private possession and inaccessible, but his draft is in ff. 531/2.

[19]   Hall, *Unpublished Scientific Papers*, pp. 187–8; and Richard S. Westfall, *Force in Newton's Physics* (New York: American Elsevier, 1971), pp. 373, 378; and *Never at Rest: A Biography of Isaac Newton* (Cambridge University Press, 1980), p. 374; see also note 2.

personally concluded on the same grounds of conceptual development that it is impossible to place it any more narrowly than in the 1670s. The watermark, however, tends to support the Halls's dating, though it is certainly insufficient evidence to compel scholars to change their opinion. The same watermark also helps to date some alchemical notes in Keynes 25 (Sotheby 26) more precisely for it nicely splits the two proposed dates, the early 1670s and the early 1680s.

Watermarks also suggest that the dates of some theological writings be reconsidered. Westfall has attributed the two versions of an essay on prophetic language in Keynes 5 (Sotheby 242) at one time to about 1680 and then more recently to the mid 1680s.[20] One of the two watermarks, horn + HG/LL on ff. I–VI, is identical to one in the manuscript of the *Opticks*, from the late 1680s or the very early 1690s; and the other, arms of the City of London + $\mathscr{AI}/\mathscr{TI}$ on ff. 1–6, is in a portion of the manuscript of the *Principia* that was written out in 1687. Thus the revised date seems fine. This example also presents us with an instance of two watermarks of the same species, arms of the City of London + $\mathscr{AI}/\mathscr{TI}$, that are not, however, the same watermark. It also appears in the manuscript of the *Opticks*, but my tracings conclusively indicate that it (with its twin) is different from the earlier one (with its twin) in the *Principia*. In another case, Westfall initially assigned the essay 'Paradoxical Questions concerning the morals & actions of Athanasius & his followers' (Keynes 10, Sotheby 268) to the 1680s and then afterwards to the late 1670s.[21] If we judge by the watermarks, though – fleur de lys/CSH on ff. 1–26 and fleur de lys/HD on ff. 27–32 – this manuscript appears to be from the early 1690s. The latter is on dated documents from June 1691 to July 1695, while the former appears on only one dated letter to Locke, on 3 May 1692.[22] Despite a single dated appearance other evidence indicates the early 1690s for this watermark. It is on some revisions to the *Principia*, Book III, Prop. VIII, Coroll. II (ULC, Add. MS 3965, ff. 270–3) that must be after 1687 and which Westfall himself

[20] Westfall, 'Newton's theological manuscripts' in Zev Bechler, ed., *Contemporary Newtonian Research* (Dordrecht: Reidel, 1982), pp. 129–44, on p. 142 [from a conference in 1977]; and *Never at Rest*, pp. 349, 804, n. 78. The manuscript is published in Newton, *Theological Manuscripts*, ed. H. McLachlan (University of Liverpool Press, 1950), pp. 119–26; the remainder of MS 5 is a different and later work.

[21] Westfall, 'Theological manuscripts', p. 142; *Never at Rest*, p. 344. The manuscript is in Newton, *Theological Manuscripts*, pp. 60–118.

[22] For the fleur de lys/HD see Newton to David Gregory [November 1691], ULC, Add. MS 3980, no. 11, in H. W. Turnbull, J. F. Scott, A. Rupert Hall, and Laura Tilling, eds., *The Correspondence of Isaac Newton*, 7 vols. (Cambridge University Press, 1959–77) vol. 3, p. 181; and Newton to Flamsteed, 9 July 1695, CCC, MS 361, no. 29, in *ibid.*, vol. 4, p. 143. For fleur de lys/CSH (which has a vertical line, with an 'X' at each end, running through the 'S' in the countermark), see Newton to Locke, 3 May 1692, Bodleian, MS Locke c. 16, ff. 149–50, in *ibid.*, vol. 3, p. 214.

dates to the early 1690s; it also forms a great part of the various drafts of the essay 'De quadratura curvarum' which was written in the fall and winter of 1691–2 (ULC, Add. MS 3962, ff. 31–40, 54–7), and it is there also combined with the same fleur de lys/HD.[23] In this case, in which the manuscript apparently was written at one time, the two watermarks overlap in time of use; but when they do not overlap; the intervals as established by dated documents must be extended, as we will see in the next example.

Newton began his 'Theologiac gentilis origines philosophicae' in the mid 1680s and continued to work on it through the 1690s. He left the work incomplete, and the papers related to it are in a state of disorder. Keynes 146 (Sotheby 295), 'Of the original of Monarchies', represents a relatively finished version of the first chapter. Westfall dates this to the 1690s, which is supported by the watermarks, horn/IR (sheets 1–6) and arms of the City of London/℃ℱ (sheets 7–13).[24] These respectively span the early 1680s to 1695 and 1696 to 1703, which requires that the dates of at least one of them be expanded, if it is assumed that the chapter was written in one brief period.[25] In fact, it turned out that the date of each could be extended beyond the termini of 1695 and 1696, which were established by dated correspondence alone. Since Newton used the horn paper for a memorandum (Keynes 26, Sotheby 45), in which he refers to a visit on 'Munday March 2d or Tuesday March 3 1695/6' by a Londoner who discussed alchemical processes with him, he must have used this paper after that date.[26] The arms of the City of London watermark appears on a draft of Newton's 'New Scheme of Learning proposed for the Mathematical Boys in Christ's Hospital' from May 1694.[27] Even had independent evidence not been available, such an association of watermarks compels one to alter the previously established dates. Clusters of the same watermark can also assist in establishing that two or more manuscripts dealing with the same material were written in the same period and are closely related. Westfall, for example, has argued on the basis of content that Newton's alchemical essay 'Praxis' started out in the early 1690s as successive sets of critical notes in Keynes MSS 21 and 53 (Sotheby 17 and 93).[28] His argument is buttressed by the use in each manuscript of the same paper with a Strasbourg

---

23   Westfall, *Never at Rest*, p. 508; Newton, *Math Papers*, vol. 7, pp. 24, 48.

24   Westfall, 'Theological manuscripts', p. 142; *Never at Rest*, p. 351, n. 55.

25   The watermark horn/IR was discussed earlier; see note 14. For the arms of the City of London/℃ℱ watermark, which has a small oval on the fold in the middle of the sheet, see the draft memorandum by Newton in June 1696, Public Record Office, Treasury Papers, Newton MS 1, ff. 8–9, in Newton, *Correspondence*, vol. 4, pp. 207–8; and Newton to Locke, 15 May 1703, Bodleian, Locke MS c. 16, f. 155, ibid., vol. 4, pp. 405–6.   26   Newton, *Correspondence*, vol. 4, p. 196.

27   ULC, Add. Ms 4005, ff. 100/1; Newton, *Math Papers*, vol. 7, p. xxii, n. 52.

28   Westfall, *Never at Rest*, p. 529.

lily watermark, which was previously unknown to me and undated. In both of these cases the evidence of watermarks was used with other sorts of evidence, and it is perilous to argue from watermarks alone. The use of watermarks, just as handwriting, is a supplement to the more fundamental historical method based on a careful reading of the documents and an analysis of their conceptual and textual elements.

In my own work with Newton's manuscripts watermarks have proved more useful for relative ordering than for absolute dating, that is, for grouping and distinguishing the numerous papers – notes, preliminary drafts, rejects, subsequent revisions, and final draft – that are related to a single text, the *Opticks*. Most of my reconstruction of the composition of the *Opticks* was carried out by traditional means, especially textual analysis. Nonetheless, the classification of watermarks not only served to confirm my reconstruction and reinforce my confidence in it, but clusters of the same watermark forced me to recognize relations between different parts of the manuscript that I perhaps would have otherwise overlooked. When clusters of the same watermark are grouped together, they indicate the composition of the *Opticks* in a striking way. I will show that each set of clusters of the identical watermark indicates that those portions were written in the same period. Newton wrote the various Books and Parts of the *Opticks* over the course of fifteen or sixteen years, but before following each step of its composition, I will present an overview of the major phases.

In about 1687 or 1688 Newton wrote the 'Fundamentum Opticae' ('The Foundation of Optics' or 'The Foundation of the *Opticks*'), which is the first draft of Book I of the published *Opticks*.[29] He started to revise this, but after writing just one page he abandoned the Latin and started anew in English, the language in which it was published in 1704.[30] In the published work this forms the two Parts of Book I, but at this time – and until shortly before publication – they were separate Books I and II. This essentially presents the theory of white light and color that he had developed in his *Optical Lectures* and optical

---

[29]   ULC, Add. MS 3970, ff. 409/10, 415/16, 394–8, 583/4, 425/6, 647/8, 407/8, 405/6, 403/4, 401/2, 399/400, 419, 422, 420/1, 411–14, 423/4, 417/18. Since the manuscript of the *Opticks* and all the other papers related to its composition are in Add. MS 3970, throughout the remainder of my account I will cite only folio numbers.

[30]   'Opticae. Liber primus', ff. 302/3, which is a revise of the opening f. 409 of the 'Fundamentum'. The watermark on ff. 302/3 is the same as that on the rest of the Books I and II in the final version of the manuscript (stage 1 in Table 2), which supports the assumption that Newton turned to the English version right after beginning in the Latin. Although Newton's name is omitted from the title page of the *Opticks: Or, A Treatise of the Reflexions, Refractions, Inflexions and Colours of Light. Also Two Treatises of the Species and Magnitude of Curvilinear Figures* (London, 1704; rpt. Brussels: Culture et Civilisation, 1966), the Advertisement is signed I. N.

correspondence in the early 1670s. Even though he took these writings as his foundation, this was a completely new exposition. The composition of Book III (II in the published work) on the colors of thin films and natural bodies was at first much easier, for he simply lightly revised the three parts of the 'Observations' that he had sent to the Royal Society in December 1675; this then comprised Part I through Part III, Proposition 8 of Book III.[31] Newton now had to decide how to conclude the *Opticks*, and this turned out to be no easy matter. He would ultimately add twelve new propositions to Part III (of which the last ten would present the theory of fits), the entire Part IV on thick plates, and a brief concluding Book IV on diffraction together with sixteen appended Queries. Book III, Parts III and IV were essentially complete by February 1692, but diffraction continued to elude Newton, and he did not write the last Book until 1703, shortly before publication. Throughout my account I will follow the division into Books and Parts of the manuscript, unless I explicitly state otherwise. It is, however, a simple matter to convert to the familiar arrangement of the published work with its one less Book: Books I and II became Book I, Parts I and II; and Books III and IV became Books II and III.

Newton's own account of the composition of the *Opticks* is somewhat vague and not altogether consistent, though we can assign Books I and II to probably 1687 or 1688, and with more certainty to no later than June 1689. The 'Advertisement' of the *Opticks* opens with a brief history of the work:

> Part of the ensuing Discourse about Light was written at the desire of some Gentlemen of the *Royal Society*, in the Year 1675. and then sent to their Secretary, and read at their Meetings, and the rest was added about Twelve Years after to complete the Theory; except the Third Book, and the last Proposition[s] of the Second, which were since put together out of scattered Papers. To avoid being engaged in Disputes about these Matters, I have hitherto delayed the Printing, and should still have delayed it, had not the importunity of Friends prevailed upon me.[32]

The 'part' written in 1675 refers to the 'Observations' on the colors of thin films and natural bodies that Newton sent to the Royal Society on 7 December

---

31 This work was not titled by Newton. In I. B. Cohen, ed., *Isaac Newton's Papers & Letters on Natural Philosophy* (Cambridge: Harvard University Press, 1958), it was published as 'Newton's second paper on color and light', pp. 177–235, while the editors of Newton's *Correspondence* assigned it the title 'Discourse of Observations'. It was first published in Thomas Birch, ed., *The History of the Royal Society of London, for Improving of Natural Knowledge, from its First Rise*, 4 vols. (London, 1756–7; rpt. Brussels: Culture et Civilisation, 1968), vol. 3, pp. 272–8, 280–95, 296–305.   32 Newton, *Opticks*, p. [iii].

1675; it was read at the Society's meetings of 20 January and 3 and 10 February 1676.[33] This formed most of Book II (and throughout this paragraph I will follow Newton's references to the Books of the published work), so that 'the rest' refers to Book I, which was written 'about' 1687–88. In the text of Book I Newton remarks that he made two reflecting telescopes 'about 16 Years ago', and in the margin of the manuscript he wrote 'Anno 1670'.[34] This would imply that he was writing Book I in 1686, a date which must seriously be doubted, since he was then very much occupied with the *Principia*, which was published in March 1687.[35] Moreover, there is no evidence that he made any telescopes at all in 1670; rather, they date from 1668, 1671, and 1672. The account in the 'Advertisement' is more reliable than the passing remark in the text, but we must take both of them only as rough recollections, in each case written about sixteen years after the event. We already know that neither watermarks nor handwriting are capable of such fine dating as to resolve the time of writing, but they are consistent with Newton's recollections. The 'Fundamentum Opticae' is written on a mixture of different papers, ten sheets with the watermark crowned horn + $\mathscr{AI}$ and six sheets of the arms of the Seven Provinces/ELF, though there are three interspersed sheets with two watermarks that cannot otherwise be dated. The crowned horn + $\mathscr{AI}$ is, of course, the watermark with the greatest span of at least fifteen years (including the 1680s) and already eliminated as useless for dating; the Seven Provinces/ELF, however, was used in 'De motu corporum in gyrum', the conceptual forerunner of the *Principia* from fall 1684, and also in a number of geometrical writings which Whiteside dates to 1693.[36] Finally, we should note that in recalling that Book I was written 'about' 12 years after the 'Observations', Newton does not enter into such a fine distinction as that between the first draft, or the 'Fundamentum', and the text of the *Opticks*

[33] 'An Hypothesis explaining ye properties of Light', which was sent to the Royal Society together with the 'Observation', was read on 9 and 16 December 1675.

[34] Newton, *Opticks*, Bk. I, Pt. I, Prop. 7, p. 76; in the manuscript Newton wrote the '1670' over '1669', f. 75r.

[35] Newton 'completed' the *Principia* in April or May 1686, but he was occupied with its revision and transcription until March 1687, as I describe at the conclusion of this paper; see Cohen, *Introduction*, pp. 68–70.

[36] 'De motu corporum in gyrum', ULC, Add. MS 3965, ff. 55–62 is published in John Herivel, *The Background to Newton's 'Principia'* (Oxford: Clarendon Press, 1965), pp. 257–74; and reproduced in a handsome facsimile in *The Preliminary Manuscripts for Isaac Newton's 1687 'Principia' 1684–1685* (Cambridge University Press, 1989), pp. 3–11. As frequently happened, the countermark deteriorated in the course of production, and some of the sheets (e.g., ff. 55/6) are missing the 'E'. See for the arms of the Seven Provinces 'Geometriae Libri Tres', ULC, Add. MS 3963, ff. 29–32, which also contains the crowned horn + $\mathscr{AI}$ (ff. 9/10) among other watermarks; see Newton, *Math Papers*, vol. 7, pp. 248–9.

itself. Surely the enterprise was under way by June 1689, when Huygens met with Newton, for he then told Huygens that a work on color would appear.[37]

Turning now to Book II, we find that Newton's remarks in the 'Advertisement' are similarly inconsistent and incomplete: He 'since' added twelve new propositions to Part III of Book II, whence I have added the 's' to 'Proposition'; and he altogether ignores the new Part IV of Book II. The manuscripts unambiguously show that he added the last ten propositions to the end of Part III all at once. Nonetheless, all the evidence supports the thrust of the account in the Advertisement, namely, that he wrote Book III and the new material for Book II after he had completed Book I and revised the earlier portion of Book II. One significant omission from Newton's account is that he does not say when he revised the 'Observations' for the *Opticks* (Book II, Parts I, II, and much of III) perhaps because it went so quickly. Let us now attempt to add some refinements to Newton's recollections.

The manuscript from which the *Opticks* was printed was afterwards returned to Newton and is now amongst his papers in the Cambridge University Library. Newton wrote out most of the manuscript himself; about a third was written by six other hands. The manuscript consists of 102 sheets of approximately pot size folded in half. The text is written on only the recto of each leaf, except for later insertions, which are often on the verso. Although many of the sheets have since split in half along their fold, only two leaves were originally half-sheets, and these are later insertions (ff. 160, 218). If the sequence of watermarks are arranged in tabular form in the order in which they appear in the manuscript (Table 1), it turns out that they tell much of the story of its composition. At first glance, it may seem that Newton and his amanuenses simply wrote out the work on whatever paper was at hand: six different watermarks were used in no apparent order. But if we look a little more carefully, we see that two watermarks predominate, arms of the City of London + $\mathcal{AI}/\mathcal{TI}$ and English royal arms + $\mathcal{AI}$/GLC, with twenty-nine and forty sheets respectively. We should observe that the former dominates Books I and II and the latter Book III, whereas Book IV consists entirely of a single watermark, arms of the City of London/EB. Indeed, it is possible to show that each book originally consisted of a single paper with the same watermark and that the other watermarks represent subsequent insertions; and moreover, that each watermark represents a particular time period (Table 2). Thus I will show that the clusters of watermarks are not arbitrary but, together with all the evidence available, indicate the order in which the *Opticks* was written. Since it consistently proved to be the case that in this one work each set of clusters with the same watermark indicated that those parts of the

[37]   See Huygens to Leibniz, 24 August 1690, in Huygens, *Oeuvres complètes*, 22 vols. (The Hague, 1888–1950), vol. 9, p. 471.

Table 1. *The watermarks of the manuscript of the* Opticks

	Folios	Number of sheets	Watermark
Bk. I, Defs. – Prop. 1	17–28	6	None
Bk. I, Props. 1–6	29–58	15	$\mathcal{TI}$
Bk. I, Props. 6–7	59–74	8	GLC
Bk. I, Prop. 7 – Bk. II, Prop. 6	75–78, 91–114	14	$\mathcal{TI}$
Bk. II, Prop. 7–11	115–30	8	LL
Bk. III, Pts. I–III, Prop. 10	131–191	$30\frac{1}{2}$	GLC
Bk. III, Pt. III, Props. 10–20	192–201	5	MC
Bk. III, Pt. IV, Obs. 1–5	202–205	2	GLC
Bk. III, Pt. IV, Obs. 5–12	206–217	6	MC
Bk. III, Pt. IV, Obs. 13; Bk. IV.	218, 219–233, 339	$8\frac{1}{2}$	EB

Notes: Shortly before publication the Books were renumbered and assumed their familiar sequence: Books I and II became Book I, Parts I and II; and Books III and IV became Book II and III.

Despite the gaps in foliation the manuscript is complete.

The watermarks are denoted by their countermarks: None = horn; $\mathcal{TI}$ = arms of the City of London + $\mathcal{AI}/\mathcal{TI}$; GLC = English royal arms + $\mathcal{AI}$/GLC; LL = horn + HG/LL; MC = horn + HG/MC (there is also a B in the lower left corner of the sheet); EB = arms of the City of London/EB.

manuscript forming that set were written in the same period, I elevated this result – somewhat grandiosely – to the principle of clustering. Of course, this is only a heuristic rule that compels the historian in each case to examine whether the identity of watermarks is an accident or a real temporal relation, but at the conclusion I will show why it is a highly probable rule that may be applied to other manuscript works. In the following reconstruction I will focus on the final manuscript of the *Opticks* without dealing with the fine structure of its composition or drafts, except insofar as they illuminate the order and time of composition. Since drafts of any portion of the text generally consist of at most a few sheets, an analysis of watermarks becomes less meaningful, though even here I have found such evidence to be consistent with the principle of clustering.

Shortly after composing the 'Fundamentum Opticae' in 1687 or 1688 Newton began to revise it. Though he now decided upon English, the text of the *Opticks* actually begins as a revised translation of the 'Fundamentum', but of course with the usual rewriting – additions, deletions, and transpositions – that one expects in a later draft. The propositions and experiments were numbered continuously throughout the 'Fundamentum', but in the *Opticks*

Table 2. *The principal stages of the composition of the* Opticks *and their corresponding watermarks*

Stage	Folios	Watermark
1	Bk. I, Props. 1–6; Bk. II, Props. 1–6 ff. 304/5, 29–58, 388/9, 470/1, 75–8, 91–114	$\mathscr{T}\mathscr{I}$
2	Bk. I, Defs. & Axioms ff. 17–28 −(304/5)[a]	None
3	Bk. II, Props. 7–11 ff. 115–30	LL
4	Bk. III, Pts. I–III, Prop. 10; Bk. IV, Pt. I, Obs. 1–12, Pt. II[b] ff. 131–91, 368/9; 202–5, 381–4, 344/5, 385/6, 306/7, 79–90	GLC
5	Bk. I, Props. 6–7 ff. 59–74 −(388/9, 470/1)	GLC
6	Bk. III, Pt. III, Props. 10–20, Pt. IV, Obs. 5–12 ff. 192–201; 206–17 −(368/9, 381–4, 344/5, 385/6, 306/7, 79–90)	MC
7	Bk. III, Pt. IV, Obs. 13; Bk. IV with Queries ff. 218, 219–33, 339	EB

Notes: a. A minus sign indicates that the folios within the parentheses were removed. b. Book IV in this state was dispersed: Part I became Book III, Part IV, and the sheets with Observations 1–5 were carried over into stage 6. Part II was removed from the manuscript for about a decade and then served as the draft for Book IV in stage 7.

Newton now divided them into two Books. At this first stage (see Table 2), after translating and revising the 'Fundamentum', the *Opticks* consisted of Book I, Propositions 1–6 and Book II, Propositions 1–6 (Figures 5 and 6), the text of which is essentially identical to that of the same propositions in Book I, Parts I and II of the published edition. The manuscript of the *Opticks* then consisted of ff. 304/5, 29–58, 388/9, 470/1, 75–8, and 91–114, which all have the same watermark, arms of the City of London + $\mathscr{A}\mathscr{I}/\mathscr{T}\mathscr{I}$ (Figure 7), and are in Newton's hand. Only Proposition 6 in Book II, which contains Newton's famous color-mixing circle, was entirely new and not in the 'Fundamentum' in some form or other.

Neither the 'Fundamentum' nor the *Opticks* at this point began with definitions and axioms as the published work does. These were added shortly afterwards. In the 'Fundamentum' five definitions were mixed in with the propositions. The only remnant of these interspersed definitions in the published *Opticks* is that on the colorific properties of light rays in Book I,

The first Book
of
Opticks

Tis not my designe here to explain the properties of light by hypotheses but to propose & prove them by reason & experiments as follows.

Propositions.

Prop. 1. Theor. 1.

Lights wch differ in colour differ also in degrees of refrangibility. the light of one natural body is more refrangibility then that of another.

The proof by Experiments.

Exper. 1. J took a black oblong stiff paper terminated by parallel sides & with a perpendicular right line drawn cross from one side to the other distinguished it into two equall parts. One of these parts J painted with a red colour & the other wth a blew. The paper was very black & the colours intense & thickly laid on that ye phænomenon might be more conspicuous. This paper J viewed through a glass prism of solid glass whose two sides through wch ye light passed to ye eye were plane & well polished & contained an angle of about 60 degrees wch J call ye refracting angle of the Prism. And whilst J viewed it, J held it before a window in such manner that the sides of yt paper were parallel to yt Prism & both those sides & yt Prism parallel to yt horizon & yt cross line perpendicular to it as that the light wch fell from yt window upon the paper made an angle wth yt paper equal to that angle wch was made wth yt same paper by yt light reflected from it to yt eye. The wall of yt chamber beyond yt the Beyond the Prism was the wall of the chamber under yt window covered over with black cloth, & yc cloth was involved in darkness that no light might be reflected from thence by yt edges of the paper to yt eye might mingle it self wth the light of the paper & obscure yt phænomenon thereof. These things

Part II, Proposition 2. I claim that the six sheets with the definitions and axioms (ff. 17–28) were inserted in place of the original opening bifolium (ff. 304/5) after the first stage of Books I and II was completed, and also that they were drafted shortly after Newton started to write out Book I. The draft of the new definitions and axioms (ff. 392/3) has the same watermark as the rest of the initial state (arms of the City of London + $\mathscr{AI}/\mathscr{TI}$), which by my principle of clustering suggests that they were written in the same period. But can we more precisely determine at what stage of the composition of the *Opticks* they were written? In fact they were drafted right after Proposition 1 was written. Definition 1 of a light ray appeared in the manuscript of the *Opticks* at the end of Proposition 1, just as it had in the 'Fundamentum', but after writing it out Newton deleted it and decided to put all the definitions and a new series of axioms at the beginning of Book I. The change of figure numbers in the manuscript show that he composed them at this point. Initially the two figures of this proposition were numbered 'first' and 'second'. They were then changed to 'seventh' and 'eighth' to accommodate the six figures in the axioms, but by Proposition 2 the text has the 'ninth' figure without having to delete 'third', and this sequence of numbering continues to the end of Book I. Why do I argue that Newton only drafted them at this time and did not then add them to the manuscript of the *Opticks*? By the clustering principle it would be most unlikely that Newton would have stopped to write them out on different paper from the rest of the text, and then resumed with the original paper. However, firmer evidence again comes from the figure numbers. The draft of the definitions and axioms lacks Axiom VI of the published *Opticks* with its four figures.[38] Hence, when it was added all the figure numbers of Book I had to be deleted and increased by four, which is just what is found in the manuscript. The changing figure numbers demonstrate that while the axioms and definitions were drafted between the writing out of Propositions 1 and 2 in the manuscript, they were not rewritten and added to the manuscript before Book I was completely transcribed, just as the watermark evidence suggested. By the same clustering principle I argue (without, however, any further evidence) that the definitions and axioms were added after Book II was transcribed. To summarize: at this second stage (see Table 2) the *Opticks* now

[38] All four sides of this bifolium, which has four figures, are fully covered with writing, and Axiom 7 (8 in the published *Opticks*) is incomplete. I assume that the remainder of this axiom with its one figure was continued on another sheet, which is now lost. This assumption is consistent with all the evidence.

Figure 5. The original opening page of the *Opticks*, which Newton removed from the manuscript when he decided to begin the book with a series of definitions and axioms: ULC, Add. MS 3970, f. 304. (By permission of the Syndics of Cambridge University Library.)

The first ~~second~~ Book
of
Opticks.
Part II.
Prop. 1 Theor 1

The Phænomena of colours in refracted or reflected light ~~are~~ are not caused
by new modifications of light variously imprest according
to the various ~~terminations~~ of light & shadow.

The proof by experiments.

Exper. 1. ffor if the Sun shine into a very dark chamber
through an oblong hole F whose breadth is ⅙ sixt or eighth part
of an inch or something less: & his beam FH do afterwards
pass first through a very ~~large~~ large Prism ABC, distant about 20
feet from the hole, & then (with its white part)
~~in some black opake~~ body whose breadth is about the fortieth
or sixtieth part of an inch & web is made in as black
opake body, & placed ~~two or three~~ at ye distance of two
or three feet from the Prism ~~○○○○○~~ in a parallel
situation both to the Prism & to ye former hole, & if
this white light be transmitted through ye hole H fall after
wards upon a white paper pt placed after ye hole H at ye
distance of three or four feet from it & ~~○○○~~ there
paint the usual colours of the Prism, suppose red at t,
yellow at s, green at r, blew at q & violet at p, you
may with an iron wire or ~~○○○○○○○○○~~ any such
like slender opake body whose breadth is about the tenth
part of an inch, by intercepting the rays at k, l, m, n or o
take away any one of the colours at t, s, r, q or p, whilst
the other colours remain upon the paper as before; or with
an obstacle something bigger you may take away any two
or three or four colours together the rest remaining;
so that any one of the colours, as well as violet may becomes outmost
in the ~~confine~~ of ye shadow ~~○~~ towards p, & any one of
them as well as red may become outmost in the confine
of the shadow towards t & any one of them may also
border upon the shadow made within the colours by the obsta
cle R intercepting some intermediat part of ye light, & lastly
any one of them ~~may~~ by being left alone may border upon
the shadow on either hand. All the colours have themselves
indifferently to any confines of shadow, & therefore the differ
ences

consisted of the first six propositions in Books I and II plus the new definitions and axioms, and ff. 304/5 were replaced by ff. 17–28 with a plain horn watermark.

I have gone into some detail over this second stage to show how watermarks raise and help to resolve questions about the composition of the *Opticks*, and how this new sort of evidence must be consistent with all the more traditional sort. Moreover, to show that such editorial minutiae have greater significance, we can explain what has seemed to some historians to be a puzzling feature of the *Opticks*. While appealing to the opening definitions and axioms as evidence for the mathematical structure of the *Opticks*, they had to concede that Newton scarcely invoked them in the text.[39] Since we now know that they were added at a late stage of composition, after much of Books I and II were nearly complete, it is not surprising that they are not an integral part of the work. This example also serves to remind us of the significance of the distinction between time of composition and that of transcription. The two can be separated by a long interval, and watermarks speak only to the time a document was written.

Book I of the *Opticks* was still far from complete. Newton had devoted the 'Fundamentum' and the *Opticks* as it then stood to formulating, demonstrating, and explicating his theory of color. He had not yet attempted to deduce the explanation of such phenomena as the rainbow, or the colors produced by prisms from the principles of his theory of color, as he had earlier in the 1670s. According to his later terminology, he had thus far adopted only the method of analysis to derive the principles of his theory and not that of synthesis or composition. To rectify this omission and to further round off the theory Newton added Propositions 7–11 to Book II (Table 2, stage 3). These eight sheets (ff. 115–30) are mostly in the hand of an amanuensis and introduce a new watermark, horn + HG/LL (Figure 1).[40] The clustering principle implies that these propositions were written at a different time from the earlier parts of Books I and II, since the watermarks are different. This is consistent with the

---

39 See, for example, Peter Achinstein, 'Newton's corpuscular query and experimental philosophy', n. 28, in Phillip Bricker and R. I. G. Hughes, eds., *Philosophical Perspectives on Newtonian Science* (Cambridge, Mass.: MIT Press, 1990), pp. 135–71, on p. 171. In arguing against Guerlac, 'Newton and the method of analysis' in Philip P. Wiener, ed., *Dictionary of the History of Ideas*, 4 vols. (New York: Charles Scribner's Sons, 1973), vol. 3, pp. 378–91, on p. 389, I. B. Cohen has observed that Newton does not in fact use the axioms and definitions; Cohen, *The Newtonian Revolution* (Cambridge University Press, 1980), pp. 13–14, 134.

40 Proposition 7 begins on f. 114r.

Figure 6. The first page of 'The second Book of Opticks', ULC, Add. MS 3970, f. 91. Shortly before publication Newton changed the 'second' Book to the 'first' and added 'Part II' between the lines. (By permission of the Syndics of Cambridge University Library.)

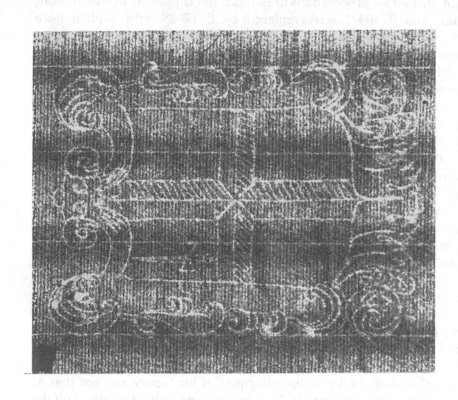

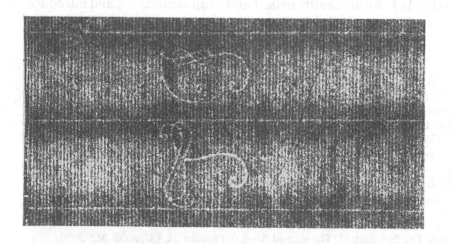

Figure 7. Beta radiographs of the watermark and countermark arms of the City of London + 𝒜𝒥/𝒯𝒥 on ULC, Add. MS 3970, ff. 113/4, a sheet from Book II, Proposition 6 of the *Opticks*. The attendant mark 𝒜𝒥 below the watermark is not included, because the plate holding the film was not large enough. (By permission of the Syndics of Cambridge University Library.)

other evidence, namely, the new material is not in Newton's hand, the nature of its content differs (theory versus application, or analysis versus synthesis), and it was not in the 'Fundamentum'. I have placed the supplement to Book II before the initial version of Books III and IV (that is, stage 3 before 4 in Table 2) by the principle of clustering alone, for I am not now aware of any other evidence that can resolve this point. The only alternative is to interchange stages 3 and 4 in Table 2, for there is no doubt that Book I, Proposition 7 (stage 5) was added after both of these, as we shall soon see. This would imply that Newton finished Books III and IV and turned to another paper for Book II, Propositions 7–11, *and* that he then returned to the first paper to add Propositions 6–7 to Book I, all within a few months. Such a sequence is certainly not impossible or even implausible, but it is more likely – and a simpler hypothesis – that he added Book II, Propositions 7–11 before he wrote Books III and IV.

Whatever the exact sequence, with the addition of these propositions to Book II Newton judged Books I and II to be essentially complete. He had numbered the folios of these Books and moved on to Book III. However, after he had written out a version of Books III and IV, he returned to Book I and made the last major addition to it, Proposition 7. This addition is more complicated than the others, for it involved adding to the end of Proposition 6 a new derivation of the sine law of refraction according to the Cartesian model, while expanding its Scholium into a proposition in its own right.[41] The aim of the Scholium was to show that the chromatic aberration of a lens (the difference of focal lengths for the extreme colors due to their unequal refractions) was as serious an obstacle to their use as he had persistently claimed. Because of the intrinsic difficulty of the observations, Newton was at first willing to accept a substantial discrepancy between the measured and predicted values of the chromatic aberration (which is proportional to the dispersive power of the glass). The discrepancy nonetheless troubled him, and he set out to eliminate it. He carried out a set of experiments to determine the chromatic aberration of his lens and the dispersive power of his glass prism.[42] This expanded into Proposition 7, 'The perfection of Telescopes is impeded by the different Refrangibility of the Rays of Light.' Newton removed from the manuscript two sheets (ff. 388/9, 470/1), which were numbered 41 and 43, and inserted eight sheets (ff. 59–74), which were numbered 41–55, bearing the watermark English royal arms + $\mathscr{AI}$/GLC with the derivation for Proposi-

---

[41]   This draft of this derivation on f. 341 has the same watermark, English royal arms + $\mathscr{AI}$/GLC, as the final version inserted into the manuscript and is published in Newton, *Math Papers*, vol. 6, p. 428.

[42]   This new proposition and its experiments are discussed in Shapiro, 'Newton's "achromatic" dispersion law: theoretical background and experimental evidence', *Archive for History of Exact Sciences*, 21(1979):91–128, on pp. 119–21.

tion 6 and most of Proposition 7. We are now at stage 5 in Table 2. Newton was able to use the last two sheets of the former Scholium to Proposition 6 for the conclusion to Proposition 7. These two sheets (ff. 75–78) have their old folio numbers 45 and 47 crossed out and 57 and 59 added in red ink. We can now understand the significance of a different watermark, English royal arms + $\mathscr{AI}$/GLC in the midst of the long run of arms of the City of London + $\mathscr{AI}$/$\mathscr{TI}$ in Books I and II. The redetermination of the dispersion of glass, which is one of the fundamental parameters of Newton's theory, in Proposition 7 necessitated changes throughout the manuscript, in particular in Book II, Proposition 10, and Book IV, Part I. The changes in these propositions show, as I just claimed, that these propositions were completed before Proposition 7 was added, or that stage 5 followed 3 and 4. We can also now explain why there are two Propositions 7 in Book I of the first edition of the *Opticks*. Before he added this Proposition 7, Newton had squeezed in another brief Proposition 7 on reflecting telescopes on ff. 77–8 in the blank space that remained at what was then the end of Proposition 6 and Book I. In the second edition this oversight was corrected and the latter of these changed to Proposition 8.

By the fall of 1691, before Newton inserted the new ending to Proposition 6 and Proposition 7 in Book I, he had already completed Books III and IV, and so the entire *Opticks*. Let us step back and try to follow the composition of the second half of the *Opticks* up to its initial finished state, for within a few months he would revise and rearrange these new Books. In briefly recounting the history of the *Opticks* in the 'Advertisement', Newton did not mention when he revised his 'Observations' for the *Opticks*, probably because the revisions were so minor. He made so few changes in the text that he was able to mark up the manuscript from 1675 (ff. 501–17) for his amanuensis to copy for the *Opticks*. This formed Book III, Parts I, II, and III, Propositions 1–8. In Parts I and II he set forth his experiments and explanations of the colors of thin films, and in Part III, Propositions 1–8, his theory of the transparency, opacity, and colors of natural bodies. As I shall soon explain more fully, there is good reason to believe that Newton undertook this revision in early 1691, give or take about six months. He did not have the text transcribed yet, for he had some problems with Propositions 8 and 9 on the physical cause of the reflection of light. A more serious problem confronting Newton, though, was in deciding how to end his book. At first he planned to follow this material with a new fourth book or part on diffraction. The earliest such draft (ff. 371/2) in fact exactly joins the end of the manuscript of the 'Observations' that served as the copy text for the manuscript of the *Opticks*: the first leaf of this draft (f. 371) was to replace f. 517 of the 'Observations' in the middle of a sentence (both were numbered 33 by Newton), and its observation numbers, beginning with 25, continue those of the 'Observations' itself. At the same time

Newton was also toying with the idea of a speculative 'Fourth Book'. This was to be more like the later Queries or his earlier 'An Hypothesis explaining ye properties of Light discoursed of in my several papers' (which in 1675 accompanied his 'Observations' to the Royal Society) than the experimentally based account of the rest of the *Opticks*. The various sketches of this projected 'Fourth Book' consist of about twenty propositions without proofs on the nature of light and its interactions with bodies and range over such topics as the physical cause of diffraction, the colors of thin films, color harmonies, and vision.[43]

Newton soon reined in his more speculative tendencies and turned to more empirical optical investigations. He continued his experiments on diffraction and also discovered an entirely new phenomenon, colored rings produced in transparent thick plates. These rings are similar to those produced in thin films ('Newton's rings'), but whereas the latter vanish after the thickness of the film increases beyond about one ten-thousandth of an inch, the former appear in plates orders of magnitude thicker (about $\frac{1}{4}$ in). After carrying out a brilliant series of experiments and calculations, Newton was able to predict exactly all their appearances by extending his earlier explanation of the colors of thin films. He now decided to end the *Opticks* with an account of his two experimental investigations and composed a Book IV with Part I on the colors of thick plates and Part II on diffraction (which Newton called 'inflexion').[44] He also worked out an ending for Part III: several paragraphs were added to the end of Proposition 8, and two entirely new propositions, 9 and 10 on the nature of reflection and refraction, were added. For a while he considered adding a physical explanation for the origin of the periodic colors of thin films and thick plates by means of the vibrations set up in bodies when light rays fall upon them (ff. 348/9), but he suppressed it, probably because he judged it too speculative. A fair copy of Book III was made by four amanuenses, whereas Newton wrote out Book IV himself. The manuscript to be joined to Books I and II then consisted of ff. 131–91 (Book III, Parts I–III, Proposition 10), ff. 368/9 (the rest of proposition 10), ff. 202–5 (Book IV, Part I, Observations 1–5), ff. 381–4, 344/5, 385/6, 306/7 (Book IV, Part I, Observations 5–12), and ff. 79–90 (Book IV, Part II); this takes us to stage 4 in Table 2. All the sheets have the same watermark, English royal arms + $\mathcal{AI}$/GLC, and by my

---

[43] See in order of composition ff. 342/1, 337/8, 335/6. I call these the 'Fourth Book' to distinguish them from the experimental Book IV soon to be described. The 'Fourth Book' is briefly discussed in Cohen, 'Hypotheses in Newton's philosophy', *Physis*, 8(1966): 163–84, on pp. 179–81; and Westfall, *Force in Newton's Physics*, pp. 379–80; and *Never at Rest*, pp. 521–2; though their interpretation of this sketch differs from mine.

[44] I describe this research in my forthcoming book, *Fits, Passions, and Paroxysms* (Cambridge University Press).

The second Book of Opticks

The fourth ~~Book Part~~

~~of Opticks~~

Part IV.

~~The fourth Part~~.

Observations concerning the reflexions &
colours of thick transparent polished plates.

There is no glass or speculum how well soever polished
but besides ye light wch it refracts or reflects regularly,
scatters every way irregularly a faint light by means
of wch the polished surface, when illuminated in a dark
room by a beam of the suns light, may be easily seen
in all positions of the eye. The phænomena are certain
scattered light, which when I first observed them seemed
very strange & surprising to me. My observations were
as follows.

Obs. 1. The sun shining into my darkned chamber
through a hole ⅓ of an inch wide, I let ye intromitted
beam of light fall perpendicularly upon a glass speculum ground concave on one side & convex on the
other to a sphere of five feet & eleven inches radius & quicksilvered over on the convex side. And holding
a white opake chart or a quire of paper at ye center of
the spheres to wch ye speculum was ground, that is at
the distance of about five feet & eleven inches from
the speculum, in such manner that the beam of light
might pass through a little hole made in y^e
middle of ye chart to ye speculum & thence be reflected
back to ye same hole; I observed upon the chart, four
or five concentric Irises or rings of colours like rain-
bows encompassing the hole much after the manner that
those wch in the fourth & following Observations of the

principle of clustering I consider them to have been written in one period. Folios 368/9, which join 191 perfectly, were removed from the manuscript afterward when Newton again revised the end of Part III. The now scattered sheets that constituted Book IV, Part I form one continuous text, which was entitled 'The fourth Book of Opticks. The First Part. Observations concerning the reflexions & colours of thick transparent plates' (Figure 8). Book IV, Part II, which was later suppressed and removed in its entirety, was entitled 'The second Part. Observations concerning the inflexions of the rays of light in their passage by the surfaces of bodies at a distance' (Figure 9). This state of the *Opticks* with its Book IV has hitherto been unknown to scholars.

The *Opticks* was now complete. If Newton intended to append his Queries at this time, there is no evidence whatsoever that he did consider this. He did not long remain satisfied with the work as it stood. Within six months of the completion of Books III and IV they were revised and essentially put into their published state. Some of the revisions were prompted by technical problems. He had encountered a five percent difference between the calculated and measured values of the diameters of the rings in thick plates, and it bothered him. Very quickly he traced the source of the error to the values for the 'wavelength' of light that he had been using for twenty years. His redetermination of this fundamental unit necessitated redoing the intricate calculations for Book IV, Part I. Moreover, after Newton had redetermined the dispersion of glass for the new Proposition 7 in Book I (stage 5 in Table 2), he also had to change the value for the index of refraction and dispersion throughout the manuscript, including the calculations for the colors of thick plates in Book IV, Part I. The new values for the 'wavelength' and index of refraction were introduced simultaneously in revision. The calculations in the initial state of Book IV, Part I were carried out with an index of refraction of 31/20 and the revision with 17/11, as required by the redetermination in Book I, Proposition 7. We can therefore place the insertion of Propositions 6 and 7 in Book I between the initial completion of Books III and IV and their revision a few months later, that is, between stages 4 and 6 in Table 2. The identity of the watermarks in Book I, Propositions 6–7 and Books III and IV suggests that the former was done shortly after the latter.

Newton, however, had a more profound reason for revising Book IV, Part I and the end of Book III, Part III. Even before he had completed the first state

Figure 8. The opening page of the original Book IV, Part I of the *Opticks* – 'The fourth Book of Opticks. The first Part' – ULC, Add. MS 3970, f. 202. When Newton decided to eliminate this Book and incorporate its first part in Book III he altered the heading to 'The fourth Part' of Book III. Shortly before publication he changed the Book numbers and altered the heading to 'The second Book of Opticks Part IV'. (By permission of the Syndics of Cambridge University Library.)

The second Part.

Observations concerning the inflexions of the rays of light in their passage by the surfaces of bodies at a distance.

Grimaldo has informed us that if a beam of the Sun's light be let into a dark room through a very small hole, the shadows of things in this light will be larger then they ought to be ~~according~~ if the rays went on by ye bodies in straight lines, & that these shadows have three ^parallel^ fasciæ ~~of colour~~ or ranks of colours adjacent to them But if the hole be too big, the phenomenon of the light wch comes from any one part of it will be obscured by ye light wch comes from ye other parts. This has been reckoned by some to proceed from y^e ^ordinary^ ~~ordinary~~ refraction of ye air but without due examination of the matter. ~~For~~ the circumstances of the phenomenon, so far as J have observed them, are as follows.

Obs. 1. J made in a piece of Lead a small hole with a pin whose breadth was the 42th part of an inch. ~~For~~ 21 of those pins laid together took up the breadth of half an inch. Through this hole J let into my darkened chamber a beam of the Suns light & found that the shadows ^placed in this beam of light^ of hairs, thred, pinns, straws & such like slender ^substances^ were considerably broader then they ought to be did ~~by y^e~~ ^y^e^ rays pass on by these bodies in right lines. And particularly a ~~mans head~~ ^hair^ whose breadth was but the 28th part of an inch ^being held^ in this light at the distance of about ~~the distance~~ 12 feet from ye hole did cast a shadow wch at the distance of four inches from ye hair was the 60th part of an inch broad, that is, above four times broader then the hair, & at the distance of two feet from the hair was about ye 28th part of an inch broad, that is, ten times broader then the hair, &

of Books III and IV, he was struggling with the idea of adding a physical explanation of the cause of the colored rings of thin films and thick plates. From the time that he had control of the phenomena of thick films, he attributed them to the same cause, namely, vibrations in the corpuscles of bodies. But was this too hypothetical to set forth in the *Opticks*? He had already decided once that it was and suppressed a draft for the conclusion of Part III (ff. 348/9) in which he invoked vibrations that are excited in the corpuscles of bodies when light particles fall on them. If the light particles encounter a compression of the vibrations at the second surface, he explained, they tend to be reflected, and if a rarefaction transmitted, thereby accounting for the alternating bright and dark colored rings. In 1675 Newton had set forth this model in great detail in the 'Hypothesis' that he sent to the Royal Society to complement his 'Observations', which contained only experiments, observations, and explanations deduced directly from them. Following his own methodological dictum, 'I shall not mingle conjectures wth certaintyes',[45] he carefully labelled the conjectural model of vibrations a 'Hypothesis' and set it forth in a paper separate from the more certain results of the 'Observations'. Twenty years later, after he had successfully extended his model of vibrations to the explanation of an entirely new phenomenon, Newton felt he had generalized it and was justified in elevating it from a hypothesis to a theory. In the 'Hypothesis' he had utilized the aether as the vibrating medium, whereas he now invoked the corpuscles of bodies for the vibrations, since in the *Principia* he had rejected such an aether. Nonetheless, the model was essentially unchanged and was capable of being developed mathematically to incorporate the newly discovered properties in thick plates. After drafting some propositions on the corpuscular vibrations for the first state of the end of Book III, Part III,[46]

---

[45] Newton, 'New theory about light and colors', *Philosophical Transactions*, 7(1671/2):3075–87, in Newton, *Correspondence*, vol. 1, pp. 92–102, on p. 100.

[46] The corpuscular vibrations in the draft (ff. 348/9) are explicitly invoked in two propositions: 'Prop. 11 The rays of light in being refracted or reflected cause an agitation in the parts of the refracting or reflecting bodies.' 'Prop. 12 The motion excited by a ray of light in its passage th[r]ough any refracting surface is reciprocal & by its reciprocations doth alternately increase & decrease the reflecting power of the surface.' Newton adopted this terminology in Book IV, Part I Observation 9 (f. 345r) where he argued that the rings in thick plates are caused 'suppose by ye mediation of

---

Figure 9. The opening page of the original Book IV, Part II, ULC, Add. MS 3970, f. 79. This Part was removed from the *Opticks* in late 1691 or early 1692. In 1703 it served as the basis for the new, last book on diffraction that Newton composed before publication. (By permission of the Syndics of Cambridge University Library.)

Newton decided against including them and instead ended that Part with Proposition 10 on the proportionality of refractive power to density in bodies. No doubt he excluded these propositions because he came to recognize that corpuscular vibrations were as hypothetical as aethereal vibrations and had no place being mingled with the more certain results of the *Opticks*. Evidently he wanted to include an explanation of the physical cause of periodic colors in the *Opticks* and pondered how to do this without violating his own methodology.

Newton resolved the problem, as was his wont, by considering the vibrations 'abstractedly' without any conjectural vibrating medium.[47] He formulated his solution, the notorious theory of fits – as he called these immaterial vibrations – in a sequence of ten abstract, mathematical propositions, from which he attempted to purge all hypothetical elements. These became Propositions 11–20 of Part III. Five sheets (ff. 192–201) with these propositions and a revision of Proposition 10 were added to the end of Part III, while the last sheet (ff. 368/9), which had concluded Proposition 10 and Part III, was removed. Newton then revised Book IV, Part I on the colors of thick plates and explained them in terms of fits, while redoing the calculations. In the midst of this revision he decided to incorporate the colors of thick plates in Book III because of their affinity to those of thin films, and this now became Part IV of that book.[48] In writing Part IV Newton was able to utilize the first two sheets with Observations 1–4 and part of 5 from the old Book IV, Part I and had to rewrite only Observations 5–12. He removed the old observations (ff. 381–4, 344/5, 385/6, 306/7) and inserted the new ones (ff. 206–17), which were written by an amanuensis, to create Part IV; stage 6 in Table 2 was now complete. The watermark on the new propositions in Part III and the rewritten portion of new Part IV are the same, horn + HG/MC, nicely supporting the principle of clustering. Book III was now complete, except for Observation 13 (ff. 217, 218), which was added later in two steps as indicated by the

some reciprocating action propagated with the rays from one surface to ye other'. In revising this passage for Book III, Part IV he replaced the reciprocations by fits and shifted this argument to the end of Observation 7 (f. 383v).

[47] When Robert Hooke had accused Newton of supporting the corpuscular model of light, Newton conceded the point and explained his way of dealing with hypothetical models: 'I knew that the Properties wch I declared of light were in some measure capable of being explicated not onely by that, but by many other Mechanicall Hypotheses. And therefore *I chose to decline them all, & speake of light in generall termes, considering it abstractedly*...without determining what that thing is...' (Newton to Oldenburg for Hooke, 11 June 1672, in Newton, *Correspondence*, vol. 1, p. 174; italics added).

[48] In the early part of his revision of Book IV, Part I, Newton referred in Observation 7 to the propositions on fits in 'ye 3^d Book', but by the time he had reached Observations 9 and 12 he referred to them as being in 'the third part of this book' (ff. 383v, 345v, 307r).

Table 3. *Newton's correspondence on eclipse colors and successive conclusions for Book III, Part III*, Opticks

Document	Content
'Fourth Book'	speed of different colors varies
10 Aug 1691, Newton to Flamsteed	enquiries about eclipse colors
27 Aug 1691, Gregory to Newton	no eclipse colors
ff. 348/9	thick plates, corpuscular vibrations, no eclipse colors
7 Nov 1691, Gregory to Newton	Cassini report; eclipse colors, red slower
ff. 339/40, 363/4[a]	draft propositions on fits
ff. 365/70	fits; red slower
24 Feb 1692, Flamsteed to Newton	no eclipse colors

Note: a. Ff. 339/40 and 363/4 were written some time between ff. 348/9 and 365/70, but there is no evidence whether it was before or after 7 November 1691.

handwriting and ink. The second step was carried out around 1703, for the countermark of the new leaf (f. 218) that was inserted matches that of the new Book IV (stage 7 in Table 2).

Thus far I have described only the successive stages of the composition of Book III, and have not attempted to provide absolute dates. Newton himself was completely vague about this, informing us in the Advertisement only that the new proposition[s] were added 'since' Book I. It is in fact possible to place the date of the greatest part of the revision of Book III in a narrow interval between 27 August 1691 and 24 February 1692, and the writing of the fair copy between the latter date and 5 May 1694. To do this, however, requires a detour into the drafts for the revision of Book III and Newton's attempt to determine whether the velocity of light varies for different colors. To avoid too extensive a digression I will present only an outline of the argument, which is presented schematically in Table 3. At the beginning of his optical research in the 1660s Newton believed that red rays are faster than violet. He reiterated his assumption of the different velocities in various sketches of the 'Fourth Book', which, I argued earlier, was written after the 'Observations' was revised for the *Opticks* but before the investigation of the colors of thick plates was undertaken.[49] In August 1691 Newton recognized that this assumption could be tested by observations of eclipses of Jupiter's moons, for when one of the satellites disappears behind the planet the slowest color should be seen last. On

[49] See Zev Bechler, 'Newton's search for a mechanistic model of colour dispersion: A suggested interpretation', *Archive for History of Exact Sciences*, 11(1973):1–37; on pp. 20–3 he discusses the 'Fourth Book', though he interprets these documents differently.

10 August he wrote John Flamsteed, Astronomer Royal, to ask if he had observed any change of color during such eclipses.[50] Flamsteed's initial reply came via David Gregory, who, after visiting Flamsteed in Greenwich, wrote Newton on 27 August that 'he said that he never observed any change of Colour in the light of appearing or disappearing satellites of Jupiter, but that hereafter he shall advert if ther be any such thing'.[51] Newton accepted Flamsteed's observation and included it in what I take to be the earliest draft (ff. 348/9) for the new propositions for the end of Part III written after the research on the colors of thick plates was well under way and the phenomenon understood.[52] If Book IV, Part I was not yet complete, it was nearly so. Gregory had not forgotten Newton's interest in eclipse colors, for on 7 November 1691 he reported to him that 'Mr Cassini assures a gentleman of my acquaintance at paris that befor the immersion the satellites appear red, and contrariwise at the emersion.'[53] This implied, contrary to Newton's earlier view, that red rays are slower than violet. Nevertheless, he accepted the validity of Cassini's observation and its implication in a Proposition 12 in a later draft (ff. 365/70) for the end of Part III, which, however, was written when the theory of fits had already been formulated.[54] More news on eclipse colors and another change of view was to follow. On 24 February 1692, Flamsteed, who had earlier promised to observe eclipse colors more carefully, now informed Newton that 'I cannot say that I ever saw any change to a blewish colour or red...'[55] Newton justly respected Flamsteed's skill as an observer and took this as the definitive judgement, for he never again adopted the assumption of different velocities for different colors.

Now let us return to the task at hand and apply this sequence of events and documents to dating the composition of Book III. The first of three attempts to end Book III as the *Opticks*, as represented by the various drafts of the 'Fourth Book', was undertaken before 27 August 1691, when Gregory wrote Newton about his visit with Flamsteed. The next attempt, when he was using corpuscular vibrations in ff. 348/9, was composed between Gregory's letter of

---

[50] Newton, *Correspondence*, vol. 3, p. 164. [51] *Ibid.*, p. 165.

[52] In Proposition 12 he states that the times of the vibration excited in the corpuscles of bodies (see note 46) by rays of different colors would follow the same ratio as the tones of an octave 'provided that all the rays of light be equally swift. For if they differed much in swiftness that difference would be discoverable by the eclipses of Jupiters satellites.' In Proposition 10 Newton already recognized that the cause of the colors of thin plates and thick films are the same. This draft is written over a letter to Newton dated 9 January 1690. [53] Newton, *Correspondence*, vol. 3, p. 171.

[54] This draft ends abruptly at the bottom of f. 370v with an incomplete proposition: 'Prop. 12. The most refrangible rays are swiftest. For the light of Jupiters Satellites is red at their immersion'; 'immersion' is the catchword.

[55] Newton, *Correspondence*, vol. 3, p. 202.

27 August and his next letter with Cassini's observations on 7 November. Book IV, Part I on thick plates had to be finished or very nearly so, since Newton was trying to end Part III with material in part deriving from that research. The last of his endings for Part III, which contains the theory of fits (ff. 339/40, 363/4), was drafted by 24 February 1692 when Flamsteed himself wrote Newton on eclipse colors, since it can be shown on textual grounds to precede ff. 365/70, which was written between 7 November and 24 February. Thus we can confidently date the conception of the theory of fits, the virtually complete final revision of Book III, Part III and Book IV, Part I, and the latter's incorporation into Book III as Part IV to the six months between 27 August 1691 and 24 February 1692. The final touches to the revision – in particular, the elimination of Proposition 12 with its assertion that red rays move slowest – and the preparation of the fair copy occurred sometime after 24 February 1692 and before David Gregory's visit to Newton in May 1694. The completion of the initial state of Books III and IV (stage 4 in Table 2) cannot be established as neatly. It was probably carried out around the time ff. 348/9 were written, that is, in the summer or fall of 1691, since on this sheet Newton was attempting to compose an ending for Part III utilizing, in part, knowledge of the cause of the colors of thick plates. When did Newton first begin his attempt to end Book III and the *Opticks*, which culminated in his work on diffraction and thick plates? I would conjecture sometime earlier in 1691, in the winter or spring, so that Newton's last active period of optical research (which also included the experiments for Book I, Proposition 7) lasted about six months to a year.

It is important to distinguish here between revising the end of Book III and actually writing out the final version, for the two need not have been contemporaneous. There is no doubt that the manuscript of the *Opticks* for Books I–III was essentially in final form by May 1694 when David Gregory spent several days with Newton in Cambridge.[56] In a memorandum dated 5, 6, 7 May Gregory noted that 'I also saw *Three Books of Opticks*' and gave a perfunctory summary of it.[57] This note has generally, albeit mistakenly, been interpreted to mean that the *Opticks* was complete by 1694.[58] Since, however,

---

[56] The watermarks of Books III and IV and their revise (stages 4 and 6 in Table 2) are of course insufficiently precise to date these sections and, in particular, to decide when between 1691 and 1694 the revise was written out. The watermark English royal arms + $\mathcal{AI}$/GLC is on dated papers from October 1690 to June 1691 and the watermark on the revise, horn + HG/MC, from October 1694 to September 1696. These dates are, however, entirely consistent with my account.

[57] Newton, *Correspondence*, vol. 3, pp. 336, 338; see also his memorandum of 4 May (pp. 312, 317), where Gregory says Newton could explain the colors of thick plates.

[58] See, for example, Shapiro, 'The evolving structure of Newton's theory of white light and color', *Isis*, 71(1980):211–35, on p. 212; and Westfall, *Never at Rest*, p. 523.

the *Opticks* was then still divided into four books, and not the three of the published edition, Book IV on diffraction was not yet finished and the *Opticks* remained incomplete. In my account of the composition of the *Opticks* I have thus far focused on the revision of Book IV, Part I and its subsequent absorption into Book III and have altogether ignored its former companion on diffraction, Book IV, Part II. Newton was evidently dissatisfied with it, and at this point removed it from the manuscript with the intention of revising it afterwards. In a fuller description of the *Opticks*, apparently written shortly after his memorandum concerning his visit with Newton, Gregory confirms this interpretation: 'I saw *Three Books of Opticks*, written in the English language ... The fourth [book], about what happens to rays in passing near the corners of bodies, is not yet complete. Nonetheless, the first three make a most complete work.'[59]

Word of the *Opticks* spread in scientific circles. In 1695 John Wallis, the Savillian Professor of Mathematics at Oxford and doyen of English mathematics, undertook a campaign to persuade Newton to publish it. Although Newton's letters to Wallis do not survive, it is apparent from Wallis's letters to him that besides his abiding fear of controversy Newton was withholding his book because he judged the last part on diffraction to be still incomplete.[60] In 1696, after spending most of his life at the University of Cambridge, Newton moved to London to assume the duties of Warden, and then Master, of the Mint. He no longer had the time and, more importantly, the interest to carry out further experiments on diffraction. By November 1702 he finally committed himself to publishing the *Opticks*.[61] From his papers he retrieved Book IV, Part II, completed more than a decade earlier, and revised it for publication. At the conclusion of the last of the eleven observations of Book IV (which was no longer divided into Parts) he explained that

> When I made the foregoing Observations, I designed to repeat most of them with more care and exactness, and to make some new ones for determining the manner how the rays of Light are bent in their passage by Bodies for making the fringes of Colours with the dark lines between them.

[59]   'Vidi ego *tres Optici libros* anglico idiomate conscriptos ... quartum de ijs quae radijs accidunt in transitu prope corporum angulos nondum absolvit. priores tamen tres opus absolutissimum constituunt.' (David Gregory, *Notae in Newtoni Principia*, Royal Society, MS 210, insert between pp. 55, 56).

[60]   Wallis to Newton, 30 April 1695, Newton *Correspondence*, vol. 4, pp. 116–17; see also pp. 100, 115, 130, 186, 188.

[61]   David Gregory recorded in his notebook that 'On Sunday 15 Nov. 1702 He [Newton] promised ... to publish his Quadratures, his treatise of Light, & his treatise of the Curves of the 2ᵈ Genre', in *David Gregory, Isaac Newton and their Circle. Extracts from David Gregory's Memoranda 1677–1708,* ed. W. G. Hiscock (Oxford: Printed for the Editor, 1937), p. 14.

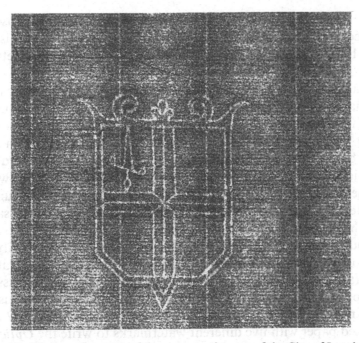

Figure 10. A beta radiograph of the watermark arms of the City of London on ULC, Add. MS 3970, f. 224, which is from Book IV of the *Opticks*, which became Book III shortly before publication. The countermark EB is divided by a chain line; a similar paper was used for the revised Latin translation of the *Opticks* in 1706, but the countermark EB is between a pair of chain lines. (By permission of the Syndics of Cambridge University Library.)

But I was then interrupted, and cannot now think of taking these things into further consideration. And since I have not finished this part of my Design, I shall conclude, with proposing only some Queries in order to a further search to be made by others.[62]

The last eight sheets, which include the sixteen appended Queries, were added to the manuscript (ff. 219–33, 339) in 1703; and a revised conclusion for Observation 13, Book III, Part IV (f. 218) replaced the original ending, which no longer survives. The *Opticks* was now complete, thereby bringing us to stage 7 of Table 2. These sheets are all in Newton's hand. Their watermark, arms of the City of London/EB (Figure 10), is also found on the sheets that were added in late 1703 to the 'De Quadratura curvarum', which was originally written in 1693 but then revised and published as the second of two mathematical essays appended to the first edition of the *Opticks*.[63] Sometime after Book IV was finished, Newton decided to combine Books I and II into

[62] Newton, *Opticks*, p. $_2$132.
[63] ULC, Add. MS 3962, ff. 1–4, 19–27; and Newton, *Math Papers*, vol. 8, p. 122.

one Book with two Parts, and then went through the manuscript and renumbered Books III and IV to II and III.[64] The *Opticks* was published by 16 February 1704 when Newton presented a copy to the Royal Society.[65]

The preceding account of the composition of the *Opticks* shows the usefulness of watermarks in reconstructing the principal stages in the writing of any manuscript. The principle of clustering, if nothing else, serves to raise questions about the sequence of writing and existence of insertions that may not otherwise be evident. The questions must, of course, be answered by other sorts of evidence. In fact, the assignment of clusters with the same watermark to the same period of composition is meaningful simply on the grounds that it is more likely than not that the same paper in two portions of the same work indicates that they were written in the same period. It is possible to develop a simple model that allows us to derive an estimate of the probability that the existence of clusters of the same watermark is due to chance rather than contemporaneity of purchase and writing. In the first place, let us assume that Newton bought the paper for the *Opticks* a few quires at a time, as is the case, and neither all at once nor in very tiny quantities. Between 1687 and 1694 Newton used paper with five different watermarks to write the *Opticks* (Table 2). To simplify the problem, let us suppose that for the *Opticks* he made one purchase of paper a year for five successive years.[66] Now, we should recall that as a consequence of the marketing process in Newton's England, the same paper would periodically appear on the market and also at any time there would be a number of other papers from different manufacturers available. In the second place, let us assume that sixteen different papers are on the market during the course of a given year. An inventory of paper offered for sale by two London merchants in 1674 lists seventeen papers of the pot size used for *Opticks*.[67] Thus, when Newton made a purchase at the stationer's shop in Cambridge, it is equally likely he would get any of these sixteen papers, though at any one time some of them (which ones, we do not know) would be unavailable.[68] Finally, since a pair of moulds lasted about four years, let us

[64] It is apparent that the change of Book numbers was made close to the time of publication, since the last book was initially headed 'the fourth Book of Opticks' and only afterwards was 'fourth' deleted and 'third' inserted (f. 219).

[65] J. Edleston, ed., *Correspondence of Sir Isaac Newton and Professor Cotes* (London, 1850; rpt. London: Frank Cass, 1969), p. lxxi, n. 147.

[66] The sixth paper for the last addition (stage 7) can be excluded. Since it was added about ten years after the rest of the manuscript was completed, it is highly unlikely that any of the earlier paper was available.   [67] See note 13.

[68] It should be noted that the feature that not all papers are available at one time does not *directly* enter the calculations, since it does not alter the probability of receiving a particular paper at any purchase. If, say, only half of the sixteen papers are on the market at any one time, the chance of purchasing a particular paper is still 1 in 16, or

assume that at the end of each year the four oldest (though we do not know which four) were replaced by four new ones; that is, four papers vanished and four appeared on the market each year, and at the end of four years the original sixteen papers would be completely replaced. To summarize the model: we assume that for the *Opticks* Newton made one purchase a year of any one of sixteen papers, and that there was an annual roll-over of four papers, so that during the course of five years thirty-two papers were on the market.

From this simplified model there is a 30 % probability that at least two of these papers are the same simply by the luck of the draw. Alternatively, the supposition that clusters of paper with the same watermark were bought at one time and written in the same period will be valid 70 % of the time utilizing no other evidence than watermarks.[69] There was, however, sufficient evidence of other sorts – as will generally be the case for any historical document – to demonstrate in all but two cases that clusters of the same watermark were written in the same period. The two cases in which the independent evidence was inconclusive were the insertion of the definitions and axioms after the completion of Books I and II (stage 2 in Table 2) rather than immediately after Book I alone; and the addition of Book II, Propositions 7–11 (stage 3) before rather than after the first state of Books III and IV. Having established the order of composition of all the other clusters on independent grounds, we can make a much stronger argument drawing solely on the probability of the sequence of watermarks. If we temporarily suspend the principle of clustering and assume that stage 2 was written between Books I and II, we can readily calculate that the probability that Newton purchased one paper, then another, and then the first one again is less than 1 in 16 (specifically, $15/16 \times 1/16$), or 6 %. This supposes that Newton bought or chose his paper in a random way, and that he did not have a separate stack of paper that he utilized for small insertions. If this reasonable alternative could be established, I would readily abandon the argument from watermarks, but in the absence of other evidence I believe that we must adopt the simpler and more probable supposition that it was added afterwards. A similar argument applies to the determination of stage 3.

$\frac{1}{8} \times \frac{1}{2}$. The periodic unavailability of paper is, however, incorporated into the model, in that it allows a typical paper to be on the market for four years rather than one. If moulds were used continually until they were worn out, so that a paper was on the market for only about one year, the pattern of watermarks would be quite different.

[69] Dennis White solved this complicated problem for me and calculated the probability. As it turned out, the realistic feature of an annual replacement of four moulds for four years is an unnecessary refinement. If it is assumed instead that there is a constant pool of thirty-two watermarks from which five purchases are made, then the probability of at least two being identical is 28 %, instead of the 29.5 % determined by the more realistic model.

The aim of this model and calculations is not to 'prove' the principle of clustering, but rather to show why it should work so well for the *Opticks*. The model, I believe, reasonably reflects the marketing of paper in Newton's day and his purchasing habits, although the specific numbers are somewhat arbitrary. Most of my estimates were conservative. For example, if I had assumed seventeen papers are for sale each year, as the inventory in fact showed, rather than sixteen, or that Newton's purchases were spread out over seven years (which is actually the narrowest documented interval) rather than five consecutive years, the principle of clustering would be valid three, four, or even five times more often than it is wrong. We can derive one more illuminating result from the model by dropping the assumption that Newton made only a single purchase of paper for a particular work each year and considering instead the totality of his purchases for all purposes. Let us also assume that there is a constant pool of thirty-two papers, rather than adopting the more complicated assumption of a roll-over of four papers each year.[70] If we suppose that he bought paper five times a year for five years, then it turns out that in the twenty-five purchases he would almost always (96% of the time) wind up with only about half of the thirty-two papers; more precisely, between twelve and eighteen of the marketed papers would not turn up in his selection. This result requires that during the course of five years there would be multiple purchases of the other watermarks, which is just what is found. (We cannot, of course, determine how many papers Newton did not buy.) This example illustrates why the principle of clustering must be restricted to a relatively small number of purchases for a single work like the *Opticks* or a related set of papers, such as those on a particular alchemical problem. Otherwise, it is virtually guaranteed that clusters of the same watermark will appear simply by chance. Finally, we should recall the assumption that each purchase involved a few quires devoted to that work. When writing a draft, such as the 'Fundamentum Opticae', or a small manuscript Newton would often use a number of papers at the same time. He would start with one, and when that ran out go to a stack of leftovers, or search for a few sheets elsewhere in his rooms, or buy more. Thus, his autograph copy of the 'Observations' begins with seven sheets with a horn and ends with two sheets with a foolscap (ULC, Add. MS 3970, ff. 501–17).

The manuscript of the *Principia* shows a similar pattern of clustering and also supports my principle, at least on the large scale on which I can test it. In his *Introduction to Newton's 'Principia'* I. B. Cohen presents a meticulous account of the composition and publication of the various editions of that great book. He describes in particular the composition of the first edition for which the printer's manuscript and other related documents – though a surprisingly

[70] See the preceding note.

Table 4. *The watermarks of the manuscript of the* Principia

Book	Folios	Number of sheets	Watermark
Bk. I	1–212	106	PT
Bk. II, Props. 1–49	213–332	60	$\mathscr{AI}$/PT
Props. 49–53	333–52	10	IV_1
Bk. III, Intro. – Prop. 6	353–62	5	$\mathscr{AI}$/PT
Prop. 6 – Prop. 41	363–440	39	$\mathscr{TI}$
Prop. 41	441–4	2	$\mathscr{AI}$/PT
Props. 41–2	445–60	8	IV_2

Note: The watermarks are denoted by their countermarks: PT = arms of Amsterdam/PT; $\mathscr{AI}$/PT = arms of Amsterdam + $\mathscr{AI}$/PT; IV = Arms of the Seven Provinces + $\mathscr{II}$/IV; $\mathscr{TI}$ = arms of the City of London + $\mathscr{AI}$/$\mathscr{TI}$; subscripts indicate different watermarks of the same species.

small number – survive. The final manuscript, now at the Royal Society of London (MS 69), consists of 460 folios in the hand of Newton's amanuensis, Humphrey Newton (no relation). Cohen is more alert to watermarks as evidence than most historians. On one occasion, for example, he invokes them in support of his account of the composition of Newton's Lucasian Lectures for 1684 and 1685, which are actually a draft of Book I.[71] When I brought the evidence of watermarks to bear upon his reconstruction of the writing of the first edition, it all either was consistent with his version or confirmed it. However, the clusters of watermarks in the manuscript of the *Principia* (Table 4) raise questions about its composition – just as it did with the *Opticks* – that are not directly addressed. As can be seen from Table 4, Book I consists entirely of one paper and the other two mostly of one sort. Like the *Opticks*, the three Books of the *Principia* were transcribed at different times, though in a much shorter period, something over a year as opposed to the sixteen years for the *Opticks*. Thus the manuscript of the *Principia* confirms what we found in the *Opticks*, namely, that different papers were used at different times.

Newton bought a large quantity of paper – at least five quires with watermark arms of Amsterdam/PT – for Book I, which was transcribed and then sent in April 1686 to Edmond Halley, his editor in London.[72] Newton had finished Book II in the summer of 1685, but he did not 'make ready' the fair copy until autumn 1686, and he did not send it to Halley until early March 1687.[73] A month later, on 14 April, Halley received Book III.[74] For the transcription of Book II Newton purchased at least three quires of paper with

[71]   Cohen, *Introduction*, p. 89.   [72]   *Ibid.*, p. 69.

[73]   Newton, *Correspondence*, vol. 2, pp. 464, 470, 472.   [74]   *Ibid.*, p. 473.

watermark arms of Amsterdam + $\mathscr{A}\mathscr{I}$/PT, and most of that book consists of this paper, except for the last ten sheets. We may ask whether Newton simply ran out of paper and then used whatever paper he could lay his hands on to complete Book II, or whether the last ten sheets are a later addition (after the first sixty sheets were transcribed), or whether those ten sheets replaced an earlier ending for Book II? For Book III Newton bought at least two quires of paper with the watermark arms of the City of London + $\mathscr{A}\mathscr{I}$/$\mathscr{T}\mathscr{I}$ (which is similar to but not the same as that used for the *Opticks*). Did Newton have the first five sheets of Book III transcribed right after completing the bulk of Book II (which perhaps then had a different ending than in its final state); or did he afterwards insert the first five sheets to replace a different beginning; and did he insert the two sheets of ff. 441–4 at the same time? Do the two short runs of different watermarks in the last ten sheets of Book III indicate later insertions, or rather than Newton had run out of paper in completing his *magnum opus*? Since I am not sufficiently familiar with the manuscripts associated with the *Principia*, I have not seriously attempted to answer these questions, or to pose still more alternatives that readily come to mind. It may well turn out that not enough evidence survives to resolve most of them. One of these questions, however, can be readily answered, and it again provides confirmation of the principle of clustering. On textual grounds Cohen and Koyré have convincingly shown that the final eight sheets of Book III with the watermark arms of the Seven Provinces + $\mathscr{I}\mathscr{I}$/IV are a later addition after Book III was completed.[75] From my own experience with the *Opticks* I am convinced that it is fruitful to pose such questions raised by clusters of watermarks, and that when they can be answered it is only by a careful analysis of the surviving papers.

At least in my own period of research, seventeenth- and eighteenth-century science, historians have unduly ignored watermarks as an independent source of evidence.[76] They supplement rather than replace any of the traditional methods in the historian's repertoire. Though my initial aim was to use watermarks for dating, it turned out that the interval of use of any paper was too long to be of great value, since most of Newton's papers can be dated otherwise to less than that interval. While they thus generally only confirm independently derived results, when these results are only tenuously established, their testimony allows greater confidence in them. Even such a fine edition as Christiaan Huygens' *Oeuvres complètes* would have gained greater

[75] Alexandre Koyré and I. Bernard Cohen, eds., *Isaac Newton's Philosophiae Naturalis Principia Mathematica. The Third Edition* (1726) *with Variant Readings*, 2 vols. (Cambridge: Harvard University Press, 1972), vol. 2, p. 715.

[76] For a pioneering use of watermarks in the history of seventeenth-century science see Stillman Drake, 'Galileo gleanings XXI. On the probable order of Galileo's notes on motion'. *Physis*, 14(1972): 55–68.

authority if watermarks had been used to assist in the dating and ordering of his numerous papers. The greatest value of watermarks is when they supply new evidence whether for dating or for analyzing the sequence of composition of a given work. If they are used to date and order all the remaining papers of one person, then it is essential that as much paper as possible be examined and an internal catalogue of watermarks be established from that examination. General catalogues of watermarks will rarely contain more than a few of those watermarks, and even if they do, they will say little about when your author used them. Such a project might prove to be useful in dating and arranging large and poorly arranged collections of papers like that of Robert Boyle. One may also judge that it is not worth the effort, since such fine historical work has been done without it. This is certainly true, but an independent source of historical evidence should not be lightly dismissed.

— III —

Newton's *Principia*

8   The critical role of curvature in Newton's developing dynamics

J. BRUCE BRACKENRIDGE

The problem of quadratures and tangents has a long history that began in antiquity and reached fruition in the form of integral and differential calculus in the nineteenth century. The first subject is concerned with the area bounded by a given curve, and the history of mathematics is rich in descriptions of the 'method of exhaustion' employed in antiquity. The modern textbook describes this problem in terms of a 'definite integral', and the name of Leibniz often appears. The second subject is concerned with the tangent to a curve, which Euclid defined (very simply) as a line touching a circle at only one point and which Archimedes (for example) determined for a spiral. The modern textbook describes this problem in terms of a 'derivative', and the name of Newton often appears. The methods and concepts of Archimedes, however, are far removed from that of a nineteenth-century mathematician. Nevertheless, there is a sense of continuity in the nature of the problem of tangents, despite the very pronounced differences in the techniques.

In modern terms, the tangent at a point to a given curve $y = f(x)$ is the derivative of y with respect to x evaluated at that point. This derivative, dy/dx, is the slope of the tangent line at that point. The curvature, however, measures the rate of change of the angle of inclination of the tangent line with respect to the arc length of the curve. In more descriptive terms, the curvature measures the rate of 'bending' of the curve at a point. If the curve is a circle, this rate of bending is uniform. Thus, the accepted method of measuring the bending of a curve is to associate it with the curvature of the circle that best approximates the curve in the immediate vicinity of the given point. The circle that represents the best approximation is called, following Leibniz, the 'osculating circle'. Such a circle has the same first and second derivatives as the curve at the given point.[1]

[1]   A rather simple method of obtaining an expression for ρ the radius of the osculating circle, i.e., the radius of curvature, is as follows: the expression for the circle is (1) $x^2 + y^2 = \rho^2$. Taking the derivative of (1) with respect to x produces $2x + 2y(dy/dx) = 0$, or (2) $dy/dx = -x/y$. Taking the second derivative of (1) produces (3) $2 + 2(dy/dx)^2 + 2y(d^2y/dx^2) = 0$. Using (1) and (2) in (3) gives $d^2y/dx^2 = \rho^2/y^3$. Solving the above expressions for ρ as a function of the first and second derivatives, one obtains the following expression: $\rho = [1 + (dy/dx)^2]^{\frac{3}{2}}/[d^2y/dx^2]$.

The historical continuity associated with the problem of tangents, however, is not to be found in the history of curvature. In a very fundamental way, curvature arises independently out of the work of Newton and Leibniz. It is traditional, however, to associate the history of curvature with the history of evolutes and involutes, particularly when a mechanical description of evolutes and involutes is employed in which the unrolling of a thread from the evolute produces the involute. It is in this sense that Huygens employed the mathematics of evolutes and involutes to produce a mechanism that gives rise to an isochronic pendulum.[2] Viewed in this way, the centers of the osculating circles of the involute curve fall on the evolute curve.

The history of the evolute is then traced back to Apollonius and his work on minimal lines in Book 5 of his *Conics*. This traditional view is summarized by Boyer: 'The concepts of radius of curvature and evolute had been adumbrated in the purely theoretical work on *Conics* of Apollonius, but only with Huygens' interest in horology did the concepts find a permanent place in mathematics.'[3] As Whiteside notes[4] and as Yoder argues very persuasively, however, neither Apollonius' work on minimal lines nor Huygens' work on evolutes are concerned directly with curvature.

> Apollonius was solving a different problem, and only through hindsight do his theorems on normals become the vestiges of curvature and evolute. In fact, startling though it may sound, Huygens' early work on evolutes also had nothing to do with curvature, and only hindsight turns his derivations of companion curves for the cycloid and conics into a method for determining curvature.[5]

Curvature in Newton's early mathematics

Newton's early work on curvature appears in his bound notebook, the *Waste Book*, in late 1664.[6] He was concerned with the problem of normals, curvature, and tangents. By May of 1665 he had developed the general formula for the radius of curvature for a function $y = f(x)$.[7] By the winter of 1670–71, Newton was writing a treatise, now called 'Methods of Series and Fluxions', that included as Problem 5 the 'Determination of the curvature of a curve at any point' and as Problem 6 the 'The quality of curvature at a point'. In this

[2] Joella G. Yoder, *Unrolling Time: Christian Huygens and the Mathematization of Nature* (Cambridge University Press, 1988), pp. 97–115.

[3] C. B. Boyer, *A History of Mathematics*, (Princeton University Press, 1985), p. 421.

[4] Isaac Newton, *The Mathematical Papers of Isaac Newton*, ed. D. T. Whiteside, 8 vols. (Cambridge University Press, 1967–81), vol. 3, pp. 163–5, note 308.

[5] Yoder, *Unrolling Time*, p. 104.

[6] Newton, *Mathematical Papers*, vol. 1, p. 245.   [7] *Ibid.*, p. 245, note 76.

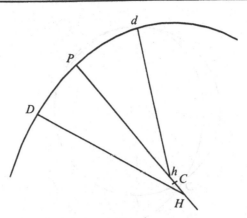

Figure 1. Newton's original method for locating the center of curvature (C), which uses three normals (DH, PC, dh) to the curve (DPd).

treatise he again obtained the general measure of curvature in Cartesian coordinates[8] and in addition also expressed it in polar coordinates.[9]

In problem 5 of 'Methods of Series and Fluxions', Newton listed five defining conditions that could serve to find the radius or center of curvature.[10] In the first method, he drew three normals to the curve, as in Figure 1. The center of curvature at the point P is the point C, where the points H and h of the normals DH and dh meet as the points d and D approach P. Newton then developed the general expression for the radius of curvature, ρ, in Cartesian coordinates:

$$\rho = (1 + y'^2)^{\frac{3}{2}}/y''$$

where y' is, in current terms, the derivative of y with respect to x.[11] Newton then provided a series of examples in which he calculated the radius of curvature of various curves, including that of the three conic sections.[12] At the conclusion of his examples he stated the following: 'Since all geometrical curves...can be referred to right-angled ordinates, I believe I have done enough. Anyone who desires more will provide it without difficulty by his own efforts, especially if, as a bonus, in illustration of the point I add a method for spirals.'[13] And he then developed an expression for the radius of curvature in

8 *Ibid.*, p. 156, note 288. The radius of curvature DC is given in terms of z, where z is dy/dx.

9 *Ibid.*, vol. 3, p. 171, note 330. The Cartesian expression $\rho = [1 + (dy/dx)^2]^{\frac{3}{2}}/[d^2y/dx^2]$ can be written in terms of a parameter t as $[x'^2 + y'^2]^{\frac{3}{2}}/[x'y'' - y'x'']$, where $x = x(t)$, $y = y(t)$, and $x' = dx/dt$, etc. It may also be expressed in terms of polar coordinates as follows: $[r^2 + r'^2]^{\frac{3}{2}}/[r^2 - rr' + 2r'^2]$ or $r[1 + (r'/r)^2]^{\frac{3}{2}}/[1 - (r'/r) + 2(r'/r)^2]$ where $r' = dr/d\theta$ and $r'' = d^2r/d\theta^2$. If $z = r'/r$, then $z' = (r''/r) - z^2$, and thus the radius of curvature is given by $r[1 + z^2]^{\frac{3}{2}}/[1 + z^2 - z']$, which is the form given in note 330 cited above with $r = y$ and $z' = z$.

10 *Ibid.*, pp. 153–4. 11 *Ibid.*, p. 157. 12 *Ibid.*, p. 159. 13 *Ibid.*, p. 169.

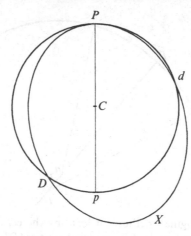

Figure 2. Newton's alternate method for locating the center of curvature (*C*), which uses three points of intersection (*d*, *D*, *P*) of the tangent circle (*Pp*) with the curve (*PX*).

polar coordinates,[14] and he again offered a series of examples in which he demonstrated how the technique could be applied to specific curves.

Newton was not content, however, with what he had done when he had 'long last...disposed of the problem'. He had employed fluxional analysis in solving these problems and neither that technique nor the type of problems themselves were familiar to his contemporaries. So he proposed 'to touch upon another [method] at once more obvious and closer to the usual methods of drawing tangents'.[15] The technique that replaced the fluxional analysis was a relationship, called Hudde's Rule, that holds for repeated roots of an equation. Schooten's second Latin edition of Descartes's *Geometry*, which Newton used extensively as early as 1664, contained Hudde's demonstration: if a polynomial had a double root, then that root also satisfies a second polynomial that is obtained by multiplying the terms of the first polynomial by an arithmetic progression.[16] When the arithmetic progression is set so that it multiplies the constant term of the first polynomial by zero, then the result is the same as that obtained by taking the derivative of the first polynomial.[17]

The first of the five defining conditions, which Newton set down as being capable of defining the radius of curvature, employed the meeting of three tangents as in Figure 1 above. This method was the traditional way to approach such problems.[18] In the last of these conditions, however, Newton introduced a new method that dealt directly with the tangent circles drawn to

[14]   *Ibid.*, p. 171.   [15]   *Ibid.*, p. 175.   [16]   *Ibid.*, vol. 1, p. 214.

[17]   This result can be demonstrated for a repeated root a as follows: if $f(x) = (x-a)^2 g(x)$, then the derivative $f'(x) = (x-a)[2(x-a)g(x)+(x-a)^2 g'(x)]$, which is $(x-a)h(x)$. Thus a is a root of $f'(x)$ as well as of $f(x)$.

[18]   Newton, *Mathematical Papers*, vol. 3, pp. 152–3, note 273.

the curve at the point at which the radius of curvature was to be determined. Figure 2 displays the circle Pp that is tangent to the curve PX at the point P. In general the tangent circle has three points in common with the curve: D, d, and the point of tangency P. The equation generated by combining the equation of the specific curve and of the tangent circle is the equation of these three common points of intersection. As the circle shrinks and the points d and D approach the point P, however, the tangent circle approaches the osculating circle at the point P (whose radius is the radius of curvature) and the roots are no longer unique; i.e., they are repeated roots and thus subject to Hudde's Rule.

The example Newton offered to demonstrate this technique was that of the parabola, given by the equation $y^2 = ax$.[19] The osculating circle centered at the points x_c and y_c is given by the equation $(x+x_c)^2+(y+y_c)^2 = \rho$, where ρ is the radius of curvature. The equations are combined by substituting $x = a^{-1}y^2$ from the first equation into the second equation and obtaining the following polynomial for the three points of intersection:

$$y^4 + (0)y^3 + (2ax_c+a^2)y^2 + (2a^2 y_c) + [a^2(x_c^2 y_c^2)] = \rho,$$

where y is now the coordinate of the intersections. The polynomial can be written in a simpler format:

$$y^4 + 0 \times y^3 + A \times y^2 + B \times y + C = 0$$

Since the roots of this equation are repeated, Hudde's Rule may be applied twice to obtain two more equations from which the coordinates of the center of curvature, x_c and y_c, can be obtained. Newton elected to multiply the first equation by the arithmetic series 4,3,2,1,0 and obtained the polynomial:

$$4 \times y^4 + 3 \times 0 \times y^3 + 2 \times A \times y^2 + 1 \times B \times y + 0 \times C = 0,$$

which reduced to the equation:

$$4y^3 + 2Ay + B = 0. \qquad (1)$$

The polynomial above was then multiplied by the arithmetic series, 3,2,1,0 which produced the polynomial:

$$3 \times 4 \times y^4 + 2 \times 3 \times 0 \times y^3 + 1 \times 2 \times A \times y^2 + 0 \times B \times y + (-1)0C = 0,$$

which reduced to the equation:

$$y^2 + (\tfrac{1}{6})A = 0. \qquad (2)$$

The coordinates of the center of curvature (x_c and y_c) can be obtained as functions of the general point y and the constant a from A and B, which are determined by Eqns. (1) and (2).[20]

Newton has provided this alternate method of solution so that those unfamiliar with fluxional analysis could employ a more 'common way of

19   Ibid., p. 177.   20   Ibid., p. 177.

operation', i.e., Hudde's Rule. The modern reader will note, however, that the polynomials obtained by Hudde's Rule can be obtained simply by taking the derivative of each of the equations.

Curvature in Newton's early dynamics

Figure 3 is the drawing from Newton's original autograph draft of the tract *On the Motion of Bodies in Orbit*, which he sent to the Royal Society in November of 1684.[21] It is used to illustrate Theorem 3 (to become Theorem 5, Proposition 6 in the *Principia*), his basic dynamical principle that is to be applied to specific problems such as that of elliptical motion. The path of the body *P* around the force center *S* is the general curve *APQ*. In the absence of a force, the body would move along the tangent line *PR*. In the presence of a force, the body moves along the curve *PQ*. The line segment *QR* therefore represents the deviation of the body from its inertial path and is thus, in some way, a general measure of the force.

Newton's first use of such a deviation as a measure of the force, however, occurs much earlier than 1684. Herivel locates its first appearance in Newton's analysis of circular motion in his writings before 1669.[22] By 1684, however, he has extended it to the analysis of other orbits as displayed in Figure 4, which is Newton's drawing for elliptical motion in Problem 2 and Problem 3 in the original draft of the *Motion of Bodies in Orbit*.[23] But this linear measure is not without difficulty for Newton. In the text, he stated that *QR* is to be constructed parallel to *SP*, the original line of force. But at the point *Q* the force is in the direction *SQ*, not *SP*, and careful inspection of Newton's drawing will show that *QR* is more nearly in line with *SQ* than with *SP*, despite Newton's textual qualification. The question of the direction of the linear deviation *QR* may appear to be resolved as *Q* approaches *P*, but the question of the mathematical rigor of Newton's treatment of the limit is still an object of controversy.[24]

Figure 5 is Newton's revised diagram for the problems of elliptical orbits that appears in his revised tract *On the Motion of Spherical Bodies in Fluids*, which Whiteside dates to December 1684.[25] In this diagram, Newton no longer drew the deviation *QR* as a linear line segment but rather drew it as a solid

[21] Isaac Newton, *The Preliminary Manuscripts for Isaac Newton's 1687 Principia 1684–1685*, intro. D. T. Whiteside (Cambridge University Press, 1989), p. 4.

[22] John Herivel, *The Background to Newton's Principia* (Oxford University Press, 1965), p. 12. See p. 192 for the dating of the text.

[23] Newton, *Preliminary Manuscripts*, p. 5.

[24] See Whiteside's notes in Newton, *Mathematical Papers*, vol. 6, p. 6, note 12, pp. 35–9, note 19, and p. 41, note 29 and Eric J. Ation, 'Polygons and parabolas: some problems concerning the dynamics of planetary orbits', *Centaurus*, 31(1989):207–21.

[25] Newton, *Preliminary Manuscripts*, p. 17.

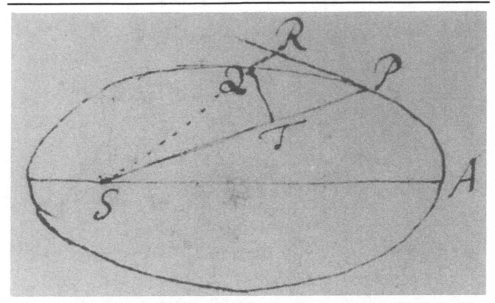

Figure 3. The diagram for Newton's basic dynamic theorem from the original tract *On Motion*, in which the force is proportional to $QR/QT^2 \times SP^2$. (From Newton, *Preliminary Manuscripts* by permission of the Syndics of Cambridge University Library.)

wedge, whose top edge is roughly parallel to SQ and whose lower edge is parallel to SP. Thus the solid wedge includes the range of inclinations that the force will have as the particle moves along the arc PQ. Clearly, Newton was attuned to and concerned with the problems evolved in the variation of the direction and magnitude of the force in the limiting process of Q going to P.[26]

Setting aside these difficulties, Newton assumed that the force can be considered to be constant over the deviation QR as the point Q approaches the point P. Under this assumption, the linear displacement QR is proportional to the constant force (i.e., a linear acceleration) and inversely proportional to the square of the time. The time is given from his first theorem (Kepler's area law); it is proportional to the area $QT \times SP$. Thus the force is measured by the ratio $QR/QT^2 \times SP^2$, which is the basic dynamic ratio employed in the first edition of the *Principia* to calculate the nature of the force for a given force center and a given orbit.

In the revised editions of the *Principia*, however, Newton added an alternate ratio as a dynamic measure of the force. Moreover, he briefly considered adapting this alternate method as the primary measure but finally held to the

[26] This variation in the drawing with the deviation QR as a solid wedge is indicated by dashed lines in A. Rupert Hall and Marie Boas Hall, *Unpublished Scientific Papers of Isaac Newton* (Cambridge University Press, 1962), p. 252. The significance of the deviation in the drawing was called to my attention by D. T. Whiteside.

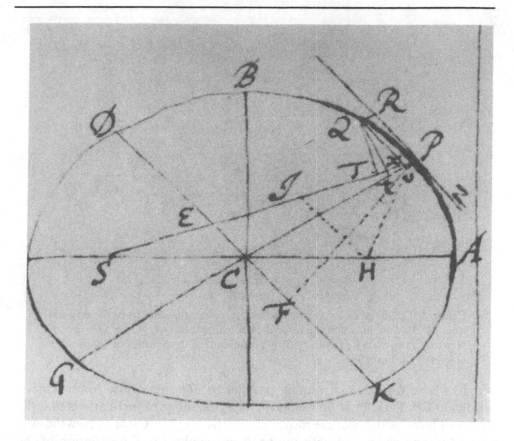

Figure 4. The diagram for Newton's solution to the problem of elliptical motion from the original tract *On Motion* (autumn 1684) with the deviation *QR* given as a single line. (From Newton, *Preliminary Manuscripts* by permission of the Syndics of Cambridge University Library.)

linear measure.[27] In developing this dynamic variant, Newton made use of his knowledge of curvature. But curvature played a role in Newton's dynamic thought long before the revised published editions. The following statement appears in his *Waste Book* of late 1664:

> If ye body b moved in an Ellipsis yn its force in each point (if its motion in yt point bee given) may bee found by a tangent circle of equal crookedness wth yt point of ye Ellipsis.[28]

Herivel refers to this statement in the context of Newton's discussion of circular motion and of the struggle to distinguish between centrifugal and

---

[27] Whiteside states that 'Newton toyed with the notion of making this derived property basic.' Newton, *Mathematical Papers*, vol. 6, p. 548, note 25. See J. Bruce Brackenridge, 'Newton's mature dynamics: revolutionary or reactionary?' *Annals of Science*, 45(1988):451–76. See p. 474 for a photograph of Newton's draft document and a discussion of the text.   [28] Newton, *Mathematical Papers*, vol. 1, p. 456.

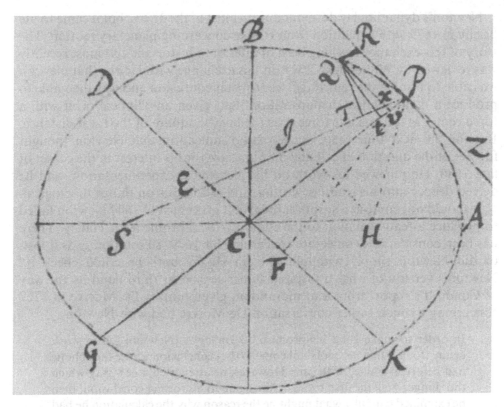

Figure 5. The diagram for Newton's solution to the problem of elliptical motion from the revised tract *On Motion* (December 1684?) with the deviation *QR* given as a solid wedge. (From Newton, *Preliminary Manuscripts* by permission of the Syndics of Cambridge University Library.)

centripetal forces. He sees this statement as 'evidence that he had actually contemplated applying the idea of centrifugal force to the problem of motion in an ellipse at a very early stage'.[29] Herivel's attention then remains fixed on the centrifugal/centripetal distinction, and he does not discuss or comment upon the role of the 'tangent circle of equal crookedness' in the problem of the motion in an ellipse. He mistakes the y^n ($=$ then) for y^t ($=$ that), which further obscures Newton's meaning. But it is precisely this reference to the osculating circle of curvature that Newton was to develop into his refined measure of force, and it is significant that one finds reference to it at this very early stage. It demonstrates the close interaction of Newton's mathematical and dynamical ideas, even in the first year of his creative life, as Whiteside makes clear in his evaluation of the statement.[30]

[29]   Herivel, *Background to Principia*, p. 60.
[30]   Newton, *Mathematical Papers*, vol. 1, p. 546, note 3.

Newton's dynamics lay dormant during most of the 1670s, but it came to life briefly in 1679 with a question from Hooke concerning planetary motion. The story of this exchange has long been of interest to historians and most recently was reviewed by Whiteside.[31] It was in this exchange with Hooke that Newton was able to resolve the centrifugal/centripetal confusion and was then able to produce a mathematical demonstration that, given an elliptical orbit with a force center at a focus, the force is as the inverse square of the focal distance. But for the next four years there was no indication that Newton thought further of the question of celestial motion, save for his interest in the comet of late 1680. He renewed his work on his alchemical experimentation, and he delved deeper into his reading, writing, and reflections on things theological.

It should not come as a surprise, then, that in August of 1684 Newton failed to produce a ready solution to the elliptical/focal problem for Halley. There has been considerable conjecture concerning his 1679/80 solution; was it lost, or did Newton simply withhold it from Halley until he could check it? Newton's version of what transpired comes second or third hand by the way of Conduitt's report from a memorandum given him by De Moivre in 1727 concerning a much earlier conversation De Moivre had with Newton.

> In order to make good his promise [to Halley]...[Newton] fell to work again, [to find that he] could not come to that conclusion wch he thought he had before examined with care. However, he attempted a new way which tho^u longer than the first, brought him again to his former conclusion, then he examined carefully what might be the reason why the calculation he had undertaken before did not prove right, and he found that having drawn an Ellipsis co[a]rsely with his own hand, he had drawn two Axes of the Curve, instead of... two Diameters somewhat inclined to one another, whereby he might have fixed his imagination to any two conjugate diameters, which was requisite he should do. That being perceived, he made both his calculations agree.[32]

The essential item to note is that there were two separate solutions. The first solution clearly depended upon the inclination of the conjugate diameters. Figure 6 is the drawing for the solution of the problem that Newton sent to Halley in London in November of 1684 and that eventually appeared in the first edition of the *Principia*. The lines *PG* and *DK* are the conjugate diameters; *DK* is drawn parallel to the tangent to the curve at *P* and *PG* is drawn from the point *P* and bisects *DK*. The angle between the conjugate diameters, therefore, is not a right angle. They must, as Newton noted, 'be somewhat inclined to one another'. The lines *BC* and *AC* are the two axes of the curve, which are at right angles, and which were drawn mistakenly for the conjugate diameters. It is the

[31] Newton, *Preliminary Manuscripts*, pp. xii–xiv.
[32] *Ibid.*, p. xv. For the source of the manuscript, see p. xix. note 27.

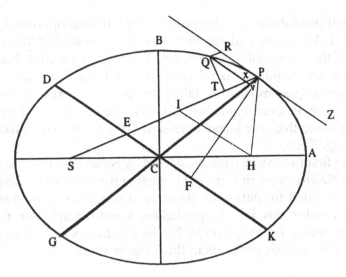

Figure 6. A drawing of the diagram for Newton's solution to the problem of elliptical motion that uses the conjugate diameters *DK* and *PG*.

line *PF* that is normal to *DK*, not *PG*. Thus, the version sent to Halley shares with the missing solution of 1679/80 a dependence upon the conjugate diameters. With the appearance of the tract sent to Halley, the history of the development of the text is on firmer ground because it is the first solution to the problem that has survived. A second version of the tract appeared shortly after the first version under the title of *The Motion of Spherical Bodies in Fluids*, and then a much expanded version appears in Book One of the first edition of *Principia*, which in turn is modified in the revised second and third editions.

But what of the alternate solution of 1684, the one 'attempted in a new way' by Newton? In contrast to his first attempt, it may not depend in any direct fashion upon the conjugate diameters because it was after 'he attempted a new way which... brought him again to his former conclusion' that he 'then examined carefully what might be the reason why the calculation he had undertaken before did not prove right' and found the faulty axes. Such an alternate solution, which does not depend upon the conjugate axes, did not appear in either of the essays on motion or in the first edition of the *Principia*. In 1690, however, Newton produced at the request of John Locke a 'simpler set of demonstrations of the properties of elliptical motion'.[33] Whiteside claims that the 'essence' of the alternate solution of 1684 may yet survive in the simplified *Demonstrations that the Planets by their gravity towards the Sun may move in Ellipses* which Newton was to send to John Locke in March 1690.[34] John Herivel has argued that the lost 1679/80 solution may well be the

[33]   Hall, *Unpublished Papers*, pp. 293–301 and Herivel, *Background to Principia*, pp. 246–54.   [34]   Newton, *Preliminary Manuscripts*, p. xvi.

prototype of this solution. In this proposal he has been opposed by A. R. Hall and by D. T. Whiteside and supported by R. S. Westfall.[35] Inspection of the diagram of the Locke solution supports Whiteside's position because it does not employ the conjugate axes. Thus, it could not have been the original solution whose reconstruction failed because the axes of the conjugate diameters were incorrectly drawn. Clearly, the Locke solution calls upon something other than the linear deviation QR and its associated measure of force, $QR/QT^2 \times SP^2$.

Consider first the special case considered by Newton in Proposition 2, which appears in Newton's manuscript but is missing from Locke's manuscript. Here Newton calculated the nature of the force of attraction at the two vertices of the ellipse's major axis, i.e., the perihelion A and the aphelion C. Figure 7 is Newton's drawing for Proposition 2, which contains only a single principal diameter. The analysis proceeds in three major steps.

Step I. From the area law of Proposition 1, the area AFE and the area CFD swept out in equal times are equal. If the time is so short that the arcs AE and CD can be replaced by the normals AM and CN, then the equal areas are given by the triangular areas $(\frac{1}{2})AE \times AF$ and $(\frac{1}{2})CF \times CD$. Thus $AE:CD::CF:AF$.

Step II. Now AM and CN are tangents to the ellipse at the two ends A and C, and EM and DN are normals to the tangents at the points M and N. Then Newton asserted:

[35] Three dates have been offered for the original composition of the manuscript upon which the Locke/Newton variation of the demonstration of motion in an ellipse rests: 1689/90, the date at which Locke received it, 1679/80, the date of the 'lost' solution generated by the Newton/Hooke correspondence on the subject, and 1684, the date of the Halley/Newton meeting in Cambridge. In 1961 Herivel first argued that the original MS must predate the first version of the 1684 tract *On Motion*; 'Newtonian Studies III. The originals of the two propositions discovered by Newton in December 1679', *Archives internationales d'histoire des sciences*, 14(1961):23–33, 'Newtonian Studies IV', *Archives internationales d'histoire des sciences*, 16(1963):14 and in 1965, Herivel, *Background to Principia*, pp. 111–17. Westfall supported Herivel's claim in 1969 and repeated in 1980; 'A note on Newton's demonstration of motion in ellipses', *Archives internationales d'histoire des sciences*, 22(1969):51–60; and *Never at Rest*, (Cambridge University Press, 1980), p. 387, note 145, and also p. 403. In response to Herivel's initial claim, the Halls argued in 1963 that there is no manifest improbability involved in setting the date at 1690 and therefore one is not entitled to set an earlier date; A. R. Hall and M. B. Hall, 'The dates of "On Motion in Ellipses"', *Archives internationales d'histoire des sciences*, 16(1963):23–8. Whiteside has set the date for the protype manuscript as August, 1684, i.e., just following Halley's visit; 'Newtonian dynamics', *History of Science*, 5(1966):104–17, note 4; Newton, *Mathematical Papers*, vol. 3 (1974), pp. 553–4, note 33, and Newton, *Preliminary Manuscripts* (1989), pp. xx–xxi, note 53.

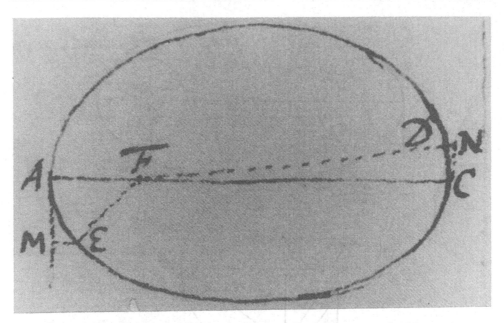

Figure 7. The diagram, which does not use conjugate diameters, for Newton's solution to the *restricted* problem of elliptical motion at aphelion and at perihelion from the tract *Demonstrations that Planets...may move in Ellipses* (1690). (From Newton, *Preliminary Manuscripts* by permission of the Syndics of Cambridge University Library.)

And because the Ellipsis is alike crooked at both ends these perpendiculars *EM* and *DN* will be to one another as the squares of the arches *AE* and *CD*.[36]

Now this statement is an echo of the statement in 1664 found in Newton's *Waste Book* previously cited above:

If the body *b* moved in an Ellipse then its force in each point...will be found by a tangent circle of equal crookedness with that point of the Ellipse.[37]

The 'alike crooked' and the 'equal crookedness' in the two statements refer to the curvature of the ellipse. The two statements differ in that the first refers to the osculating circle only at perihelion and aphelion, while the second refers more generally to the osculating circle at any point. One should not, however, allow that difference to obscure the possible connection. In the second statement Newton pointed out, in effect, that an element of the path in the vicinity of any point can be replaced by the element of the osculating circle at that point. In the first statement, he selects aphelion and perihelion to be the

[36]   Herivel, *Background to Principia*, p. 249.   [37]   *Ibid.*, p. 60.

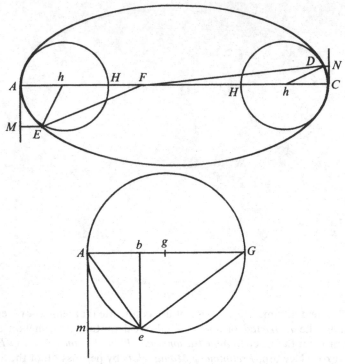

Figure 8. A drawing of the diagram in Figure 7 with the osculating circles inserted above and the details of a tangent circle restructured below.

two points of interest. These two statements were sufficient for Newton to claim that the normals *EM* and *DN* are proportional to the square of the arcs *AE* and *CD*; i.e., $EM:DN::AE^2:CD^2$. He offered no demonstration of this result other than the statement given above. Figure 8 is Newton's drawing with the osculating circles *AEH* and *CDH* inserted and below an enlarged drawing of a circle *AeG*, which could be constructed tangent to the curve above at the point *A* (or at point *C*). Following the outline provided by Newton in Lemma 11 of the *Principia*, the dependence of the square of the arcs *AE* and *CD* upon the normals *EM* and *DN* can be demonstrated as follows. Triangles *Abe* and *AeG* are similar because they contain the common angle *eAb* and right angles *Abe* (by construction) and *AeG* (chords of a circle). Thus $Ae:Ab::AG:Ae$, or since $Ab = me$, then $Ae^2/me = AG$. Then as the tangent circle shrinks down to the osculating circle at point *A*, the diameter *AG* approaches the unique value of *AH*, i.e., the diameter of the osculating circle at the point *A*. Moreover, as the point *e* approaches the point *A*, the chord *Ae* can be replaced by the arc *Ae*. This result, which depends upon the properties of the circle of curvature, is clearly obvious to Newton, who defends it in Lemma 11 simply with a parenthetical expression, '(from the nature of the circles)'.[38] Thus as Newton stated, 'because the Ellipsis is alike crooked at both ends' then $EM:DN::AE^2:CD^2$.

Step III. Now Newton employed the normals DN and EM as the linear deviations and hence as proportional to forces in given times.

> Tis by the force of the attractions that the bodies are drawn out of the tangents from M to E and from N to D and therefore the attractions are as those distances MR and ND.[39]

Thus, from Step III the forces are proportional to the deviations ME and ND, which in turn from Step II are proportional to the squares of the arcs AE and DC, which in turn from Step I are proportional to the focal distances FE and FD. Thus, as Newton stated,

> The attraction at the two ends of the Ellipsis shall be reciprocally as the squares of the body in those ends from that focus.[40]

Thus, Newton achieves the solution to this special case without the explicit use of conjugate diameters. He has used properties of the circle of curvature at the two special points, but he has not employed it in any general fashion. In any event, whether or not Newton in 1690 had in mind his insight of twenty-five years before, there can be no doubt that, because he did not allude to conjugate diameters in it, this part of the Locke proof could not have been the first solution and could well have been the alternative one concocted by him in 1684.

In Proposition 3, which follows the special case of elliptical motion at aphelion and perihelion discussed above, Newton treats the general case of elliptical motion at any point under a central force directed toward the focus. As it was for the diagram for Proposition 2 (Figure 7), Newton's diagram for

---

[38] Isaac Newton, *The Mathematical Principles of Natural Philosophy*, trans. Andrew Motte, intro. I. Bernard Cohen, 2 vols. (London, 1729, rept. London: Dawsons, 1968), Bk. I, Lemma 11, vol. 1, p. 51. The enlarged drawing of the osculating circle also provides the basis for the following contemporary analysis in which the angles θ_1, and θ_2 in radians are given as $\theta_1 = CD/r$ (where $r = Dh$) and $\theta_2 = AE/r$ (where $r = Eh$). The normals to the tangents, DN and EM are given as: $DN = r - r\cos\theta_1 = r(1 - \cos\theta_1)$ and $EM = r - r\cos\theta_2 = r(1 - \cos\theta_2)$. Now the function $(1 - \cos\theta)$, called the versine, can be evaluated for small angles (which is the situation here when E approaches A and D approaches C) by employing the expansion for the cosine function; i.e., $\cos\theta = (1 - \theta^2/2! + \theta^4/4! - ...)$. To the first order of approximation then, the normals DN and EM are given by the square of the angles θ_1 and θ_2, which in turn are given by the arcs AE and CD.

[39] *Ibid.*, p. 249.   [40] *Ibid.*, p. 248.

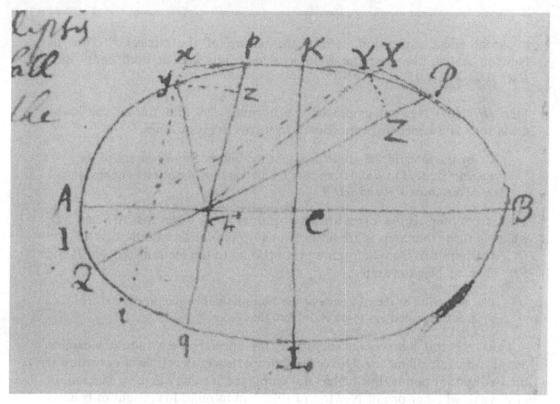

Figure 9. The diagram, which does not contain conjugate diameters, for Newton's solution to the *general* problem of elliptical motion at any point from the tract *Demonstrations that Planets...may move in Ellipses* (1690). (From Newton, *Preliminary Manuscripts* by permission of the Syndics of Cambridge University Library.)

Proposition 3 (Figure 9) still does not contain an explicit rendering of the conjugate diameters. In contrast to the simplified solution for the special case considered in Proposition 2, however, the solution for the general case in Proposition 3 is much more complicated. Since this solution was sent by Newton to Locke in response to the latter's appeal for a 'simpler solution', it is tacitly assumed that in some meaningful way, Proposition 3 is 'simpler' than the proof in the *Principia*. But careful inspection of this solution challenges that assumption of simplicity. D. T. Whiteside[41] has pointed out that the demonstration depends on four scarcely self-evident propositions in Apollonius' *Conics*, or their post-Apollonian equivalents referred to in the demonstration by Newton as 'shewed by all those who treat of yᵉ Conic sections'.[42] Since the demonstration of each of these four propositions would have presented to Locke a mathematical challenge of the sort he was seeking

---

[41]   D. T. Whiteside in personal correspondence, 14 December 1990.

[42]   Herivel, *Background to Principia*, p. 251 or Hall, *Unpublished Papers*, p. 298.

to avoid, simplicity is achieved only if Newton ordains that their truth be accepted without demonstration. But that is not the way in which Newton developed the solution and his devotion to mathematical rigor in his presentation of the lemmas undercuts any claim that the solution was generated solely or primarily in response to a plea for simplicity. If the four propositions are stated as simple lemmas, however, then Whiteside holds that a general solution follows in quite a simple fashion. (See Appendix A for his reconstruction of Newton's demonstration cast in the format of four lemmas, a corollary, and a proposition.)

But what sense can one claim that the proof of Proposition 3 does not employ conjugate diameters? As noted above, they do not appear explicitly in the diagram for that proposition. They do appear, however, in the diagram for Lemma 3 that proceeds Proposition 3 and are therefore employed in its solution. Thus, one can claim that Newton's demonstration of Proposition 3 is free of conjugate diameters only in the sense that he casts them out of the main frame of his argument of Proposition 3 and relegates them to his lemmatical subroutines. This solution lies somewhere between the solution in Proposition 2, which does not employ conjugate diameters in any fashion, and the solution in the first edition of the *Principia*, which employs them directly.

There are two distinct methods, therefore, employed by Newton to obtain a solution to the Kepler problem. The first method employs the conjugate diameters and the linear deviation QR in the form of the ratio $QR/QT^2 \times SP^2$. It has its beginning in the early work on circular motion, is employed in the 1679/80 solution, is attempted first unsuccessfully and then successfully in 1684, and provides the basis for the two early tracts on motion and the first edition of the *Principia*.

The second method employs the curvature to obtain the basic relationships and does not directly use the conjugate diameters. It has its beginning in Newton's early work on curvature, is employed as the alternate solution of 1684, and contains the 'essence' of the solution sent to Locke in 1690. It does not appear in the first edition of the *Principia*.[43] This argument, therefore, supports Whiteside's dating of 1684 for the prototype of the Locke/Newton MS and rejects Herivel's claim that it might have been the 'lost' MS of 1679/80.

[43] The concept of curvature is basic to Lemma 11 in all editions, but in the first edition this lemma is applied only to the problems of circular and spiral motions (Propositions 7 and 9) and not to the problems of elliptical motion (Propositions 10 and 11). In the revised editions curvature is employed explicitly, in the form of Lemma 11, to all problems (Propositions 7-13).

Curvature in Newton's revised dynamics

Following the publication of the first edition of the *Principia*, Newton immediately began to correct and revise it. In the early 1690s, however, he went far beyond correcting mathematical and verbal errors and began to produce what Whiteside calls a 'grand scheme of revision.'[44] In these radical revisions to the format of the first edition of the *Principia*, Newton proposed three distinct but interrelated methods for producing solutions to problems of motion such as the Kepler problem. In contrast to the first edition, in which curvature appeared only in a minor role, the proposed radical revisions promoted curvature to a major position. The first of the new methods employed the linear deviation QR and the area law as in the first edition. In the revised editions, however, they were used to develop a 'comparison theorem' in which the ratio of the forces at any two points was given instead of the linear dynamics ratio of the first edition.[45]

The second method again began with the linear deviation QR and it developed the linear dynamics ratio, $QR/QT^2 \times SP^2$, that appeared in the first edition. The basic supporting arguments for this ratio, however, are changed from Lemma 10, in which curvature does not appear, to Lemma 11, in which curvature is a central element.[46] Otherwise the form of the solutions are as they appear in the first edition.

In dramatic contrast, however, Newton developed a third method that followed as a corollary to the second method and which depended explicitly upon curvature. This new dynamics ratio is given by $1/SY^2 \times PV$, where SY is the normal to the tangent through the center of force S, and PV is the chord of the circle of curvature drawn from the general point P on the curve through the center of force S to its terminus on the circle of curvature V. It is in this third method that the use of curvature reaches fruition. Newton developed a lemma in which the *latus rectum* was related to the chords of the circle of curvature for a conic drawn through its center and focus.[47] Then there followed a problem solution in which the force necessary for motion in a conic about a given center was given in two or three lines (in contrast to the full page solution required for a linear dynamics ratio).

These dramatic revisions of the early 1690s did not all appear in the published revised second edition of 1713. In his original proposed revisions, Newton had introduced an entirely new numbering scheme for the propositions and lemmas to set off the three new methods. Each theorem was numbered as was each set of problem solutions. Thus, most of the theorems,

---

[44] Newton, *Mathematical Papers*, vol. 6, p. 568, note 1. [45] *Ibid.*, pp. 570–80.

[46] J. Bruce Brackenridge, 'Newton's unpublished dynamical principles: A study in simplicity', *Annals of Science*, 47(1990): 3–31.

[47] Newton, *Mathematical Papers*, vol. 6, pp. 583–84.

problems, and lemmas that were to be carried over from the first edition would have had to be renumbered. Even more challenging, all the internal references would have had to be found and revised. Apparently the enormity of this task was prohibitive. In the preface to the first edition of the *Principia*, Newton apologized for not revising the numbering scheme as he added items to his first manuscript: 'Some things found out after the rest, I chose to insert in places less suitable, rather than change the number of the propositions and the citations.'[48] It is a position he maintained in the revised editions of 1713 and 1726, in which he retained the numbering format of the first edition of the *Principia*. Thus, the three distinct methods of the proposed but unpublished revisions were scattered about the published revisions.

In the published revised edition, Newton retained Proposition 6 as the repository for his basic dynamical theorems. The linear dynamics ratio of the first edition (method 2 above) appeared again but with a primary reference to Lemma 11 (and hence to curvature). But more dramatic was the addition of the alternate circular dynamics ratio (method 3 above) as a corollary to Proposition 6 and of its application as an added alternate solution to the problem solutions that precede the Kepler problem (i.e., circle/circumference, spiral/pole, and ellipse/center). The comparison theorem (method 1 above), which held the status of an independent theorem in the unpublished revision, appeared in the published revision simply as a corollary to the problem solution in Proposition 7. It is this theorem that Newton used to provide the alternate solution for the Kepler problem (i.e., ellipse/focus), although he could have used the alternate dynamics ratio as for the other problems. Clearly, he wished to demonstrate the full extent of all three methods within the confines of his old numbering scheme. Unfortunately, the scattered presentation of these revised methods in the published editions has provided as much confusion as it has produced clarification for the readers of the revised editions.

Newton's outline for a solution for the inverse problem

Curvature played even yet another role in the published editions. The problems that Newton addressed in the opening three sections of the *Principia* were what the eighteenth century called the 'direct problem', i.e., given the path and the force center, find the nature of the force necessary to maintain that motion. Given the knowledge from Kepler of the elliptical orbits of the planets, what was unknown was the nature of the force. From the perspective of the nineteenth and twentieth centuries, which accept the law of universal gravitation, it is the 'inverse problem' that is of major interest; i.e. given the

---

[48] Newton, *Mathematical Principles*, (Motte) vol. 1, pp. v–vi.

force and the force center, find the path. Newton also struggled with the inverse problem in Section 8 of the *Principia*, but in the opening three sections he was primarily concerned with solutions to the direct problem.

Even in the opening sections, however, Newton did not entirely ignore the inverse problem. He provided an outline of a proof of the inverse problem that employed his solution of the direct problem. In the first edition Newton appeared to assume that the solution of the direct problem provided by itself a solution of the inverse problem. In Propositions 11, 12 and 13 he had demonstrated that for elliptical, hyperbolic, and parabolic motion with the force directed toward a focus, the magnitude was as the inverse square of the distance. In Corollary 1 of Proposition 13 in the first edition, he made the following claim:

> It is a consequence of the three most recent propositions that, should any body *P* depart from the place *P* following any straight line *PR* and with any velocity whatsoever, and if it be instantaneously snatched up by a centripetal force which is proportional to the square of the distance from its center, this body shall move in some one of the conic sections.[49]

This statement is an improvement over the initial version in *On Motion* sent to Halley in 1684, in which only the problem of the ellipse was presented and the other two conic sections were not discussed. Nevertheless, as it stands in the first edition, it is unsatisfactory, and it was criticized by Johann Bernoulli before the publication of the revised second edition.[50] Newton's solution to the direct problem of conic/focal motion in the 'three most recent propositions', i.e., Propositions 11, 12, and 13, does not, by itself constitute a solution for the inverse problem. As Whiteside puts it, 'The hidden assumption here made that no curve other than a conic may, in an inverse-square force-field centered on *S*, satisfy all possibilities of motion at *P* wants – for all its manifest plausibility – an explicit, rigorous justification, and Newton was later fairly criticized by Johann Bernoulli for merely presupposing its truth without demonstration.'[51]

---

[49]   Isaac Newton, *Philosophiae Naturalis Principia Mathematica* (London, 1687, rept. London: Dawsons, 1953), p. 55. The translation from Newton, *Mathematical Papers*, vol. 6, p. 147.

[50]   For Whiteside's discussion of Bernoulli's criticism of the issue and of Newton's response to it by way of John Keill's directed correspondence, see Newton, *Mathematical Papers*, vol. 6, pp. 146–9, note 124 and pp. 348–51, note 209. See also note 73 on pp. 56–7 and note 38 on p. 556. In private correspondence, Whiteside has more recently reviewed his analysis of the outline of the inverse problem as set forth in Corollary 1 to Principia, Book 1, Propositions 11–13. Hopefully, it will soon appear in the public domain. See also E. J. Aition, 'The inverse problem of central forces', *Annals of Science*, 20(1964):81–99.

[51]   Newton, *Mathematical Papers*, vol. 6, pp. 147–8, note 124.

The revised second edition, however, contains the following extension to Corollary 1 of Proposition 13:

> And the contrary. For the focus, the point of contact, and the position of the tangent being given, a conic section may be described, which at that point shall have a given curvature. But the curvature is given from the centripetal force and the bodies velocity given: and two orbits mutually touching one the other, cannot be described by the same centripetal force and the same velocity.[52]

Newton intended for this extension to serve as an outline for the solution to the inverse problem, and apparently it satisfied Johann Bernoulli, for in 1719 he wrote to Newton,

> Gladly I believe what you say about the addition to Corollary 1, Proposition 13, Book I of your incomparable work, the *Principia*, that this was certainly done before these disputes began, nor have I any doubts that the demonstration of the inverse proposition, which you have merely stated in the first edition of the work, was yours; I only said something against the form of that assertion, and wished that someone would give an analysis that led *a priori* to the truth of the inverse [proposition] and without supposing the direct [proposition] to be already known. This indeed, which I would not have said to your displeasure, I think was first put forward by me, at least so far as I know at present.[53]

Whiteside claims that Newton could have employed a general polar curve with given curvature and thus could have computed the elements of the general conic.[54] But as Newton did not explicitly produce such a solution, there have been continued challenges to the validity of his outline. Most recently, claims have been made that Newton's solution, even as given in outline in the corollary above, is radically flawed and contains a gross, irreparable fallacy.[55] But now, as in the past, every critic produces a series of defenders. (See Appendix B for the details of one recent skirmish.)

Independent of the validity of his outlined proof, it is of interest to note that here, as elsewhere in the *Principia*, Newton assumes on the part of the reader a background in the mathematics of curvature. In Lemma 11, the reference to

[52] Newton, *Mathematical Principles*, (Motte) Bk. I, Proposition 13, vol. 1, p. 85.

[53] *The Correspondence of Isaac Newton*, ed. H. W. Turnbull, J. F. Scott, A. R. Hall and Laura Tilling, 7 vols. (Cambridge University Press, 1959–77), vol. VII, p. 78. This letter was called to my attention by D. T. Whiteside.

[54] See Newton, *Mathematical Papers*, vol. 6, pp. 146–9, note 124 for Whiteside's presentation of the details of a potential solution of the inverse problem.

[55] R. Weinstock, 'Dismantling a centuries-old myth: Newton's *Principia* and inverse-square orbits', *American Journal of Physics*, 50(1982):610–17, and 'Long-buried dismantling of a centuries-old myth: Newton's *Principia* and inverse-square orbits', *American Journal of Physics*, 57(1989):846–9.

curvature appeared in only a parenthetical expression, even though curvature was primary to the relationship developed in the lemma. In the final corollary to Proposition 7, Problem 2, the extension of the analysis on circular motion to the comparison theorem and general motion was defended by a final single sentence that called upon curvature. And here in Proposition 13, Newton again assumes that he need not reproduce his extensive work on curvature.

A detailed solution of the inverse problem

It is more satisfying, perhaps, to produce a solution to the indirect problem that does not explicitly employ the solution to the direct problem. There is no evidence that Newton actually did the calculation, but Whiteside claims that he was capable of such a solution and has demonstrated the validity of this claim.[56] Looking back at his earlier work on curvature, however, one sees there in detail what only appears above in outline. It is instructive, therefore, to flesh out these details in contemporary terms by taking each element from Newton's earlier work and developing from them a proof of the inverse problem.

Central to the analysis is Newton's expression for the radius of curvature (ρ) in polar coordinates (r, θ), which he developed in the early 1670s:[57]

$$\rho = r(1+z^2)^{\frac{3}{2}}/[(1+z^2)-z'],$$

where z is the slope of the curve $(1/r)(dr/d\theta)$ and z' is $dz/d\theta$. The derivative z' can be expressed in terms of $(1+z^2)$:

$$z' = dz/d\theta = (dz/dr)(dr/d\theta) = (r/r)(dr/d\theta)(dz/dr) = rz\,dz/dr$$
$$= (r/2)\,d(1+z^2)/dr.$$

Thus, the expression for the radius of curvature (ρ) can be written:

$$\rho = r(1+z^2)^{\frac{3}{2}}/[(1+z^2)-(r/2)\,d(1+z^2)/dr],$$

Two other general expressions may be employed that only require the force to be directed toward a given center (and hence that the motion be in a plane). Figure 10 is derived from the diagram for Proposition 6, Theorem 5 in the revised edition of the *Principia*. The general curve is *APB*, the tangent to the curve is *ZPY*, the center of force is *S*, and *PVX* is the circle of curvature at the point *P* with the radius $\rho(= PO)$. The first expression is obtained from Proposition 1, i.e., Kepler's area law, and is:

$$rv \sin(\alpha) = K,$$

where r is the radius *SP*, v is tangential velocity at *P*, the angle α is between the radius *SP* and the tangent *ZPY* (the complement of the angle θ shown in

56   Newton, *Mathematical Papers*, vol. 6, pp. 148–9, note 124. In what follows, I develop a solution for the inverse problem in the notation currently employed by physicists.
57   Newton, *Mathematical Papers*, vol. 6, p. 171, note 330.

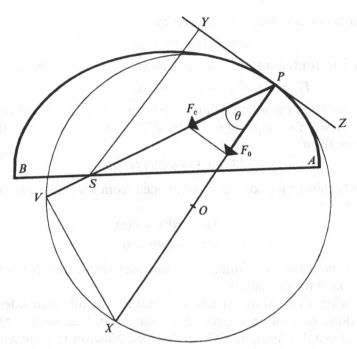

Figure 10. A drawing of a portion of the revised diagram for Newton's basic dynamic theorem from the second and third editions of the Principia with the osculating circle inserted.

Figure 10), and K is a constant proportional to the area swept out per unit time. The second expression obtains from Newton's replacement of motion along an incremental arc of the general curve at point P with uniform circular motion along an incremental arc of the circle of curvature at the same point. It is given as:

$$F_0 = F_s \cos(\theta) = F_s \sin(\alpha) = v^2/\rho,$$

where F_s is the force directed toward the center of force S and ρ is the radius of curvature. The component of force F_0 provides the centripetal circular acceleration v^2/ρ as required in Proposition 4 of the *Principia*. Newton has explicitly employed this uniform circular replacement in Lemma 11, Corollary 3; and in Proposition 7, Problem 2, Corollary 5; and he has implicitly employed it in other discussions. Combining the two expressions, the radius of curvature (ρ) can be written as:

$$\rho = v^2/F_s \sin(\alpha) = K^2/F_s r^2 \sin^3(\alpha),$$

The angle α can be expressed in terms of the slope z as:

$$z = (1/r)\, dr/d\theta = \operatorname{ctn}(\alpha) = \cos(\alpha)/\sin(\alpha),$$

and thus the expression $(1+z^2)$ is given as:

$$(1+z^2) = 1+\cos^2(\alpha)/\sin^2(\alpha) = 1/\sin^2(\alpha).$$

Thus the radius of curvature (ρ) is given by:

$$\rho = K^2(1+z^2)^{\frac{3}{2}}/F_s\,r^2,$$

which when substituted into the original curvature equation produces:

$$F_s = (K^2/r^2)[(1+z^2)/r - (\tfrac{1}{2})\,\mathrm{d}(1+z^2)/\mathrm{d}r].$$

If one now expresses the relationship in terms of the radius r and the angle θ instead of Newton's slope $z(=1/r\,\mathrm{d}r/\mathrm{d}\theta)$, it then takes on the more contemporary form:

$$F_s = (K^2/r^2)[1/r + \mathrm{d}^2(1/r)/\mathrm{d}\theta^2],$$

which is the 'orbital equation' that is obtained from the equations of motion in polar form, i.e.,

$$F_r = m(\ddot{r} - r\dot{\theta}^2) = F(r),$$
$$F_\theta = m(r\ddot{\theta} + 2\dot{r}\dot{\theta}) = 0$$

by using the chain rule to eliminate the time derivative and the constant of motion $K = mr^2\dot{\theta}$ to eliminate $\dot{\theta}$.[58]

Thus, Newton's curvature equation is mathematically equivalent to the polar equations of motion expressed as the orbital equation. Moreover, the circular dynamics ratio, $1/YS^2 \times PV$, which Newton introduced into the revised edition, is also identical with the orbital equation. From Figure 10,

$$PV = 2\rho\cos(\theta) = 2\rho\sin(\alpha) \text{ and } SY = r\cos(\theta) = r\sin(\alpha).$$

Thus, the circular dynamics ratio is $1/SY^2 \times PV = 1/[r^2\rho\sin^3(\alpha)]$, which is the orbital equation:

$$F_c = 1/[r^2\rho\sin^3(\alpha)] = (K^2/r^2)[(1+z^2)/r - (\tfrac{1}{2})\,\mathrm{d}(1+z^2)/\mathrm{d}r]$$
$$= k^2/r^2[1/r + \mathrm{d}^2(1/r)/\mathrm{d}\theta^2].$$

Whiteside has previously demonstrated that the linear dynamics ratio, $QR/QT^2 \times SP^2$, when expanded in a Taylor series, reduces to the orbital equation.[59] It comes as no surprise, therefore, that the circular dynamics ratio also is given by the orbital equation. In this latter case, however, the relationship is direct, and no expansion is required since it arises out of the curvature equation and hence out of the equivalence of the first and second derivatives of the curve and its osculating circle.

Having demonstrated the general equivalence of the orbital equation and the curvature equation, it is instructive to consider the special case of the inverse square force and then to solve the curvature equation for the polar equation of the general conic. The curvature equation can be written as:

$$\mathrm{d}(1+z^2)/\mathrm{d}r - 2(1+z^2)/r = -(2/K^2)(r^2F).$$

[58]   See any textbook on mechanics such as Keith R. Symon, *Mechanics* (Reading, Mass.: Addison-Wesley, 1971) 3rd edn, pp. 124–6. The orbital equation appears as Eqn. (3.222).   [59]   Newton, *Mathematical Papers*, vol. 6, p. 42, note 30.

If the force F is given as c/r^2, and $(1+z^2)$ is written as the function $f(x)$, the equation reduces to:

$$df/dr - 2f/r = -2A,$$

where $A = c/K^2$. This equation is a first order linear differential equation whose complementary solution f_c satisfies the equation $df_c/dr - 2f_c/r = 0$ and is given by $f_c = Cr^2$, where C is an arbitrary constant. The particular solution, f_p, is given by $2Ar$, and thus the full solution f is:

$$f = f_c + f_p = 2Ar + Cr^2.$$

Substituting $(1+z^2)$ for f (as defined above) and $(B^2 - A^2)$ for C (where B is arbitrary and A is given above), the following is obtained:

$$z = r[B^2 - (1/r - A)^2]^{\frac{1}{2}} = (1/r)\,dr/d\theta.$$

Solving for the angle θ, one obtains the following relationship:

$$\theta = (1/r^2)\,dr/[B^2 - (1/r - A)^2]^{\frac{1}{2}} = \cos^{-1}[(1/r - A)/B] - \varepsilon,$$

where ε is a constant of integration. Solving for $1/r$, the polar equation of the general conic is obtained:

$$1/r = A + B\cos(\theta + \varepsilon).$$

Thus, as Newton has claimed, given the curvature from the force, the path is uniquely determined. Whether Newton could have produced a version of this proof, as Whiteside claims, is a matter of personal conviction. I, for one, have no doubt that he could. But he need not have done so, for the outline provided in Corollary 1 of Proposition 13 is adequate.

Conclusion

Newton's conception of curvature as an important devise in the analysis of equations has long been recognized. He developed independently an expression for the radius of curvature in both Cartesian and polar coordinates. His writings on the subject are extensive, including a large number of examples in which conic sections are prominent. It is also important to note, however, the central role that curvature plays in his dynamics. The two explicit measures that he developed for the force, the linear dynamics ratio $QR/QT^2 \times SP^2$, and the circular dynamics ratio, $1/SY^2 \times PV$, have been demonstrated to be equivalent to the orbital equation, which is the combined polar equations of motion. Newton does not employ these relations in the form of a second order differential equation. Nevertheless, he demonstrates that he is aware of the relationship between the force and the curvature. And the outlines that he provides for solutions to both direct and inverse problems are substantiated both in his extensive work with curvature and with the correspondence to contemporary solutions. The ultimate reduction of these early dynamic forms to their modern equivalents clearly must not be attributed to Newton. What is

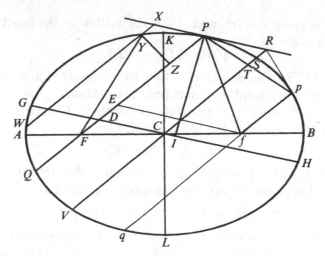

Figure 11. A composite figure for the lemmas and Proposition 3 for the Locke manuscript.

true, however, is that the modern form of the second order differential equation, $F = m\,\mathrm{d}^2a/\mathrm{d}t^2$, which the textbook calls 'Newton's second law', resides implicitly in Newton's dynamic expression of curvature. Moreover, the evidence is clear that such ideas were in his thoughts on dynamics as early as 1664, that they appeared in his solutions of 1684, and that they reached fruition in his unpublished revisions of the 1690s. Newton made the role of curvature quite clear in these unpublished revisions, even to the extent of proposing a lemma with the heading, 'To find the curvature of curves'. In the *Principia*, however, the critical role of curvature was not made clear save for the few explicit statements such as found in Corollary 1 of Proposition 13 that appear in the outline for the inverse problem. Moreover, references to curvature such as those found in Lemma 11, in Proposition 7, and in Proposition 13 require the reader to supply the details of the underlying analysis of curvature. Nevertheless, as one looks back from the vantage point of three hundred years of successful analysis, the central and fundamental role of curvature for Newton in his revised dynamics can be seen as the forerunner of modern dynamics.

Appendix A: restructured 'simplified' solution to the direct problem

D. T. Whiteside has suggested that if one accepts the following four propositions from Apollonius' *Conics*, then the four lemmas and the corollary give rise to a simplified version of the proof found in the Newton and the Locke manuscripts (see note 41). The figures found in Lemma 1, Lemma 3, and Proposition 3 of the Newton manuscript are replaced by a single figure (see Figure 11). Comparison with the original drawing for Proposition 3 (see Figure 9) will show that the point now labeled W appeared as I in the original and the point E in the drawing for Lemma 3 (not shown) is now labeled as I. The point now labeled W in the figure is a general point on the ellipse of axes AB, KL, center C and foci F, f. The tangent XP at P meets VCS, which is parallel to QFP, at R; qfp, which is parallel to QFP, meets the ellipse in p, and T is the meet of Pp with CS; GCH and fE, each parallel to XPR meet PG in D and E respectively, with PI perpendicular to GH. WYX, meeting the ellipse in W and Y, and PR in X, is any parallel to QP, with YZ perpendicular to QP.

Lemma 1. Since $FC = Cf$, there is $FD = DE$.[60] Whence, since PI bisects angle FPf (from Apollonius, *Conics* III, Prop. 48 – which Newton omits to mention),[61] and so $PE = Pf$, there is $PD = (\frac{1}{2})(PF + (PE \text{ or})Pf) = (\frac{1}{2})AB$.

Lemma 2. From the property 'shewed by all those who treat of y^e Conic sections'[62] (from Apollonius, *Conics* III, Prop. 42) there is $CR/CS = CS/CT$.[63] Here $CR = PD = (\frac{1}{2})AB$ (by lem. 1), $CS = (\frac{1}{2})VS$ and $CT = (\frac{1}{2})(PF + (pf \text{ or})FQ)$[64] $= (\frac{1}{2})PQ$. Accordingly, $VS^2 = AB \times PQ$.

Lemma 3. 'A Property of the Ellipse [*lege* any conic] demonstrated by all that write of the conic sections.'[65] (from Apollonius, *Conics* III, Prop. 16)[66] $WX \times YX/PX^2 = CS^2/CH^2$, that is, $(VS/GH)^2$.

[60] From similar triangles FDC and FEf, $FD/FC = FE/2FC$ or $2FD = FE$ and thus, $FD = DE$.

[61] Proposition 48 'With the same things being so, it must be shown that the straight lines drawn from the point of contact to the points produced by the application make equal angles with the tangent.' Apollonius, *Conics III* in *Great Books of the Western World* (GBWW) ed. M. J. Adler, 54 vols. (Encyclopedia Britannica, 1952), vol. 11, p. 787.

[62] Newton, *Preliminary Manuscripts*, p. 244. See also, Herivel, *Background Principia*, p. 251.

[63] Proposition 42: 'If in an hyperbola or ellipse or circumference of a circle or opposite sections straight lines are drawn from the vertex of the diameter parallel to an ordinate, and some other straight line at random is drawn tangent, it will cut off from them straight lines containing a rectangle equal to the fourth part of the figure to the same diameter.' Apollonius, *Conics* III in *GBWW*, vol. 11, p. 780. On the line preceding the final statement on page 783, 'rect. HF,FM = sq.FG', where comparison of diagrams show that $HF = CR$, $FG = CS$, and $FM = CT$. Thus $CR/CS = CS/CT$.

[64] In the trapezoid $FfpP$, where $FC = Cf$ and FP, CT, and fp are parallel, CT is the mean of PF and fp. Thus, $CT = (\frac{1}{2})(PF + pf)$, where further, $pf = FQ$.

[65] Newton, *Preliminary Manuscripts*, p. 244.

[66] Proposition 16 'If two straight lines touching a section of a cone or circumference of a circle meet, and from some point of those on the section a straight line is drawn parallel to one tangent and cutting the section and the other tangent, then as the squares on the tangents are to each other, so the area contained by the straight lines between the section and the tangent will be to the square cut off at the point of contact.' Apollonius, *Conics* III in *GBWW*, vol. 11, p. 747.

Lemma 4. 'They have also demonstrated that all Parallelograms circumscribed about an Ellipsis are equall [in area]'.[67] (from Apollonius, *Conics* VII, Prop. 31)[68] Whence $2PI \times GH = AB \times KL$.

Corollary. Since, from similar triangles, $PX/YZ = PD/PI =$ (by Lem. 1) $AB/2PI$, there is $WX \times YX/YZ^2 = (PX/YZ)^2(WX \times YX/PX)^2 =$ (by Lem. 3) $(AB/2PI \times VS/GH)^2 =$ (by Lem. 4) $(VS/KL)^2 =$ (by Lem. 2) $(AB \times PQ/KL^2)$ and therefore $YX/YZ^2 = AB/KL^2 \times PQ/WX = (1/L) \times PQ/WX$, where, by definition, $L = $ *latus rectum* $= KL^2/AB = 2CK^2/AC$.

Proposition. Since by Proposition 1 [Newton's generalization of Kepler's area law] the time of orbit over arc PY is measured by the area (PFY), where f is the force which at P acts along PF to F, in the limit as arc PY becomes vanishingly small, there will be $Y = (\frac{1}{2})ft^2$, where $t \propto (PFY) = \Delta PFY = (\frac{1}{2})PF \times YZ$, and therefore, $f \propto (1/PF^2)$ Lim $(XY/YZ^2) = (1/PF^2)(AB/KL^2) = (1/PF^2)(1/L)$, since Lim $(PQ/WX) = 1$ and $(AB/KL^2) = (1/L)$. Thus, the force f is given proportional to the inverse square of the distance PF.

Appendix B: Newton's outline for a solution to the inverse problem

The objections to be discussed in what follows were put forth by R. Weinstock in two papers that appeared in the American Journal of Physics in 1982 and 1989. I shall not refer to the considerable contributions of D. T. Whiteside on this question, cited previously in this paper, nor to his more recent review of the subject, not yet published. Rather, I shall discuss Weinstock's criticism and a recent rebuttal of it by B. Pourciau.

In his first paper, Weinstock points out what he sees as a gross irreparable fallacy in the outline. In a second paper he notes that there were a number of objections raised to his initial claim and proposes to overcome all such objections by further asserting that Newton 'begins the purported proof that inverse-square force implies conic-section orbit with the hypothesis of a particle moving in a conic-section orbit!' Thus, Weinstock claims that Newton violates a 'simple principle of elementary logic', in the light of which 'all possible doubt of the immanence of fallacy evaporates.'[69] Pourciau agrees with the claim, made in

[67] Newton, *Background to Principia*, p. 244.

[68] This statement also appears as Lemma 12 in the *Principia*. See Newton, *Mathematical Papers*, vol. 6, p. 34, note 14 for Whiteside's discussion of it.

[69] R. Weinstock, 'Dismantling a centuries-old myth' and 'Long-buried dismantling of a centuries-old myth' (see note 55). In these two articles Weinstock claims that Newton is guilty of gross logical fallacies in his outline of a proof. Moreover, he claims that his case against Newton is so strong 'that no competent person who is acquainted with the facts should refuse to credit'. I must admit, however, that despite spending considerable time reading Weinstock's papers, more time discussing these papers with colleagues in the history of science, and even more time in correspondence with Weinstock himself, I still do not understand, much less agree with, his criticism of

Weinstock's first paper, that a gap exists in the logic of Newton's outline. He disagrees with Weinstock, however, in the import of such a lapse. Once the gap is identified, Pourciau sees the filling of it as intuitively obvious and easy to prove. He points out, in fact, that in another section of the *Principia* Newton has proved the statement that is needed to fill the gap. Moreover, he rejects categorically the claim, made in Weinstock's second paper, that Newton has created a circular argument in violation of a basic principle of logic and he demonstrates how the solution to the direct problem fits logically into the outline for a solution to the inverse problem.[70]

The original statement in the first edition '*If any body goes from the place P...urged by the action of a centripetal force, that is reciprocally to the square of the distance of the places from the center*' can be seen to express the supposition that there exists some *actual motion* with the following five attributes: (1) initial position P, (2) initial speed V, (3) tangent PR, and a force (4) directed to a fixed

Newton. Moreover, Weinstock appears to see a conspiracy on the part of historians of science to evade or hide this perceived gross fallacy of Newton's (see notes 9 and 11 in the first paper and p. 848 of the second). He sees his thesis as being 'judged unacceptable by a few men who had been cultivating the *Principia* orchard since long before I [Weinstock] wandered in to pluck a wormy apple from one of its trees'. I therefore encouraged a colleague, Bruce Pourciau (a mathematician with no ties to the community of historians of science) to read these papers and to correspond with Weinstock and then to explain to me the point that is so obvious to Weinstock and so elusive to historians of science. Weinstock readily agreed to the exchange and after a number of letters Pourciau was convinced that a 'minor gap' could be found, namely that Newton 'forgets to show that there is a motion on his constructed conic with a central acceleration but this GAP is easy to prove and *intuitively obvious*'. Such a gap, however, is far from a gross fallacy in logic as claimed by Weinstock, and once the gap is restored, Newton's outline provides a method for a solution to the inverse problem.

[70] Bruce Pourciau, Department of Mathematics, Lawrence University, Appleton, Wisconsin. The version of Newton's outline to follow in this paper came from personal correspondence and conversations with Pourciau: now see 'On Newton's proof that inverse square orbits must be conics', *Annals of Science*, 48(1991). Other recent commentaries on Weinstock's claim can be found in Eric J. Aiton, 'The solution of the inverse-problem of central forces', *Archives internationales d'historie des sciences*, 38(1988):pp. 274–5 and Herman Erlichson, 'Comment on "Long-buried dismantling of a centuries-old myth: Newton's *Principia* and inverse-square orbits", by Robert Weinstock [*Am. J. Phys.* 57,846–9(1989)]', *American Journal of Physics*, 58(1990):pp. 884–6. In Newton, *Mathematical Papers*, vol. 6, p. 350, Whiteside refers to an extensive debate that prefigures the modern controversies. In addition to the corollary to Proposition 13, there is the question of the role Proposition 17 plays in the logic of the inverse-problem. On the one hand, the paper by Kruse (*Acta Eruditorum*, Oct. 1718, pp. 454–66) raises the issue of circularity in this proposition, while on the other hand, Euler in *Mathematica* (1736) (*Opera Omnia*, series II, vol. 1, p. 221) claims that with this proposition Newton solved the inverse-problem. Weinstock reacts to Euler's claim with the comment, 'Even Euler – the overwhelmingly great Euler – reports on the *Principia* without having made certain of what he reports on!' (Weinstock, 'Long-buried dismantling of a centuries-old myth', p. 848).

center with (5) magnitude c/r^2. Now the challenge is to find a *conic motion* with the same five attributes and then to invoke a *uniqueness principle* that will unite the two motions and hence demonstrate that the actual motion under a central inverse-square force is the conic motion.

One can begin the task of constructing the conic motion by computing at point P the curvature κ of the actual motion following Newton's directions, '*but the curvature is given from the centripetal force and the bodies velocity given*'. Having then determined the curvature at the general point P for the actual motion, Newton's outline employs it to construct a conic section that has the focus at the force center and that passes through the point P with a curvature κ and a tangent PR. '*For the focus,…being given…a conic section may be described…given curvature*'.

At this point one has only a conic curve. The proper motion must be imposed, i.e., the motion must sweep out equal areas in equal times as directed in Proposition 1. Newton does not explicitly include this step in his outline. This missing step is the gap that Pourciau identifies as the fallacy that Weinstock has in effect called to attention. Contrary to Weinstock's claim that the outline is thus beyond repair, Pourciau sees it as a gap that, once called to the reader's attention, is readily filled. Newton, for example, demonstrates in Section 6 of the *Principia* how such motions are to be found in given orbits, the elliptical/focal orbit being one example that he solves.

With the gap filled, the conic motion displays four of the five attributes of the actual motion. What remains to be demonstrated is that the central force for the conic motion is an inverse-square. But this attribute is precisely what Newton has demonstrated in the preceding three propositions, i.e., that for motion on a conic section with the force directed toward a focus, the magnitude of the force goes as c/r^2. Moreover, the curvature of the conic motion was constructed from the curvature of the actual motion, and thus the proportionality constant c is the same for both motions. In this fashion, the solution to the direct problem is employed in a logically sound fashion to aid in the demonstration of the inverse problem.

Finally one must employ a *uniqueness principle* to demonstrate that the actual motion and the conic motion are one. Simply stated, it is that at most one motion can satisfy the five conditions above. Thus, since the conic motion shares all of the five attributes of the actual motion, then the actual motion is identical to the conic motion. Or as it appears in Newton's outline , '*two orbits* [about the same force center] *mutually touching one the other cannot be described by the same centripetal force and the same velocity*'. Thus, as Newton claimed '*From the three last propositions* [plus the revisions, i.e., a conic construction, a conic motion (gap), and a uniqueness principle] *it follows, that…the body will move in one of the conic sections, having its focus in the center of force: and the contrary*'.

9 Newton and the absolutes: sources

A. RUPERT HALL

Let us start with the received wisdom as stated by Richard S. Westfall: 'If Newton's conception of absolute space drew upon Gassendi, so also it drew upon the Neoplatonism of Henry More, especially in its assertion that extension is a disposition of being *qua* being.'[1] To unpack this rather cryptic utterance we may turn back to an earlier article by Westfall on 'The foundations of Newton's philosophy of nature', which is in fact an account of Newton's juvenile essay, *Quaestiones quaedam philosophiae*. Here Westfall refers (in a footnote) to 'the major influence on the *Quaestiones*' of 'Gassendist atomism'. Later in the same footnote Westfall adds: 'When Newton wrote *De Gravitatione et Aequipondio Fluidorum*, which the Halls date between 1664 and 1668, he drew the discussion of absolute space and time directly from Gassendi's *Syntagma Philosophicum*'. However, a few lines before this he had also written 'I am unable to say how much of the influence [of Gassendi upon Newton] came via [Walter] Charleton and I have not found any notes in the *Quaestiones* that repeat Gassendi's words verbatim as the first *Quaestio* repeats Charleton.'[2]

Thus while Newton's knowledge of Gassendi's philosophical ideas mediated by Walter Charleton's *Physiologia Epicuro-Gassendo-Charltoniana* (1654) is accepted by Westfall, as perhaps by all modern scholars, it cannot confidently be claimed that Newton had spent much time on Gassendi's very large, and then recently published, tomes. In Newton's library as reconstructed by John Harrison, Gassendi is represented only by a late (1682) edition of *Institutio Astronomica*, and Charleton not at all.[3] We might not expect a necessarily imperfect catalogue to include all the books read by Newton as a young man, but I incline to agree with D. T. Whiteside's suggestion, borne out by Newton's notebooks, that though as a Cambridge student Newton was a wide reader, he

---

[1]   Richard S. Westfall, *Never at Rest: A Biography of Isaac Newton* (Cambridge University Press, 1980), p. 304.

[2]   *Idem*, 'The foundations of Newton's philosophy of nature', *British Journal for the History of Science* 1(1962): 171–82, on pp. 172–3, note.

[3]   John Harrison, *The Library of Isaac Newton* (Cambridge University Press, 1978), p. 147.

was also an impatient one.[4] At this stage he found the world of nature too enthralling to wish to lose himself in a world of paper. I know nothing of the availability of the great Lyons 1658 edition of Gassendi's writings in Cambridge five years later, but I doubt whether Newton, having gone through Charleton, would have been inspired to measure how exactly he had represented (or misrepresented) the metaphysics of the recent French philosophers.

One must also remember that Newton's early reading did not necessarily stamp his mind for ever. To give only two examples: he was almost immediately to reject Charleton's notions about colours (though the idea of light as a stream of particles, and that of vision as the penetration of these particles into the eye, were to last longer); and before long he would shake off Charleton's scepticism about the 'illuminated *Sons of Hermes*, who boast to have, if not attained to the bottom of the mystery, yet out done the endeavours of all other Sects of Philosophers, in profounding it.'[5] It is the historian, not Newton, who finds the positive influence standing out in letters of gold in books that Newton read.

If we look into Charleton's rather naive pages on space and time in his *Physiologia* we find (as is to be expected) Aristotle to be his chief enemy in philosophy, Epicurus his major friend. Charleton argues (1) that a vacuous space can be imagined; (2) that the three spatial dimensions can be introduced by imagination within such a space; (3) Descartes was in error in making an equation of 'empty', 'nothingness', 'non-being' and 'dimensionless'; (4) *space* and *time* are more general than *substance* or *accident*, and are therefore embraced by neither of these terms; (5) *space, pace* Descartes, is present within the boundaries of a vacuum where:

> (indeed) nothing *Corporeal* doth intercede,...that is, either Substance or Accident; but yet there doth intercede *something Incorporeal*, such as we understand by *Spatium, Intercapedo, Distantia, Intervallum, Dimensio,* which is neither Substance nor Accident.

From these arguments Charleton concludes that we can properly 'discriminate Dimensions into' (a) corporeal, pertaining to physical bodies, (b) spatial, pertaining to space without bodies.[6]

---

[4] Newton did not quite finish reading Hooke's *Micrographia* (A. Rupert Hall and Marie Boas Hall, *Unpublished Scientific Papers of Isaac Newton* (Cambridge University Press, 1962, 1978), p. 413) and hence did not note Hooke's suggestion (*Micrographia*, 1665, p. 245) that the Moon has a gravitating power.

[5] R. H. Kargon (ed.), *Physiologia Epicuro-Gassendo-Charltoniana*. Reprinted from the London Edition of 1654. (New York and London: Johnson Reprint Corporation, 1966), p. 184.   [6] *Ibid.*, pp. 62–7.

Charleton in these pages has already defended the right of the philosopher to imagine the destruction of the world by God. Now he maintains that in this event infinite space would remain, as it had existed before God created the world as a small finitude placed within infinity; further, 'these immense Spaces are absolutely *Immoveable*' and no part of them has any respect to the matter that may fill it at any time. I do not altogether follow Charleton's meaning when he expands this last point: I think it is that though the dimensions of a portion of space remain the same whether it be empty or filled with matter, the dimensions of space are independent of those of body. Charleton now draws a distinction that he might better have inserted earlier: the word *incorporeal* may signify either (a) absence of body, and of 'any positive Nature capable of Faculties and Operations' but not of dimension; or (b) absence of body but not of

> a true and germane *substance*, to which certain *Faculties* and Operations essentially belong; and in that sense it is adscriptive properly to God, Angels, the Souls of men, &c. spiritual Essences.[7]

It is interesting that Newton could have taken the concept of 'active incorporeal space' from either Charleton or More. But the former, unlike the latter, also held that it is not impious to believe that space is '*improduct* by, and *independent* upon God'.

If (following him still further) we thus detach ideas of space and dimension from those of body, we can equally do so with our notions of *place*, the location of body. So if we say that an angle, though immaterial, is *here*, 'in a determinate place', this is simply to say that it is seen in a certain part of space. Because it is immaterial, the angel's form is able

> to penetrate the Dimensions of any [of] the most solid Bodies, so that the whole [spiritual] substance of an Angel may be *simul & semel*, altogether and at once in the same place with that of a stone, a wall, the hand of a man, or any other body whatever, without any necessity of mutual *Repugnancy*.[8]

This too might well have been written by More! Charleton here means that though material dimensions cannot interpenetrate, the dimensions of space are infinite and can interpenetrate; therefore a place can be occupied by both matter and spirit.

As with space, so with time: it owes nothing to body and is independent of motion:

> if it be any thing at all it seems to be the *Twin-brother* of *Space*, devoyd of all relation to Corporiety, and absolutely independent on the Existence of any Nature whatever ... the understanding must deprehend Time to

---

[7]   *Ibid.*, p. 68.   [8]   *Ibid.*, p. 71.

continue to be what it ever was, and is, whether there be any Motion or Mutation in the World, or not, nay, whether there be any World or not...Motion holds no relation to Duration, nor *è converso*, Duration to Motion, but what is purely *Accidental* and *Mental*, i.e. imagined by man in order to his commensuration of the one by the other.[9]

The former is Newton's absolute time, the latter Newton's relative time. In this instance Charleton notes his departure from the ancient atomist tradition. We need not pursue him into the discussion of eternity.

Turning now to the youthful Newton of *De gravitatione et aequipondio fluidorum*, there are certainly traces of a reading of Charleton's *Physiologia*, though nothing to indicate (as I see it) that Newton had the book open before him. There are many ways in which Newton's discussion of space, place, motion and matter differs significantly from Charleton's. The essay opens with a long and specific critique of Descartes's metaphysics, largely ignored by Charleton. Hence Newton writes a good deal about *extension* – not considered by Charleton – and about *motion* likewise. Descartes's conceptualisation of these two technical terms led Newton into areas of debate not touched upon by Charleton, though Newton's rebuttal of Descartes is certainly made in a Charletonian spirit. Newton introduces geometry into the 'corporifaction' of space, again going beyond Charleton in a significant manner. I do not think that we can find in *Physiologia* Newton's important idea that

The parts of duration and space are only understood to be the same as they really are because of their mutual order and position; nor do they have any hint of individuality apart from that order and position which consequently cannot be altered.[10]

Still less is Newton's emphasis on the transcendental character of time and space typical of Charleton, though the latter is no less strong a deist and creationist than Newton. For Newton, space and duration are effects arising 'from the first existence of being', that is, from God, the origin of all being. He silently spurns Charleton's acceptance of the independence of space and time from the creation. Charleton at least for the purpose of philosophical debate can conceive of spaces where God (temporarily) is not; Newton could not conceive of such a deprivation. If Charleton allowed the filling of both empty space and matter by a spiritual substance, Newton's view in *De gravitatione* seems to be closer to that of Henry More – an author not noticed by Charleton – that spiritual substance is omnipresent. Charleton tends to introduce God into his arguments as an instrument in the exercise of the philosophical imagination: 'Let us suppose that God...' Newton always remembers God as the cause of all things and therefore the final term of explanation.

[9]   *Ibid.*, p. 73.   [10]   Hall & Hall, *Unpublished Scientific Papers*, p. 136.

I suppose it would be possible to regard *De gravitatione* as an essay in which the principles of the two recent English philosophers whom we have seen to have some thoughts in common were applied by Newton in critical evaluation of the earlier parts of Descartes's *Principia philosophiae*. (In parenthesis, it is to be noted that Newton at this time, and perhaps always, seems to have paid no attention to Henry More's own critique of Descartes in his letters to the French philosopher, printed (in revised form) in More's *Philosophical Works* of 1662.) But this seems a sterile approach, minimising the strong elements of Newton's individuality and especially his mathematical character.

Henry More's writings have become far better known for their influence upon Newton than have those of Walter Charleton – for example, in Westfall's biography.[11] Newton's atomism has been attributed to More, as have his Platonism (as distinct from Hermeticism)[12] and especially his ideas of space and time.[13] The debatable belief that More was the *éminence grise* behind Newtonian metaphysics has been buttressed by the even less plausible assertion that the two men were intimate personal friends. They were from the same county, the same school, the same University. They were certainly well acquainted. All this does not make them intimate friends. The disparity in their ages – twenty-eight years – was considerable; their intellectual developments, their interests and approaches to the study of the natural universe were no closer. True, they shared a common interest in the interpretation of prophecy and of the early history of philosophy, but these were ordinary enough concerns in the late seventeenth century. Such interest, rather than the themes of Newton's youth, may have brought them together. One work on the former topic (*A plain and continued exposition of the several prophecies or divine visions of the Prophet Daniel*, 1681) was presented by More to Newton, who also owned five other publications of or about More's late theological studies (that is, *post* 1680).[14] It is well known that the unique letter in which More wrote of Newton is of this period (18 August 1680) and is concerned with Daniel. More's personal allusions to Newton:

> the manner of his countenance which is ordinarily melancholy and thoughtful [was] then mighty lightsome and chearfull and…in a manner transported…Mr Newton has a singular Genius to Mathematicks, and I

[11] Westfall, *Never at Rest*, p. 97; R. H. Kargon, *Atomism in England from Hariot to Newton* (Oxford: Clarendon Press, 1966), p. 120.

[12] J. E. McGuire, 'Neoplatonism and active principles: Newton and the *Corpus Hermeticum*' in Robert S. Westman and J. E. McGuire, eds., *Hermeticism and the Scientific Revolution* (Los Angeles: Clark Memorial Library, University of California, 1977).

[13] Alexandre Koyré, *From the Closed World to the Infinite Universe* (Baltimore: Johns Hopkins University Press, 1957), especially Ch. VI.

[14] Harrison, *Library*, p. 196.

take him to be a good serious man ... he has free leave from me to enjoy his own opinions. We have free converse and friendship, which these differences [about the prophecies] will not disturb ...[15]

suggest to me, besides the slightly patronising tone of an elder scholar of established reputation for a younger one, the degree of acquaintance that arises between the few senior men (other than Heads of Houses) permanently resident in the University and also deeply involved in the same subject, though of differing opinions. More's letter notably lacks the warm emotional tone which he reserved for friends of his soul like Sir John Finch and Francis Mercury van Helmont.

Newton's collection of More's books was quite extensive, beginning with *Philosophical Poems* (1647) which was bound in one volume with *An Antidote against Atheism* (1653) and *The Immortality of the Soul* (1659). Newton did not own *Philosophical Writings* (1662) containing More's letters to Descartes, still less Clerselier's earlier publication of them. Sending to Newton on Boyle's behalf his 'new Book of Effluviums' in September 1673, Henry Oldenburg asked him to deliver other copies to Barrow and More. A unique allusion to More in Newton's own correspondence (February 1685) records Newton's bizarre attempt to enrol the septuagenarian philosopher in an abortive Cambridge Experimental Philosophy Society.[16] That the air-pump elsewhere noted by Newton at More's College (Christ's) had any connection with him seems unlikely since it is not mentioned in More's Will along with other philosophical apparatus. In this Newton was remembered – the only Cambridge resident outside Christ's to be so – and Newton presumably attended More's funeral in the early days of September 1687. It would be as interesting to know Newton's opinion of More as an experimenter in pneumatics, as to know More's of the philosophy of the *Principia*. Would More have rejoiced in Newton's version of a non-mechanical gravitation?

Whereas older writers made much of More's powerful influence upon Newton, more recent ones conscious of the paucity of solid evidence warn against its exaggeration.[17] The editors of the *Quaestiones quaedam philosophiae* tell us that More 'was not a major influence on Newton's views concerning infinity and extension' and that, being directly acquainted 'with many of the ancient sources of atomism on which More himself drew' the maturer Newton at any rate demonstrated a good deal of independence from More, despite the oft-quoted youthful memorandum

[15]   Marjorie Hope Nicolson, *The Conway Letters* (New Haven: Yale University Press, 1930), pp. 478–9.

[16]   H. W. Turnbull (ed.), *The Correspondence of Isaac Newton*, vol. I (Cambridge University Press, 1959), p. 305; *ibid.*, vol. II (1960), p. 415.

[17]   Westfall, *Never at Rest*, p. 349.

And y^t Matter may be so small as to be indiscerpible The excellent Dr Moore in his booke of ye soules Immortality hath proved beyond all controversie.[18]

It would indeed be hard if everyone had his judgements of twenty anchored upon his years of discretion. In his published writings Newton's singular use of More's name disowns any metaphysical connection, disowns indeed what some have seen as a permanent influence of More upon Newton. Was it not unreasonable (Newton complained anonymously in 'An Account of the Book entituled *Commercium Epistolicum*' (1715) that

> Mr Newton should be reflected upon for not explaining the Causes of Gravity and other Attractions by Hypotheses; as if it were a Crime to content himself with Certainties and let Uncertainties alone;

what was worse, the editors of the *Acta Eruditorum* had told the world that

> Mr *Newton* denies that the Cause of Gravity is Mechanical and that if the Spirit or Agent by which Electrical Attraction is performed be not the Ether or *subtile Matter* of *Cartes*, it is less valuable than an Hypothesis, and perhaps may be the Hylarchic Principle of Dr. *Henry Moor*...[19]

Evidently, while denying that the cause of gravity is mechanical in the sense of 'Cartes' or Leibniz, Newton was still enough of a mechanical philosopher to have as little wish to be linked with the hylarchic principle or Spirit of Nature as had Boyle and Hooke in the previous generation.

It is generally recognised that the chief innovation that Newton introduced into the mechanical philosophy – indeed, transforming its character – was the concept of force or more broadly that of natural powers and active principles as the effective agents of the phenomena of Nature, being the ultimate causes of the motions of particles.[20] He thus undermined the concepts of the conservation of motion in the Universe and of the derivation of all motion (or change) from antecedent motion, both fundamental to the Cartesian philosophy. The influence of More upon Newton in these matters has been

18 J. E. McGuire and M. Tamny, *Certain Philosophical Questions: Newton's Trinity Notebook* (Cambridge University Press, 1983), pp. 319, 340; A. R. Hall, 'Sir Isaac Newton's Notebook, 1661–1665', *Cambridge Historical Journal*, 9(1948):239–50, on p. 243. It is worth remarking that no mean part of Newton's notes on *The Immortality of the Soul* (1659) – the only work of More's known to have been read by Newton in these early years – relates to fields in which neither More nor Newton were ever distinguished, for example: 'When a snaile creepes a gale of spirits circuit from her head downe her back to her taile & up her belly to her head againe' [*Immortality*, p. 203]; McGuire and Tamny, *Trinity Notebook*, p. 418.

19 A. Rupert Hall, *Philosophers at War* (Cambridge University Press, 1980), p. 313 reprinting *Philosophical Transactions* 29(1715):223.

20 Marie Boas Hall, *The Mechanical Philosophy* (New York: Arno, 1981), reprinting from *Osiris* 10(1952):412–541, on pp. 509–10.

stressed by McGuire.[21] Whether such ideas as these were, in the last resort, more central to Newton's sense of the deep reality of Nature (as they appear to be) than the alternative aetherial hypotheses that he proposed at various times is another issue not to be considered here, and perhaps incapable of resolution. What is clear is that More and Newton alike insist that space (or something immaterial filling space) can be active (or the locus of activity) while matter is necessarily inert and passive. More finds Descartes in error in his attachment of the idea of dimension or space to the idea of matter, as though to measure were always to take the length of something and not to take the interval between two things:

> For it is plain that not so much as our Imagination is engaged to an appropriation of this *Idea of Space* to corporeal *Matter* ... and therefore it may as well belong to a *Spirit* as a *Body*. Whence as I said before, the *Idea* of God being such as it is, it will both justly and necessarily cast this ruder notion of *Space* upon that infinite and eternal *Spirit* which is God.[22]

More's opinion, shared as we have seen by Charleton, is that the ideas of distance and dimension have no necessary connection with the idea of matter, and that space and intervals within it would remain if all matter were abstracted from the world. Thus arguing, More arrives at the partial identity of space and God which is singular to himself:

> if after the removal of *corporeal Matter* out of the World, there will still be *Space* and *distance*, in which this very matter, while it was there, was also conceived to lye, and this *distance Space* cannot but be something, and yet not corporeal, because neither impenetrable nor tangible, it must of necessity be a substance Incorporeal, necessarily and eternally existent of it self: which the clearer *Idea* of a *Being absolutely perfect* will more fully and punctually inform us to be the *Self-subsisting* God.[23]

Now More was surely conscious of the outrageous boldness of the statement that the Universe of matter is contained within God, whose being forms the space within which matter exists, and also that his list of the twenty attributes common to space and to God did not include many attributes normally assigned to the deity, such as infinite power, infinite goodness, infinite knowledge, and infinite wisdom, not to say emotions found in the Scriptures such as anger or pity. God must be more than space, indeed more than the Universe which is more than space. This More recognised in weakening the initial force of his assertion; none of the common qualities of God and space, he added,

---

[21] McGuire, 'Neoplatonism'.

[22] Henry More, *An Antidote against Atheisme* (London, 1652), revised in his *Collection of severall philosophical writings* (London, 1662), p. 163 quoted by Koyré, *Closed World*, p. 135.   [23] More, *loc. cit.*, p. 164; Koyré, *Closed World*, p. 137.

appears to concern the divine life and activity, but simply his bare essence and existence.[24]

On this point Newton was to depart somewhat from More, while agreeing with him that space is more than empty nothingness.

Thus, against the Cartesian statement of the philosophical absurdity of the idea of a space empty of matter, More maintained the theological absurdity of the idea of a space empty of spirit. For spirit must be where matter is not. (That a volume *apparently* filled with matter can also contain spirit was obvious to any atomist.) If the Universe were full of matter to the exclusion of spirit, as the mechanists believed, natural philosophy could be nothing but materialism. Further, More held that Descartes's homologisation of matter, extension and space might serve as grounds for the denial of divine creation. (For if no matter then no space, and if no space then no God.) So long as Descartes argued

> that extension was the essence of matter (not minds) and that matter was
> a substance, his teaching could not be rid of the implication that matter
> was uncreated and also that God and minds are nowhere.[25]

For if spirit is excluded from the material creation, God is excluded from the Universe. In his later writings Henry More applied the label 'nullibists' to the mechanists of the Cartesian tradition that had once seemed to him so promising (of Gassendi, the reviver of Epicurus, More had always a poor opinion), 'nullibists' because although professing belief in God, their universe had no room for him. Although in More's letters addressed to Descartes in 1648–49 there had been besides admiration reasoned metaphysical objections and others questioning the internal logic of some of Descartes's propositions in the *Principia philosophiae*, in his later, strongly anti-Cartesian writings More's criticisms seem to be, in effect, correlatives of his conceptions of God's infinity and ubiquity. They are theological notions given a metaphysical dress, certainly not principles formulated (as were the metaphysical presuppositions advanced by Descartes, Leibniz and Newton) as foundations of a natural-philosophical system. More was both more careless in his language about space and more bold in deifying it than was Newton who (in his General Scholium to the *Principia* (1713)) did not go beyond the assertion that 'by existing always and everywhere, [God] constitutes duration and space'. He is

---

[24] Henry More, *Enchiridion Metaphysicum* (London, 1671), Part I, chap. 8, sect. 14, quoted by J. E. McGuire, 'Existence, actuality and necessity: Newton on space and time', *Annals of Science*, 35(1978):463–508, on p. 480.

[25] W. von Leyden, *Seventeenth Century Metaphysics* (London: Duckworth, 1968), p. 232.

not duration and space, but he endures and is present.[26] I take Newton to be saying (unlike Charleton and More) that the infinity of time and the infinity of space exist because God exists and is infinite (in time and in space and in all attributes); 'constituit' I read as having the sense 'established, created, ordered', and so time and space are themselves to be thought of as elements of the divine creation.

The clearest legacy of Newton's early reading of More in the General Scholium is his firm insistence that God is really present throughout his creation; Newton was as much an enemy as More himself of the nullibists:

> As each and every particle of space exists for ever, and each and every indivisible moment of duration exists everywhere, surely the Creator and Lord of all things will not be *never* and *nowhere*.

We may by analogy, Newton continues, affirm that as a man's soul impalpably pervades his body, so God pervades his Universe:

> He is omnipresent not as a *virtue* only, but as a *substance*, for a virtue cannot exist without a substance. In him the Universe is contained and is moved, but without mutual interaction, for just as God feels nothing from the motions of bodies, so bodies feel no resistance from the omnipresence of God.[27]

How God is able to permeate and govern the Universe we can no more tell than a blind man can understand the nature of colours but (Newton claims) the relation of God to his Universe is not that of man's soul to his body, which demands the intermediacy of a brain and nervous system. God has no need of instrumentalities in the vacuity of space; and though (in the cognate passage of *Opticks*, Query 28) Newton also spoke of space as the *sensorium* of God (or the quasi-*sensorium*) wherein he perceives all things intimately in their immediate presence to himself, Newton only meant to emphasise (*pace* Leibniz) that God does not stand aside from his creation, being in this respect precisely unlike the watchmaker who has fabricated a watch and set it aside as perfect.[28]

---

[26] 'Non est aeternitas vel infinitas, sed aeternus et infinitus; non est duratio vel spatium, sed durat et adest. Durat semper & adest ubique, & existendo semper & ubique, durationem & spatium, aeternitatem & infinitatem constituit.' (*Principia*, 1713), p. 483.

[27] *Ibid.*, my translation. That God should feel nothing from the motion of bodies, is puzzling, unless we read 'feel' as implying human physiological sensations, such as pain. It would be self-contradictory for Newton to say that God 'has no awareness of' the motions of bodies. In Query 28 of *Opticks* Newton writes of God seeing the things themselves intimately, perceiving them, and comprehending them.

[28] So in the late editions of *Opticks*; see the Dover, New York, 1952 reprint of the London, 1936 edition, p. 370.

If in search of further understanding of the development of these instances of that interaction between natural philosophy and divinity which Newton regarded as fitting we go back again to *De gravitatione et aequipondio fluidorum* we perceive even more clearly than in the published Scholium the influence of Henry More, but his is not the *sole* influence there. Like him, and for similar reasons, Newton criticised Descartes's definition of motion as translation, that is, the displacement of a body from one context to another. In Descartes's view a body fixed in its context is at rest, even though the context moves. Newton in *De gravitatione* thought such a relativist definition of motion 'absurd'. If Descartes were right

> not even God himself could define the past position of any moving body accurately and geometrically now that a fresh state of things prevails, since in fact, due to the changed positions of the bodies, the place does not exist in Nature any longer.[29]

Therefore (and this is a new point made by Newton) the possibility of a geometrical account of motion demands the existence of an absolute, unchanging system of coordinates against which not relative motion only but true or absolute motions can be measured. In Westfall's words: 'To avoid the Scylla of relativity, Newton embraced the Charybdis of absolute space.'[30] His choice was justified by a transcendental concept of absolute time and space emanating from God. In *De gravitatione* the concept of absolute space was made to depend more heavily upon the infinity of God than it was to do in the published General Scholium. Extension – that is, space – Newton wrote, 'is not substance; on the one hand because it is not absolute in itself, but is as it were an emanent effect of God, or a disposition of all being' and on the other hand because space does not affect substance [matter] or minds. Nor is it an accident inherent in some subject, and it is certainly not nothingness.[31] Space is infinite and motionless. It is a 'disposition of being *qua* being':

> No being exists or can exist which is not related to space in some way. God is everywhere, created minds are somewhere, and body is in the space that it occupies... space is an effect arising from the first existence of being, because when anything is postulated, space is postulated.

Equally, Newton goes on, the existence of being postulates the existence of time, for

> the quantity of the existence of God was eternal, in relation to duration, and infinite in relation to the space in which he is present... If ever space had not existed, God at that time would have been nowhere.[32]

[29]   Hall & Hall, *Unpublished Scientific Papers*, p. 130.

[30]   Richard S. Westfall, *Force in Newton's Physics* (London: Macdonald, 1971), p. 339.

[31]   Hall & Hall, *Unpublished Scientific Papers*, p. 132.   [32]   *Ibid.*, pp. 136–7.

Similarly God could not be said to be eternal if it were possible to think of a time without God, or God not existing in time. Unlike him, created things occupy only limited regions of time and space. Space is indifferent to the presence of matter: 'space is no more space where the world is, than where no world is', by which Newton means that God's creation of matter did not also create space (as Descartes's theory required) since God and space antedated the material creation. There is therefore a differentiation by this argument also between space and void, or nothingness; here Newton by implication approached tentatively towards More's partial identification of space with God.[33]

At any rate, in *De gravitatione* as in later published works Newton followed the example of More (and, it seems to me, Charleton) in rejecting the Epicurean concept of space as the void, that is, nothingness, the negation of everything. In the words of Pierre Gassendi:

> the void [*Inane*] is devoid of mass, untouchable, incapable of action, of feeling and of resistance, [whereas] body is endowed with mass, is tactile and capable of action and feeling and resistance.[34]

Space as the void can have no ontology; space as More and Newton conceived it possessed (as McGuire has elaborately demonstrated in several papers) a complex ontology. I therefore conclude that McGuire was mistaken in suggesting, in an early paper, that Newton's thinking about space 'may well have been influenced' by reading Gassendi's *Syntagma philosophicum*. Temporarily, McGuire seems to have forgotten that Newton was a Christian philosopher, and that as the Abbé Conti wrote to Nicolas Rémond

> Si vous demandez ce qu'il y a dans les espaces vuides de matiere, les newtonistes vous repondent, que comme l'espace est une proprieté de la divinité, il y a tout ce qui accompagne la divinité...[35]

a passage quoted by McGuire only two pages after his suggestion that Newton shared the Epicurean concept of space as 'the *Inane*'!

Like More, Newton in *De gravitatione* finds it unnatural to seek an ontology for space that is not primarily theological. When, however, he turns to the characteristics of space he seems to hear another voice, for now (unlike More)

---

[33] *Ibid.*, pp. 137–8. Cf. also: 'Although space may be empty of body, it is not in itself a void; and *something* is there, because spaces are there, although nothing more than that.'

[34] J. E. McGuire, 'Body and void in Newton's *De systemate mundi*: Some new sources', *Archive for History of Exact Sciences*, 3(1966):206–48, on p. 226 quoting from Pierre Gassendi, *Syntagma Philosophicum*, Lyons 1658, vol. I, p. 231; my translation. I have rendered *passio* by 'feeling' but perhaps 'reaction' might be preferable.

[35] McGuire, 'Body and void', p. 228.

he considers space geometrically, as dimension or the source of dimension. And he finds that we cannot form a satisfactory concept of motion without first forming appropriate ideas of space. As his thinking about space tends towards the philosophical rather than the theological, Newton seems to recall the lectures of Isaac Barrow, an older scholar whose impression upon Newton was at least as great as that of Henry More.[36] Whiteside and others have pointed out that as Barrow exercised no educational responsibility towards Newton as an undergraduate, we know nothing of the student's acquaintance with the professor until Barrow introduced Newton into the world of learning and yielded to Newton his own Chair. In this respect Newton's first years in Cambridge are a blank. Newton took no extant notes from Barrow's lectures, delivered at Cambridge in 1664 (OS) and subsequent years, but (in my view) it would be as strange to imagine that Newton was not an auditor of them, as to deny that Barrow in later years wrote of Newton as his own protégé.[37]

The message of Barrow's first few lectures, whose sharpness and clarity contrast with More's diffuse expositions, is that physical science is best treated mathematically, above all geometrically, and he explains why geometrical knowledge is certain. (It could hardly be denied that Newton took this precept to heart.) Barrow is a realist: the geometer is not concerned with imaginary or ideal lines or figures that exist only in his mind, though it is true that the magnitudes he is concerned with are perceived by the eye of reason rather than the eye of light:

[36]   Isaaci Barrow, *Lectiones Mathematicae XXIII*; *in quibus Principia Matheseos generalia exponuntur*: *habitae CANTABRIGIAE A.D. 1664, 1665, 1666* (London, 1683), reprinted in W. Whewell (ed.), *Mathematical Works of Isaac Barrow* (Cambridge University Press, 1860), vol. 1, pp. 5–414 (but Whewell prints 27 lectures); translated by John Kirkby as *The Usefulness of Mathematical Learning explained and demonstrated* (London, 1734; 23 lectures). Very different views have been taken of Barrow's philosophical impact upon Newton. E. W. Strong, 'Barrow and Newton', *Journal of the History of Philosophy* 8(1970):155–72 finds it very significant in *De gravitatione*. However, in relation to *Quaestiones quidam philosophiae* McGuire and Tamny affirm that on topics 'of a specific nature, the influences of Galileo, Barrow, Wallis, Euclid, and Digby are certainly present. But their contributions to Newton's general intellectual outlook are minimal, if judged in comparative terms.' (McGuire and Tamny, *Trinity Notebook*, p. 139). It is possible that the writing of the notes edited by them preceded Barrow's course of lectures beginning in March 1664, *De gravitatione* is surely later.

[37]   Newton more than once admitted that 'its probable that Dr Barrows Lectures might put me upon considering the generation of figures by motion, tho I not now remember it' (D. T. Whiteside, *Mathematical Papers of Isaac Newton*, vol. I, Cambridge 1964, p. 344 note 4; see also *ibidem ante* p. 11 note and p. 150). If, with Whiteside, we take this allusion to be to the third of Barrow's *Lectiones mathematicae XXIII*, delivered probably in March or April 1664 – reading (with him) the date '14 March 1664' attached to the Prefatory Lecture as indicating the New Style year – then we have proof positive that these *lectiones* were indeed attended by Newton.

each and every geometrical shape that can be comprehended is really present in every particle of matter whatsoever; is present, I repeat, in actuality and in perfection though it does not appear to the senses.[38]

Thus mathematical physics is not only impeccable logically in its internal construction, but actually corresponds to the reality of Nature.

Mathematics and mathematical physics are concerned with magnitudes, and magnitudes occupy and measure out space. What then is space, Barrow asks?[39] Is it distinct from magnitude? To think of it as uncreated and eternal, so as to be independent of God, is abhorrent to both reason and piety. Some, like Descartes (and Aristotle before him) reason a way out of this difficulty by identifying space with extension, that is, material magnitude. But space is infinite while matter is not; there is space between worlds, and God can create new worlds still in the deep infinity of space. Moreover, motion necessarily takes place in space, each magnitude occupying its own portion; here Barrow clearly differs again from Descartes. He concludes that space does really exist, distinct from magnitude, and is what exists when magnitude is absent. It is interponibility ('betweenness'), the potentiality of which magnitude is the actuality.[40] More had said something similar.

Despite a certain resemblance in their deductions of ideas of space from the infinity and ubiquity of God, Barrow and More have little enough in common. Notably, Barrow does not share More's faith in the infinity of material worlds, nor does he fill space with spirit. More argued along Aristotelian lines that spirit and space are related as substance and accident, that is, that space inheres in spirit; he cannot accept space as pure dimension or potential dimension. Barrow like Newton later rejected More's spirit, even though he seems reluctant to admit that any space *within* a world can be proved to be empty of matter. In other words (though there is at this point some difficulty in following the discussion, as Barrow admitted to his auditors) he seems to conceive of a single geometric idea of space, which is common to all worlds, which exists where worlds are not, and which is the potential locus of all material existence. His space exactly anticipates Newton's conception of

---

[38] Whewell, *Barrow*, vol. 1, pp. 84–5; Kirkby, *Mathematical Learning*, p. 76; my translation.

[39] Whewell, *Barrow*, *Lectio* X, vol. 1, p. 149; Kirkby, *Mathematical Learning*, p. 163.

[40] 'Space is not something actually existing and actually distinct from quantified things, nor does it have actual dimensions proper to itself or separate from the dimensions of magnitudes. What is it, then? What does this riddle signify? I do not much like my answer, nor dare I hope that you will be satisfied by it, but since the rule requires me to reply I say that space is nothing other than a pure putative potentiality, a mere capacity, a "placement" or (forgive the term!) an "interponibility" of any magnitude.' Whewell, *Barrow*, vol. 1, p. 158; Kirkby, *Mathematical Learning*, p. 176. The latter is (as usual) to be read with care, rendering *potentia* as 'Power'!

absolute space and is not theologically conceived in the manner of Henry More.

Just as space existed before the creation of worlds, so did time; just as space exists outside worlds, so does time. There is, declares Barrow, a great analogy and affinity between space and time:

> For as space is to magnitude, so time appears to be to motion, so that time is, as it were, the 'space' of motion.[41]

And just as space is the potentiality of magnitude, so time is the potentiality of existence. The regular circular motions of the heavens are used by us to measure time, but absolute time has no dependence upon motion:

> the quantity of time is dependent neither upon motion nor upon rest; time flows with the same even tenour whether the Universe moves or is at rest, whether we sleep or wake.[42]

Suppose the stars to be at rest, the whole Universe to be motionless and changeless, yet in this repose of all things time would continue. It is only in order to *measure* time that we must call on the aid of motion, choosing the most uniform and perfect that we can. For we can perceive nothing unless our senses are affected and we know the measure of time only because our senses can perceive motion. We commonly take the celestial motions to be the most uniform accessible to us, and hence to be the primary measures of time, but this is not strictly true; the primary and original measures of time are rather

> those motions near to us which we observe with the senses and which are subject to our experiments, since by means of these we are able to judge of the regularity of the heavenly movements. Not even the Sun is the arbiter of time or a reliable witness, except in so far as his truthfulness is assured by mechanical timekeepers.[43]

(I assume that Barrow here alludes to the relatively recent discovery of the equation of time.) Barrow adds that as we have no proof of the constancy of the Earth's revolutions through the millennia, we cannot be sure that the days of the past or the years of Methuselah were identical in absolute measure to our days and years.

Although Charleton had anticipated Barrow in the distinction between absolute and relative time, and perhaps others before Charleton, Barrow's is a richer and more precise discussion. Nothing of this kind is to be found in Henry More.

Newton's mature views on these matters were given in the Scholium following the Definitions at the opening of his *Principia* (1687).[44] Like Barrow,

---

[41] Whewell, *Barrow*, vol. 1, p. 165 (*Lectio* X *ad fin.*); Kirkby, *Mathematical Learning*, p. 184. Compare Barrow's *Lectiones geometricae* (London, 1670) in Whewell, *Barrow*, vol. 2, 161ff. [42] Whewell, *Barrow*, vol. 2, p. 161. [43] *Ibid.*, p. 164.

[44] I. Newton, *Philosophiae Naturalis Principia Mathematica* (London, 1687), pp. 5–11.

he asserts that absolute time is a uniform flow, or duration, without relation to anything external; relative time is what is measured by means of motion. We can approximate to absolute time by a mathematical correction of solar time (as Barrow had indicated), but [though we improve mechanical timekeepers indefinitely] we cannot be sure that we shall ever have access to 'an equable motion, whereby time may be accurately measured', in such a way as to be able to attain to absolute time:

> All motions may be accelerated or retarded, but the flowing of absolute time is not liable to any change. The duration or continuance of the existence of things remains the same, whether the motions are swift or slow, or none at all; and therefore this duration ought to be distinguished from what are only sensible measures thereof [that is, relative time].[45]

Similarly absolute space depends upon nothing external and is always and everywhere similar and motionless. And as the order of the instants of time is unchanging, so is the order of the parts of absolute space. Space and time cannot, as it were, move within themselves. Newton means that just as the instant in the succession of absolute time is unique, so is any place defined by the coordinates of absolute space, and only a movement with respect to such places and time is an absolute movement.

Just as relative time is measured by motion, so we measure our relative space with yardsticks and for this purpose define a relative space with respect to some arbitrarily chosen centre of coordinates, for example the centre of the Earth. But if the Earth moves in absolute space

> any space of our air, which relatively and with respect to the earth remains always the same, will at one time be part of the absolute space into which the atmosphere passes, at another time it will be another part of the same, and so, absolutely understood, it will be continually changed [in position].[46]

Hence,

> Absolute and relative space are the same in shape and magnitude, but they do not always remain numerically the same.[47]

For what is constant in relative coordinates is changing in absolute coordinates, and *vice versa*. We note here again, as with Barrow, the geometric nature of the discussion.

An absolute place is defined by absolute spatial coordinates, a relative place by any coordinate system we choose to adopt. The absolute motion of a body is equivalent to the vectorial sum of its relative motions. Thus, in Newton's example, a seaman moves relatively to his ship, which sails in relative motion

[45] *Ibid.*, p. 7; translation by A. Motte, rev. F. Cajori (ed.), *Sir Isaac Newton's Mathematical Principles*...(Berkeley: University of California Press, 1946), p. 8, modified.   [46] *Principia* (1687), p. 5; Motte–Cajori, p. 6 modified.

[47] *Ibid.*

to the Earth, which may be presumed to possess an absolute motion about the Sun. But we cannot be sure that the presumption of the absolute immobility of the Sun and the stars is correct, Newton reminds us, and so the absolute motion of the seaman cannot be determined with confidence.

Because we have no confident knowledge of either absolute space or absolute time, we are forced to employ only relative measures of these quantities. However, Newton points out for the first time, though we cannot always complete the calculus for determining an absolute motion, we can assuredly point to some instances of it. Newton then goes on to explain how an absolute rotary motion is distinguished by the manifestation of a force of recession from the axis of rotation, but as this conception and its development are unique to him I shall omit further consideration of them here.

From one brief passage in the Scholium (already quoted) where Newton writes of a material body occupying successive positions in absolute space as it moves absolutely, thus raising the issues of (a) the distinction between a 'part' of space and the 'part' of the body filling it, and (b) the universal conformity between these two distinct sets of 'parts', so that (c) we have to suppose that the 'parts' of space are protean in that (unlike the parts of matter) they are infinitely variable in form, dimension and position, we are led to recall the elaborate discussion of just these metaphysical questions in *De gravitatione*. E. W. Strong has pointed to this discussion, echoed in the *Principia* Scholium (however faintly), among other features of *De gravitatione*, as indicative of Barrow's influence upon Newton:

> less acute metaphysically than Barrow is in distinguishing the 'capacity' of space as potential from occupancy or filling of space as actual, young Newton so abstracts from the 'dispositions and properties of bodies' as to conceive the remainder – the 'uniform and unlimited stretching out of space in length, breadth and depth' – to be actually existent, whether or not (and for God, not) any bodies existed in the first place from which to do the abstracting.[48]

Strong has also argued that Newton's criticism of Descartes in *De gravitatione* and defence of alternative metaphysical notions of space and body were stimulated by the Lucasian Professor:

> The correspondence of arguments tendered by Barrow and Newton indicates that the tenth lecture [in Barrow's course, delivered in January 1665/6] constituted a source from which Newton gleaned principal objections to Descartes's *Principia Philosophiae*.[49]

Before Newton (and perhaps before Charleton and More) Barrow charged Descartes with the error of following a priori rather than experimental reasoning, and with finding

[48]   E. W. Strong, 'Barrow and Newton', p. 168.   [49]   *Ibid.*, p. 156.

no place in his system for an inward Spirit eminently able to employ matter and motion by design and not merely mechanically.[50]

This as early as 1652 in Barrow's MA thesis. Strong also makes the link with Barrow run right through to the General Scholium in the *Principia*:

> The pupil is following upon his master in asserting in the General Scholium that God is 'not duration or space...'[51]

quoting at length the passage where Newton dilates upon God as universal, thinking, percipient and substantial, discussed above.

This article by E. W. Strong (of 1970) was, I think, the first to look into Barrow's *philosophical* influence upon Newton, though Toulmin had given a hint of it long before. My own opinion is that the article makes too much of Barrow's influence as affecting *De gravitatione* and the General Scholium. As far as the *Principia* Scholium on space and time is concerned Strong's claims are wholly just; only I would myself lay greater emphasis on the fact that Barrow's metaphysics was set out in a *mathematical* context, and that Newton (like the professor) was conscious that its concepts were not only those appropriate for a geometer, but infused with mathematical notions. Barrow was not only instructing his class in metaphysics, but in what we should now call 'foundations of mathematics'. Newton's approach follows Barrow's. Consider the following passage from *De gravitatione*, where Newton geometrises space; it is (he says) everywhere divided by surfaces infinite in extent and number, which also constitute boundaries of all kinds and in all possible positions:

> hence there are everywhere all kinds of figures, everywhere...circular, elliptical, parabolical and all other kinds of figures, and those of all shapes and sizes, even though they are not disclosed to sight. For the material delineation of any figure is not a new production of that figure with respect to space, but only a corporeal representation of it, so that what was formerly insensible in space now appears to the senses to exist. For thus we believe all those spaces to be spherical through which any sphere ever passes, being progressively moved from moment to moment, even though a sensible trace of the sphere no longer remains there. We firmly believe that the space was spherical before the sphere occupied it, so that it could contain the sphere; and hence that as there are everywhere spaces that can adequately contain any material sphere, it is clear that space is everywhere spherical. And so of other figures.[52]

This remarkable, strange passage is not only a statement about space (that it is an entity, not mere nothingness) but also a statement about figures, and the

[50]   *Ibid.*, p. 157.   [51]   *Ibid.*, p. 170.
[52]   Hall & Hall, *Unpublished Scientific Papers*, p. 133.

geometer's idea of figures. In a highly Platonic vein, Newton is saying that every imaginable figure already exists in space, provided of course that it is not self-contradictory. As Barrow had said before, geometry is not just a game.

If we omit the drawing of the ontological distinction between rectilinear and circular motions peculiar to Newton, everything that Newton writes in the Scholium is a clear expression in terse language of what Barrow had declared to his Cambridge audience twenty years before. Like Barrow's, Newton's reasoning is commonsensical and a-theological, some of it springing straight from the kinematical tradition of Galileo. Like Barrow, Newton endeavours to extend the presuppositions of pure geometry, to move as it were from the arbitrary three dimensions of the mathematician's world to the real and permanent dimensions of space extending through the continuum of time. To me, Barrow's treatment and Newton's in the General Scholium read as close companion-pieces so that we can scarcely doubt the influence of the former upon the latter. Henry More's opinions, though parallel in certain respects, belong to a different realm of thought and argument.

Despite the filiations from *De gravitatione* to the *Principia*, it is evident that Newton's thoughts about space and time underwent a certain hardening and clarification as he grew older and his thinking became more mature. Barrow's influence seems to have survived better than More's, though the latter (whom we may fairly regard as the founder of natural theology in England) is still to be found in the General Scholium. We have no way of tracing the development of Newton's ideas about space and time through the twenty years (or so) separating the unpublished essay from the printed book, still less of accounting for it. Some of the change may perhaps be associated with Newton's increasing confidence in the combination of mathematical analysis with mechanical principles, a combination swiftly changing its character in the early 1680s as Newton took up the concept of force. This confidence was of course strengthened by his success in dealing with the problems of motion in the pre-*Principia* drafts of late 1684 and early 1685. Newton's experience of philosophical controversy in the 1670s had taught him that it was desirable to avoid metaphysical pronouncements altogether; his subsequent transform-ation of the mechanical philosophy persuaded him that it was possible, by adhering to a mathematical, axiomatic form of exposition as strictly as possible, to avoid exposing himself in theological metaphysics in such a way as would lead to his identification with the Cambridge Platonist philosophers. He made it his business in the first *Principia* to speak as a geometer, just as Barrow had done twenty years before in his mathematical lectures.

If Newton so reasoned in the mid-1680s, the long years of preparation for the second edition of the *Principia* (1713) gave him courage to express his thoughts with greater freedom – most openly, of course, with younger friends like David Gregory. The Scholium on space and time remained essentially the same from first to last, but Newton now added a theological conclusion to the whole work in the form of a General Scholium. Here Newton plainly asserted (as by this time he had already in *Opticks*) that the complex motions of the heavens cannot arise from mechanical causes alone; the whole perfection of the system of the world could only spring from the design and authority of a being both intelligent and powerful, who rules all things not as a 'soul of the world' but as the master of the Universe. On this Scholium we have already touched. One cannot read it without recalling Henry More.

In considering the sources of the variety of philosophical arguments available to Newton when writing the *Principia* in the mid-1680s I think it is necessary to be eclectic. Newton is not readily categorised as a follower of Boyle, of Gassendi, of Charleton, of More or of Barrow as though, in particular aspects of his thought (such as his examination of our concepts of space and time) he fell predominantly under a single influence. Yet Newton learned from each of the authors I have named, and especially from Descartes. Historians of ideas more commonly write about differences between philosophers than about the concepts or arguments they mutually accept as valid. Historians, like philosophers, make it their business to draw distinctions. But, in the transmissions of ideas, what an English reader of the 1660s might have learned from a number of books was more likely to impress him than the idiosyncratic opinion of a single author. Further, the historian may easily misjudge an author's position in his own time by supposing it to be more (or less) original than it really was; it is impossible to understand Henry More, for example, if one imagines that his Christian spiritualism was as rare among philosophers in his own day as in ours.

One of my objects in this paper has been to suggest that in a fair number of respects we may find the legacy inherited by Newton to consist of ideas common to his sources, such as Barrow, Charleton and More.[53] Rather than opposing each other at every point, as the refined analyses of modern historians may imply, and therefore forcing Newton to choose between them – or permitting the historian to choose a posteriori for him – the books that he particularly noted, or quoted, offered many common ideas which he embraced and developed. In each author (Barrow least) he found matter to be rejected, or

[53] Obviously I differ from Howard Stein, 'On the notion of field in Newton, Maxwell, and beyond', *Minnesota Studies in Philosophy of Science* 5(1970) who contends (p. 396) that because Barrow was not concerned to use concepts of space and time as a base-line for mechanics, 'Here, as in other areas, Newton's debt to Barrow has been exaggerated'. He also finds it 'unlikely that More could ever have been much help on most of the subject matter of *De gravitatione*' (p. 398) and doubts that theological or metaphysical notions are necessary to the formulation of the idea of absolute space (p. 431 note). Neither am I wholly convinced by Robert Palter, 'Saving Newton's text', *Studies in History and Philosophy of Science* 18(1987).

particular points to admire, but on many things their voices spoke as one. It is obvious that a certain kind of criticism of Descartes was common to all these authors, and it is taken up by Newton also. All were theistic atomists, all doubtful of strict mechanism as a philosophy of Nature, all accepted in some measure the idea of 'active space'.

This seems to me the most important point of all, since it is the key to Newton's own philosophy of Nature. If the Cartesian concepts of the Universe as a plenum, and of the exchange of motion as the sole cause of active phenomena within it, are rejected then (1) a distinction between space and matter must be postulated and (2) a new cause of active phenomena must be found. English mechanical philosophers like Walter Charleton and Robert Boyle who were sceptical of Cartesianism as an integral system nevertheless failed to surmount these difficulties. Newton did. He asserted (as his sources had hinted) that dimension is not determined by matter but by space; and that activity is not within matter but in space. (To be a little more precise, since Newton did not deny that the momentum of the wind drives round the windmill's sails, he denied that a space-filling fluid of the Cartesian kind is the source of activity.) To me, Newton's concept seems quite unlike that of the alchemists, which was that of an *active matter*: an idea that Newton regarded as philosophically deplorable. In Newton's view of things, an atom of matter in isolation is inert; an atom of matter in space is endowed with forces or active principles by the space in which it exists. That the resulting phenomena could not appear were the atom not present to be so endowed is clear, but it is not itself the origin of the forces or principles and still less (for the occurrence of any phenomena without two atoms at least is unimaginable) are atoms the cause and agents of the force-interactions between themselves, acting through space, which are the basic phenomena of physics. To be more precise again, it seems to me that for Newton the forces or active principles causing motions in the atoms do not reside in space *simpliciter*, that is, the interval between atoms, empty nothingness, *intercapedo* Charleton calls it. Space is no void, Newton wrote, and it is because it is more than a void that it can be active. Here Newton's early reading of Charleton and More, and Barrow's lectures, were together of crucial importance for each of these three scholars rejected not only Cartesian extensionalism but the Epicurean *Inane* or void. From them Newton learnt that space has an ontology. The alternative to Descartes's metaphysics of extension was not Epicurean vacuity but *active space*. It was to be long (as Whiteside has shown) before Newton needed to recall and transform this concept but it was latent in his mind, available for use when it became essential to his own internal need to understand how matter in space could move without impact or magic.[54]

[54]   D. T. Whiteside, 'Newton's early thoughts on planetary motion: a fresh look', *British Journal for the History of Science* 2(1964):117–37, especially pp. 120, 128–9; also *idem*

Though the aetherial theory revived by Newton from 1713 onwards may seem at first sight to conflict with this concept of *active space*, as Newton's aetherial ideas of the 1670s would have done, it really is not so. A space-filling fluid, disproved mathematically in the *Principia*, scouted in Query 28 of *Opticks* as a vehicle of light, is not at all revived by Newton in Queries 17 to 24. The Newtonian aether which performs such useful functions in these lately-added Queries is of a quite different character, 7×10^5 times more elastic than air, above 7×10^5 times more rare. By Newton's own principles (*Opticks*, Book II, Part III, Prop. 8) such a medium must be exceedingly porous – its mass not effectively resisting the tails of comets – that is, its particles must be very widely separated and there must therefore be an exceedingly strong repulsive force between them. Newton's aether, which is not a kinematic aether like Descartes's, or Fatio de Duillier's, or Bryan Robinson's, by no means obviates force and action at a distance between particles: it merely changes their character (without making that any less remarkable!). In other words, active space alone can explain the activity of the aether.

Of course I do not mean to deny the qualitative differences between the various elements in the legacy of ideas about time and space that was received by Newton from his English forerunners. Where Newton speaks geometrically, one seems to hear Barrow's voice; where Newton strikes a transcendental note, More's voice is echoed. And Charleton seems to have prepared Newton for these stronger influences. But whereas Barrow is trying to adjust the space of Euclidean geometry to serve the Universe, Newton's scholium is an essay on the difference between things as they are and things as they seem. Then again (as J. E. Power has expressed it):

> While More liked the thought of absolute space because it brought the spiritual so close to experience, Newton postulated absolute space in view of other needs. Unlike More...Newton considered it the subject of the highest importance in the system of the universe.[55]

Unlike Barrow, More made no contribution to Newton's study of dimensionality, nor to his distinction between the absolute (true) and the relative (useful), which perhaps he would not have comprehended.

In the recent past scholars have also considered two further questions: (1) Did Newton really need metaphysical sources? (2) Did Newton (as author of the *Principia*) need theology? The answer to the first to some extent responds to the second, and therefore for brevity's sake I limit myself to it. The pioneer

'Before the *Principia*: the maturing of Newton's thoughts on dynamical astronomy, 1664–1684', *Journal for the History of Astronomy* 1(1970):5–19.

[55]   J. E. Power, 'Henry More and Isaac Newton on absolute space', *Journal of the History of Ideas* 31(1970):289–96, on pp. 290–91.

sceptic in this matter was Stephen Toulmin (1959), in a long and weighty essay. Toulmin's point, simplified, is that for Newton to adopt the concepts of absolute time and space was scientifically reasonable, indeed necessary; what he did not need was a lot of theological vapouring that had been attached to these concepts by his precursors, whom he read, and whose model he followed. Let us fan away the theological vapour, says Toulmin, and we shall see that the definitions of absolute time and absolute space are to be adopted as *axioms*, necessary foundations of mechanics:

> The distinction between absolute, mathematical and relative, sensible space, time and motion can be interpreted consistently as a logical rather than a metaphysical distinction, and the theory as a whole justified on purely dynamical grounds.[56]

To ask whether absolute space really exists is absurd, asserts Toulmin.

His opinions run counter to those previously advanced by other scholars. Toulmin's main object of criticism was Alexandre Koyré, who had argued (1) that the conclusions of the Scholium on space and time settle metaphysical issues fundamental to Newtonian science: 'Newton's...natural philosophy stands or falls with concepts of absolute time and absolute space'; and (2) that these concepts are identical with those of Henry More, and may be presumed to derive from his writings.[57] Similar ideas had been canvassed in a previous generation, with less firmness, by E. A. Burtt.[58] But Burtt, who brought Barrow under review as well as More – and it is a weakness in Koyré's book that Barrow's name never figures in it[59] – also put the question (somewhat in Toulmin's style) why Barrow did not

> drop the absolutistic terminology and treat space and time as purely relative to magnitude and motion, inasmuch as practically that is how they must always be treated?[60]

Burtt did not, however, also address this question to Newton's shade, of whom it might be asked with even greater cogency.

If Koyré is to be interpreted as saying: no More means no Newtonian physics, then his claim would be absurd. Newton could have derived the notions of absolute time and space from Charleton and Barrow, or even from Euclid. And – as everyone emphasises – Newton was far from identifying

---

[56]   S. E. Toulmin, 'Criticism in the history of science: Newton on absolute space, time and motion', *Philosophical Review* 68(1959): 1–29, 203–27, on pp. 7, 214.

[57]   A. Koyré, *Closed World*, p. 160.

[58]   E. A. Burtt, *Metaphysical Foundations of Modern Physical Science* [rev. edn 1932, reprinted] (London: Routledge & Kegan Paul, 1949), especially pp. 254–63.

[59]   Barrow's name is not indexed. Burtt regularly couples Boyle, Barrow and More as intellectual influences upon Newton.

[60]   Burtt, *Metaphysical Foundations*, p. 152.

space with God or filling it with an hylarchic spirit. On the other hand, Toulmin could hardly expect to be credited if he were arguing that universal axioms about space and time (whether or not applied to the science of mechanics) are not metaphysical in character. Or that such axioms do not deserve some justification by metaphysical reasoning, if the natural-philosophical system to which they relate is to possess more than logical purport. (And that metaphysics and theology were strongly related in the seventeenth century can hardly be in question.) Toulmin's discussion therefore seems to be rather out of sympathy with the intellectual spirit of Newton's age, as von Leyden has pointed out.[61] Certainly Toulmin regarded More's philosophy as bizarre.

Burtts' question seems curiously naive. Given that we suppose the three-dimensional space of Euclidean geometry to extend throughout the Universe, as I suppose was always believed in Newton's time and afterwards, it is impossible not to ask whether, of the infinity of co-ordinate systems defining limited portions of such universal space, there is not one that is privileged because its point of reference is fixed and unchanging? Newton made it plain that he thought it worthwhile to ask such a question in order to determine whether the Earth is the centre of such a privileged system, or whether it truly moves in space. (It is significant that – like Toulmin – he regarded the question whether the Sun and stars really move or are fixed as indeterminable.) And similarly of time: it is one of the most obvious metaphysical speculations to distinguish for oneself time (as measure) from universal duration (an independent invariable). Further, if the geometer can conceive of the ideal line or circle, perfect in the way no mark traced by ruler or compass can be perfect, then the philosopher can as easily conceive of an ideal space and time. All these are absolutes, geometrical or philosophical. To frame a logically composed system of mathematical physics without taking absolute space and time as the system of reference would, in Newton's time, have been like framing a geometry by measuring lengths of thread and weighing paper cut-outs. If physics were to be mathematised – that is, subjected to an absolute logical system – it was only logical to treat time and space universally as other dimensions were treated universally, that is, in absolute terms. In this, of course, Barrow preceded Newton, but these considerations were not within More's horizon.

I have here argued – and in its essentials the notion seems self-evident – that the geometer favours a geometrical metaphysics, the theologian godly metaphysics, while the scholar (like Charleton) weighs one authority against another. I have suggested that the young Newton was moulded by such varied

61   W. von Leyden, *Seventeenth Century Metaphysics*, pp. 259–60.

(but by no means wholly antithetic) trends in his reading, and was far from being one man's pupil. In writing about his development, I have hinted, too much subtlety and too many fine distinctions may obscure the main outlines, for though Newton was a sharp and subtle thinker as well able to cry *distinguo* as any schoolman, he believed that in all respects truth is always clear and robust. The search for the truth of things is not a matter of selecting one shade of grey amid many hues. Perhaps part of the appeal of mathematics to his mind lay in its certainty of reasoning; one wonders if he took in Barrow's long encomium of

> Mathematics, which effectively exercises the attentive mind, not purposelessly leading it astray nor wretchedly tormenting it with thorny subtleties, involved problems and argumentative speeches; mathematics, conquering without battle, triumphing without ostentation, compelling without violence, ruling absolutely without loss of liberty...demonstrating everything clearly, advancing swiftly, offering no false shadow of knowledge but veritable knowledge itself to which the mind firmly adheres as soon as it is grasped, and will never afterwards renounce of its own accord or be driven from it by any force...Mathematics, finally, supported by principles that are evident to the mind and confirmed by experience, evincing certain conclusions, unravelling pleasant problems, producing astonishing results, which is the fruitful parent of all arts, the unshakeable foundation of the sciences, the plentiful fountain of benefits to human concerns...[62]

[62] Barrow, *Lectiones*, Oratio Praefatoria; Whewell, *Barrow*, vol. 1, 19–20; Kirkby, *Mathematical Learning*, xxviii–xxix. My rather free translation. Barrow's sentence runs to 73 lines.

10 Newton's ontology of the force of inertia

ZEV BECHLER

Newton's physics has sometimes been interpreted as presupposing two kinds of motion in the world, i.e., natural and enforced. The natural motions are inertial and are described in the first law, and the rest are enforced and are described in the second and third laws.

The distinction, according to this interpretation, between these two kinds of motion, besides the kinematic features, lies in their causal features, and these entail the logical structure of Newton's physical explanations. According to this interpretation, natural, inertial motions are unenforced, and so uncaused (they are natural to matter as such), and therefore cannot and need not be causally explained. Hence physical explanation can be logically offered only for enforced, accelerated motions, which is done by identifying their causal forces.

I will argue that this interpretation is false. Firstly, it trivialises the scientific revolution in the seventeenth century, for the simple reason that the classification of motion into natural vs. enforced is an Aristotelian classification, in fact, the foundation of Aristotle's physics and of his logic of explanation. Consequently, there will be no difference of principle but only of detail between systems of physics which accept this classification.

In fact, if inertial motions are defined by their being natural, uncaused and unenforced, then Aristotle's natural motions are his inertial motions. In his physics these inertial motions include the down–up straight motions of free fall and free rise of the four sublunar terrestrial elements, and the circular motions of the fifth element. All these need no explanation beyond a mention of their nature, while every other kind of motion must be explained by the efficient causes that force bodies to move against their nature.

The ontological focus of this interpretation is the non-separateness of the force of inertia. It presupposes that inertia is only a phenomenon and does not denote a force as an entity separate from some motion. Thus the view that since inertial motion is 'natural', it cannot and need not be explained within Newton's physics, is tightly linked with the ontological view that inertial motion is not caused by any force, and finally that the force of inertia is identical with mass.[1]

Hence it would follow that Newton's physics is Aristotelian in its basic conception of motions and of explanation, and the scientific revolution was strictly a reshuffling of motions within the old traditional classifications.[2]

Incidentally, another absurdity ensues on current attempts to incorporate

[1]   Most of the current accounts of Newton's physics belong to this kind. Following Koyré's analysis of Galileo's and Descartes' theories of inertia as strictly relational theories (e.g. his *Galileo Studies*, Brighton: Harvester Press, 1978, p. 130), his view was applied to Newton, e.g. J. W. Herivel, *The Background of Newton's Principia* (Oxford University Press, 1965), ('Herivel, *Background*'), p. 28, I. B. Cohen, 'Newton's attribution of the first two laws of motion to Galileo', in *Atti del simposium internationale* (1964) '...*Galileo nella storia e nella filosofia della scienza*' (Florence: Marzocco, 1967), p. 147, 'History and the philosopher of science', in *The Structure of Scientific Revolutions*, ed. F. Suppe (Urbana: University of Illinois Press, 1974), pp. 308–60, on p. 330, n. 68, *The Newtonian Revolution* (Cambridge University Press, 1980), pp. 191–3, R. S. Westfall, *Force in Newton's Physics* (London: McDonald, New York: American Elsevier, 1971), ('Westfall, *Force*'), pp. 445–6, B. Ellis, 'The origins of Newton's laws of motion' in *Beyond the Edge of Certainty*, ed. R. G. Colodny (Pittsburgh University Press, 1965), pp. 29–68, on pp. 41–43, N. R. Hanson, 'A response to Ellis's conception of Newton's first law', in *ibid.*, pp. 69–74, on p. 71, and also his 'Newton's first law: a philosopher's door into natural philosophy', in *ibid.*, pp. 6–28, on pp. 18–21, which exemplifies the development of this interpretation of Newton's law of inertia as a counter-factual statement. The ultimate conventionalist outcome of this interpretation is fully exhibited in B. Ellis, 'The existence of forces', *Studies in History and Philosophy of Science* 7(1976): 171–85. Herivel has suggested that inertial force be viewed as 'potential' (in *Background*, p. 28) but this is, for him, merely to deny its reality, i.e., to deny that it has any effect in a uniform state. My view is that the force of inertia in Newton is fully factual since it is always fully actual and active and is independent of circumstances. In this I am in full accord with Gabbey's general trend in his 'Force and inertia in the 17th century: Descartes and Newton', in *Descartes: Philosophy, Mathematics and Physics,* ed. S. Gaukroger (Brighton: Harvester Press, 1980), pp. 230–320, ('Gabbey, *Force & inertia*'), on p. 287, even though I do not quite see why he relapses at times and seems to accept the 'ontological equivalence of motion and rest' in Newton (e.g. p. 277) or agrees that Newton 'equates force (*vis impressa* and the resisting *vis inertiae*) with the quantity of change of state' (p. 287). My view of the importance of the ontology involved in Newton's technical work was shaped mainly by McGuire's discoveries, see note 36. The current interpretation identifies forces with their effects since it holds the ontological and dynamical equivalence of rest and motion in the new inertial physics. Inconsistency is obtained (e.g., in Gabbey, *ibid.*) if this ontological equivalence of rest and inertial motion is combined with the assumption of the objective existence of an inertial force as an entity of any definite magnitude.

[2]   See, for example, A. Koslow, 'The law of inertia: some remarks on its structure and significance', in *Philosophy, Science and Method*, eds. S. Morgenbesser, P. Suppes, and M. G. White (New York: St Martin's Press, 1969), pp. 549–67, on p. 552, about the close similarity of Aristotle's and Newton's laws of inertia.

Aristotle's notorious law of free fall and free rise within his system. The law says that velocity is proportional to weight, and in the attempts to explain how such a manifestly false law could be seriously declared by Aristotle, it is routinely suggested that weight acts as a force and so the law of enforced motion explains natural motions. Thus, Aristotle's physics becomes as mechanistic as any seventeenth-century physics, all amicably agreeing that no-force (or cause)–no-motion is an essential tenet in Aristotle's physics.

In both these current accounts, the logical difference between seventeenth-century physics and Aristotelian physics dissolves. First Newton's inertial motion becomes an Aristotelian natural motion, and then Aristotle's natural free fall motion emerges as an enforced, gravitational-like motion.

The problem of space and the laws of motion

I pass now to a more sharply focused discussion of the status of space and motion in Newton's physics, conceived now rather as an informative explanation of all motions in the world. Elsewhere[3] I have analysed in some detail the consequence of interpreting Newton's physics in this way and I now summarise this analysis.

One important tenet of Newton's informationist view of explanation is the rejection of the notion that to be real, entities must act, or derivatively, that they must be somehow testable or verifiable in order to be admitted into a physics. This is, fundamentally, a rejection of the principle of the identity of indiscernibles for it implies that two entities – such as states or properties of a substance – could be different from each other without this ever being discernible for us. Typically, such entities in Newton's physics are points of space and of time, and inertial states of motions. Two inertial states of a given mass may be different yet their difference, just as their very existence, cannot be discerned by us.

So, even though Newton offered his bucket argument in connection with this discernibility problem, it can be shown that the bucket cannot be regarded as an argument for the discernibility of absolute space, or even as a proof of the reality of absolute motion as a relation between body and space. Briefly, the argument is that if the reality of motion is not denied from the start, then

---

3   I have developed this contrariety between what I called the Platonic and the Aristotelian traditions and ontologies in the history of scientific thought in my *Newton's Physics and the Conceptual Structure of the Scientific Revolution* (Dordrecht: Kluwer Academic Press, forthcoming) ('Bechler *Newton's Physics*'), and *Potentiality, Toleration, and Quantum Jumps: A Study in Modern Aristotelian Philosophies of Nature* (1991). A detailed analysis of Aristotle's philosophy of nature as a foundation of the Aristotelian tradition may be found in my *Non-Informative Knowledge: Aristotle's Philosophy of Nature, Science, and Mathematics* (forthcoming).

every explanation of observed relative motion of two masses must presuppose either that both, or that one, or that if neither then the observer, is in real motion. So both the relativist and the absolutist presuppose that only, and each case of, relative motion reveals real motion. The crucial difference is that only the absolutist identifies real with absolute, i.e., relative to absolute space. Hence Newton's bucket could not be an argument to prove the reality of motion, for this is a presupposition shared by the relativist too. Moreover, in his arguments Newton employs the change of shape, but this can be valid only if shape is presupposed to be absolute, i.e., relative to absolute space. Consequently, the bucket must be interpreted as an argument internal to the absolutist only. It presupposes the absoluteness of motion and of space, and proves no more than that given these assumptions and this ontology absolute motion may be discerned in the case of the motion of one single macroscopic body, even though only by means of relative motions that occur within it. Moreover, Newton's argument presupposes a complex ontology of the inertial force which no relativist can possibly accept.

In his explanation of the ways in which absolute motion may occasionally become observed, Newton refers to the relative motions which necessarily result from absolute motions as the 'properties' of absolute motions, i.e., the essential accidents which result from the nature of absolute motion. The other factors which may similarly make it observed he calls 'causes and effects' of absolute motion. These are the forces which are involved in accelerated motions, both as their causes and as their effects.[4]

The 'effects' of motion and Newton's inertial forces

Now, why does the relative motion of a cluster's components and the ensuing change of its shape entail that the cluster as a whole is in absolute motion (which is, in general, different from the relative motion of its components)? The answer to this depends on Newton's theory of that brand of forces which are the *effects* (rather than causes) of absolute motion. As he declared, these are, besides the 'properties' and the causes of true motion, the signs by which absolute motion can be distinguished from mere relative motion. What are they, exactly?

There is no doubt that they are real and so absolute forces, because they, in their turn, *cause* deformations. But since they are also inertial forces, this creates some doubt about their reality if Newton's physics is interpreted as based on an Aristotelian-type principle of inertia. For such a principle would declare that uniform-velocity motion is not caused at all, so that no force of

---

4   Isaac Newton, *Principia*, trans. A. Motte, ed. F. Cajori (Berkeley & Los Angeles: University of California Press, 1962), ('*Principia*'), pp. 8–9.

inertia exists, and consequently that there is no ontological difference at all between the concepts and the states of 'rest' and 'uniform velocity', or between 'force of inertia' and 'mass in uniform motion' or sometimes even 'mass'.

Obviously, if forces of inertia are not absolute, Newton's argument from their existence to absolute motion is in trouble, and so is his whole conception of absolute space and time. I shall now proceed to show that this Aristotelian interpretation has no compelling textual evidence, argue that it fails to understand Newton's arguments for absolute motion, and present an alternative interpretation which solves this difficulty and illuminates another.

Newton's early notes on the mechanics of collision and of circular motion show that 'force of a body in motion', be this motion inertial or circular, denoted an inner force which is distinct from its motion and is close to Descartes' conatus and to the medieval impetus.[5] The close conceptual connection to impetus may be clearly seen in the following attempt, which even though Newton crossed it out, still tells the same story as the formulation which replaced it:

> The force which the body has to preserve it self in its state shall be equal to the force which put it into that state; not greater for there can be nothing in the effect which was not in the cause, nor less for since the cause only loses its force only by communicating it to its effect there is no reason why it should not be in the effect when it is lost in the cause.[6]

He then replaced it by a definition of 'motion', yet this new definition depends again on that of the 'force' which generates or annihilates it:

> A body is said to have more or less motion as it is moved with more or less force, that is, as there is more or less force required to generate or destroy its whole motion.[7]

The 'force of a body to preserve itself in its state' can now be quantitatively equated with 'its motion' on the one hand and with the sum of forces which actually have accelerated it to its present 'state' on the other hand. The conglomerate concept 'force of motion' which Newton will use synonymously with momentum, will be thus connected to the 'force of inertia' as two aspects of the same entity: the 'force of motion' will indicate the force stored up in the body as the sum of forces which acted on it in the past and as itself a possible cause in the future when it will collide with other bodies. Yet another aspect of the 'force of inertia' will now indicate its absolute force for velocity

---

[5]   See Herivel's arguments for this view, in his *Background*, p. 28.

[6]   *Ibid.*, p. 160.   [7]   *Ibid.*, p. 157.

preservation (absolute in the sense that the velocity is reckoned in the absolute space and time frame) and thus as the cause of its self-motion.

This same view is expressed also in a somewhat later paper which Newton entitled 'The Laws of Motion' (around 1670), in which he first uses the concept of absolute space in order to define velocity, motion and the force of motion. Of special importance is the fact that he uses 'motion' and the 'force to persevere in that motion' to denote two *distinct* entities, even though they are measured by the same quantity, namely, the 'factus' $m \cdot v$:

> But *the motion it self and the force to persevere in that motion* is more or less accordingly as the factus of the body's bulk into its velocity is more or less. And that force is equivalent to that motion which it is able to beget or destroy.[8]

Thus, 'the force to persevere in that motion' is an absolute entity, but is measured by the particular velocity it preserves in absolute space and by the 'body's bulk'. It is evident, therefore, that in absolute space, the force of inertia (which is identical with this force of persevering in a given motion) preserves and so causes the absolute velocity of the body. This is fully consistent with the notion which we saw in his previous paper, that this force of inertia is equal to the sum of forces which had generated this absolute motion, and Newton partly repeats this conception here in the last sentence of the passage quoted.

We may view the present passage in the light of Newton's subsequent explicit insistence (in the Scholium to the definitions, *Principia I*) to distinguish between entities and their measures and also between causes and their effects.[9] Accordingly, the force of motion is distinct from the motion it generates, as a cause is distinct from its effect, even though the effect is, as a rule, the only measure of the cause. The same consideration then holds for the 'equivalence' of the 'force' with 'the motion it is able to beget or destroy', for here the equivalence is between the measure and the entity it measures, or of cause and effect.[10]

---

[8] *Ibid.*, p. 208.

[9] 'Relative quantities are not the quantities themselves whose name they bear, but those sensible measures of them (either accurate or inaccurate), which are commonly used in place of the quantities themselves.' (*Principia*, p. 11). '*Def. 18*: The representatives of times, spaces, motions, speeds and forces are any quantities whatsoever proportional to the things represented.' (Herivel, *Background*, p. 312).

[10] Hence the embarrassment in having to explain the fact that Newton never eliminated the *vis insita* from his physics. The only period in which he dispensed with it altogether was his *Waste Book* period, prior to 1669, the date of his anti-Cartesian revolt. From then on, all his writings on mechanics are full of it. And even though the formulations of *Law 1* in the *Lectiones De Motu* and the final *Principia*, do not refer to it directly,

The distinctness of inertial force from body and from motion

An argument has been suggested for the 'dynamic identity' of rest and inertial motion in Newton's physics from the fact that both need the same amount of external force in order to 'beget or destroy' the same change of velocity Δv (for a given mass and time).[11] Now, one fact is indisputable, and that is that Newton here introduces the concept of absolute space in order to declare an absolute distinction between inertial motions with different velocities, and then between the forces (of inertia, obviously) which preserve each of these different velocities. Hence there is no dynamic equivalence in the sense that all bodies of the same mass have the same inertial force in them independent of their absolute velocities: no dynamic equivalence, since each such body needs *another* inertial force to preserve its absolute velocity.

Hence, also, the fact that for the same mass the same amount of force is needed to generate in a given time the same velocity change Δv irrespective of its initial absolute velocity cannot be used to argue that the forces of inertia are the same irrespective of the absolute velocity. Also the fact that these various forces of inertia are *not differently* opposed to this external force does not indicate that they are all the same force. This automatic adjustment of the inertial reaction, whereby a body adjusts its resistance to an external attempt to change its velocity so that it is exactly equal to the external force, is simply a strange property of bodies. This fact most surely does not show that all bodies moving uniformly are 'dynamically equivalent' even though all bodies, irrespective of their masses and velocities, resist a given external force to exactly the same amount.

Just as the force of inertia is not identical but merely proportional to the absolute velocity of the body, so too is it distinct from but proportional to its mass. This can be easily seen by the difference in the exerted forces in the collision reaction to the same oncoming moving body, when the exerting mass is at absolute rest and when it is in absolute motion. For if the force of inertia were the mass, its exerted force in the two cases would have to be the same whereas in fact it depends on the difference of the absolute velocities of the two colliding bodies. Hence, combining this consideration with the previous one

---

Definition 3 of the *Principia* rectifies this with a vengeance. All the formulations of *Law 1* between 1669 and the *Lectiones* make this *vis insita* their central causal explanation of states of motion and later on of rest.

[11] The notion of dynamic equivalence seems to be derived from Reichenbach's term 'dynamic relativity' (e.g. his *Selected Writings 1909–1953*, Dordrecht: Reidel, 1978, vol. 2, p. 62) which Koyré adopted for his Aristotelian interpretation of Galileo and Descartes (e.g. see his conclusion that inertial motion must be a 'relational state' in his *Galileo Studies*, p. 130), from which it passed on to us, as in Westfall 'dynamic identity' e.g. *Force*, p. 347.

leads to the conclusion that the force of inertia is distinct from either the mass or the velocity, but is proportional to their 'factus' $m \cdot v$ in the appropriate circumstances. That is, for bodies at rest the force of inertia is proportional to the mass but in moving bodies it is proportional to the mass and the velocity.

Clear evidence that this is regarded by Newton as an inner conatus or impetus which pushes the body on its way and so is an entity distinct from the motion itself, is Newton's early treatment of circular motion. The key notion here was that a body moves circularly because there is an external force which acts against the inertial force of motion. Once these two are pitched against each other they counteract until some equilibrium results in the compromise of some curved motion. It seems that at least until 1681, Newton held this view of the 'force of motion', and that he assumed that the exact nature of the curve depended on the exact ratio of the inner 'force of motion' to that of the external force.[12]

The consideration of the curved motion is clear enough to clinch the case for this interpretation of inertial forces. For if Newton had meant to interpret the force of motion as identical with the actual motion, then he would not have had to explain, nor would he have been able to explain, curved motion as a resultant of two forces. Since neither of these two component forces is equal to the resultant quantitatively, nor is either actualised in the actual motion, it must be concluded that they are distinct from the actual resultant motion. This resultant Newton calls 'the force with which the body is moved', whereas the inner component he calls 'the force by which the body endeavors from the center', and the external component he calls the 'continued checking or reflection of it [the moving body] from the tangent line in every point of the circle'.[13] These are three distinct forces, and Newton shows this by arguing that in the motion of a ball inside a hollow cylinder, it is the external force of the cylinder that causes the ball to keep 'it equidistant from m' (the center) for otherwise, if the cylinder 'should cease to check it, it would continually move in the [tangent] line obliquely from the center'. This proves, he argues, that the ball 'continually presses' itself upon the cylinder, so it follows that the ball has an 'endeavor from the center'. This is formulated as a conclusion of the argument, showing that two component forces are involved as causes:

> Hence it appears that all bodies moved circularly have an endeavor from the center about which they move, otherwise the body OC [the ball] would not continually presse upon edf [the cylinder].[14]

But both the internal endeavor and the external force are different from the resultant they generate, since he goes now to prove that 'the whole force by

---

[12]   See D. T. Whiteside 'Before the *Principia*: the maturing of Newton's thoughts on dynamical astronomy, 1664–1684', *The Journal for the History of Astronomy* 1(1970):5–19, on p. 14.   [13]   Herivel, *Background*, p. 147.   [14]   *Ibid.*

which the body *CO* [the ball] endeavors from the center in half a revolution' is more than double 'the force with which it is moved', that is, the resultant. The proof is as follows.

The endeavor of the ball from the center is instantaneously destroyed at each point of its path by the external force of the cylinder. Hence they are equal in magnitude and opposite in direction at each point. But this external force thus destroys an internal force of magnitude *mv* and generates, at the end of half a revolution, another internal force of magnitude *mv*, thus expending in sum at least 2*mv* measures of external force, which therefore gives also the sum of the instantaneous internal endeavors of the ball. Hence the sum of the internal endeavor is

> more than double to the force which is able to generate or destroy its motion, that is to the force with which it moves.[15]

The conception is clearly that external forces generate or destroy motion, internal 'endeavour' conserves what motion is thus generated as far as it can (hence it sometimes is only an endeavor), and the 'force with which the body moves' is the actual resultant. The distinction between them is first the potentiality of the external force and internal endeavor (potentiality since the ball moves neither outward nor inward) as against the actuality of the resultant. But the distinction is also that between the causes and their resultant effect. One point is clear, each of these forces points in another direction, and so each is distinct.

Newton's law of inertia is conceived and formulated in terms of forces and is a clear continuation of impetus physics. Nor do I see any fundamental change in this as Newton's thought evolved. Thus, the fact that an analysis which began from the principle of inertia arrives at the description of curved motion as an equilibrium of forces, is in my view only the logical route for such an ontology of separate forces and is strange only on the Aristotelian interpretation, in which forces are ontically identical with motion, and statics and dynamics are the classificatory non-causal descriptions of two distinct natural kinds.[16]

The splitting of the force of inertia

Moreover, not only are the component forces distinct from the resultant motion, they even go on existing while the resultant motion takes place, for

[15] *Ibid.*

[16] I take it as natural to Newton the Platonist to have from the start an informationist ontology of forces and their separate effects. Within such an ontology there is no difference whatsoever between statics and dynamics, and so Newton's crossed-out axiom in his *Waste Book* surely 'seems to dissolve kinematics in a universal dynamics' (Westfall, *Force*, p. 347), just as his treatment of curved motion makes 'the distinction of dynamics from statics … most difficulty' (*ibid.*, p. 353).

surely they are not eliminated and created anew after causing each elementary-infinitesimal circular motion.

This is also the point where Newton's force of inertia starts a splitting process. The force of inertia acts here both as the force which pushes the body along a straight line, or in the direction of the tangent, and also pushes it against the external force. Since this external force is centripetal, the inertial reaction to it is centrifugal, and in the case of a circular route is at a right angle to the inertial tangent force. This is what is said explicitly in the following draft definition for the *Principia* (drafted around 1684). It defines the 'internal, inherent and essential force' of a body as that by which it conserves its rest or uniform motion, and then continues:

> It is proportional to the quantity of the body, and is actually exercised proportionally to the change of state, and in so far as it is exercised it can be said to be the exercised force of the body, of which one kind is the centrifugal force of rotating bodies.[17]

This is of course the only possible analysis of the force situation in the case where the external centripetal force is a push contact force.[18] This is, for example, the case of the ball revolving inside a cylinder, and of the planetary motion in the Cartesian world. The centrifugal force of the ball acts on the cylinder and that of the planet acts on the ambient particles, and is a split-off from the inertial force, exactly as is the case in every collision situation. What is peculiar in Newton's early analysis is, however, that he regards his centrifugal force as acting on the ball itself. But then again, this is peculiar only in comparison with his later analysis of the forces in collision, where this is called the 'exerted force' of the body and it acts only on the other body. The early analysis, on the other hand, is not peculiar at all if we remember that the centrifugal force of the ball is part of its force of inertia, and so should act on the ball itself exactly as the force of inertia and as any external force does.

This inertial genealogy of the centrifugal force may explain also the nature of the equilibrium that is reached in curved motion. Even though the centrifugal force acts on the ball itself, it may be discounted when the resultant which acts on the ball is calculated, simply because it is an inertial and not an accelerative force. Only in so far as it acts on the cylinder is it accelerative since then it is external. The fact that an equilibrium state is also a case of acceleration should not bother Newton, therefore, mainly because the equilibrium consists in part of inertial components.

The case of the ball inside the cylinder illustrates a case of exact equilibrium

[17] Herivel, *Background*, pp. 306, 311; see also the later draft on p. 320.

[18] This might have been the reason for Herivel's view that Newton could not possibly have accounted for the effects of rotation without his employment of centrifugal forces; *ibid.*, p. 64.

between the external centripetal force and the internal centrifugal force, since this latter internal force is a reaction to the external force and so is exactly equal to it. The case of a heavy body on the equator of the earth, say, illustrates a non-equilibrium situation, since here the centrifugal internal force is not a reaction, or is not excited by the gravitational external force, and so may be smaller than gravity. Newton, in an early paper that documents the discovery of the inverse square ratio (prior to 1669), calculates that 'the force of gravity is many times greater' (Newton arrived at the figure of 350) than the 'endeavor from the center'.[19]

Curved motion as an equilibrium

I have argued that the centrifugal force is an unavoidable entity in the analysis of curved motion which is caused by mechanical centripetal pushing, if the force of inertia is postulated as a force which resists any change of motion. The accepted solution to the ensuing equilibrium, which places the centrifugal force in the external, pushing body, ignores the fact which is presupposed by Newton's physics, that the force of inertia acts on the body in which it originates. Thus, an inertially moving body A is acted upon by its own force of inertia which keeps it going, even though on colliding with a body B, the exerted force of inertia acts as an impressed external force on B. However, at the moment of collision this force acts also on A. Thus, the true solution must be that in Newton's physics a curved motion under mechanical centripetal push or constraint is a semi-equilibrium state, and the nature of the curve depends on the exact relation between the internal centrifugal and the external centripetal forces acting *on the body in motion*. On the other hand, in a curved motion which is not the result of a mechanical push, there cannot be any centrifugal force exerted, and then the radial force acting on the body in motion will be the external centripetal force. The situation then will not be an equilibrium.

It seems that the first time it occurred to Newton to regard curved motion as a non-equilibrium situation, and hence as not caused by mechanical push action, was under the influence of Hooke in 1679. Newton's solution of the problem of the motion of a free falling body under the action of the central gravitation force of the earth, as he wrote in his answer to Hooke, was that the body will 'circulate with an alternate ascent and descent made by its *vis centrifuga* and gravity alternately over-balancing one another'. And then, all of a sudden, one passage later the whole notion of a *vis centrifuga* disappears, and he proposes an account which corresponds to Hooke's 'theory of circular motions compounded by a direct motion and an attractive one to a center', as Hooke had previously described it:

[19]   *Ibid.*, pp. 192–8.

For the motion of ye body is compounded of ye motion it had at *A* towards
M and of all ye innummerable converging motions successively generated
in every moment of its passage from *A* to *G*.[20]

Here, one component of the motion (towards *M*) is the inertial motion along
the tangent, and the other component is radially centripetal. This kind of
analysis is valid only if the centripetal component is not caused by any contact
action, and so this may be taken as the first appearance of Newton's
gravitation as action at a distance.[21]

This should serve also as a clinching argument concerning the mode of
action of gravitation in the *Principia*. The fact that Newton considers there to
be no centrifugal but only tangential inertial forces and centripetal gravitation,
means that contact push force is tacitly rejected and hence that gravitation is
taken here as action at a distance. This argument should then be supplemented
by Cotes' argument from the application of the third law of motion, to
overcome any remaining doubt.[22]

Inertial force as a causal agent and the transformation of internal and external forces

This equal existential footing of inertial and external forces in Newton's
physics stems from their being equally causes of motions. Since he regarded his
physics as a science of causes, his three basic laws of motion describe the causal
effectivity of each kind of force. To the purely inertial forces he devotes *Law 1*,
and its explicit formulations in the drafts to the *Principia* leave no doubt as
to their causal and ontological role. In the early tract (circa 1669) *De
Gravitatione* he had formulated it in this manner very clearly as a reaction to
Descartes' conception:

> Force is the causal principle of motion and rest. And it is either an external
> principle that generates and destroys or otherwise changes impressed
> motion in some body; or it is an internal principle by which existing motion
> or rest is conserved in a body, and by which any being endeavours to
> continue in its state and opposes resistance.[23]

Nothing essential changes, when, more than ten years later, he drafts the first
version of *Law 1*, split into a definition and a hypothesis:

[20] H. W. Turnbull, J. F. Scott, A. Rupert Hall, and Laura Tilling, eds., *The Correspondence of Isaac Newton*, 7 vols. (Cambridge University Press, 1959–77), vol. 2, pp. 307–8.

[21] The fact that Newton suddenly abandoned the equilibrium conception at this point has been systematically missed by commentators.

[22] See the detailed treatment in Bechler, *Newton's Physics*, chapt. 11.

[23] A. R. Hall and M. B. Hall, eds., *The Unpublished Papers of Isaac Newton* (Cambridge University Press, 1962), p. 148.

> *Definition 2*: [That] by which a body endeavours to persevere in its motion in a straight line [I call] the force inherent in a body [*vim … corpori insitam*].
> *Hypothesis 2*: Every body under the sole action of its inherent force moves uniformly in a straight line to infinity unless anything extrinsic hinders it.[24]

This he repeats in a second version he drafts, and then in a third in which Hypothesis 2 becomes *Law 1*:

> *Law 1*: By its inherent force alone a body will always proceed uniformly in a straight line provided nothing hinders it.[25]

And in some other drafts, which are probably revisions to the third version of the tract *De Motu*, he repeats the same conception:

> *Definition 12*: The internal and inherent and essential force of a body [*corporis vis insita innata et essentialis*] is the power by which it perseveres in its state of rest or moving uniformly in a straight line…[26]

Thus, the 'internal and inherent and essential force' of the body is the cause of its continuous rest or uniform velocity, and it is proportional both to its mass and to this absolute velocity. Indeed, only the externally impressed force changes the absolute velocity of the body, but even so there is some kind of transformation which converts internal forces into external ones, and thus dissolves the apparently ontological distinction between them. This is what happens when the internal force is 'exercised' or 'exerted'. On some occasions of enforced change of motion, the inertial force is partly transformed into an impressed external force which acts on the other body if it is in touch with it. Thus are explained changes of motion on collision, where no acceleration prior to the collision exists, and thus no external forces are at work at all in the area. The collision creates the external forces out of the internal forces, and these work to change the motions.[27]

The laws of motion and the classification of forces

Close upon the final draft of the *Principia* Newton regarded the centrifugal force as such an exerted inertial force. Even though later on he decided to suppress any explicit theory about it, implicitly it stayed in its most important

---

[24] Herivel, *Background*, pp. 257–8.   [25] *Ibid.*, p. 299.   [26] *Ibid.*, p. 309.

[27] Thus I reject, and show here why, Westfall's thesis that in the *De Motu* drafts Newton's 'definition of inherent force of a body…was in the process of a revision which would render it useless in the determination of absolute motion' (*Force*, p. 442). The definition was 'useless' for that purpose from the start, and it never underwent any revision. Along with this I reject Westfall's thesis that *Law 1* underwent any change of essence, and that it ever described in Newton anything other than 'the concept that uniform motion is the product of a uniform force' (*ibid.*). As regards the transformation of forces from internal to external and the difficulties such a reading may raise, Gabbey's 'Force and inertia' should be consulted, especially pp. 273–6.

role, namely, as the ultimate proof for the absoluteness of motion. For it was this transformed inertial force that appears in the *Principia* as the 'effects' by which true motion can be distinguished from merely relative. The following definition from a revision to the tract *De Motu* connects the normal inertial force with the exerted inertial force that arises as the effect of an enforced change of motion:

> *Definition 13*: The force of a body arising from its motion is that by which the body endeavours to preserve the total quantity of its motion. It is commonly called impetus and is proportional to the motion, and according to its kind is absolute or relative. The centrifugal force of rotating bodies is to be referred to the absolute kind.[28]

The first sentence identifies the effect-force with the cause-force, and since the cause-force of uniform motion is the force of inertia, so too is the effect-force. The second sentence identifies it with the 'impetus' and the third with the centrifugal force. Now, 'impetus' is defined (in a twin draft penned at the same time) as the 'exerted' 'internal force of matter', that is, as the externalised force of inertia.[29] The body exerts or externalises its force of inertia on one occasion only: whenever its motion is checked by another body, which it then endeavours to resist. After defining 'the internal force of matter', Newton adds:

> In fact the body only exerts this force [*Exercet vero corpus hanc vim solummodo*] in changes of state made by another force impressed on it, and its exercise is *resistance* and *impetus* which are distinct only in relation to each other: being resistance in so far as the body opposes itself to an impressed force, and impetus in so far as the body by yielding with difficulty attempts to change [*conatur mutare*] the state of the other body.[30]

The same conception is described in the parallel definition of the previous draft:

> *Definition 12*: The internal and innate and essential force of a body is a power by which it perseveres in its state of rest or uniform right line velocity, and it is proportional to the quantity of matter, and is in fact exerted [*exercetur vero*] in proportion to the change of state, and in so far as it is exerted it can be said to be the exerted force of the body, of which one kind is the centrifugal force of rotating bodies.[31]

This definition, together with the definition of an impressed force (the following Definition 14), was conceived by Newton as sufficient to entail the third law of motion. Here appears also the detailed description of the transformation of the internal force of inertia into an external impressed force:

[28] Herivel, *Background*, p. 311.   [29] See 'Definition 3', *ibid.*, p. 318.
[30] *Ibid.*, p. 318.   [31] *Ibid.*, p. 311, translation modified.

> *Law 3*: As much as any body acts on another, so much does it experience
> in reaction...In fact this law follows from Definitions 12 and 14 in so far as
> the force exerted by a body to conserve its state *is the same as* [*sit eadem
> cum*] the impressed force on the other body to change its state, and the
> change of state of the first is proportional to the first force and of the
> second to the second force.[32]

This completes Newton's classification of forces. It turns out that they are
either internal, described by *Law 1*, or external, described by *Law 2*, or
transformed from internal into external, as described by *Law 3*. Centripetal
force is external, but centrifugal is internal transformed into external
'impetus'. The main obvious difference in their causal action is the fact that
external forces only change motion and internal forces only conserve motion.
But as far as the body is in motion both are vectorial in essence, and as a result
of the impetus exertion process, even inter-transformable.

The mechanism of inertial deformation – the bucket

This is one important methodological reason why Newton had to work with
infinitesimals. It is a fundamental assumption of the *Principia* that force causes
motion which covers in the first infinitesimal moment a distance proportional
to the magnitude of the force, and this assumption applies equally to inertial,
impressed, and exerted forces. It is by this assumption, and only by it, that the
application of the composition of forces described in corollary I (to the laws of
motion) to the crucial first theorems of the *Principia* is legitimised. Here
Newton composes the distances covered in the infinitesimal moment in which
the forces act, and thereby makes the heterogeneity of these forces irrelevant.
Put otherwise, working out the problem for equal time intervals eliminates the
apparent incompatibility between the two laws $f \propto m\Delta v$ and $f \propto m\Delta v/\Delta t$. And
since v is proportional to s for equal time periods, it follows that $f \propto \Delta s$ for any
kind of impressed, external force. But since $v \propto s$ for a constant v, the rule $f \propto
\Delta s$ applies to inertial, internal forces as well. As a result, for equal and
infinitesimal time moments, forces are proportional to the velocities or the
distances covered and can be vectorially replaced by either of these. This is
what Newton implicitly announces in his parallelogram law, and this is also
what allows him to compose two forces which are as heterogeneous as internal
and external forces.[33]

Seen in this context, it becomes clear that the essential property of forces, as
they are actually employed in Newton's physics, is the fact that they cause

[32]   *Ibid.*, p. 313, corrected.
[33]   See D. T. Whiteside, 'Newtonian dynamics', *History of Science* 5(1966): 104–17 and
          The Mathematical Papers of Isaac Newton, ed. D. T. Whiteside, 8 vols. (Cambridge
          University Press, 1967–81), vol. 6, p. 94.

absolute motion. That one kind causes only acceleration and the other only constant velocity becomes systematically secondary since this heterogeneity is eliminated from the calculations constructed within the infinitesimal proportions machinery.

But in fact, once the essential feature of force – all forces – is seen as being the cause of absolute motion, the inertial and impressed forces become truly 'identical in their relations to motion', as Westfall noted, and this is exactly what the parallelogram of forces presupposes. Westfall asked in regard to the three kinds of forces: 'Could they be compared in this way [in the parallelogram of force]? In fact the word 'force' [vis] was the only factor common to concepts that were otherwise utterly disparate.'[34] However, since what is common to the three kinds of forces is the fact (and not the name) that they are forces, that is, causes of absolute motion, Westfall is wrong. Also, to state, as he does, that the parallelogram assumes that both components cause acceleration or that both cause constant velocity is a mistake. Nothing more than that both components cause absolute motion needs to be assumed.[35]

It is this fundamental ontology of forces that stands at the basis of Newton's argument for the absoluteness of motion from deformation. For what causes the deformation of the accelerated body are its own inertial forces which it exerts *on itself*, in reaction to some externally impressed force. In the two spheres version, as a reaction to the cord's external impressed centripetal force, the spheres exert a centrifugal force. However, this exerted force acts on the cord and is able to extend it centrifugally only because it acts on the spheres themselves first, and they in their turn pull the cord.

This mechanism is very clearly the one responsible for the deformation of the water's shape in the bucket argument. In reaction to the external, impressed centripetal force of the bucket, the water activates an exerted, inertial, centrifugal force on the bucket. But the water is able to act on the bucket only because it first acts on itself.

Newton's argument from deformation to absolute motion passes, therefore, through the deformation's cause, the exerted inertial centrifugal force. Since there is no impressed external force around that acts centrifugally and that can thus account for the deformation, Newton infers that this is caused by an

---

[34] Westfall, *Force*, p. 434.

[35] Another matter that worried historians is Newton's well-known systematic indifference to the time rate change of velocity in his definitions of impressed force and its measure (such as *Law 2*). This worry is now well understood since we can see that Newton's indifference here is merely another facet of his indifference to the alleged heterogeneity of inertial and impressed forces. Whiteside pointed out (*Math Papers*, vol. 6, p. 94, n. 5) that this worried Newton himself who wrote that 'the exertion of innate force to preserve a state is not comparable to the force impressed to change it' in a note he added to *Definition 3* in a revise of the tract *On Motion*. But Newton then cancelled this note!

exerted inertial force. Since an exerted inertial force is the 'effect' of a change of absolute motion, its existence proves the change and hence the presence of absolute motion in the water.

Inertial force as a force at a distance

An apparent result of this line of reasoning is now that brute inanimate matter can act where it is, without the need of an immaterial mediating agency. Contact action would then seem to be a direct action. However, we know that contact action is in fact the action of the *vis insita* or *vis inertia* of mass. It is not mass as such that changes the motion of the impinging body. But if so, then Newton's concept of matter forbids not merely direct action at a distance but also direct action at contact. It is always forces that act on mass, never mass on mass.

There is, therefore, no real difference between gravitational force as the initiation of a new motion, and collision as the redistribution of existing momentum. Both may be regarded equally as either, and the difference between active and passive principles should be re-examined.[36] It is quite a problem to try and formulate the exact point where gravitation, allegedly one of the active principles, initiates new motion. Surely this does not occur at any point in the present career of the universe since the conservation of momentum and the laws of mechanics are a necessary condition for Newton's proof that gravitation is a real force.

The pronouncements Newton made on the direct action of the 'small particles of bodies' may be accounted for now by interpreting them as a description of the action of the non-material forces and spirits which inform them:

> Have not the small particles of bodies certain powers, virtues or forces by which they act at a distance, not only upon the rays of light for reflecting, refracting and inflecting them, but also upon one another for producing a great part of the phenomena of nature?[37]

The last clause might not refer only to chemical, electric and magnetic action. I see no evidence that this is intended to exclude collision action, as this is an action at short range, similar to those mentioned above. Just as he reinterprets

---

[36] Much of the work has been done in J. E. McGuire's publications, mainly 'Transmutation and immutability: Newton's doctrine of physical qualities', *Ambix* 14(1967):69–95, 'The origins of Newton's doctrine of essential qualities', *Centaurus* 12(1968):233–60, 'Force, active principles and Newton's invisible realm', *Ambix* 15(1968):154–208, 'Newton on space, time and God', *The British Journal for the History of Science* 11(1978):114–29.

[37] Isaac Newton, *Opticks* (New York: Dover, 1952), p. 376, Query 31, in a passage added in 1717.

the interaction of light and matter by suggesting it is not a contact collision action as it seems to be, so he might reinterpret all contact collision action. He had prepared for this in his *Principia* analyses of *vis inertia* as a many-faced entity, distinct from mass as such.

Newton's ontology entails, in effect, that the absoluteness of the force of inertia is its separateness from matter. Such separateness implies that Newton's conception of essence denotes no more than minimal necessary conditions so that an essential property is such only that cannot be logically denied to the substance. In this sense it is identical with Descartes' concept, and leaves only extension as the essence of matter. The rest of its properties are accidents which are superadded to it according to God's will, and one of these is the force of inertia.[38] This separateness of the force of inertia accords well with Newton's texts throughout his writings. This view is supported by his concept of the absoluteness of space and motion. In his bucket argument he makes crucial use of the separateness of the force of inertia by an implicit theory of its transformability from an internal to an external, impressed force. Only if this implicit separateness theory is taken into account can his bucket argument be consistently understood as a valid argument, but then it must be taken not as a refutation of the relationist but as internal to the absolutist, with the heavy baggage of an ontology of forces and their vicissitudes. Only thus can it emerge as non-circular, but then the separateness of the force of inertia is unavoidable.

[38] See Bechler *Newton's Physics*, chapt. 11/11, 12, 13.

11 Newton's *Mathematical Principles of Natural Philosophy*: a treatise on 'mechanics'?

ALAN GABBEY

Acknowledgement

I am indebted to the editors of this *Festschrift* for some salutary observations on an earlier draft, and in particular to Alan Shapiro for inviting me to take a closer look at Barrow's *lectiones mathematicae* and Newton's *lectiones opticae*.

Many will find the question posed in my title trivial, and dismiss it with an unqualified affirmative. After all, I will be reminded, Newton's *Principia* was described as a treatise in mechanics by his contemporaries, it has occupied an honoured place in every history of mechanics since the eighteenth century, and physicists in particular will remind me that the three laws of motion preceding Book I appear in every modern mechanics textbook. So what is the point of the question, I will be asked, and why the scare quotes round 'mechanics'?

It is true that many of Newton's contemporaries and immediate successors saw the *Principia* as a treatise in 'mechanics'. The review in *Bibliothèque universelle* (March 1688), written probably by Locke, begins with the comment:

> If those who work in mechanics [*les méchaniques*] understood perfectly the rules of geometry ... they would never fail to achieve their end, and they could give to their works all the exactness and perfection that mathematicians are capable of imagining ... Mr Newton proposes to himself the same end, and takes the same path in this treatise, explaining in the first two Books the general rules of natural mechanics [*les regles génerales des Mechaniques naturelles*], that is to say the effects, causes and degrees of weight, lightness, elastic force, the resistance of fluids, and the powers called attractive and impulsive ...[1]

[1] Quoted in I. Bernard Cohen, *Introduction to Newton's 'Principia'* (Cambridge University Press, 1971), p. 146, note 9, On the other hand, Halley in his review referred only to 'the Principles of Natural Philosophy' and did not use the term 'mechanics' or its cognates at all, except to say that 'the powers of all sorts of Mechanical Engines are demonstrated' from the corollaries following the laws of motion. *Philosophical Transactions*, No. 186 (Jan–March 1687), pp. 291–7. Reproduced in facsimile in *Isaac Newton's Papers and Letters on Natural Philosophy and Related Documents*, ed. and gen. intro., I. Bernard Cohen, assisted by Robert E. Schofield (Cambridge University

The highly critical reviewer (possibly Pierre-Sylvain Régis) for the *Journal des sçavans* conceded that the *Principia* is 'a mechanics, the most perfect that one could imagine', his reason being that 'it is not possible to make demonstrations more precise or more exact than those he gives in the first two books on lightness, on springiness, on the resistance of fluids, and on the attractive and repulsive forces that are the principal basis of physics'.[2] In 1710 Berkeley referred to the *Principia* as 'a treatise of Mechanics, demonstrated and applied to nature, by a philosopher of a neighbouring nation, whom all the world admire'.[3] Seventy years later, Lagrange wrote that it was 'la mécanique' which 'became a new science in Newton's hands, and his *Mathematical Principles*, which appeared for the first time in 1687, was the epoch of this revolution.'[4]

Yet Lagrange does not discuss the 'natural philosophy' he omits from Newton's title, and talk of a 'new science' of mechanics invites searching questions about the identity of old one(s) that Newton replaced or transformed. One wonders if Berkeley assumed there could be a mechanics that did *not* apply 'to nature' and if the reviewer in the *Bibliothèque universelle* intended a contrast between 'natural mechanics' and some other kind. And one wonders how the reviewer in the *Journal des sçavans* reviewed the relation before and after Newton between mechanics, whether or not precise and exact, and physics.

It is also true that most historians of science have described Newton's revolutionary work as a treatise in 'mechanics'. Surveys of the history of mechanics, whatever their scale, typically include a chapter or section on Newton's *Principia* or at least on the definitions, laws of motion and their corollaries in Book I. Mach's view was that Newton 'completed the formal

---

Press, 1958), pp. 405–11. The reviewer in *Acta Eruditorum* mirrored Newton's usages of 'mechanica' and its cognates, without saying anything of particular significance for my purposes in this essay. *Acta Eruditorum* (June 1688), pp. 303–15.

[2] *Journal des Sçavans*, vol. 16 (2 August 1688) pp. 237–8. Quoted from I. Bernard Cohen, *The Newtonian Revolution. With Illustrations of the Transformation of Scientific Ideas* (Cambridge University Press, 1980), p. 96. See also Cohen's précis of this review in his *Introduction*, pp. 156–7, and Paul Mouy, *Le développement de la physique cartésienne 1646–1712* (Paris: Vrin, 1934), pp. 256–8.

[3] *Treatise concerning the Principles of Human Knowledge* (1710), sect. 110. George Berkeley, *A New Theory of Vision, and Other Writings*, intro. A. D. Lindsay (London: Dent, 1972, first pub. 1910), pp. 168–9, note 2. In the second edition of the *Treatise* (1734) the description reads: 'a certain celebrated treatise of mechanics: in the entrance of which justly admired treatise, time, space and motion, are distinguished into absolute and relative, true and apparent, mathematical and vulgar'.

[4] Joseph-Louis Lagrange, *Mécanique analytique. Edition complète, réunissant les notes de la troisième édition* [1853] *revue, corrigée et annotée par Joseph Bertrand, et de la quatrième édition* [1888] *publiée sous la direction de Gaston Darboux*, 5th edn [ed. Albert Blanchard], 2 vols. (Paris: Albert Blanchard, 1965), vol. 1, p. 208. The first edition appeared in 1788. Except where otherwise stated, all translations are my own.

enunciation of the mechanical principles now generally accepted'.[5] According to Dugas, 'Galileo and Huygens had emancipated mechanics from a scholastic straightjacket. However, the task of constructing an organised corpus of principles in dynamics remained. This was the work of Newton, who set his seal on the foundations of classical mechanics at the same time that he extended its field of application to celestial phenomena. Newton's work in mechanics is called *Philosophiae naturalis principia mathematica ...*'[6] Dijksterhuis claims that by creating order out of the chaotic conceptual legacy of the Peripatetics and other predecessors, Newton had axiomatized mechanics.[7] Historians today still categorize the *Principia* as a treatise on mechanics[8] (or dynamics: see below), though few of them go as far as Mach or Dugas in their characterization of its achievements.

However, such modern categorizations of the *Principia* are largely equivocal in taxonomic intention if they are intended to be understood with reference to conceptions of 'mechanics' in the seventeenth century. Or they are anachronistic if it is assumed without further ado that Newton's treatise can be described in post-Eulerian or post-Lagrangian language. To my knowledge there has not been a studied examination of the precise referents of the term 'mechanics' in Newton's day,[9] or of the precise nature of Newton's contribution to mechanics, as opposed to the natural philosophy whose mathematical principles he was explicitly concerned to create on the bases of

[5] Ernst Mach, *The Science of Mechanics: a Critical and Historical Account of its Development*, 6th edn, with revisions through the 9th German edition (1933), trans. Thomas J. McCormack (La Salle, Illinois: Open Court, 1974), p. 226. In today's scientific community, and among some historians of science of the older school, the *Principia* is assumed to contain the earliest complete formulation of the principles of 'classical mechanics'. The liveliest antidote to this piece of mythology is still the work of Clifford Truesdell: see essays II, III, and V in his *Essays in the History of Mechanics* (Berlin, New York: Springer-Verlag, 1968), pp. 85–137, 138–83, 239–71.

[6] René Dugas, *A History of Mechanics*, trans. J. R. Maddox, preface Louis de Broglie (London: Routledge & Kegan Paul, 1957), p. 200.

[7] E. J. Dijksterhuis, *The Mechanization of the World Picture*, trans. C. Dijkshoorn (Oxford University Press, 1969, 1st pub. 1961), pp. 464–5.

[8] I include myself in this blanket censure. See my 'The case of mechanics: one revolution or many?', in *Reappraisals of the Scientific Revolution*, eds. David C. Lindberg, Robert S. Westman (Cambridge University Press, 1990), pp. 493–528. There I raise several doubts about 'the Newtonian Revolution' in mechanics, but I fail to examine the notion of 'mechanics' itself in that context.

[9] By contrast, in the case of the Renaissance (including the early seventeenth century), splendid work has been done by Wallace and Laird. William A. Wallace, *Galileo and his Sources: the Heritage of the Collegio Romano in Galileo's Science* (Princeton University Press, 1984), especially Chapter 4.4, pp. 202–16. W. R. Laird, 'The scope of Renaissance mechanics', *Osiris* (2nd series) 2 (1986): 43–68; 'Giuseppe Moletti's "Dialogue on Mechanics" (1576)', *Renaissance Quarterly* 40 (1987): 209–23.

new patterns of mathematical thought and new laws of motion. Similar questions may be asked, just as pressingly, about his contributions to 'dynamics', another term that historians use indiscriminately with 'mechanics' to describe his work, despite Newton's own contemptuous dismissal of this borrowing from the Greek that Leibniz made part of our scientific language. But that is another story.[10]

So the question in my title does have a point, though it has no single or simple answer. Mechanics as a discipline underwent radical changes in nature during the period covered by the prehistory, publication and early reception of the *Principia*. Correspondingly, the term 'mechanics' as used by writers of the time often carried equivocal senses. We should not therefore expect the interpretation and descriptive application of the term to be always straightforward matters, whether in Newton's writings or in those of his contemporaries. If we look afresh at Newton's purpose in the *Principia* within this perspective, we find that it both was and was not a treatise on mechanics, depending on whether 'mechanics' is understood in an emerging new sense which the *Principia* played a major role in shaping, or in an older sense that was not Newton's concern, though its subsequent evolution at the hands of others owed much to the principles and methods he set out in 1687. At the same time, the *Principia* was a revolutionary text in 'natural philosophy', the discipline traditionally held to be distinct from (though related to) 'mechanics' in that older sense.

The background to these aspects of the *Principia* is the transformation of mechanics in its twin role as an *ars*, distinct from *scientiae* such as natural philosophy, and as a *scientia media*, participating in both mathematics and natural philosophy, into mechanics as an integral and foundational part of a programme for the general *mathematization* of natural philosophy, out of which eventually developed mathematical physics as understood today. Since antiquity, the mechanical or manual *artes* had been concerned with the construction and operation of machines and other artifacts designed to move terrestrial things *contra naturam* and for human ends. On the other hand, in Peripatetic manuals the traditional definitions of *philosophia naturalis* (equivalently *physica* and *physiologia*) typically employed the phrase 'the science of natural bodies, *in so far as they are natural*'. Physics did not deal with artificial things *qua* artificial, so it did not share the concerns of the *artes mechanicae*. The first line of the *Mechanica* of Pseudo-Aristotle alludes to the distinction:

10   See for example Pierre Costabel, 'Newton's and Leibniz's dynamics', trans. J. M. Briggs, in *The Annus Mirabilis of Sir Isaac Newton 1666–1966*, ed. Robert Palter (Cambridge, Mass.: MIT Press, 1970), pp. 109–16, on p. 111, or my 'Force and inertia in the seventeenth century: Descartes and Newton', in *Descartes: Philosophy, Mathematics and Physics*, ed. Stephen Gaukroger (Sussex: Harvester Press; New Jersey: Barnes & Noble, 1980), pp. 230–320, on pp. 230–43.

'Remarkable things occur in accordance with nature [κατὰ φύσιν], the cause of which is unknown, and others occur contrary to nature [παρὰ φύσιν], which are produced by skill for the benefit of mankind.'[11]

However, already in Pseudo-Aristotle's *Mechanica*, as in Book 8 of Pappus's *Collection*, mechanics was also a theoretical discipline that dealt mathematically with problems relating to the use of machines. With the rediscovery of the *Mechanica* in the Renaissance, mechanics became a developed *scientia media* (like optics and astronomy), operating 'midway' between mathematics and physics in the sense that physical and empirically accessible objects were treated in a mathematical way. Mechanics was therefore 'subalternated' to both mathematics and physics, though in the Renaissance and early seventeenth century there was disagreement as to whether it should be seen as subalternated principally to one or other of the two parent sciences.[12] For a statement that typifies how mechanics was generally understood in the early seventeenth century, we may consult the article on μηχανικὴ in Rudolf Goclenius's *Lexicon philosophicum Graecum*:

> Mechanics is the art through which are constructed, not only on mathematical and physical grounds but also through manual arts, machines and instruments that are necessary, agreeable and convenient to human society. Mechanics is λογικὴ (rational) or theoretical, or χειρουργικὴ manual (otherwise practical).[13] Pappus of Alexandria, Mathematical

11 Aristotle, *Minor works*, trans. W. S. Hett, Loeb Classical Library, vol. 14 (London and Cambridge, Mass., Harvard University Press, Heinemann, 1980, 1st edn 1936), p. 331. On the *Mechanica* see Paul Lawrence Rose and Stillman Drake, 'The Pseudo-Aristotelian *Questions of Mechanics* in Renaissance culture', *Studies in the Renaissance*, 18(1971): 65–104. François de Gandt, 'Les *Mécaniques* attribuées à Aristote et le renouveau de la science des machines au XVIe siècle', *Les Études Philosophiques* (1986):391–405.

12 Laird, 'The scope of Renaissance mechanics'; William A. Wallace, 'Traditional natural philosophy', in *The Cambridge History of Renaissance Philosophy*, eds. Charles B. Schmitt, Quentin Skinner, Eckhard Kessler, assoc. ed. Jill Kraye (Cambridge University Press, 1988), pp. 201–35; *Galileo and his sources, passim*.

13 'Mechanica est λογικὴ (rationalis) seu theoretica, vel χειρουργικὴ (aliâs practica) manuaria', and further down, 'Mechanica est duplex, rationalis & manuaria magis. Rationalis (λογικὴ) est, quae mathematicis & physicis rationibus potissimum perficitur. Constat hypothesibus & theorematis...' Recently Cohen has noted Newton's use of the term *mechanica rationalis* in the *Principia* (Preface to the 1st edn) and asks 'Is this the earliest usage of this term?' This entry in Goclenius's *Lexicon* proves that the term was in (probably common) use during the first decade of the seventeenth century, at the latest; in any event, 'rationalis' would be the natural Latin rendering of Pappus's λογικὴ in any period. I. Bernard Cohen, 'Newton and Descartes', in *Descartes: il Metodo e i Saggi. Atti del Convegno per il 350° anniversario della pubblicazione del Discours de la Méthode et degli Essais*, eds. Giulia Belgioioso, Guido Cimino, Pierre Costabel, Giovanni Papuli, 2 vols. (Rome: Istituto della Enciclopedia Italiana, 1990), vol. 2, pp. 607–34, on p. 617, note 51.

collections, Book 8. The latter is so called because it is done through the operation of hands.

Theoretical mechanics is done by using geometrical, arithmetical, physical, and astronomical reasonings...

Mechanics is two-fold: rational and (not least) manual. Rational mechanics is done especially through recourse to mathematical and physical reasonings. It consists in hypotheses and theorems...[14]

These disciplinary divisions explain why in the Renaissance and early seventeenth century the authors of the Peripatetic manuals of *philosophia naturalis* (or *physica*) did not publish treatises on mechanics. The treatises on mechanics were all written by mathematicians, not by 'scholastic' advocates of *physica Peripatetica*. Conversely, relatively few of the writers on mechanics published any natural philosophy. To see this one only has to glance at representative lists of those who are best known for their writings on mechanics (rational and practical) and those who are known for their treatises on Peripatetic philosophy, including in particular natural philosophy. They form distinct and minimally overlapping groups. On the one hand, we have Baldi, Benedetti, Besson, Commandino, Galileo, Giovanni di Guevara, Guidobaldo del Monte, Maurolico, Ramelli, Stevin, Tartaglia, Valerio, Zonca, to cite the better known authors. On the other hand, we have scholastic authors such as Abra de Raconis, Burgersdijk, the Conimbricenses, Eustache de St Paul, Keckermann, Magirus, Ruvio, Suarez, or Toletus.

Descartes' *Principia philosophiae* (1644) is revealing in this connection. In traditional histories of mechanics it is a – for some *the* – key text in the period from Galileo to Huygens and Newton, yet there is no hint of a theory of machines in this famous non-mathematical *summa* of the principles of natural philosophy. But that is wholly in keeping with Descartes' purpose, which was to establish laws of *nature* and display their explanatory roles within broad domains of mostly *natural* phenomena. To find Descartes' theory of the five simple machines, that is, his *mécanique(s)* or *mechanica*, we must turn to his correspondence with Mersenne and Constantin Huygens.[15] Viewed within the

[14] Rudolf Goclenius, *Lexicon philosophicum Graecum, opus sane omnibus philosophiae alumnis valde necessarium cum perspicientia philosophysici sermonis plurimum etiam ad cognitionem rerum utile. Accesit adiicienda Latino lexico sylloge vocum & phrasium quarundam obsoletarum...*(Marburg, 1615), pp. 141–2. Two years earlier Goclenius published his larger *Lexicon philosophicum, quo tanquam clave philosophiae fores aperiuntur...*(Frankfurt, 1613). Both *lexica* were published in facsimile as one volume by Olms (Hildesheim) in 1964.

[15] I examine this problem at length in 'Descartes' physics and Descartes' mechanics: chicken and egg?', which was presented at the conference celebrating the 350th Anniversary of Descartes' *Discourse on Method*, San Jose State University, 15–17 April 1988 (proceedings to be published by Oxford University Press, ed. Stephen Voss).

disciplinary framework I have outlined, Descartes' *Principia philosophiae* was of a piece with the Peripatetic manuals he intended his treatise to replace in Europe's colleges and universities.

More significantly, Galileo (like Benedetti) was as much a natural philosopher as he was the author of *Le meccaniche* (*c.* 1600), and he succeeded in bridging some of the disciplinary gap between *philosophia naturalis* and *mechanica* by mathematizing one of the most intractable phenomena of the natural world, the fall of heavy bodies. Through the deployment of mathematics and *ars mechanica* to yield experimentally-derived data *de motu* he created powerful methods for exploring the operations of the *natural* world. Newton's was by no means the first attempt to bring to light the mathematical principles of natural philosophy.

Partly through the impact of Galileo's achievements in this area, and partly as a result of other developments, including aspects of Descartes' programme in natural philosophy, the era of Cartesianism saw a marked evolution in the application of mathematics to natural philosophy, in which the nature of mechanics in particular began to undergo significant transformations. The evolution is exemplified in Isaac Barrow's *lectiones mathematicae* (1664–66), some of which the undergraduate Newton attended.[16] In Lecture II (Spring 1664), 'Of the particular division of the mathematical sciences', Barrow passes in critical review earlier taxonomies of the mathematical sciences, notably that of Geminus (for whom mechanics was 'the art of discovering the weight and motive power of any massy body'), and more recently those of Biancani, Hérigone, Guldin, and other 'Moderns'. Most previous writers on the matter shared a false distinction between the intelligible and the sensible (the objects of mathematics being both), and misjudged some of the taxonomic divisions themselves. Barrow's own judgement is that 'mixed mathematics' (another term for *scientiae mediae*)

> ought all to be taken as parts of natural science, being the same in number
> with the branches of physics ... For magnitude is the common affection of
> all physical things, it is interwoven in the nature of bodies, blended with all
> corporeal accidents, and well-nigh bears the principal part in the
> production of every natural effect ... there is no part of this [physics] which
> does not imply quantity, or to which geometrical theorems may not be
> applied, and consequently which is not some way dependent on geometry;
> I will not except even *Zoology* itself ...[17]

[16] Richard S. Westfall, *Never at Rest: a Biography of Isaac Newton* (Cambridge University Press, 1980), pp. 99–101.

[17] Isaac Barrow, *The usefulness of mathematical learning explained and demonstrated: being mathematical lectures read in the publick schools at the University of Cambridge ... To which is prefixed, the Oratorical Preface of our learned author, spoke before the University on his being elected Lucasian Professor of the Mathematics* (London, S.

In particular, Barrow notes, since weight and the forces that sustain or elevate weights are both determined jointly by bulk, position, and distance from the centre of motion,

> the disciplines concerning the rest or motion of these weights...are ascribed to the mathematics...of which kind are statics (i.e. the science of weights), mechanics, centrobarycs, horrhopics ['the art of balancing'], and the like. And to these may be referred automatopoetics, or the art of making machines moveable, not by an extrinsecal impulse, but from an internal gravity, or elastic virtue within themselves. Yea, because no local motion as to its duration, impetus, intention, direction, and all kinds of differences can be otherwise estimated in itself, or compared with another motion, but from the spaces...it passes through, therefore most parts of physics...are to be accounted mathematical: from whence accrues a plentiful crop of sciences to the mathematics...[18]

Barrow's 'mixed mathematics' was of very broad scope indeed. The plentiful crop of sciences he saw accruing to mathematics, however diverse their subject matter, comprise virtually the whole of physics, which thereby becomes redefinable as the mathematical treatment of corporeal magnitudes moving in whatever way under forces of whatever kind. Within that framework the foundations of all practical and theoretical mechanical disciplines, though they might bear the traditional names Barrow retains for them, become identical with the principles of natural philosophy.

The vision of a generalized mathematical treatment of natural philosophy is also evident in Newton's *Lectiones opticae*, delivered during 1670–72 on his appointment to the Lucasian Chair of Mathematics at the University of Cambridge in succession to Barrow. In Lecture 3 there is a significant digression in which he explains why he is using mathematics in the investigation of colours, 'which are not supposed to pertain to mathematics', that is (we may gloss) not supposed to according to Peripatetic natural philosophy. Mathematics is necessary for the study of colours because of the inseparable bond between the properties of colours and those of refractions. Besides, he continues,

> were I not discussing refractions, and were an inquiry into them not the occasion for undertaking to explain colours as well, the generation of colours involves so much geometry, and knowledge of them is secured by so much that is evident to the understanding, that even on their account I could still have begun to extend somewhat the boundaries of mathematics. For just as astronomy, geography, navigation, optics and mechanics are held to be mathematical sciences though they deal with the heavens, the earth, ships, light and local motion, which are physical things, so although

Austen, 1734), pp. 21–3. The original Latin *Lectiones mathematicae* was first published in 1683.   [18]   *The usefulness of mathematical learning*, pp. 23–6.

colours belong to physics, nevertheless scientific knowledge of them must be considered mathematical, in that they are treated through mathematical reasoning. Indeed, since they seem to be among the most difficult things that a philosopher needs to know accurately, I hope to show, by way of example, how valuable mathematics is in natural philosophy...[19]

In his *Mechanica: sive, de motu, tractatus geometricus* (1670–71), John Wallis defines mechanics simply as 'the geometry of motion'. He distinguishes between *artes mechanicae*, the illiberal arts, and doing something 'mechanically' in geometry, that is, using instruments for geometrical operations. However, in his treatise, as he informs the reader, he will not be using *mechanica* in either of these senses: 'we understand it to be the part of geometry that deals with [local] motion, and investigates, apodictically and using geometrical reasoning, the force with which such and such a motion takes place.' He follows this with a brief historical note on the etymology of the term *mechanica*.[20] In keeping with his definition of mechanics, Wallis's treatise includes treatments of motion in general, the fall of bodies, the balance, centres of gravity, the simple machines, composition of motions, accelerated and retarded motions, projectiles, percussion, collision theory and centres of percussion, elasticity, hydrostatics, and other topics. Especially significant in this context is Proposition 7 of Part I, Chap. 1: 'Effects are proportional to their adequate causes', which Wallis claims in a scholium 'opens the way to go from purely mathematical speculation to physics; or rather it forms the link between the two'.[21]

Clearly a new sense of the scope and nature of mechanics was emerging during the 1660s, at least in England, with Wallis as probably the foremost 'technical' advocate in print[22] of a new kind of universal mechanics dealing with forces of whatever kind in nomological relations to their associated motions. Less technically involved or equipped, but possibly as influential in this respect with a wider audience, was Robert Boyle, whose *Usefulness of mechanical disciplines to natural philosophy* (1671) begins:

> To prevent the danger of stumbling (as they speak) at the threshold, I shall begin this discourse with advertising you, that I do not here take the term

[19]   *The Optical Papers of Isaac Newton*, ed. Alan E. Shapiro (Cambridge University Press, 1984–), vol. 1, p. 86.

[20]   Pars prima, Cap. I ('De motu generalis'). *Opera Mathematica*, 3 vols. (Oxford, 1693–1699), vol. 1, p. 575.   [21]   *Opera Mathematica*, vol. 1, p. 584.

[22]   Newton had not yet published anything *de motu corporum*, but of course he had already done much important work in that area. See John Herivel, *The Background to Newton's Principia: a Study of Newton's Dynamical Researches in the Years 1664–1684* (Oxford: Clarendon Press, 1965). Richard S. Westfall, *Force in Newton's Physics: the Science of Dynamics in the Seventeenth Century* (London, New York: Macdonald, American Elsevier, 1971), chapters 7 and 8. *The Mathematical Papers of Isaac Newton*, ed. D. T. Whiteside (Cambridge University Press, 1967–1981), vol.6.

mechanicks in that stricter and more proper sense, wherein it is wont to be taken, when it is used only to signify the doctrine about the moving powers (as the beam, the leaver, the screws, and the wedge,) and of framing engines to multiply force: but I here understand the word mechanicks in a larger sense, for those disciplines, that consist of the applications of pure mathematicks to produce or modify motion in inferior bodies: so that in this sense they comprise not only the vulgar staticks, but divers other disciplines, such as the centrobaricks, hydraulicks, pneumaticks, hydro-staticks, ballisticks, etc., the etymology of whose names may inform you about what subjects they are conversant.[23]

Boyle had a sharp understanding of what was involved in the new mechanics and the new natural philosophy that Barrow and Wallis were advocating. He goes on to note that 'the phenomena afforded us by these [mechanical] arts ought to be looked upon as really belonging to the history of nature in its full and due extent. And therefore as they fall under the cognizance of the naturalist, and challenge his speculation; so it may well be supposed, that being thoroughly understood, they cannot but much contribute to the advancement of his knowledge, and consequently of his power'. Accordingly,

> ... there are several doctrines about physical things, that cannot be well explicated, and some of them not perhaps so much as understood, without mechanicks.
>
> That which emboldens me to propose a thing, that seems so paradoxical, is, that there are many phaenomena of nature, whereof though the physical causes belong to the consideration of the naturalist, and may be rendered by him; yet he cannot rightfully and skilfully give them without taking in the causes statical, hydrostatical, &c. (if I may so name them) of those phenomena, i.e. such instances as depend upon the knowledge of mechanical principles and disciplines.[24]

Boyle's seemingly paradoxical coupling of natural philosophy and mech-anical principles had been legitimated by Barrow and Wallis, and was to find effective expression in Newton's *Principia*. Yet the *Principia*, though coming long after the *Lectiones opticae*, has its own seemingly paradoxical juxta-positions. If we read it without reference to later senses of 'mechanics' or its cognates,[25] we find that it is not free of equivocal usage of these terms, since it offers both negative and affirmative replies to the initial question of my title. Rather than a signal for puzzled concern, such equivocalness is a sign of the

---

[23]   *The Works of the Honourable Robert Boyle in five volumes. To which is prefixed the life of the author*, ed. Thomas Birch (London, 1744), vol. 3, p. 162.

[24]   *Ibid.*, p. 163.

[25]   Later senses of the term, including those currently in use today, derived to an important degree from the laws of motion in Book 1 and their implications. But that is not germane to the problem of the changing nature of 'mechanics' during the seventeenth century.

disciplinary changes I have been trying to describe, so we should not be surprised to find in Newton's text vestiges of earlier states of affairs co-existing with elements of new patterns of thought.

A negative response to the question in the title is implied by Newton in Book 1. In the Scholium following the sixth corollary to the 'Axioms, or Laws of Motion', he illustrates the wide applicability of the three laws, including their employment (as he imagines) by Wren, Wallis and Huygens in their respective solutions to the problem of hard-body collision. After half-a-dozen pages on his experimental confirmation of Law 3 using colliding bodies of different materials, Newton digresses to note that the operation of all machines (*instrumenta mechanica*) depends on the principle that the speeds of equipollent agents vary inversely as their forces. Then he begins to wind up the Scholium with a remark that remained unchanged in all three editions of the *Principia*: 'But to treat of mechanics [*mechanica*] is not my present purpose. I just wanted to show the great extent and certainty of the third law of motion.[26] In a similar spirit he introduced Corollary 2 (on the resolution and composition of forces), whose great extent and certainty are abundantly confirmed 'from mechanics [*ex mechanica*]', and upon which indeed 'depends the whole of mechanics [*mechanica tota*] as demonstrated by authors in various ways'.[27] Furthermore, at the beginning of Book 3 he pauses to remind the reader:

> In the preceding books I have laid down the principles of [natural] philosophy, not however the philosophical but just the mathematical principles that permit inquiry into philosophical matters. The principles that concern most particularly [natural] philosophy are the laws and states [*conditio*] of motions and forces...[28]

There is no explicit mention here of mechanics as the overriding concern of the first two Books of the *Principia*. On the evidence of these passages, read *per se*, 'mechanics' for Newton was *the theory of machines*, and mechanics in this sense was not therefore the main business of the *Philosophiae naturalis principia mathematica*.

This was also Roger Cotes's view, to judge by his preface to the second edition (1713) of the *Principia*. It is significant that his laudatory presentation

---

26 Book I, Law 3, Corollary 6, Scholium. *Sir Isaac Newton's Mathematical Principles of Natural Philosophy and his System of the World*, trans. Andrew Motte, ed. Florian Cajori (Berkeley, Los Angeles: University of California Press, 1947), pp. 27–8 (translation modified). (Henceforth abbreviated as *Principles* (Motte–Cajori).) *Isaac Newton's Philosophiae Naturalis Principia Mathematica*, 3rd edn. (1726) with variant readings, eds. A. Koyré, I. B. Cohen, Anne Whitman, 2 vols. (Cambridge University Press, 1972), vol. 1, p. 72. (Henceforth abbreviated as *Principia* (KCW).)

27 *Principia* (KCW), vol. 1, pp. 57, 59.

28 *Principles* (Motte–Cajori), p. 397 (translation modified). *Principia* (KCW), vol. 2, p. 549.

took the form of a favourable comparison of the treatise and its methods with the approaches of two classes of 'those who have undertaken to treat of physics [*physica*]', that is, of certain other *natural philosophers*, not authors of *libri mechanicorum*. Predictably, into the first class fall Aristotle and the Peripatetics, and into the second fall the Cartesians (though Cotes does not identify them by name). Equally predictably, there is a third class of natural philosophers 'who possess experimental philosophy'. 'From some select phenomena they deduce by analysis the forces of nature and the more simple laws of forces; and from thence by synthesis show the constitution of the rest. This is that incomparably best way of philosophizing, which our renowned author most justly embraced in preference to the rest'.[29] This 'context of comparison' provides important supplementary proof of the descriptive accuracy of the title Newton chose for his treatise. Cotes did not compare the *Principia* with the mechanical treatises (say) of the Renaissance mathematicians or engineers for the very good reason that there was no such comparison to make. They and Newton were doing quite different things.

As for an affirmative answer to my initial question, Newton as good as gives one in his 1687 Preface, to which I return later, and elsewhere in the *Principia*. Proposition 52 of Book 2 shows that for a fluid set in rotational motion by a sphere rotating within it, the periodic times of the parts of the vortex are as the squares of their distances from the centre of motion. Corollary 5 of this proposition states that if another rotating sphere is placed within the first vortex, the two spheres will eventually revolve about a common centre, and will recede from each other, unless they are held in check by some external force.

> Afterwards, if the constantly impressed forces, by which the globes continue in their motions, should cease, and everything be left to act according to mechanical laws [*omnia legibus mechanicis permitterentur*], the motion of the globes will languish by degrees (for the reason assigned in Cor. III and IV), and the vortices at last will quite stand still.[30]

Corollary 11, in a similar context, also appeals to 'leges mechanicae' (adj.), as does the scholium to Proposition 53 showing that in a Cartesian planetary vortex the speed of flow of vortical matter through a cross-section should vary inversely as the cross-sectional area, a result that conflicts with the astronomical data.[31]

[29] *Principles* (Motte–Cajori), pp. xx–xxi. *Principia* (KCW), vol. 1, pp. 19–20.

[30] *Principles* (Motte–Cajori), p. 391 (translation modified as indicated in main text). *Principia* (KCW), vol. 1, p. 540.

[31] *Principles* (Motte–Cajori), pp. 392, 395–6. *Principia* (KCW), vol. 1, pp. 542, 546. The wording of these appeals to 'leges mechanicae' remained unaltered throughout the three editions of *Principia*.

Now in the Motte–Cajori translation of Corollary 5 and of the Scholium, 'leges mechanicae' (in the appropriate grammatical forms) is given as 'the laws of mechanics', but in Corollary 11 it appears as 'the mechanical laws', which to my mind is the correct translation of the expression, apart from the superfluous definite article (compare my modified translation of the above excerpt from Corollary 5). 'The laws of mechanics' connotes the *axiomata* of a distinct though not explicitly defined discipline, whereas 'mechanical laws' connotes those laws that govern the material universe as understood according to the mechanical philosophy. I think Newton had the latter sense primarily in mind in Corollaries 5 and 11. Once again I find support in Cotes's 1713 preface, in the passage where he observes, alluding to the Cartesians, that 'those who take the foundations of their speculations from hypotheses, even if they then proceed with the greatest accuracy in accordance with mechanical laws [*leges mechanicae*], may indeed be said to have produced a fable that is charming and perhaps judicious, but a fable it will remain for all that.'[32]

However, the conclusion to draw is that just as the mechanical laws here are those of Descartes or other Cartesians, so in the case of Newton's two corollaries the implied laws can scarcely be other than the 'Axiomata sive leges motus' of Book 1, especially in the light of their Corollary 2 and the scholium to Corollary 6. The laws of motion therefore double as laws of mechanics. As for the Scholium to Proposition 53, there the reference to 'leges mechanicae' (adj.) is to one of Benedetto Castelli's theorems on flowing water.[33] So the use of 'leges mechanicae' in the two corollaries and the scholium to Proposition 53 implies that Newton would have been happy to describe his critique of the vortex theory, a major concern of the *Principia*, as an exercise in 'mechanica'.

Other texts point to an affirmative response to my question. In one of the Queries to the *Opticks* (4th edition, 1730) we read that 'as in Algebra, where affirmative Quantities vanish and cease, there negative ones begin; so in Mechanicks, where Attraction ceases, there a repulsive Virtue ought to succeed'.[34] Since the mathematical investigation of one variety of attraction

---

[32] *Principia* (KCW), vol. 1, p. 20. My translation, Motte–Cajori's being quite unsatisfactory.

[33] The inverse relation between cross-sectional speed of flow and cross-section was given by Benedetto Castelli in his *Della misura dell'acque correnti* (Rome, 1628; later eds. 1639, 1660). Thomas Salusbury published an English translation of the 1660 edition in his *Mathematical Collections and Translations* (London, 1661), vol. 1, pt. 2. I assume that Castelli's treatise, or at least some of its content, was known to Newton.

[34] Isaac Newton, *Opticks, or A Treatise of the Reflections, Refractions, Inflections & Colours of Light*, foreword A. Einstein, intro. E. Whittaker, preface I. B. Cohen, table of contents D. H. D. Roller (New York: Dover, 1979), Query 31, p. 395. In the 1706 *Optice*, where Query 31 first appeared (as Query 23), this passage reads: 'Et sicuti in algebra, ubi quantitates affirmativae evanescunt & desinunt, ibi negativae incipiunt; ita in mechanicis, ubi attractio desinit, ibi vis repellens succedere debet.' *Optice: sive*

was the highpoint of the *Principia*, this remark implies that that investigation too was an exercise in mechanics. The manuscript entitled 'The Elements of Mechanicks', written about or after the mid-1690s, begins with eleven 'theoremes' (minus the demonstrations) in which Newton states his three laws of motion; the parallelogram rule of composition and resolution; the 'inertial' property of the centre of gravity of an isolated system; the law of falling bodies and the parabolic motion of water spouts and projectiles; the formulae for centrifugal and centripetal force; Kepler's laws as consequences of motion under a inverse square central force; the resistance to bodies moving in a medium; hydrostatical laws; laws of vibratory and pendular motion. Appearing as Theorem no. 3, immediately following the third law, is a version of the principle of 'virtual work' stated loosely in terms appropriate to machines: 'The forces wth wch the parts of one & ye same engine move at one & ye same time are reciprocally proportional to their velocities'. And the second of the six 'Problemes' at the end of the manuscript is the traditional headache of mechanicians and engineers: 'With any given force to move any given weight'.[35] It is clear from this document that the elements of mechanics comprise both the theory of machines and the theoretical content of the *Principia*.

The same holds for the Preface to the first edition of the *Principia*. Newton opens the Preface by reporting Pappus on the ancients' views of the status and nature of *mechanica*. Pappus had written that mechanical science (μηχανικὴ θεωρία), 'takes almost first place in dealing with the nature of the material elements of the universe. For it deals generally with the stability and movement of bodies, and their motions in space, inquiring not only into the causes of those that move in virtue of their nature, but forcibly transferring [others] from their own places in a motion contrary to their nature...' According to Pappus, Hero's school had divided mechanics into a theoretical and a manual part:

de reflexionibus, refractionibus, inflexionibus & coloribus lucis libri tres. *Authore Isaaco Newton, Equite Aurato, Latine reddidit Samuel Clarke...Accedunt Tractatus duo ejusdem authoris de speciebus & magnitudine figurarum curvilinearum, Latine scripti* (London, 1706), p. 338.

[35] Cambridge University Library, MS Add. 4005, ff. 23–5. Isaac Newton, *Unpublished Scientific Papers of Isaac Newton: a Selection from the Portsmouth Collection in the University Library, Cambridge*, ed. A. R. Hall and M. B. Hall (Cambridge University Press, 1978 (1962)), pp. 165, 168. Note that the traditional problem of moving a given weight with a given force is explained near the end of the Scholium to the Laws of Motion: *Principles* (Motte–Cajori), p. 27; *Principia* (KCW), vol. 1, p. 72. The Halls' surmisal (p. 165) that this undated manuscript 'is certainly later in date than the *Principia*' is confirmed by the watermark of ff. 23–4, which shows that the paper was used by Newton between the mid 1690s and the early 1700s. I am indebted to Alan Shapiro for this information (private communication); see his paper elsewhere in this volume.

'the theoretical part is composed of geometry, arithmetic, astronomy and physics, the manual of work in metals, architecture, carpentering and painting and anything involving skill with the hands.'[36]

Newton begins the summary of Pappus with an explanation of his purpose in the *Principia*:

> Since the ancients (as we are told by Pappus) set the highest value on mechanics [*mechanica*] in the investigation of natural things, and since the moderns, abandoning substantial forms and occult qualities, have undertaken to apply mathematical laws to the phenomena of nature, it seemed fitting in this treatise to cultivate mathematics in so far as it concerns [natural] philosophy.[37]

Newton can bracket the two apparently disparate reasons because it is mathematics that is common to the rational mechanics of antiquity and to the moderns' programme of mathematizing natural philosophy. The *Principia* is primarily a treatise in mathematics, which explains why Newton does not say in the Preface that he has cultivated *mechanics* is so far as it concerns philosophy, or even that he has cultivated natural philosophy in so far as it can be handled mathematically.

Newton continues with the distinction between rational and practical mechanics (compare the excerpt above from Goclenius) and he comments on contrasts between the geometrical and the mechanical, concluding that 'geometry is founded in mechanical practice, and is nothing but that part of universal mechanics [*mechanica universalis*] which accurately proposes and demonstrates the art of measuring'. Then he lets us see why we would be right to describe the *Principia* as a treatise in a new kind of rational mechanics, and as we read we recall the passages quoted earlier from Wallis and Barrow:

> But since the manual arts are employed chiefly in moving bodies, it happens that geometry is commonly referred to their magnitude, and mechanics to their motion. In this sense rational mechanics will be the science of motions [*scientia motuum*] resulting from any forces whatsoever, and of the forces required to produce any motions, accurately proposed and demonstrated. The ancients cultivated this part of mechanics with respect to the five powers that relate to the manual arts; they scarcely considered gravity (not being a manual power) other than in relation to moving weights with those powers. However, being concerned with

---

[36] *Selections illustrating the History of Greek Mathematics*, ed. Ivor Thomas, 2 vols., Loeb Classical Library (London: Heinemann; Cambridge, Mass.: Harvard University Press, 1967–1968), vol. 2, pp. 615–19. For the Greek text with Hultsch's Latin translation see *Pappi Alexandrini Collectionis quae supersunt e libris manu scriptis edidit Latina interpretatione et commentariis instruxit Fridericus Hultsch*, 3 vols. (Berlin: Weidmann, 1876–1878), vol. 3, pp. 1022–9.

[37] *Principles* (Motte–Cajori), p. xvii (translation modified). *Principia* (KCW), vol. 1, p. 15.

philosophy rather than arts and writing about natural not manual powers, we investigate especially those things that relate to gravity, levity, elastic force, the resistance of fluids and suchlike forces, whether attractive or impulsive. And on that account we offer this work of ours as the mathematical principles of philosophy. For the whole burden of philosophy seems to consist in this – from the phenomena of motions to investigate the forces of nature, and then from these forces to demonstrate other phenomena ...[38]

It is no accident that this Preface shares material with several parts of drafts (c. 1693) for an unfinished treatise on geometry,[39] where Newton is more forthcoming on the relation between mathematics and mechanics, and where one feels even more strongly the presence of Barrow and Wallis. The drafts merit closer study than I have space to give them here, but a couple of passages will suffice to illustrate Newton's position:

> I say that geometry from its nature is nothing other than the more select part of mechanics, indeed I say that nearly all the mathematical sciences originated from mechanics. It is mechanical to count different things placed before one and to count continued parts of the same thing, and to carry out arithmetical operations on paper. It is mechanical to observe the motions of the stars. It is mechanical to measure refractions of light, and make telescopes. It is mechanical to treat of musical instruments, and so on. These are all practical mechanics, and speculations grounded in them are rational mechanics. Each time some part of mechanics is seen to be worth treating separately, another name is given to it and only what is then left of mechanics maintains the proper name.[40]

> ...therefore geometry and mechanics are joined together in the closest unity, and to speak more truly, they are nothing other than two parts of the one and the same science. For what is geometry other than the art of measuring lengths and shapes produced by agreed manual operations? And what is mechanics other than the art of determining and measuring the distances [longitudines] and shapes that moved bodies[41] describe in given times?[42]

Recently Cohen has written that 'Newton wanted his readers to be certain that he himself had constructed mathematically a system of rational mechanics, whereas Descartes had been more concerned to expound a general

[38] *Principles* (Motte–Cajori), pp. xvii–xviii (translation modified). *Principia* (KCW), vol. 1, pp. 15–16.

[39] Whiteside, *Mathematical Papers*, vol. 7, pp. 248–561 (Books 1 and 2), especially pp. 286–99, 338–45 (on mechanics and geometry).

[40] *Mathematical Papers*, vol. 7, p. 341.

[41] *corpora mota*: note that the perfect participle passive connotes motion caused *ex alio*, that is motion under a force, which is what Newton has in mind.

[42] *Mathematical Papers*, vol. 7, p. 339–40, note 6 (cancelled passage).

philosophy of nature.[43] On comparing the gloss on Pappus in the Preface with the rest of the treatise, Newton's readers could see that he had greatly expanded the boundaries of 'rational mechanics' as it had been understood since antiquity, and no doubt this was Newton's intention, but they could also see that he did not *call* the result 'mechanica rationalis'. To develop Cohen's remark, the contrast is not just between Newton's 'rational mechanics' *per se* and Descartes' natural philosophy *per se*, but rather between two approaches to the same discipline of natural philosophy, one a wholly descriptive exercise in one version of the mechanical philosophy, the other a mathematical exploration of a different version of the mechanical philosophy.

Yet the result of that exploration plainly *was* rational mechanics, given the definition in the preface of rational mechanics as 'the science of motions resulting from any forces whatsoever'. The *Journal des sçavans* review was hostile, but the reviewer's hostility to the *Principia* was a Cartesian reaction consequent upon a shrewd identification of Newton's purpose and methods: 'these demonstrations cannot be considered as other than mechanical, since the author himself admits...that he has treated their principles not as a physicist [*physicien*], but simply as a geometer'. At the beginning of Book 3, the reviewer noted, Newton again admits the same thing, 'where he tries nonetheless to explain the system of the world'; but he does this 'only with hypotheses that are arbitrary for the most part, and which consequently serve as the foundation only for a treatise on pure mechanics [*un traité de pure Mécanique*]'.[44] The reviewer concluded that to bring to complete perfection his already perfect 'mécanique', all Newton had to do was to produce 'a physics as exact as the mechanics'.

Cohen is surely justified in surmising that 'Newton would have been furious to find that his presentation of the System of the World in Book 3 was described as having been based on a set of introductory "hypotheses" that are for the most part arbitrary and hence not to be considered as the basis of a true physics.'[45] Yet were it recast in more neutral language, possibly with a spot of *emendatis emendandis*, the review in the *Journal des sçavans* would not come across as a wholly unfair account of Newton's *Principia*. Newton often distinguished between physical questions and mathematical procedures, the preference being given to the latter where (as was frequently the case) the former could not be resolved in ways amenable to experimental check.[46] But

---

[43]   Cohen, 'Newton and Descartes' (1990), p. 617.

[44]   Cohen, *Introduction*, pp. 156–7 (original text: my translation). See also Cohen, *The Newtonian Revolution*, pp. 96–9; Mouy, *Le développement de la physique cartésienne*, p. 256.   [45]   Cohen, *Introduction*, p. 157.

[46]   See for example Book 1, Section 11, introductory paragraph and Prop. 69, Scholium; Book 2, Prop. 4, Scholium and Prop. 23, Scholium. *Principles* (Motte–Cajori), pp. 164, 192, 244, 302. *Principia* (KCW), vol. 1, pp. 266–7, 298, 359, 431. For a

the reviewer in the *Journal des sçavans* saw in that approach something that was characteristic of texts in pure (and therefore rational) mechanics, since they too dealt geometrically with the relations between motions and forces, and couched their findings in the form of definitions, axioms, propositions, corollaries, and one or two hypotheses.

By the definition given by Pappus and Newton, 'rational mechanics' must include the theory of machines (the five mechanical powers), yet we saw that Newton explicitly excluded this kind of inquiry from the programme of the *Principia*, equivocally (or inadvertently?) calling it 'mechanica' without further qualification, while at the same time choosing not to include 'mechanica' in the title of the treatise.[47] But that should not be a surprise. Newton's *Principia* was a transitional and truly 'pivotal' text. It was not a manual on the *artes mechanicae*, nor was it a *liber mechanicorum*. It was assuredly and obviously a mathematical treatise, but whether it was a revolutionary treatise in natural philosophy or a revolutionary new departure in rational mechanics depends on whether we choose to look backwards from 1687 to the achievements in natural philosophy of Descartes and the Peripatetic tradition, or whether we choose to look forward, while remembering Galileo and Huygens, to the great syntheses of Euler and Lagrange.

detailed examination of the important relation between mathematics and physics in Newton's thought, notably in the *Principia*, see Cohen, *The Newtonian Revolution*, Part 1, Chapter 3: pp. 52–154.

[47]   It is worth noting that 'mechanica rationalis' substituted for 'philosophia naturalis' in the title would have seemed odd to the earliest readers of the *Principia*. 'Mechanicae rationalis principia mathematica' would have been pleonastic, since rational mechanics, unlike natural philosophy, could be nothing other than mathematical. The title of Newton's *Principia* was well chosen.

12 The review of the first edition of Newton's *Principia* in the *Acta Eruditorum*, with notes on the other reviews

I. BERNARD COHEN

Introduction: the reviews of the first edition of the *Principia*

The publication of Newton's *Principia* was a notable event. For historians, with the perspective of a long backward view, Newton's *Principia* marks both the high point of the Scientific Revolution and the maturity of scientific thought about motion. It is the foundation stone of the new science of 'rational mechanics', to use the name introduced by Newton in the Preface to the *Principia*, or the new science of 'dynamics', to use the name invented by Leibniz.[1] To paraphrase Galileo, it was a 'new' work on a very 'old' subject, motion.[2]

In any consideration of the novelty of the *Principia* and its profound restructuring of the most fundamental of the exact sciences, the question must naturally arise of the reaction of the scientific community to Newton's presentation of his ideas. Did Newton's contemporaries at once recognize the *Principia* as a truly significant book? Did it create any stir? Was there any immediate recognition of its novel features? Did Newton's fellow scientists accept or reject his concepts?

The contemporaneous reaction to the *Principia* may be gauged in various ways. First of all, there are recorded opinions in the correspondence of such scientists as Huygens and Leibniz.[3] Much of this discussion centered on

---

[1] In a manuscript note, Newton wrote a brief account of the development of dynamics, beginning: 'Galileo began to consider the effect of Gravity upon Projectiles. Mr. Newton in his Principia Philosophiae improved that consideration into a large science. Mr. Leibnitz christened the child by [a] new name as if it had been his own, calling it *Dynamica*.' See Supplement I to I. Bernard Cohen, *Introduction to Newton's 'Principia'* (Cambridge: Harvard University Press; Cambridge: Cambridge University Press, 1971, 1978). See, further, Eric J. Aiton, *Leibniz: a Biography* (Bristol: Hilger, 1985).

[2] *Discorsi e dimostrazioni matematiche, intorno à due nuove scienze* (Leyden, 1638) (the 'Two New Sciences'), Third Day, introduction.

[3] On the reactions of Leibniz and Huygens, determined from their correspondence, see Alexandre Koyré, *Newtonian Studies* (Cambridge: Harvard University Press, 1965), chapter 3, suppl. A.

Newton's apparent use of the concept of 'attraction' and the seemingly consequent adoption of the ancillary concept of action-at-a-distance.[4] A second domain in which to seek for reactions is in published books or articles discussing, criticizing, emending, or recasting some parts of the *Principia*. An example is Jacob Hermann's presentation of a version of Newton's second law in a form very much like that familiar to generations of college students,[5]

$$G = M \, \mathrm{d}V : \mathrm{d}t$$

'where G stands for the weight [*pondus*] or gravity [*gravitas*], however variable, of the mass M'. Yet a third measure of the contemporaneous reaction is to examine the book reviews.

The *Principia* gave rise to four very different book reviews. These appeared in the *Philosophical Transactions*, the journal of the Royal Society of London, then edited and published by Edmond Halley, the *Bibliothèque Universelle*, the *Journal des Sçavans*, and the Leipzig *Acta Eruditorum*. Of these, the most influential was probably the one published in the *Acta*. The reason is that the Leipzig *Acta* were widely read all over the Continent; the reviews published in the *Acta* were on the whole fair and impartial; the journal itself was dedicated to the 'republic of letters' without a particular bias – that is, the *Acta* were not a bastion of conservative Cartesianism. In the *Acta*, furthermore, the information concerning new work in science as well as in other fields of learning was superior to that in many other journals; there were both book reviews and articles presenting the results of original research. Finally, this particular review was long and very detailed, running to thirteen printed pages.[6]

The author of the review in the Leipzig *Acta*

All of the reviews except Halley's were published anonymously. There is now good evidence that the one in the *Bibliothèque Universelle*, in French, was written by the philosopher John Locke.[7] The author of the one in the *Journal*

---

4   *Ibid.*, chapter 3, suppl. B.

5   Jacob Hermann: *Phoronomia, sive de Viribus et Motibus Corporum solidorum et fluidorum libri duo* (Amsterdam: apud Rod. & Gerh. Wetstenios, 1716), p. 57. See I. B. Cohen, *The Newtonian Revolution* (Cambridge University Press, 1980), pp. 143–6.

6   *Journal des Sçavans* (2 August 1968):128–9; *Bibliothèque Universelle* 8 (March 1688):436–50; *Philosophical Transactions*, no. 186 (January, February, March 1687):291–7; *Acta Eruditorum* (June 1688):303–15.

7   On Locke's authorship of this review, see Rosalie L. Colie: 'John Locke in the Republic of Letters', in *Britain and the Netherlands*, ed. J. S. Bromley and E. H. Kossmann (London: Chatto & Windus, 1960):pp. 111–29; James L. Axtell: 'Locke,

des Sçavans has never been positively identified, although Paul Mouy suggested that he may have been Pierre-Sylvain Régis, a strict Cartesian philosopher.[8] For a long time the authorship of the review in the *Acta* was shrouded in mystery. There was always the possibility that it might have been written by Otto Mencke, the editor of the *Acta*, but Mencke was a classicist and philosopher whose talents were literary and humanistic. He was professor of theology at the University of Leipzig. With only a limited knowledge of mathematics and physics, it is doubtful whether he could have understood the *Principia* well enough to have written such a review.

New evidence has recently enabled the author of this book review to be identified as Christoph Pfautz (1645–1711), a professor of mathematics at the University of Leipzig.[9] He wrote three 'dissertations': on parallax (1668), on eclipses considered from arithmetical and geometrical principles (1674), and on solar eclipses (1684).[10] There is little biographical information available concerning Pfautz. Poggendorff's presentation is very brief, mentioning his professorship at the University of Leipzig and giving his birth and death dates – 11 Oct. 1645, Leipheim bei Ulm; 2 August 1711, Leipzig – and the titles and dates of his three publications. The recent history of the editorship of the *Acta*, in which Pfautz's authorship of the review is ascertained, gives volume and page numbers of Pfautz's notes and reviews in that journal over many decades.[11]

Poggendorff derived his information about Pfautz from Christian Gottlieb Jöcher's four-volume *Allgemeines Gelehrten-Lexicon* (Leipzig, 1750–51). Further details are given in the supplement, Johann Christoph Adelung's *Fortsetsung und Ergänzungen zu Christian Gottlieb Jöchers Allgemeinem Gelehrten-Lexicon* (Rotermund).[12] From these volumes we learn, additionally,

Newton and the Elements of Natural Philosophy', *Paedagogica Europaea*, 1 (1965):235–44; *idem*, 'Locke's review of the *Principia*', *Notes and Records of the Royal Society of London* 20 (1965):152–61.

8   Paul Mouy: *Le Développement de la Physique Cartésienne, 1646–1712* (Paris: Librairie Philosophique J. Vrin, 1934), p. 256. On the physics of Régis, see *ibid.*, pp. 145–67.

9   A. H. Laeven: *De 'Acta Eruditorum' onder Redactie von Otto Mencke* (Amsterdam: APA – Holland University Press, 1986), p. 285. I am grateful to Domenico Bertoloni Meli for bringing this book to my attention.

10   See J. C. Poggendorff: *Biographisch-Literarisches Handwörterbuch zur Geschichte der Exacten Wissenschaften* (Leipzig: Verlag von Johann Ambrosius Barth, 1863).

11   See Laeven, *De 'Acta Eruditorum'*, suppl. 5, pp. 349–50.

12   A photo-reprint of the original four volumes has been made by Georg Olms Verlagsbuchhandlung (Hildesheim, 1961); the biography of Pfautz appears in vol. 3, p. 1484. Although the first volume of Adelung's revision was published in 1784, vol. 5 (containing the biography of Pfautz) was not published until 1816. This set has also been reprinted by Georg Olms (Hildesheim, 1961). In the original 1751 volume, a work by Carl Pfautz (a physician) on the blood from an incision is mistakenly attributed to

that Pfautz's father (also Christoph Pfautz) was a preacher. A final source of information is the first volume of *Miscellanea Lipsiensia Nova*, which contains an 'oratio' on Mencke written by his son, edited with notes by Mencke's grandson. This 'oratio' contains a longish note about Pfautz, from which we learn that Pfautz and Mencke were closer than ordinary academic colleagues. Pfautz married Mencke's sister Catharina (1657–1703); they had four sons and two daughters.[13] From these data we learn that when Pfautz wrote his review of the *Principia* he was forty-two years of age.

From the correspondence of Mencke, the editor of the *Acta*, and from the notes of his grandson, we may learn that Pfautz and Mencke were closely associated in many ways. They were members of the same 'collegium' in the university.[14] Additionally, in 1684, five years before the publication of the *Principia*, Pfautz and Mencke went on a trip together to Holland and to England. In England they met the classicist and philosopher Thomas Gale and also Robert Boyle. They went on to Oxford and there met John Wallis. There is no mention of a visit to Cambridge or a meeting with Isaac Newton.[15] Although Newton later corresponded with Mencke, and obtained a set of the *Acta*, his correspondence does not mention Pfautz, nor is there a discussion of Pfautz's review.[16]

Pfautz was a logical choice to be the reviewer of the *Principia*. He was a professional mathematician interested in astronomy. He was a close associate of the editor, Mencke, and was a rather regular reviewer of the *Acta*. Another virtue of Pfautz as reviewer was that he seems to have had no personal 'axe to grind'; he does not, for example, appear to have been overly committed to the Cartesian physics and cosmology.

The very existence of four book reviews in three languages (English, French, and Latin) is one gauge of the immediate recognition of the significance of the *Principia*. Another is the fact that Mencke was willing to publish Pfautz's review, even though it was very long and discussed the contents of the book in

---

Christoph Pfautz; this error is corrected in the later edition. In this 1816 edition, there is mention of a copper-plate portrait, but no information is given concerning it.

[13]   Johann Burkhard Mencke, 'De Vita et in Literas meritis Ottonis Menckenii ... Oratio', *Miscellanea Lipsiensia Nova*, vol. 1 (1742), pp. 644–737, with notes by Friedrich Otto Mencke; a note about Pfautz occurs on p. 709. Pfautz was in correspondence with the astronomer Hevelius; see *Actorum Eruditorum ... Supplementa* 9 (1719): 360–1.

[14]   Mencke, 'Oratio'.

[15]   On this trip see Mencke, 'Oratio' p. 709 note. Of course, at this time Newton was not yet the celebrity that he was to become later; there was no reason for them to have sought a visit.

[16]   See *The Correspondence of Isaac Newton*, ed. H. W. Turnbull, J. F. Scott, A. Rupert Hall, and Laura Tilling, 7 vols. (Cambridge University Press, 1959–77), letters 414, 430, vol. 3, pp. 270–1, 291–3.

considerable detail. As we shall see in a moment, Pfautz himself was fully aware that his review was more detailed and more extensive than usual and he apologized to his readers for this fact.

The principal novel features of the *Principia*

From today's historical vantage point, the major novelties in the *Principia* seem obvious. They begin with the opening 'Definitiones', primarily the new concepts of mass and of centripetal force, plus the three measures of force (the absolute, the accelerative, and the motive); the 'Axiomata sive Leges Motus', primarily the second law and also the third; and Newton's lemmas on the method of limits (or 'first and last ratios'). The second and third sections of Book One displayed – in Propositions 1, 2, and 3 – the significance of the law of areas, showing that this law is a necessary and sufficient condition for a centrally directed force continually acting on a body with an initial component of inertial motion. Newton's result was astonishing. For the first time it was shown that there is a connection between the law of areas and the law or principle of inertia.[17] Then, in Prop. 6 Newton introduced a radically new dynamical measure of a force. In Prop. 11 Newton showed that motion along an ellipse implies a centrally directed force that must be as the inverse square of the distance; subsequent propositions showed that it is the same force in the case of a parabola and a hyperbola.[18] Sections Four and Five are geometric, devoted to properties of conic sections.[19] Later parts of Book One explore the

[17] Although this result is not mentioned in any of the four contemporaneous reviews, its significance was at once apparent to Leibniz, as we learn from his manuscript annotations on first reading the *Principia*. I gained this information from Domenico Bertoloni Meli, whose researches on Leibniz have cast much new light on the relations between Leibniz and Newton. See his 'Leibniz's excerpts from the *Principia Mathematica*', *Annals of Science*, 45(1988):477–505.

[18] The converse, that an inverse-square force implies an orbit that is a conic section, was stated by Newton (Corol. 2 to Prop. 13, Book One) without proof. In the second edition of the *Principia*, he sketched a proof. See Isaac Newton, *The Mathematical Papers of Isaac Newton*, ed. D. T. Whiteside, 8 vols. (Cambridge University Press, 1967–81), vol. 6, pp. 146–9, note 124. The validity of Newton's proof has been called into question by Robert Weinstock. See J. Bruce Brackenridge's paper in this volume for an account of Weinstock's arguments and subsequent debate over it.

[19] Since Sections Four and Five are, in major part, concerned with the geometry of conics, they can be considered loosely to have been a legitimate part of the *Principia*. These two sections (especially Section Five) do not, however, consist of materials necessary for the development of the theory of elliptic or parabolic orbits and were introduced by Newton probably because he had some results available and written up and wished to publish them. We know this is the case for Section Five, for which there exists a pre-draft. For details, see my *Introduction*, pp. 96–8. On the mathematical

dynamics of force and motion for two-body systems, for some three-body systems, for bodies with spherical and even certain non-spherical shapes, and concludes with Newton's analogy of the physics of light paths (as in sound) and the dynamics of particle motion.

In Book Two Newton explored various conditions of resistance to motion, introduced a form of the calculus (Lemma Two[20] on 'moments'), investigated the motion of pendulums, developed the mathematical physics of simple wave motion, and concluded with a demonstration that any theory of vortices is inconsistent with Kepler's law of areas. The inescapable consequence is that the Cartesian system of the heavens must be false.[21]

In the first edition of the *Principia*, Book Three – on the System of the World – began with a set of 'Hypotheses', which Newton later divided into 'Phaenomena', 'Regulae Philosophandi', and some non-provable statements which remained 'Hypotheses'.[22] In this presentation, it is shown that the orbital motion of Jupiter's satellites and the orbital motion of the planets (in the Copernican system) demand an inverse-square force.[23] This is followed by Newton's elegant and convincing demonstration that the terrestrial force which we call 'gravity' extends to the moon (Prop. 4 and scholium)[24] and this

significance of these two geometrical sections, see D. T. Whiteside's commentary in Newton, *Math. Papers*, pp. 230–98.

[20] In this lemma, Newton develops his method of finding the derivative and the integral of quantities raised to powers and of polynomials. In the first two editions, he referred rather generously to the work of Leibniz saying that the latter's method 'hardly differed from mine, except in his form of words and symbols'. To this statement there was added in the second edition (1713) the words 'and the concept of the generation of quantities'. In the third edition (1726), which was published after the priority dispute, a new scholium was introduced and this reference to Leibniz was eliminated.

[21] A number of scientists, Lagrange among them, believed that the whole purpose of Book Two was to prove that the Cartesian vortical cosmology is false.

[22] For details see I. B. Cohen, 'Hypotheses in Newton's Philosophy,' *Physis* 8(1966):163–84. The opening of Book Three in the first edition contained nine 'Hypotheses'. In the second edition Hypothesis I and Hypothesis II became the first and second 'rules'. Hypothesis III on the transformation of bodies was eliminated. Hypotheses V–VI–VII–VIII–IX became Phaenomena I–II–III–IV–V. Hypothesis IV ('The center of the system of the world is at rest') was moved to the place where it was needed in the text and there became Hypothesis I. In the second edition there was an additional third rule which had not appeared in the first edition. Also there was a Hypothesis II which had not been so labeled in the first edition. The final (fourth) rule was not added until the third edition (1726).

[23] In fact, Prop. 1 of Book Three deals with the forces with which the satellites of Jupiter are attracted by Jupiter, while Prop. 2 introduces the forces by which the sun attracts the planets so as to keep them in their orbits. Prop. 3 introduces the force exerted by the earth on the moon.

[24] Prop. 4 and the scholium which follows are devoted to a proof that the force of gravity (which we know on earth as the cause of weight) extends to the moon. Even the

leads to the dramatic conclusion that the force which acts between the planets and the sun, between planets and their satellites, and between one planet and another (force of perturbation) is gravity. The force is seen to be universal because it is applicable to all masses; it is directly proportional to the product of the masses and inversely proportional to the distance between any two masses.

In Newton's system of the world, the sun and moon exert a force of gravity on the waters of the ocean, to produce the tides.[25] A major topic, occupying about half of Book Three, is the physics and motion of comets. Comets are shown to be a 'kind of planet' and therefore must move according to the same laws of dynamics as planets and their satellites, i.e., in orbits having the shape of conic sections. Although Newton was aware that many comets move in elliptical orbits, he worked out a mode of using parabolas in the computation of cometary positions.

One of the most brilliant parts of Book Three is Newton's theoretical linking of the shape of the earth and the phenomenon of precession (Prop. 39, Book Three) and his analysis of the moon's motion. Newton's principles of mechanics indicate on theoretical grounds that the shape of the earth must be a oblate spheroid. The reason is that in this case the moon will exert a greater gravitational pull on the near bulge than on the far bulge, with the net effect of an endeavor to change the tilt of the earth's axis (an endeavor to make the axis more nearly perpendicular to the plane of the earth's orbit); any such endeavor to change the orientation of the axis of a spinning body must produce precession.

Newton addressed the problem of the moon's motion by a simple and direct application of his new theory of universal gravity. In this theory the moon's motion is considered primarily in terms of the centrally directed gravitational attraction of the earth and the perturbing gravitational attraction of the sun. For the first time in the long history of astronomy the moon's motion became subject to study on the basis of physical causes and ceased being an example of trial and error celestial geometry. Newton was not fully successful in this endeavor and he revised and expanded this discussion in the second edition.[26]

Cartesians who were not willing to accept the theory of attraction could not fault Newton's argument. For evidence concerning the reaction of Cartesians to this particular part of Newton's work, see Mouy, *La Physique Cartésienne*, pp. 252–3.

[25] Newton was aware that he could not account for all aspects of tidal phenomena by mere solar and lunar gravitational forces and so he introduced considerations of channels through which tidal waters might ebb and flow.

[26] See D. T. Whiteside: 'Newton's lunar theory: from high hope to disenchantment' *Vistas in Astronomy* 19(1976):317–28. See, further, Craig B. Waff, *Universal gravitation and the motion of the moon's apogee: the establishment and reception of Newton's inverse-square law* (unpubl. diss., Johns Hopkins University, 1976); Philip P.

But even if Newton did not solve all the problems of the moon's motion by gravitational rational mechanics, and had to introduce some arbitrary geometric schemes, his analysis shifted the study of this subject to a new and wholly different theoretical plane[27] and enabled his successors to complete the work he had inaugurated.

Pfautz's review in the *Acta*: Books One and Two

We may judge the quality of Pfautz's review by seeing how many of the fundamentals of the *Principia* he presented. He begins: 'It has seemed best to the distinguished author, a first-rate mathematician of our time, to emulate the zeal and industry of the ancients and moderns together in promoting natural science, and to exhibit for the public good an outstanding example of it in this work of profound learning'. He then proceeds to summarize Newton's Preface, paraphrasing Newton's definition of 'rational mechanics' as 'the science of motions due to any forces, and of forces required for any motions'. He follows Newton in observing that in the three books comprising the *Principia*, Newton 'has undertaken to discuss matters relating to gravity and levity, elastic force, the resistance of fluids, and other forces of this sort – whether attractive or repulsive – and indeed the motion of bodies'. He then says that Newton has applied these principles and 'general propositions' to 'a clear model of the system of the world', setting forth 'by what forces of gravity bodies tend toward the sun and toward the other plants, in agreement with the celestial phenomena, and how the motions of the planets, moon, comets, and the sea follow from this'.

Continuing with his summary–paraphrase, Pfautz turns to the Definitions and Axioms. He mentions the definitions given by Newton of 'quantity of matter and of motion', of the 'inherent [or innate: Lat. *insita*], impressed, and centripetal force of matter', and of the 'absolute, accelerative, and motive quantity of centripetal force'. He does not give the reader either extracts or paraphrases of Newton's definitions of these quantities, nor does he comment on their possible novelty or originality, but merely echoes Newton to the effect that these are among the 'less commonly used terms' and therefore are to be defined. He then summarizes the scholium on space and time, stressing

Chandler, 'The *Principia*'s theory of the motion of the moon's apse', *Historia Mathematica* 4(1977): 405–10.

[27]   For further information see I. B. Cohen: *Isaac Newton's Theory of the Moon's Motion (1702), with a Bibliographical and Historical Introduction* (Folkestone: Wm. Dawson & Sons, 1975). See, also, Eric Forbes, 'Tobias Mayer's contributions to the development of lunar theory', *Journal for the History of Astronomy*, 1(1970), 144–54; and *The Euler–Mayer Correspondence: A New Perspective on Eighteenth-Century Advances in Lunar Theory* (London: Macmillan, 1971).

Newton's distinction between 'relative' or 'measured' and 'absolute' quantities. No critical comments are made concerning Newton's views on absolute and relative space and time.

The 'Axioms'[28] or 'Laws of Motion' are treated at greater length. Each of the three laws of motion is quoted, as are the corollaries. Thus the parallelogram law is stated and also Newton's result that 'the common center of gravity of all bodies acting on one another (external actions and impediments excluded) either is at rest or moves uniformly straight forward'. The scholium following the laws is paraphrased at great length.

It is evident that Pfautz is making a careful and complete paraphrase or summary or 'epitome' (as we shall see Newton call it) of the *Principia*, much like an extended and detailed analytical table of contents. This is his method, once again, in dealing with Section One, on 'first' and 'last' ratios, where he paraphrases each of Newton's lemmas concerning limits and also Newton's statements concerning the reasons why he has rejected the method of indivisibles. As a trained practicing mathematician and a university teacher of mathematics, Pfautz was able to grasp the principal features of Newton's treatment of limits and to appreciate the significance of the purely mathematical features of Newton's presentation.

This brings Pfautz to Section Two and Proposition 1, the beginning of Newton's development of rational mechanics. On this score Pfautz says that he had 'decided to present to the reader' the arguments of the first two books 'arranged in order', and with 'as much abridgment as possible'. But, as he was 'making a synopsis of this material, which extends through various categories and relations of bodies and motions combined with one another in various ways', the 'writing grew to such a bulk that it far exceeded the limit which I had set for myself'. And thus he 'had to fear that the annoyance caused to the reader by such a long description of general propositions, so varied and not adapted to concrete things, ... might be greater than the loss caused by a disconnected review skimming only the main topics'.

From this point on, the presentation takes on more of a general summary character, in which Pfautz will not present the content of the *Principia* proposition by proposition and section by section. Adopting this more general tone, Pfautz observes that in the first two books

> motions of bodies of every kind are discussed – spherical and non-spherical, ascending and descending, projected, pendulous, fluid, and agitated by any kinds of forces; motions in a straight line, in curved lines;

[28] Newton used both 'axioms' and 'laws' in the heading, 'Axiomata sive Leges Motus', but each of them was then called 'Lex' and not 'Axioma'. In fact this is the only occasion in the *Principia* in which Newton uses the actual term 'axiom' (in the nominative plural 'Axiomata').

motions circular, spiral, in conic sections – concentric with the center of forces and eccentric – in moving or unmoving orbits; progressive motions; motions propagated through fluids; likewise, the centripetal, absolute, and accelerative forces of motions; times, velocities, and the increase and decrease of the latter; centers, areas, places, apsides, spaces, mediums, and the densities and resistances of mediums.

These, we are told, are all 'set forth in an investigation worthy of so great a mathematician'. Pfautz adds that interspersed among the propositions there are important lemmas on geometry, especially the geometry of conics, and also scholiums giving 'philosophical' (i.e., physical) examples of the abstract mathematical principles being expounded so that the latter will not appear sterile.

Following an extended summary of the many different topics explored by Newton, together with some of the most important results of the first two books, Pfautz reaches the conclusion of Book Two. He fully appreciates the significance of Newton's demonstration that the speed of planets in their orbits, moving 'more slowly in their aphelia and more swiftly in their perihelia' is 'the opposite of what ought to happen according to the mechanical laws of vortices'. He apparently can find no fault with Newton's ringing conclusion that the planetary speeds in Cartesian vortices contradict the celestial phenomena, even though he does not comment on the significance of the dreadful blow that Newton has dealt to Cartesian physics.

In retrospect Pfautz has done a remarkable job of presentation, reducing Newton's rich treatment of rational mechanics and natural philosophy to a limited compass. But it will seem odd to a critic of today that Pfautz has not told his readers in clear and unambiguous terms that in Props. 1–2–3 of Book One Newton has found the significance of the area law of Kepler and has shown that this law is related to the principle of inertia (Newton's Axiom or Law One). That is, Newton has shown that area law is a necessary and sufficient condition for a centrally directed (centripetal) force, directed in fact at the same center with respect to which the equal areas have been computed. Nor has he said specifically that in the succeeding Prop. 11 Newton has found the significance of Kepler's law of elliptical orbits in their implication that the force directed to the focus varies as the inverse-square of the distance.

Pfautz's account of Book Three on the system of the world

Turning to Book Three, Pfautz observes that Newton 'lays down as foundation for the book' a set of 'hypotheses, partly physical, partly relying upon astronomical observations'. After summarizing these 'hypotheses', he gets to the heart of Book Three. Here, he says, Newton uses these astronomical data and the mathematical results of the first two books to 'demonstrate' that the forces by which the circumjovial planets, the primary planets, and the moon are continually drawn off from their rectilinear motions and kept in their orbits are from their gravitation toward Jupiter, the sun, and the earth, and are reciprocally as the squares of the distances from the center of Jupiter, of the sun, and of the earth; that all bodies gravitate toward each of the planets, and that their weights toward any single planet, at equal distances from the center of the planet, are proportional to the quantity of matter in each one (from which he further infers that the weights of bodies do not depend upon their forms and textures, that a vacuum is necessarily given...).

Here is an admirable summary statement of some of the principal new aspects of the Newtonian system of the world.

Pfautz also takes note of Newton's finding that 'gravity is of a different nature from magnetic force' and that 'gravity occurs toward all bodies universally and is proportional to the quantity of matter in each individual body'. Furthermore, Pfautz paraphrases in summary the primary features of Newton's system of the world:

that if it is posited that the matter in two globes gravitating toward each other is everywhere homogeneous in regions equidistant from the center, then the weight of either globe on the other is reciprocally as the square of the distance between their centers...;

that, proceeding downwards from the surfaces of the planets, gravity decreases as nearly as possible in the ratio of the distances from the center;

that the motions of the planets in the heavens can be conserved for a very long time;

that the common center of gravity of the earth, sun, and all the planets is at rest;

that the sun is agitated by a continual motion, but never recedes far from the common center of gravity of all the planets, which accordingly he contends should be considered the center of the world;

that the planets move in ellipses, having their focus in the center of the sun, and that by radii drawn to that center they describe areas proportional to the times (where he speaks about the perturbations of motions by the mutual gravity of the planets);

that the aphelia and nodes of the orbits are at rest (where from the imperceptible parallax of the fixed stars he asserts that they cannot produce perceptible effects around us);

that the diurnal motions of the planets are uniform and that the libration of the moon arises from its diurnal motion;

that the axes of the planets are less than the diameters drawn perpendicularly to the axes (from which he gathers that the diameter of the earth between the poles is to the diameter at the equator as 689 to 692, and therefore that, since the earth's mean semi-diameter taken according to the recent measurement by the French is 19615800 Paris feet, the earth here is 85200 feet or 17 miles higher ...) ...;

that the equinoctial points regress and that the axis of the earth by oscillating in its individual revolutions is twice inclined toward the ecliptic and twice restored to its former position;

that all the lunar motions and all the inequalities of the motions follow from the principles laid down;

that the ebb and flow of the sea arise from the actions of the sun and the moon, with the actions of the luminaries in conjunction and opposition working together toward the greatest effect of the motion but in the quadrature impeding each other and bringing about the smallest tide...

In this extract we can see how Pfautz carefully drew out of Newton's text not only the primary features of Newton's system of the world but also many important details of Newton's presentation. These include Newton's results concerning the relative densities of the planets, the motions of the planets, the shape of the earth as an oblate spheroid, the precession of the equinoxes as a gravitational consequence of the earth's shape, the variation in weight with latitude, the inequalities of the motion of the moon explained by gravitational forces, and the ebb and flow of the sea also explained.

The review ends with a number of pages summarizing Newton's findings about comets. Here is a portion of Pfautz's summary in his own words.

Concerning comets: that they are higher than the moon (the evidence being the lack of diurnal parallax) and they are in the region of the planets (their annual parallax proving this)...

He infers that the comets shine with light reflected from the sun and that in the circumsolar region many more are seen than in the region opposite to the sun.

Further, from their motion, in every way very free, preserved for a very long time even contrary to the motions of the planets, he concludes that the aether has no resistance...

He concludes that comets are a kind of planet, returning by continual motion into orbit and that they move in conic sections (ellipses but so like a parabola that parabolas can be used in their place without sensible error) having their focus in the center of the sun and that by radii drawn to the sun they describe areas proportional to the times.

Pfautz observes that Newton found 'an analogy with the planets' in that those comets 'are smaller which revolve in orbits that are smaller and closer to the sun'.

Pfautz's review ends abruptly, without a proper conclusion, by referring to Newton's graphic method for determining the parabolic orbit of a comet from three observations. So, too, did the *Principia* end abruptly in the first edition;

the famous concluding Scholium Generale did not appear until the second edition (1713).[29]

Some further aspects of Pfautz's review

Pfautz seems to have introduced almost all of Newton's major innovations and he referred to many topics which I have not mentioned here. Among them are Newton's study (in Sect. Fourteen, Book One) of motion of particles that follow paths very much like that of light being refracted, Newton's method of the calculus (Lemma Two, Book Two) along with a mention there of Newton's reference to Leibniz, the problem of the solid of least resistance and the design of ships,[30] and much else. He did not, however, explicitly call the reader's attention to Newton's actual methods of demonstration and there are no diagrams. And, as noted earlier, he did not specifically note that Newton (Props. 1–3, Book One) had shown the physical significance of the law of areas as a necessary and sufficient condition for a centrally directed force. Nor did he particularly stress Newton's other great discovery, the significance of the law of elliptical orbits (Prop. 11, Book One) in implying that the centrally directed force varies as the inverse-square of the distance.[31] Newton's discovery of the significance of Kepler's law of elliptical orbits appears only in relation to its use in Book Three on the system of the world.

The review contains no adverse critical comments and the reader is left in no doubt concerning the admiration of the anonymous reviewer for Newton's magnificent achievement. Every careful reader of Pfautz's presentation would gain a very good idea of the contents and scope of Newton's *Principia* – and to a greater degree than by reading any of the other three reviews.

That Pfautz fully understood some of the main principles of Newton's

[29] Newton evidently intended to have the first edition conclude with a general discussion of the forces of nature, including those on the microscopic level as well as those on the macroscopic level (i.e., those which are observable in ordinary physical events and in the phenomena of the heavens). This soon became speculative and even hypothetical and was rejected by Newton before a final version was completed. Newton's intentions were first discovered by A. Rupert Hall and Marie Boas Hall, who published the extant drafts in their collection, *Unpublished Scientific Papers of Isaac Newton* (Cambridge University Press, 1962).

[30] See I. B. Cohen, 'Isaac Newton, the calculus of variations, and the design of ships: an example of pure mathematics in Newton's *Principia*, allegedly developed for the sake of practical applications', in R. S. Cohen, J. J. Stachel, and M. M. Wartofsky, eds., *For Dirk Struik* (Dordrecht/Boston: D. Reidel, 1974), pp. 169–87.

[31] Newton had shown in Prop. 11 of Book One that the elliptical orbit implies that the centripetal force acting on a planet must vary as the inverse-square of the distance. Prop. 4 of Book One had dealt with a special case of the inverse-square law for uniform square motion.

dynamics is made clear in his discussion in Book Three of the motion of the planets. For he says that, using what he has 'said in the preceding books', Newton 'demonstrates' that 'the forces by which the circumjovial planets, the primary planets, and the moon are continually drawn away from rectilinear motions and are kept in their orbits' are the result of 'their gravitation toward Jupiter, the sun and the earth'. It was a primary feature of Newtonian dynamics thus to explain curved or orbital motion by this combination of a rectilinear (inertial) component of motion and a continual acceleration or falling toward a center as a result of a centripetal force, which in the case of planets and satellites is the force of gravity. Thus Newton dismissed from physics the ambiguous and misleading notion of a centrifugal force.[32]

Leibniz's use of Pfautz's review in writing the *Tentamen*: was Leibniz's knowledge of Newton's science based solely on Pfautz's review while he was writing the *Tentamen*?

Pfautz's review achieved a special importance in 1689, when Leibniz referred to it in one of three articles published in the *Acta*: the 'Tentamen de Motuum Coelestium Causis' (Essay on the Causes of the Motions of the Heavenly Bodies).[33] In this work, Leibniz set forth an alternative explanation to Newton's.[34] Leibniz explained that while in Rome he had 'come upon an account of the celebrated Isaac Newton's Mathematical Principles of Nature' in the June 1688 issue of the *Acta*.[35] Although the topic of the causes of celestial motions was at that time 'far removed from my present line of thought', he wrote, his reading of the account of Newton's book recalled to mind some

---

[32] Newton learned from Hooke, during correspondence in 1679/80, how to analyze orbital motion as a combination of two components: an accelerated motion of falling towards the sun and an inertial motion along the tangent; see my *Newtonian Revolution*, pp. 241–258. I believe this indebtedness to Hooke was first brought to scholarly notice in R. S. Westfall, 'Hooke and the law of Universal Gravitation', *British Journal for the History of Science* 3(1967):245–61. See, further, Patri Pugliese, 'Robert Hooke and the dynamics of motion in a curved path', in *Robert Hooke, New Studies*, ed. Michael Hunter and Simon Schaffer (Woodbridge [Suffolk]: Boydell Press, 1989), pp. 181–205.

[33] *Acta Eruditorum*, February 1698, 82–96. Leibniz's *Tentamen* is reprinted in Leibniz's *Mathematiche Schriften*, ed. C. I. Gerhardt, 7 vols. (Berlin: Verlag von Ascher & Comp., 1849–63), vol. 6, pp. 135–44.

[34] For an analysis of Leibniz's explanation of the causes of the orbital motion of the heavenly bodies, see Eric J. Aiton, 'The celestial mechanics of Leibniz: A new interpretation', *Annals of Science* 20(1964):111–23. See, further, Aiton's *Leibniz*.

[35] It should be noted that Leibniz has used an incorrect title. Instead of *Philosophiae Naturalis Principia Mathematica*, Leibniz has referred to Newton's book as if it were *Naturae Principia Mathematica*.

336

work he had done in Paris twelve years earlier. At that time he had decided not to publish his ideas on this subject, waiting until he could have the opportunity 'to make a more careful comparison of the geometrical laws with the most recent observations of astronomers'. But now, after reading about Newton's work, he had been stimulated to write up and to publish his own very different explanations.[36] Leibniz's own system for the planets combined what he called a 'harmonic circulation' and a 'paracentric circulation', an explanation that on first reading seems confusing[37] but that has been carefully elucidated for us by E. J. Aiton.[38]

Newton never believed that Leibniz's theorems, published in the *Acta* in 1689, had truly been found earlier and independently of the *Principia*. In fact, he even declared that Leibniz's theorems '(Errors and Trifles excepted) are Mr. *Newton's* (or easy Corollaries from them)'. In another document, in which Newton referred to Pfautz's review as 'an epitome' of the *Principia*, Newton said that Leibniz read the review and then proceeded to compose the three essays for the *Acta* 'as if he himself had discovered the principal propositions of Newton concerning these matters, and had done so by a different method, and had not yet seen Newton's book'. According to Newton, even if Leibniz had not seen the *Principia* (which Newton personally doubted), he ought – before writing his articles for the *Acta* – nevertheless to have examined the *Principia* 'before he published his own thoughts concerning these same matters'.[39]

[36] Although Leibniz does not say so specifically, he evidently wanted to set forth his own ideas on the cause of planetary motion so they could be discussed as an alternative to Newton's; he did not want to lose credit for his originality and priority.

[37] Thus, for example, Alexandre Koyré (*Newtonian Studies*, Appendix A to chapter three) misinterpreted Leibniz's concept of a 'harmonic vortex' and as a result gave an incorrect view of Leibniz's system. As Aiton pointed out, *Annals of Science* 21(1965):204–5, 'The velocity of circulation of a body moving in a "harmonic vortex" is the trans-radial velocity and not, as Koyré supposes, the velocity of the body along its path.'

[38] See E. J. Aiton, 'The celestial mechanics of Leibniz in the light of Newtonian criticism', *Annals of Science* 18(1962):31–41; *The Vortex Theory of Planetary Motions*, (London: Macdonald; New York: American Elsevier, 1972); and also 'A new interpretation'.

[39] See Cohen, *Introduction*, pp. 153–4, and the anonymous review of the *Commercium Epistolicum*, the account of the dispute between Newton and Leibniz concerning priority in the invention of the calculus. This 'Recensio Libri' was actually written by Newton himself and published anonymously as 'An Account of the Book entituled *Commercium Epistolicum Collinii & aliorum, De Analysi promota*, published by order of the Royal-Society, in relation to the Dispute between Mr. Leibnits and Dr. Keill, about the Right of Invention of the Method of Fluxions, by some call'd the Differential Method', *Philosophical Transactions* 19(1715):173–224. On this subject see A. Rupert Hall, *Philosophers at War: The Quarrel between Newton and Leibniz*

For two centuries there had been no way of telling whether or not Leibniz had actually seen the *Principia* before writing his three articles, whether or not he had derived his information about Newton's results exclusively from Pfautz's review. When Leibniz's annotated copy of the *Principia* turned up a few years ago, there was hope that a definitive statement could be made about Leibniz's honesty in this matter, one way or another. Unfortunately, the annotations did not help to decide the question since they were consistent with either a reading of the *Principia* by Leibniz after writing the *Tentamen* (and the other two articles) or before writing them.[40] More recently, however, the profound investigations of Domenico Bertoloni Meli have uncovered some hitherto unnoticed notes by Leibniz.[41] By a combination of careful analysis and deep historical insight, Bertoloni Meli has been able to identify these notes, to assign a date to them, and to determine their significance. They leave no doubt that Leibniz had been carefully reading the *Principia* before writing and publishing the three articles in the *Acta*.

A possible influence of Pfautz's review on Newton: the reference to God in the First Edition of the *Principia*

Although there seems to be no direct evidence that Newton read Pfautz's review, there is a significant alteration in the *Principia* that may be related to Pfautz's presentation. Newton's correspondence with Mencke shows that he obtained a set of the *Acta*.[42] He was certainly aware of the existence of the review since he referred to it in a critical response to Leibniz, as we have seen, calling it an 'epitome', in many ways an exact characterization. In this case Newton was responding to a remark made by Leibniz (in his *Tentamen*) to the effect that he had not yet seen the *Principia*, that his knowledge of Newton's work was based solely on the account of the book that he had encountered in

(Cambridge University Press, 1980), where Newton's 'Recensio Libri' is published in facsimile on pp. 263–314.

[40] Leibniz's annotations have been transcribed and published by E. A. Fellmann; G. W. Leibniz, *Marginalia in Newtoni Principia Mathematica* (Paris: Librairie Philosophique J. Vrin, 1973 – Collection des Travaux de l'Académie Internationale d'Histoire des Sciences).

[41] Bertoloni Meli's masterly study of Leibniz's mathematical physics, based primarily on manuscript sources not hitherto discovered and used, is scheduled to be published by Oxford University Press. His findings concerning Leibniz's reading of Newton's *Principia* before writing his *Tentamen* are forthcoming, 'Public claims, private worries: Newton's *Principia* and Leibniz's theory of planetary motion', *Studies in History and Philosophy of Science*, esp. sect. 3 and n. 36.

[42] See John Harrison, *The Library of Isaac Newton* (Cambridge University Press, 1978), pp. 67, 83 (§ 7).

the *Acta*. But Leibniz used the more customary word 'recensio' in referring to Pfautz's review. Very likely, then, Newton would have seen the review since he was aware that it took the form of an 'epitome'. If his sole knowledge of the review had been Leibniz's reference in the *Tentamen*, he would most likely have called it – following Leibniz – a 'recensio'. In fact, when Newton's anonymous review of the *Commercium Epistolicum* was reprinted in a Latin translation in the second edition of the *Commercium* (London 1722), it was there titled 'Recensio Libri'.

Most authors would seek out and read avidly all reviews of their works and so it might be supposed that Newton would have read all the reviews of the different editions of the *Principia* and of his other writings. But Newton was not like 'most' authors and I have found no direct evidence that Newton ever did fully read the four reviews of the first edition of the *Principia*, although it would be most surprising if he had not read Halley's review in the *Philosophical Transactions*.

We shall see below that there is some indirect evidence that Newton had seen the review in the *Journal des Sçavans*. There is similar indirect evidence that would indicate a strong probability that Newton had also read Pfautz's review in the *Acta*. This evidence is related to a very interesting question in Newtonian scholarship: Newton's interest in religion. For a long time it was believed by many scholars and scientists that Newton turned to theology only late in life, so that this concern could be taken as a sign of senility or dotage. This was the line taken by such figures as Biot and Laplace, who found evidence for their point of view in two facts. The first is that the first edition of the *Opticks* (1704) does not conclude with a discussion of God or of any religious theme. These topics appear only at the conclusion of the final Query (no. 31) in the later editions. The second is that the first edition (1687) of the *Principia* ends with the discussion of comets. Only in the second (1713) and third (1726) editions does this work conclude with a discussion of the attributes of God and the strong statement that the system of the world could not have come into being without the 'dominion and counsel' of a wise creator.

Biot and Laplace knew in detail only the final version of the *Principia*, and the third and final edition concluding in the General Scholium. There was no critical edition – giving the changes made by Newton from edition to edition – until only a few years ago. Hence Newtonian commentators were unaware that religious issues were of importance in Newton's thought long before he wrote the *Principia*.[43] There was also ignorance of the fact that the religious

---

[43] See R. S. Westfall, 'Newton's theological manuscripts', in Zev Bechler, ed., *Contemporary Newtonian Research* (Dordrecht: D. Reidel Publishing Company, 1982), pp. 129–44; and *Never at Rest: a Biography of Isaac Newton* (Cambridge University Press, 1980). See, further, James E. Force and Richard H. Popkin, *Essays on the*

theme had appeared in the first edition of the *Principia* and had been removed by Newton while preparing the second edition.

Pfautz was fully aware of Newton's reference in the *Principia* to God and the creation of the universe and he called attention to this passage in his review. He wrote that on page 415 of the *Principia* there is a discussion of the distances, masses, and densities of the planets,

Ex quo Deum concludit collocasse Planetas in diversis a Sole distantiis, ut pro ratione densitatum suarum a Sole calorem reciperent.	From which he concludes that God placed the planets at different distances from the Sun, so that they would receive heat from the Sun according to the proportion of their densities.

Pfautz was referring to Corol. 5 to Prop. 8 in Book Three. In the first edition of the *Principia*, these corollaries were quite different from those that appear in the later editions;[44] in particular Newton eliminated some of his calculations concerning the densities of the planets and the moon and he reduced the number of corollaries from five to four. In Corol. 4 in the first edition, Newton computed the densities of the planets Saturn, Jupiter, and the earth and developed a rule for the ratio of their densities. This led him in Corol. 5 to conclude:

Collocavit igitur Deus Planetas in diversis distantiis à Sole, ut quilibet pro gradu densitatis calore Solis majore vel minore fruatur.	God therefore placed the planets at different distances from the Sun, so that each one may, according to the degree of its density, enjoy a greater or smaller amount of heat from the Sun.

It is possible that Newton read Pfautz's review and came upon this single reference to God's role in establishing the universe. Although the sentence in question is not very prominent in the *Principia*, occurring in the middle of Corol. 5, its significance is aggrandized by the fact that it is the only sentence referring to either God or the creation of the universe. Pfautz obviously considered this sentence to be sufficiently important to be paraphrased in his presentation. If Newton did in fact read through the review, he might very well have decided that it had been an error to introduce the creating God in the

---

Context, Nature, and Influence of Newton's Theology (Dordrecht: Kluwer Academic Publishers, 1990).

[44] See Alexandre Koyré, I. B. Cohen, and Anne Whitman, eds., *Isaac Newton's 'Philosophiae naturalis principia mathematica'. The Third Edition (1726) with Variant Readings*, 2 vols. (Cambridge: Harvard University Press, 1972).

midst of his scientific presentation. In any event, in his own interleaved copy of the *Principia*, Newton crossed out this reference to God and changed the beginning of the sentence from

Collocavit igitur Deus Planetas in diversis distantiis à Sole...

to

Et collocandi erant Planetae in diversis distantiis a Sole...

which, in the second and third editions, was slightly rewritten so as to become

In diversis utique distantiis a sole collocandi erant planetae...

That is

The planets, of course, had to be set at different distances from the Sun so that each one might, according to the degree of its density, enjoy a greater or smaller amount of heat from the Sun.[45]

John Locke's review

We may gain some appreciation of Pfautz's review by comparing it with the one written by John Locke for the *Bibliothèque Universelle*.[46] Locke's review is also a summary of the *Principia*, but of a very different kind. Unlike Pfautz, who actually summarized (even if in paraphrase) the contents of the laws, propositions, and lemmas, Locke (if, as it seems, he was the reviewer) merely translated into French the headings of the different sections of Books One and Two, commenting on a proposition or proof here and there and then gave a summary of Book Three. In fact, he actually apologized to the reader for not having translated a set of Newton's headings in Book Three, as he had done in Books One and Two. He should not be blamed, he wrote, since the fault was not his but the author's. Newton had not divided Book Three into sections as he had done in Books One and Two.

It is typical of the lack of attention paid to physics in this review that the contents of the laws of motion are neither stated nor summarized. The only reference to the laws of motion occurs in a remark that Newton has followed the method of the geometers, prefacing the work with 'several definitions and axioms concerning motion'. But Locke does summarize in full Newton's proof

---

45   For details, see I. B. Cohen, 'Isaac Newton's *Principia*, the scriptures, and the divine providence', in Sidney Morgenbesser, Patrick Suppes, and Morton White, eds., *Philosophy, Science and Method* (New York: St. Martin's Press, 1969), pp. 523–48.

46   See notes 6, 7. All extracts have been translated from the French.

(at the end of Book Two) that Descartes's system of vortices is inconsistent with Kepler's laws, even reproducing Newton's diagram.

Most of this presentation of Book Three is devoted to the introductory 'Hypotheses' which contain numerical data.[47] There is even a detailed reproduction of Newton's table of the mean distances and periods of the satellites of Jupiter and of the mean distances of the major planets from the sun.[48] There is but a single sentence about the tides, which does not mention either the term 'attraction' or 'gravitation' but merely states: 'Le flux & le reflux de la Mer procèdent de l'impression du Soleil & de la Lune.'

No reader would learn anything from this review concerning what we today would consider the main feature of the Newtonian system of the world in Book Three, namely, the way in which the force of universal gravity binds the system together. Nor would anyone learn from this review how Newton had dealt with the dynamics of Keplerian motion, nor how he had shown that the force of gravity extends to the moon. The sense of majesty conveyed by Pfautz's presentation, however uncritical it may be, is completely missing here.

The closest approach to a statement of the law of universal gravity is that 'there is weight [*pesanteur*] in all bodies, according to the quantity of matter which is in each one of them, and the effort [*effort*] made by weight on all equal particles of a body is reciprocally as the square of the distance between the positions of these particles'. I doubt whether any reader would be much enlightened on the subject of Newton's celestial mechanics by the reviewer's general statement about the way in which, he says, 'Newton undertakes, in Book Three, to explain the system of the world.' He does so, the reader is informed, by determining 'the degrees of weight [*pesanteur*] which carry bodies toward the sun, or toward any planet, and which being known serve him to find the cause of the motion of planets, comets, the moon, and the sea'.

We know that Locke was not sufficiently mathematically literate to follow Newton's mathematics. The review provides evidence that he could not really appreciate the logical structure of Newton's rational mechanics. It is well known that Newton's disciple, J. T. Desaguliers, recorded how he had been told 'several times by Sir Isaac Newton himself' that Locke was unable to determine 'whether the mathematical Propositions in Sir Isaac's Principia were true', and that he turned to Huygens for a critical judgment. On being told that he could 'depend upon their certainty' he then 'took them for granted, and

---

[47] See note 22.

[48] These data appeared in Hypothesis V (the satellites of Jupiter) and in Hypothesis VII (on the distances of the planets and the earth from the sun). Hypothesis V listed the periodic times of the four satellites, as determined from observations, followed by their distances from the center of Jupiter according to observations made by Cassini, Borelli, Townley, and Flamsteed. These data were then compared to the distances as determined by Kepler's third law from the periodic times.

carefully examined the Reasonings and Corollaries drawn from them' so as to become 'master of all the Physics'.[49] An analysis of the review would indicate that any such degree of mastery would have been a later acquisition.

On another occasion,[50] I have suggested that perhaps Newton had a real person in mind when he provisionally made a curious addition to the introduction to Book Three of the *Principia*. In this introduction, as printed in all three editions, Newton said that he would hardly advise every reader to study all the propositions of the first two books, since such an effort would 'cost too much time, even to readers of good mathematical learning'. Rather, he advised his readers to concentrate on the Definitions, the Laws of Motion, and the first three sections of Book One. In his personal copy of the first edition, in which he recorded proposed emendations for a second edition, Newton wrote:

> Those who are not mathematically learned can read the Propositions also, and can consult mathematicians concerning the truth of the Demonstrations.[51]

Newton did not publish this addition, explicitly contrasting those who are *not* mathematically learned with the previous mention of readers who *are*. In writing this remark he may very well have been thinking of the example of John Locke.

The review in the *Journal des Sçavans*

I have mentioned that the reviewer in the *Journal des Sçavans* remains unidentified.[52] Whoever he was, he seems to have recognized some of the philosophical implications of the *Principia* and to have appreciated the quality of Newton's mathematics. This author, however, takes a position of strict Cartesianism, which produces a very limited stance from which to examine the *Principia*. The reviewer is obviously so concerned with philosophical issues that he scarcely troubles to take note of Newton's technical solution of major problems in physics and astronomy.[53] He especially takes Newton to task for 'arbitrary' suppositions which have not been proved.[54] He declares that the demonstrations of the *Principia* are 'only mechanical' (i.e., belonging only to

49   J. T. Desaguliers, *Experimental Philosophy*, 3rd edn (London, 1763), vol. 2, p. viii.

50   Cohen, *Introduction*, p. 148.

51   Translated from the Latin. See Koyré *et al.*, *The Principia with Variant Readings*, vol. 2, pp. 548–9.   52   See note 8. Extracts are translated from the French.

53   In fact there is good evidence, as we shall see, that he had not read the text fully.

54   The reviewer does not at this point specify which of Newton's suppositions he has found to be 'arbitrary', in the sense of not having been proved. Perhaps he had in mind that the whole work was premised on a set of 'axioms', the designation Newton gave to the laws of motion, 'Axiomata sive Leges Motus'.

the realm of mechanics: 'mécaniques'), a feature that Newton is said to have recognized himself. In evidence he cites 'the end of page four and the beginning of page five', where – he continues – Newton says that 'he has not considered' the principles 'as a Physicist, but only as a Geometer'.

The reviewer is referring to the 'Definitions' at the start of the *Principia*, the final paragraph of Newton's discussion of the three 'measures' of centripetal force – the 'absolute quantity', the 'accelerative quantity' and the 'motive quantity'. Newton's own statement is an elaboration of a sentence occurring a few lines earlier, to the effect that the concept of force which he is using is 'purely mathematical', since he is 'not now considering the physical causes and sites of forces'. The paragraph in question reads as follows:

> I use the words 'attraction', 'impulse', or expressions for any sort of propensity toward a center interchangeably and indiscriminately, considering these forces not from a physical but only from a mathematical point of view. Therefore, let the reader beware of thinking that by words of this kind I am anywhere defining a species or mode of action or a physical cause or reason, or that I am attributing forces in a true and physical sense to centers (which are mathematical points) if I happen to say that centers attract or that centers have forces.[55]

We should take note, as the reviewer did not, that Newton does not say that throughout all three books of the *Principia* he is not considering 'the physical causes and sites of forces' but rather that 'now' [*jam*], during the mathematical development of results which he will later apply to physical situations, he is not concerned with physical causes. Newton's references to 'sites' as well as 'causes' arises from the fact that he is considering a 'system' consisting of a mass point moving about a mathematical center of force, even though he knows that in physical nature forces originate in and are directed toward bodies.

In Book One Newton does begin with a purely mathematical intellectual construct, a 'system' consisting of a body (a mass point) moving about a mathematical center of force. It is, in fact, only in such a system that Kepler's area law and law of elliptical orbits are absolutely true.[56] After developing the mathematical properties of this system, Newton advances to a two-body system, in which the center of force becomes the second body. The reason, as

---

[55] This and succeeding extracts are quoted from a new translation of the *Principia*, made by I. B. Cohen and Anne Whitman, to be published by the University of California Press.

[56] A good part of Book One and part of Book Two of the *Principia* is, in fact, devoted to the mathematical elaboration of the emendations required for Kepler's laws in systems of two or more bodies; see I. B. Cohen, 'Newton's theory vs. Kepler's theory and Galileo's theory', in *The Interaction between Science and Philosophy*, ed. Yehuda Elkana (Atlantic Highlands: Humanities Press, 1974), pp. 299–338.

Newton explains in the beginning of Sect. Eleven (Book One) is that in nature 'attractions are always directed toward bodies'. Hence Newton will

> now go on to explain the motion of bodies that mutually attract one another, considering centripetal forces as attractions, although perhaps – if we speak in the language of physics – they might more truly be called impulses. For here we are concerned with mathematics; and therefore, putting aside any debates concerning physics, we are using familiar language so as to be more easily understood by mathematical readers....

Critics have not accepted Newton's explanation that he is 'here concerned with mathematics' and therefore is 'putting aside any debates concerning physics'.[57]

After developing propositions concerning the motions of two bodies mutually attracting one another, Newton turned to the very important question of the relations between mathematical laws and principles and those pertaining to physics or natural philosophy. The nub of his argument is as follows:

> Mathematics requires an investigation of those quantities of forces and their proportions that follow from any conditions that may be supposed. Then, coming down to physics, these proportions must be compared with the phenomena, so that it may be found out which conditions [or laws] of forces apply to each kind of attracting bodies. And then, finally, it will be possible to argue more securely concerning the physical species, physical causes, and physical proportions of these forces.

We shall see in a moment that this statement defines a manner of doing science, one that I have called the 'Newtonian style'.

On considering the magnitude of Newton's achievement, the reviewer cannot help but admit that even though Newton's work is merely a 'mechanics', it is 'the most perfect one that could be imagined, since it is not possible to make demonstrations more precise or more exact than those which he [Newton] gives in the first two books on levity [lightness], on springiness, on the resistance of fluid bodies, and on the attractive and repulsive forces that are the principal basis of Physics'. This laudation is tempered, however, by a fundamental Cartesian criticism. The latter appears strongly in some further remarks about the general character of Newton's book. Newton, according to the reviewer, 'confesses' once again, at the beginning of Book Three, that his

[57] Since the reviewer did not cite this passage, in which Newton defines the mathematical level of discourse of the first two books of the *Principia*, I assume that he had not encountered it, that he had not read the whole text of the *Principia* carefully. Most likely, he concentrated on the 'Definitions', the 'Axioms or Laws of Motion', a few theorems about elliptic motion, and the first part of Book Three on the 'System of the World'. This was Newton's own suggested way to read the *Principia*, given forth at the beginning of Book Three.

principles are being considered merely from the point of view of geometry and not physics.

In introducing Book Three, Newton gave a final presentation of his method. In the preceding two books, he wrote, he had 'presented principles of natural philosophy that are not, however, philosophical but strictly mathematical', principles 'on which natural philosophy can be based'. That is, these 'principles are the laws and conditions of motions and of forces which especially relate to natural philosophy'. He had, however, shown how these principles may be used in natural philosophy in some 'philosophical scholiums' dealing with such basic topics in natural philosophy as 'the density and resistance of bodies, spaces void of bodies, and the motion of light and sounds'. Now, in Book Three, his goal was 'to explain the system of the world from these same principles'.[58]

If one follows Newton's procedure carefully, as stated at the start of Book Three and as exemplified throughout Book One, Newton is seen to proceed to develop the mathematical properties of systems that become increasingly more complex, advancing from a single body (particle or mass point) and a center of force to a system of two bodies. Then he will introduce the question of three and even more bodies, finally getting to the physical properties of such bodies, e.g., their shape and constitution and physical constitution. Of course, it would be apparent to readers that Newton's procedure is primarily to develop mathematically the properties of a system which later can be applied to the physical or external world of nature. And it is in order to keep the reader constantly aware that this is his aim, that he introduces from time to time some scholiums in which he indicates how his results may be applied to physical situations. Finally, in Book Three, he will apply these mathematically developed results to the System of the World.

I have called this method of procedure the Newtonian style.[59] It is in this sense that the *Principia* was a revolutionary book, being marked off from many centuries of traditional thought about physical problems. That is, Newton gets down to the derivation of mathematical results without taking time off for metaphysical or ontological discussions of the nature of force or its mode of action. Once he has obtained his results, he applies them to real problems and – so to speak – gets on with the job of producing a celestial mechanics in the same mode as terrestrial mechanics without getting bogged down in philosophical problems concerning foundational problems. We may

[58] The omission of this discussion reinforces the assumption (see the previous note) that the reviewer had not carefully read the book he was reviewing.

[59] On the Newtonian style, see my *Newtonian Revolution*. I have given an abbreviated presentation of the main features of the Newtonian style in *Some Truer Method: Reflections on the Heritage of Newton*, ed. Frank Durham and Robert D. Purrington (New York: Columbia University Press, 1990), pp. 15–57.

see this new style of mathematical physics in the work of Euler and Lagrange, reaching a high classical peak in the *Mécanique Céleste* of Laplace. To a Cartesian critic of Newton, however, this was not the way to proceed. Because Newton had avoided the philosophical questions that had loomed so central in the scientific thought of Descartes and which still obsessed his followers, Cartesians could only conclude that his book was not concerned with nature or 'physics' but only with mathematics or 'mechanics'.

The literature of the history of science contains no clearer statement of the gulf between the new science emerging from the Scientific Revolution and the traditional approach, which was heavily laden with philosophical assumptions, than is found in the review of the *Principia* in the *Journal des Sçavans*. Not that Newton was free of philosophical assumptions, some of which – as in the case of the nature of space and time – were presented in the *Principia*. But Newton, like Laplace and like the succeeding generations of mathematical physicists, did not become derailed by preliminary philosophical discussions of basic concepts. Rather, as we have seen, he explored the mathematical properties of systems of forces and particles or bodies without entering into discussions of the nature of forces, the modes of operation of forces, or the existence of forces. In the end, he produced a consistent and coherent system of mathematical physics that, as he was later to say in the General Scholium, serves fully to explain the motion of our seas and of the heavenly bodies. The same system also explained many aspects of terrestrial physical phenomena. As Newton said, in the General Scholium (1713), 'Satis est', 'That is enough'!

There is one passage in the review of the *Journal des Sçavans* that may particularly attract our attention. It occurs in the conclusion, where the reviewer faults Newton's attempts 'to explain the System of the World.' Newton has erred because he has proceeded 'only by hypotheses which are, for the most part, arbitrary and which therefore can serve as foundation only for a treatise in the realm of pure mechanics'. The reviewer obviously has in mind the set of Hypotheses introducing Book Three in the first edition.[60] The reviewer does take note that Newton introduces 'the principle that all the planets gravitate mutually toward one another' and that he applies this principle to explain the 'inequality of the tides'. But this principle is 'arbitrary', since 'it has not been proved'; therefore any demonstration based on it, as in the explanation of tides, can 'only be mechanics' and not 'physics'.

The reviewer concludes by recommending that Newton now undertake to compose 'a work as perfect as possible' by producing 'a Physics as exact as his Mechanics'. To do so, Newton would have only to substitute 'true motions in place of those which he has supposed'. Who can doubt that Newton would

---

[60] See note 22.

have been especially provoked by these comments, especially the statement that he had based his system of the world on hypotheses. We know of Newton's long-standing abhorrence of having any of his work called a 'hypothesis' or of the charge that his physics was to any degree hypothesis-based. In 1672 he had taken umbrage at the suggestion made by Pardies that his theory of light and colors was a hypothesis. He replied that if he had not known that the 'properties of light' he had found were 'true', he 'should prefer to reject [them] as vain and empty speculations, than acknowledge them as my hypothesis.'[61] In 1704 Newton opened his *Opticks* with the declaration that his 'Design in this Book is not to explain the Properties of light by Hypotheses, but to propose and prove them by Reason and Experiments'. Nine years later, the second edition of the *Principia* would conclude in a General Scholium containing the famous slogan: '*Hypotheses non fingo*!' 'I feign no hypotheses.'[62]

Soon after the publication of the review in the *Journal des Sçavans*, that is, in the early to mid 1690s, Newton eliminated the rubric of 'HYPOTHESES' at the beginning of Book Three of the *Principia*. As had been mentioned, he inaugurated a series of revisions in which he converted these largely into 'PHAENOMENA' and 'REGULAE PHILOSOPHANDI'.[63] The new designations were a more accurate characterization and we may wonder why Newton had ever called them 'hypotheses', but that is a separate question. Here we are concerned only with Newton's alterations of the foundation of Book Three. Who can doubt that a reading of the allegation that the subject of the *Principia* was mechanics and was not relevant to nature ('physics'), that this work was based on arbitrary principles, would have had a strong impact on Newton! In that event, this review may have been the immediate cause for the change from 'hypotheses' to 'phenomena' and 'rules'.

Halley's review in the *Philosophical Transactions*

Halley's review, in the *Philosophical Transactions*,[64] although brief by comparison with Pfautz's and Locke's, did nevertheless convey to the reader a real sense of Newton's accomplishment. Furthermore, as a geometer as well

[61] Newton to Oldenburg, in reply to Pardies, 13 Apr. 1672, *Newton Correspondence*, vol. 1, p. 140.

[62] This translation was suggested by A. Koyré on the basis of a comparison of the English *Opticks* (London, 1704) and the Latin translation, *Optice* (London, 1706).

[63] See note 22.

[64] See note 6. Halley's review is published in facsimile (with an introduction by Robert E. Schofield) in *Isaac Newton's Papers & Letters on Natural Philosophy and Related Documents*, ed. I. Bernard Cohen and R. E. Schofield (Cambridge: Harvard University Press, 1958; revised edn, 1978), V. § 2.

as astronomer, Halley was able to appreciate the contributions to mathematics embodied in the great work. Although Halley does not present anything like the wealth of detail given by Pfautz, he does call attention to such mathematical novelties as Newton's extensive use of infinite series.[65]

While there can be no doubt that Halley's review in the *Philosophical Transactions* was in some ways intellectually superior to the others, most readers would be easily aware that it was not an impartial evaluation.[66] The *Philosophical Transactions* was the organ of the Royal Society and Newton's book had been published under the Society's sponsorship. The title-page boldly displayed 'IMPRIMATUR. S. Pepys, Reg. Soc. Praeses', a declaration of the official approval given above the name of Samuel Pepys, the learned diarist, in his capacity as President of the Royal Society. Additionally the printer of the *Transactions* was Joseph Streater, who was also the printer of the *Principia*.[67] Finally, the review was written and signed by Halley, who was the editor and publisher of the journal, and the man who had seen the *Principia*

[65] See I. B. Cohen, 'Halley's two essays on the *Principia*', in *Standing on the Shoulders of Giants, a Longer View of Newton and Halley*, ed. Norman J. W. Thrower (Berkeley/Los Angeles: University of California Press, 1990), pp. 91–108.

[66] Yet it must not be thought that Halley's presentation was pure laudation. In referring to Newton's system of the world, Halley wrote that the results 'agree with the Phenomena of the Celestial motions, as discovered by the great Sagacity and Diligence of *Kepler*'. And, in discussing Book Three, Halley referred to the '*principal phenomena* of Nature', chiefly 'the *Hypotheses* of Kepler'. That this was critical of Newton may be seen in the fact that Newton does not introduce the name of Kepler in reference to the law of equal areas or the law of elliptical orbits in Book One, where they are first used; nor is Kepler mentioned by name in Book Three when Newton introduces these two laws and applies them to the system of the world. Only the third or harmonic law is ascribed by Newton to Kepler.
Halley not only corrected Newton by introducing Kepler's name as the discoverer of the first two laws, but did so in reference to Kepler's 'great Sagacity and Diligence'. Yet it is to be doubted whether many of the readers of Halley's review would have recognized that this statement had a slightly critical tinge.

[67] The heading of Halley's review announced the *Principia* as published in '4to. Londini. Prostat apud plures Bibliopolas.' The original title-page reads, in part, 'Londini, Jussu Societatis Regiae ac Typis Josephi Streater. Prostat apud plures Bibliopolas.' That is, 'London, [published] by order of the Royal Society and printed by Joseph Streater. For sale at many book-sellers.' The variant title-page (found on a cancel in a number of copies) contains a slightly different version, '...ac Typis Josephi Streater. Prostant Venales apud Sam. Smith ad insignia Principis Walliae in Coemiterio D. Pauli, aliosque nonnullos Bibliopolas.' That is, in addition to having been printed by Joseph Streater, the *Principia* is said to be sold 'by Sam. Smith at the Arms of the Prince of Wales in St Paul's Church-yard.' This may be compared with the printer's notice (in English) in the *Philosophical Transactions*, 'Printed by J. Streater, and are to be sold by Samuel Smith at the Princes Arms in St Paul's Church-yard.' On the two title-pages, see the bibliographical essay by William Todd in vol. 2 of the edition cited in note 44.

through the press.[68] Then, as if this were not enough, the *Principia* contained an introductory ode signed by Halley, praising Newton's creative genius and declaring the scientific magnificence of the *Principia*. 'Nearer the gods', Halley's ode proclaimed, 'no mortal can approach'.

Conclusion

A comparison and contrast of the four reviews shows that Pfautz's presentation gave readers the most complete account of the subjects treated by Newton in the *Principia*, the only one in fact that really succeeded in presenting the whole range of topics covered by Newton, together with his principal results. The only other attempt to present the whole range of topics was Locke's review. But since this was merely a set of translations into French of the headings of the sections, it did not really give the reader any true sense of Newton's methods and results, in the sense that Pfautz did. For example, the reader would learn nothing about Newton's treatment of Kepler's first two laws of motion from Locke's translation of the headings of Sect. Two (containing Props. 1–2–3 on the area law) and of Sect. Three (containing Prop. 11 on the elliptical orbits).

I have mentioned Locke's feeble attempt to deal with universal gravity, which may be contrasted with Pfautz's elegant paraphrase of Newton's own statements concerning the inverse-square of the distance and the two masses involved. Furthermore, Pfautz understood – as Locke evidently did not – how gravity explains the perturbing action of the sun on the two-body system of the earth and the moon and also explains the action of the sun and the moon in producing the tides in the oceans.

Halley's review was very different from either Pfautz's or Locke's. He did not try to be encyclopedic but was rather very selective. A feature that differentiates his review from Pfautz's is that he stressed such mathematical novelties of the *Principia* as Newton's extensive use of infinite series. No doubt Halley was a more perceptive mathematician than Pfautz. On the other hand Halley's goal differed greatly from that of Pfautz, who wanted to present a detailed account of all the topics introduced by Newton.

[68] In an Advertisement, printed at the conclusion of his book review, Halley explained why 'the Publication of these *Transactions* has for some Months last past been interrupted'. He asked the reader 'to take notice that the care of the Edition [i.e., editing the *Principia* for publication, reading the proofs, and seeing the volume through the press] of this Book of Mr. *Newton*' had 'lain wholly upon the Publisher [i.e., Halley himself]'. This, together with 'some other pressing reasons' had prevented him from getting the *Transactions* out on time. Halley told his readers that in seeing to the publication of the *Principia* he 'conceives he hath been more serviceable to the Commonwealth of Learning' than in editing and publishing the *Transactions*. On Halley's role in the publication of the *Principia*, see my *Introduction*.

Finally, the anonymous reviewer in the *Journal des Sçavans* did not really attempt to give his readers any measure of Newton's accomplishment. His primary purpose, as we saw, was to criticize Newton for basing a system of mechanics on the notion of attraction. Yet even this reviewer did inform his readers that Newton had introduced an explanation of the phenomena of the tides based on the concept of a gravitating force.

One of the few Newtonian scholars to mention Pfautz's review was W. W. Rouse Ball. In his *Essay on Newton's Principia*,[69] Rouse Ball disparaged Pfautz's review by saying that 'it is, and purports to be, little more than a synopsis of the contents'. In late nineteenth-century Cambridge, it must have seemed an easy task to have made such a synopsis or epitome. In that same volume, Rouse Ball himself produced an admirable and useful summary outline of the Principia. But Rouse Ball for the most part merely translated the section heads and translated or paraphrased (or modernized) the statements of the definitions and laws and the statement of each proposition, adding some new material from manuscript and other sources from time to time. He omitted the scholiums and most of the corollaries.

Rouse Ball was a Cambridge mathematician who had studied the *Principia* as a student and who had taught some parts of this book to his own students. Accordingly, he was familiar with the subject matter of the book itself as well as the general subjects of rational mechanics and celestial mechanics. He therefore tended to read the *Principia* as an early form of a later science which he knew well. But Pfautz was in a very different position. Although he was trained in mathematics and was adept at some parts of mathematical astronomy, he would have found much of Newton's work new and very different from anything he had ever encountered before. To make a careful and well balanced presentation of so revolutionary a work on first encounter is a vastly more difficult undertaking than to do so when one has studied the book with the advantage of the hindsight of mature later developments of the subjects. We may observe in this regard, furthermore, that Pfautz introduced no major errors nor did he distort Newton's thought.

Rouse Ball, on the other hand, was so unable to read Newton's book in its own terms that he could not help but revise Newton's text to make it agree with the rational mechanics of his own day. Thus, when Rouse Ball came to the second law of motion, he could not allow Newton's own words to stand as he had written them. It was 'obvious' to him that Newton's statement of the second law is incorrect as stated by Newton, that Newton's presentation differs in one major feature from the second law that Rouse Ball had learned and had taught and that every physicist still learns today.

[69] W. W. Rouse Ball, *An Essay on Newton's Principia* (London: Macmillan and Co., 1893; A photo-reprint, with an intro. by I. B. Cohen, New York: Johnson Reprint Corporation, 1972).

Newton says that the change in 'quantity of matter' or momentum is proportional to the 'force impressed', whereas every beginning student of physics knows that it is the *rate of change of momentum* or the *change in momentum per unit time* that is proportional to the 'force'. It is obvious to anyone who reads the *Principia* that Newton knew our form of the second law and actually used it again and again. Rouse Ball concluded, therefore, that Newton must have erred in his initial statement of the law. Accordingly, he corrected Newton's fault by adding the missing phrase 'per unit of time' so as to make the first law read:

> *Law* 2. The change in momentum [per unit of time] is proportional to the moving force impressed, and takes place in the direction in which the force is impressed.[70]

It apparently never occurred to Rouse Ball that perhaps a mathematician and physicist as acute as Newton might have meant what he said, and all the more so since the statement of the second law went unchanged through three editions. Had Rouse Ball tried to find out what Newton had in mind, there can be little doubt that he quickly would have discovered that Newton used two different versions of the second law. They use two different concepts of 'force', one of which is ours and the other is not.[71] In the statement of the second law in the Definitions, the force in question is not a continually acting force such as gravity but rather an instantaneous blow or impact, in which case Newton's form of the law is quite correct.[72]

---

[70]   *Ibid.*, p. 77.

[71]   On Newton's concept of force, see R. S. Westfall, *Force in Newton's Physics* (London: Macdonald; New York: American Elsevier, 1971); also I. B. Cohen, 'Newton's second law and the concept of force in the *Principia*', in Robert Palter, ed., *The Annus Mirabilis of Sir Isaac Newton, 1666–1966* (Cambridge: The MIT Press, 1970), pp. 143–85.

[72]   Rouse Ball was not alone in his reading of Newton's second law. For example, A. B. Webster, in his *Dynamics of Particles and of Rigid, Elastic, and Fluid Bodies* (Leipzig: B. G. Teubner, 1904), p. 22, quotes the second law in Latin from Newton's *Principia*, followed by a reasonably accurate English translation – 'Change of motion is proportional to force applied, and takes place in the direction of the straight line in which the force acts.' Then, without further justification or comment, Webster declares, 'By change of motion is meant acceleration.'
There is, of course, a fundamental difference between the case of Rouse Ball and of Webster (and other mathematicians and physicists). Webster (like others) was writing a textbook for students of mathematics and physics, whereas Rouse Ball was producing a historical monograph and was the author of other historical works including a *Short History of Mathematics* which went through many editions and a learned monograph on *The Study of Mathematics at Cambridge*. Sad to relate, this error has crept into the writings of historians of science. For example, Max Jammer's *Concepts of Force* (Cambridge: Harvard University Press, 1957), p. 124, quotes the second law in the original Latin, followed by this comment: 'The English translation

A generation earlier, J. C. Maxwell had fully appreciated this aspect of Newton's second law. In his *Matter and Motion* (1877), Maxwell stated this law in Newton's own words:

> Law II. – Change of motion is proportional to the impressed force, and takes place in the direction in which the force is impressed.[73]

He noted that by 'motion, Newton means what in modern scientific language is called Momentum', that is, 'the quantity of matter moved is taken into account as well as the rate at which it travels'. He further observed that by 'impressed force he means what is now called Impulse', in which 'the time during which the force acts is taken into account as well as the intensity of the force'. An additional comment explained that the 'word impulse was originally used to denote the effect of a force of short duration, such as that of a hammer striking a nail'.[74]

The 'cautionary' example of Rouse Ball's reading of Newton's second law highlights a major difference between the task of reading the *Principia* from the vantage point of post-Newtonian physics and of doing so, as Pfautz did, when the subject was still new and strange. It is true that what Pfautz produced was a summary or epitome of Newton's *Principia*, but we must remember that he had to do so with no previous guides. Pfautz had to make his way alone through a very long complex and difficult technical treatise, one that in many ways is not presented in the most orderly fashion for understanding on a first unaided reading. That he accomplished this mission in so complete a fashion without gross error or distortion, selecting the main topics and results for his presentation, must seem all the more remarkable when we remember that he was probably only the third person in the world to read the *Principia* from cover to cover.[75]

Rouse Ball's treatment of the second law may help us to appreciate all the more the worth of Pfautz's generally complete and always sound 'epitome' of Newton's rational mechanics and physical principles of the system of the world. It is an honor to be able to make his name generally known to the world of Newtonian scholarship.

> of the second law, although faithful to the Latin original, may easily be misinterpreted; what Newton meant when saying *mutatio motus* should be rendered in modern English as "rate of change of momentum".'

[73] J. C. Maxwell, *Matter and Motion* (London: Society for the Promotion of Christian Knowledge, 1877; 2nd edn, with appendixes by Joseph Larmor, 1924), p. 32, §44.

[74] *Ibid.*, p. 37, §49, 'On Impulse'.

[75] The first two, of course, were the author and Halley (editor and proof-reader). Neither Locke nor the reviewer in the *Journal des Sçavans* appears to have read the whole of the *Principia*. I don't count among 'readers' the amanuensis who wrote out the text for the printer, Humphrey Newton, described by Newton as an 'Emanuensis who understood not Mathematicks'. See my *Introduction*, pp. 294, 299–301.

13 Newton, Cotes, and $\sqrt{\sqrt{2}}$: a footnote to Newton's theory of the resistance of fluids

D. H. FOWLER

Book II, Section VII of Newton's *Principia*, on the resistance of fluids, is a layered mixture of experiment (in the mind, in various laboratories, and in St Paul's) and theoretical analysis, all contrived to fit a discernible palimpsest of changing conclusions. As a piece of natural philosophy, the result is, by common consent, most unsatisfactory; but it gives us an unusually rich glimpse into the workings of the mind of one of the greatest of natural philosophers.[1] From this rich and heavy plum-pudding, I propose to extract just one small raisin, from Proposition XXXVI,[2] and savour some of its delights.

This proposition, crucial to Section VII, concerns the velocity of water flowing from a hole in a cylindrical vessel. At different times, and in the light of different experiments, some of his own, others proposed by different correspondents, Newton makes the result equal to the velocity the water would acquire by falling either through its depth in the vessel, or through half of that depth. Now the outflowing velocity can be deduced by measuring the amount of water that flows from a hole of measured size. But, Newton observes:

> [The particles of the effluent water] meet in a stream whose diameter is a little smaller below the hole than at the hole itself; its diameter being to the diameter of the hole as 5 to 6, or as $5\frac{1}{2}$ to $6\frac{1}{2}$, very nearly, if I measured those diameters rightly. I procured a thin flat plate, having a hole pierced in the middle, the diameter of the circular hole being five eighth parts of an inch. And that the stream of running water might not be accelerated in falling, and by that acceleration become narrower, I fixed this plate not to the bottom, but to the side of the vessel, so as to make the water go out in the

---

[1] For a description, see R. S. Westfall, *Never at Rest* (Cambridge University Press, 1980), p. 707 ('No part of the *Principia* had been more imperfect....Not much of anything in Section VII was satisfactory.') to p. 712 ('No one would care to deny that it [sc. Newton's treatment of the topic] was a virtuoso performance. Few would care to dignify it with the title of scientific investigation.')

[2] In the numbering of the second and third editions; the corresponding proposition XXXVII of the first edition is very different. See *Isaac Newton's Philosophiae Naturalis Principia Mathematica: The Third Edition (1726) with Variant Readings*, 2 vols., eds. A. Koyré and I. B. Cohen (Cambridge University Press, 1972), vol. 1, pp. 478–87, vol. 2, pp. 777–9.

direction of a line parallel to the horizon. Then, when the vessel was full of water, I opened the hole to let it run out; and the diameter of the stream, measured with great accuracy at the distance of about half an inch from the hole, was twenty-one fortieth parts of an inch. Therefore the diameter of this circular hole to the diameter of the stream was very nearly as 25 to 21.[3]

Newton was a consummate experimenter, but could he really arrive at such a measurement of 21/40 inch for a stream of water? Something is afoot, and Cotes – who, in seeing the second edition through the press, is constantly suggesting revisions and corrections – knows perfectly what is going on:

I have received Your Letter with the inclosed Paper & am very well satisfied with Yr solution of the difficulty which I formerly proposed to You concerning the Velocity of the effluent water. I find that 25 & 21 express the proportion of $\sqrt{\sqrt{2}}$ to 1 as nearly as it is possible for so small numbers to do it, whence it is probable yt the exact proportion of the diameter of the Hole to ye diameter of ye Stream is that of $\sqrt{\sqrt{2}}$ to 1, & then ye proportion of 44 to 37 will be much nearer the truth than yt of 25 to 21.[4]

Can Cotes really be proposing that Newton repeats the experiment with the 5/8 inch orifice and measure the stream, to check if it is more nearly 185/352 inch, since this gives

$$\text{(diameter stream)/(diameter hole)} = \frac{185}{352}\bigg/\frac{5}{8} = 37/44?[5]$$

Surely not; what Newton and Cotes are presenting are various approximations to $\sqrt[4]{2}$, namely 6/5, $6\frac{1}{2}/5\frac{1}{2}$ (or 13/11), and 44/37, that they are using to suggest the measurements fit Newton's theory closely, conveniently forgetting how Newton has constructed his theory to fit these observations, just as in the first edition he constructed a different theory to fit a different conclusion apparently suggested by different observations. But how did they come by these approximations?

At this point, a footnote somewhere in the output of the Newton industry

[3] Adapted from the translation of Andrew Motte (*Sir Isaac Newton's Mathematical Principles*, tr. A. Motte, rev. F. Cajori, London, 1729, rept. University of California Press, 1934), vol. 1, pp. 338–9.

[4] Cotes to Newton, 31 March 1711, *The Correspondence of Isaac Newton*, eds. H. W. Turnbull, J. F. Scott, A. R. Hall, and L. Tilling, 7 vols. (Cambridge University Press, 1959–77), vol. 5, no. 829, p. 107; also see Newton to Cotes, 24 March, *ibid.* no. 826, p. 103. The description and most of the quotations so far can be found in Westfall, *Never at Rest*, though the figure given there does not quite correspond either to Newton's figure or Westfall's text, and this makes Newton's already difficult argument even more difficult.

[5] So much more outrageous than Westfall suggests (*Never at Rest*, p. 711): 'Possibly Newton feared he could not with a straight face claim to have "measured" a proportion of 44 to 37 "with great accuracy".'

should provide the answer. Since I have been unable to find it so far, I present it here immediately: These ratios are convergents and intermediate convergents of the continued fraction expansion of $\sqrt[4]{2} = [1,5,3,1,1,40,\ldots]$, and, because of the large term of 40, the convergent $[1,5,3,1,1] = 44/37$ will be an unusually good approximation. Moreover Cotes describes at great length, in his 'Logometria' (1714), one method of evaluating such approximations.

Tom Whiteside, to whom my footnote is dedicated in admiration, respect, and friendship, knows the explanation so well that he probably hasn't inserted it in one of his own densely annotated volumes, spidery handwriting in ballpoint pen, only because it is so blindingly obvious; indeed the ingredients for a complete mathematical and historical discussion can already be found in his thesis. Or he may already have written it in a letter, perhaps several pages of erudition in reply to some casual question – the bread that he has cast upon waters like this has fed an army of historians of mathematics. But, because the underlying mathematics has been unfashionable for some years now, many mathematicians may be unable to give, or even to understand, this explanation; therefore it has slipped from view of historians of mathematics also, so many of them can give neither the explanation nor its context. The rest of this article will therefore fill in some of this background.

Pocket calculators and computers have revolutionised calculation in our lifetime, just as logarithms did in the seventeenth century, and I have learned from Tom not to spurn the advantages these instruments give us. Today multiplication and division, powers and roots, are cheap, quick, accurate, and universally available. A moment's button-pushing gives

$$\sqrt[4]{2} = 1.1892\ldots \qquad \frac{1}{0.1892\ldots} = 5.2852\ldots \qquad \frac{1}{0.2852\ldots} = 3.5061\ldots$$

$$\frac{1}{0.5061\ldots} = 1.9757\ldots \qquad \frac{1}{0.9757\ldots} = 1.0249\ldots \qquad \frac{1}{0.0249\ldots} = 40.1815\ldots \quad \text{etc.}$$

Unravelling this, we see that

$$\sqrt[4]{2} = 1 + 0.1892\ldots = 1 + \frac{1}{5.2852\ldots} = 1 + \cfrac{1}{5 + \cfrac{1}{3.5061\ldots}} = 1 + \cfrac{1}{5 + \cfrac{1}{3 + \cfrac{1}{1.9757\ldots}}}$$

$$= 1 + \cfrac{1}{5 + \cfrac{1}{3 + \cfrac{1}{1 + \cfrac{1}{1.0249\ldots}}}} = 1 + \cfrac{1}{5 + \cfrac{1}{3 + \cfrac{1}{1 + \cfrac{1}{1 + \cfrac{1}{40.1815\ldots}}}}} \quad \text{etc.}$$

Now 1/40.1815... is very small so, if we ignore it, we will make a very small error buried right at the bottom of this last complicated continued fraction.[6] Therefore

$$1 + \cfrac{1}{5 + \cfrac{1}{3 + \cfrac{1}{1 + \cfrac{1}{1}}}} = \frac{44}{37}$$

is likely to be a very good approximation of $\sqrt[3]{2}$. And, similarly, if we ignore or adjust some earlier remainders, we get

$$1 + \frac{1}{5} = \frac{6}{5}, \quad 1 + \cfrac{1}{5 + \cfrac{1}{1}} = \frac{7}{6}, \quad 1 + \cfrac{1}{5 + \cfrac{1}{2}} = \frac{13}{11}, \quad 1 + \cfrac{1}{5 + \cfrac{1}{3}} = \frac{19}{16}, \quad 1 + \cfrac{1}{5 + \cfrac{1}{3 + \cfrac{1}{1}}} = \frac{25}{21}.$$

These are all likely to be less good but improving approximations.

We can describe the process so far as follows. Given a number, record its integer part, replace the original number by the reciprocal of its fractional part, and so continue forward; and, for our example of $\sqrt[3]{2}$, the quotients 1,5,3,1,1,40,... will be produced. Also, it would be useful to have a more convenient way of evaluating the complicated fractions that arise than working them out case by case, as I have just done. Here is such a procedure, described at great length:

> Having made these calculations [of the quotients 1,5,3,1,1,40,...] we must set out two columns of ratios, one of which contains the terms which have a ratio greater than the true one, and the other the terms which have a ratio less; beginning the computation with the ratios 1 to 0 and 0 to 1, which are most remote from the true one; from there deducing the remaining ratios which approach ever closer to the true one. Let the terms 1 & 0 be multiplied by the first quotient 1, and let the result 1 & 0 be written under the terms 0 & 1; addition will produce the ratio $1+0$ and $0+1$, or 1 to 1. Let the terms of this [ratio] be multiplied by the second quotient 5, to make 5 & 5, which are added to the terms 1 & 0; and the ratio $5+1$ and $5+0$, or 6 to 5 will be obtained. Let the terms of the ratio be multiplied by the third quotient 3, to make 18 & 15, which added to the preceding terms 1 & 1; and the ratio 19 & 16 will be obtained. Let the terms to this ratio be multiplied by the fourth quotient 1, to make 19 & 16, and added to the preceding terms 6 & 5, and the ratio 25 to 21 will be obtained. Let the terms of this

6   This name 'continued fraction' is first found in John Wallis, *Arithmetica Infinitorum* (1655), Proposition CXCI and its Scholium, in connection with Brouncker's expression for $4/\pi$ which, he writes, has a *denominator continuae fractus*, a 'successively fractioned denominator'.

be multiplied by the fifth quotient 1, to make 25 & 21 to be added to the preceding 19 & 16, and the ratio 44 to 37. Let the terms of this ratio be multiplied in the same way by the sixth quotient 40, making 1760 & 1480 to be added to the preceding 25 & 21, to produce the ratio 1785 to 1501, and so continue as far as you wish, transferring alternate factors to alternate columns. When this is done, we will have the following ratios greater than the true one: 6 to 5, 25 to 21, 1785 to 1501, etc. The ratios less than the true one will be 1 to 1, 19 to 16, 44 to 37, etc. And these are the principal and primary ratios[7] by which the proposed ratio is continually approached.

Ratios greater than the true one		*Ratios less than the true one*	
		0	1
1	0 × 1	1	0
5	5	1	1 × 5
6	5 × 3	18	15
19	16	19	16 × 1
25	21 × 1	25	21
1760	1480	44	37 × 40
1785	1501	&c.	&c.
&c.	&c.		

But if we require the entire series of all the ratios greater than the true one, which are such that no ratio greater than the true one, which comes closer to the true one, can be described in lesser terms; and if we likewise require the entire series of all the ratios less than the true one, which are such that no ratio less than the true one, which comes closer to the true one, can be described in lesser terms; then besides these primary ratios which we have just discovered, we can obtain other secondary ratios.[8] These occur when the quotient is greater than one. These are found by replacing the multiplication by the quotients which occurred above by repeated addition of terms, as many times as there are units in the quotient. Thus, because the first quotient was 1, the terms 1 & 0 are added once to the terms 0 & 1; and

---

[7]   These 'principal and primary ratios' are now known as convergents. I do not know who first used this obvious name, but this quick procedure for evaluating them can already be found in Daniel Schwenter, *Deliciæ Physico-Mathematicæ* (Nuremberg, 1636) and in the third edition of his *Geometria Practica Nova* (1741); Wallis (see the previous note) gives an equivalent procedure for evaluating them in the slightly more complicated situation where not all of the numerators are unity. See D. T. Whiteside, 'Patterns of mathematical thought in the later seventeenth century', *Archive for History of Exact Sciences*, 1 (1951): 179–388, on p. 195, n. 42. The early history of all these features will be summarised below.

[8]   These 'secondary ratios' are now generally known as intermediate convergents; the name *fractions intermédiaires* seems to have been first used in Joseph-Louis Lagrange, *Additions aux Éléments d'Algèbre d'Euler*. See *Œuvres de Lagrange*, ed. J.-A. Serret, 14 vols. (Paris: Gauthier–Villars, 1867–92), vol. 7, pp. 5–179, on pp. 29ff.

the sum will give the ratio 1 to 1. These last terms 1 & 1, since the second quotient was 5, are added five times to the ratio 1 & 0, and this will give the ratios 2 & 1, 3 & 2, 4 & 3, 5 & 4, 6 & 5. These terms 6 & 5, since the third term was 3, are added three times to the terms 1 & 1, and the sum will give 7 & 6, 13, & 11, 19 & 16. ...

Ratios greater than the true one			*Ratios less than the true one*	
1	**0** × 1		**0**	**1**
1	1		1	0
2	**1**		**1**	**1** × 5
1	1		6	5
3	**2**		**7**	**6**
1	1		6	5
4	**3**		**13**	**11**
1	1		6	5
5	**4**		**19**	**16** × 1
1	1		25	21
6	**5** × 3		**44**	**37** × 40
19	16		&c.	&c.
25	**21** × 1			
44	37			
69	**58**			
&c.	&c.			

And so one can proceed as far as seems convenient. When this operation is at length completed, the entire series of all the ratios greater than the true one will be 1 to 0, 2 to 1, 3 to 2, 4 to 3, 5 to 4, 6 to 5, 25 to 21, 69 to 58, 103 to 95, etc., and similarly the entire series of all the ratios less than the real one will be 0 to 1, 1 to 1, 7 to 6, 13 to 11, 19 to 16, 44 to 37, &c.

The application of these useful approximations is widespread, wherefore I have given a somewhat prolix exposition of their derivation, by the method which seems to me simplest and easiest. The celebrated men Wallis and Huygens[9] have dealt with the same argument, slightly differently.

This quotation has been translated and adapted to fit the numbers of this particular example of $\sqrt[4]{2}$ from Cotes' own 'Logometria'.[10] Cotes' description

---

9 The contributions of Wallis and Huygens will be summarised below.

10 Published in *Philosophical Transactions of the Royal Society*, 29(March 1714):5–47, and reprinted verbatim in his posthumous *Harmonia Mensurarum* (1722); the translation is adapted from R. Gowing, *Roger Cotes, Natural Philosopher* (Cambridge University Press, 1983), pp. 147–9. Let me emphasise: the first sentence of the passage will be quoted below; I have altered the numbers in Cotes' text to fit this example here; and I have modified the tables, introducing a displacement of one line in the columns of the first table, and setting the approximations of both of them in bold type, to clarify his procedure.

is, on his own admission, prolix, and even more dramatically so when compared with the uncompromising terse symbolic style of the rest of the article. But Cotes knows a valuable trick, and seems to think it worth setting it out at length. He has learned the trick from Wallis and Huygens; but in his article, he has just used it to uncover another remarkable and unexpected phenomenon that I shall describe below.

I recommend those who are unfamiliar with this process for generating the sequence of quotients, such as 1,5,3,1,1,40,... for $\sqrt[4]{2}$, to programme their pocket calculating machines to do it and explore some examples. Square roots will exhibit a periodic behaviour that I have proposed might have been discovered by the mathematicians around Plato, and which might have played a not inconsiderable part in the developments of pre-Euclidean Greek mathematics.[11] Further explorations of square roots (try $\sqrt{19}$, $\sqrt{21}$, $\sqrt{22}$, $\sqrt{(11/2)}$, $\sqrt{(23/2)}$, $\sqrt{(23/3)}$,... but beware that round-off errors will eventually corrupt the result) will reveal refinements of this pattern that intrigued Euler, Lagrange, Legendre, Galois and others. The example of π was explored at great length by Wallis, who used a different process to generate the 'entire series' of 494 intermediate convergents and convergents that can be deduced from van Ceulen's 35 place decimal value of π, and his resulting table (which is analogous to Cotes' second table) fills eight large folio pages of his *Treatise of Algebra* (1685), pp. 48–55. The reader can easily generate the expansion $\pi = [3,7,15,1,292,1,1,...]$ using a pocket calculating machine (though most machines will not have enough accuracy to evaluate the expansion correctly thus far); Wallis, in effect, implicitly generates 34 terms of this expansion, though I have argued that he seems unaware of the relationship between his method and the continued fraction process.[12]

Cotes is clearly aware that large terms in the continued fraction expansion will correspond to specially good approximations by the convergent which truncates the expansions just before this term. (This explains why 22/7 is a good approximation to π, and 355/113 is a very good approximation indeed.) Thus, truncating the expansion of $\sqrt[4]{2}$ to [1,5,1,3,1,1] will generate the approximation 44/37 which is 'much nearer the truth than yt of 25 to 21', although the terms of the fraction are not much larger.

In 'Logometria', Cotes gives, correctly, the values of e and e^{-1} to twelve

---

11 See my *The Mathematics of Plato's Academy: A New Reconstruction* (Oxford University Press, 1987); these proposals are based on a different formulation of the algorithm which uses only the positive integers and Euclidean-style geometry and proportion theory. Further details of many aspects of the algorithm, including a sketch history of the theory of continued fractions up to today, can be found in this book.

12 See my 'Approximation technique, and its use by Wallis and Taylor', *Archive for History of Exact Sciences*, 41(1991):189–233.

decimal places; perhaps this is the first printed value of e, though surely other people must have already calculated it. Then, in Scholium 3 to Proposition 1, he does what every mathematician should do when coming across a new number: he evaluates its continued fraction expansion, thus stumbling on a rich new mathematical phenomenon. Here is the beginning of the quotation above, this time with Cotes' own example:

> The major term 2.71828 &c. [= e] should be divided by the minor 1, or again the major 1 by the minor 0.367879 &c. [= e^{-1}] and once more the minor by the number which is left, and this again by the last remainder, and so continue forward; and the quotients 2, 1, 2, 1, 1, 4, 1, 1, 6, 1, 1, 8, 1, 1, 10, 1, 1, 12, 1, 1, 14, 1, 1, 16, 1, 1 &c. will be produced. Having made these calculations, we must set out two columns of ratios, ...

and we have reached the passage given above, in adapted form. Several comments are in order.

First, this and other examples of this new kind of regular pattern that Cotes has discovered here, but about which he says nothing, will appear in Euler's works from 1737 onwards, and Euler's work combines with Lambert's, from 1761 onwards, to give proofs of it; then Hurwitz will give further explanations of this kind of phenomenon in the 1880s. It is clear that Cotes has induced the pattern, seventeenth-century style, because a twelve-decimal-place approximation to e will only yield twelve terms of the expansion, at which point it becomes corrupted with round-off error; the expansion that Cotes gives, up to 1,16,1,1, requires knowing twenty-one places of e.[13] But nobody seems to acknowledge Cotes as the discoverer of this remarkable behaviour of e.

Second, Cotes describes the continued fraction expansion in terms of division, but it is clearly division with remainder, more akin to subtraction than division and much more amenable to hand-calculation than the pure division process I described above. Also his casual description (contrast the prolix description that follows, quoted above) seems to indicate that he expects

[13]   In general, n decimal places of a number will usually provide slightly more information than n places of the continued fraction expansion. This is a subtle consequence of the probabilistic behaviour of continued fractions first discovered by Gauss and rediscovered by Kuz'min and others. For an explanation, see D. Shanks, 'A logarithm algorithm', *Mathematical Tables and Other Aids to Computation*, 8(1954):60–4, where Shanks gives a continued fraction algorithm which he describes as 'quite unlike anything previously known to the author and [which] seems worth recording because of its mathematical beauty and its adaptability to high-speed computing machines' [note the date of the paper!]. His algorithm is a rediscovery of a method previously described by Brook Taylor ('A new method for computing logarithms', *Philosophical Transactions of the Royal Society* 30(1717):618–22) and by David Rittenhouse, a self-taught eighteenth-century American mathematician. Further details are given in my 'Approximation technique' (see the previous note).

his readers to be familiar with it. And so they should be, for it is the so-called Euclidean algorithm, to be found in *Elements* VII 1–2 and X 2–3. The calculation for e proceeds as follows:

$$e = 2.7183\ldots = 2 \times 1 \qquad\quad + 0.7183\ldots$$
$$1 \qquad\quad = 1 \times 0.7183 \quad + 0.2817\ldots$$
$$0.7183\ldots = 2 \times 0.2817\ldots + 0.1548\ldots$$
$$0.2817\ldots = 1 \times 0.1548\ldots + 0.1269\ldots$$
$$0.1548\ldots = 1 \times 0.1269\ldots + 0.0280\ldots$$
$$0.1269\ldots = 4 \times 0.0280\ldots + 0.0150\ldots$$

etc.

We can immediately convert this subtraction process into the division process I described first by rescaling each line, dividing the second line by 0.7183..., the third line by 0.2817..., etc. So the two algorithms yield the same sequence of quotients. Or we can, as Cotes suggests, start with 1 and $e^{-1} = 0.3679\ldots$

$$1 \qquad\quad = 2 \times e^{-1} \qquad + (1 - 2 \times e^{-1})$$
$$= 2 \times 0.3679\ldots + 0.2642\ldots$$
$$0.3679\ldots = 1 \times 0.2642\ldots + 0.1036\ldots$$

etc.

and this will again give the same sequence of quotients; again, we have just rescaled each line of the calculation.

Next, I think it is appropriate to pose the question as to why Cotes gives here such a prolix description of the calculation of the convergents, especially as very similar procedures can already be found, as he says, in Wallis' and Huygens' works. My guess, and it is only a guess, will be connected with my final question: Cotes' article provides a convincing explanation of how he derived the approximation 44/37, but how did Newton generate his approximations 6/5, $6\frac{1}{2}/5\frac{1}{2}$, and 25/21? To this final question, I have no plausible answer; to the best of my knowledge, there is no overt or covert sign of continued fraction techniques anywhere else in Newton's works,[14] and, in this

---

[14] I have surreptitiously checked this opinion with Tom Whiteside and he concurs, with the possible exception of one place where Newton could have used the continued fraction process: see *The Optical Papers of Isaac Newton*, ed. A. E. Shapiro, Cambridge University Press, 1984–), vol. 1, p. 181, n. 29. Newton there gives the approximation to 8097/5240 (= [1,1,1,5,36,2,6]) by 17/11 (which is the convergent [1,1,1,5] which terminates before the large term of 36), but there is no other indication that he did use continued fractions, and any sufficiently efficient procedure could yield this approximation. Earlier, in his 'New theory about light and colors' (*Philosophical Transactions of the Royal Society* (1672):3075–87, on p. 3077), Newton gave an approximation to this same fraction by 31/20 which is not a convergent, 31/20 =

passage, the way that he writes $6\frac{1}{2}/5\frac{1}{2}$ seems to indicate that he is using some *ad hoc* procedure, such as observing that $6/5 > \sqrt[4]{2}$ (because $6^4 > 2 \times 5^4$) while $7/6 < \sqrt[4]{2}$, so $6\frac{1}{2}/5\frac{1}{2}$ will lie between them and, since $6\frac{1}{2}/5\frac{1}{2} > \sqrt[4]{2}$, it will be a better approximation than $6/5$. But I cannot make any suggestion about $25/21$ beyond the general remark that any efficient method for generating approximations by rational numbers will give the convergents, since these are the best possible such approximations, in some precise sense.

So why did Cotes give such a detailed description? He may well have been writing his 'Logometria' around the same time that Newton's revision of the *Principia* had reached the section on the resistance of fluids, for he sent the completed paper to Newton for approval on 25 May of the following year, 1712, and acknowledged Newton's reply on 10 August,[15] and in both letters he presses Newton to suggest corrections and emendations. Is it possible that he was trying to interest Newton in the technique and in his discovery of the behaviour of e? I like to think so, though I know of no other evidence to support this.

Cotes' contribution comes at the end of what may be described as the first, exploratory phase of the modern development of the theory of continued fractions. The second phase is then initiated by the work of Euler (whose first known reference to the subject appears in a letter of 1731 to Goldbach in connection with the Ricatti differential equation;[16] soon thereafter it spreads throughout his work), and continues with Lambert, Lagrange, Legendre, Gauss, and Galois. After this, all that was needed for the subject to enter on its modern phase were some advances in notation, especially an efficient index notation, and further developments in analysis and geometry.

This is therefore an appropriate point to attempt to summarise the features that are to be found in those early authors, from Bombelli to Cotes, who deal with matters closely related to continued fractions. These features will be grouped under the following headings:

I   Some description of a *layout* of the fractions.

II   *Simple continued fractions*, with all numerators unity (and all denominators positive integers; other possibilities do not enter before Euler and Lagrange).

III   *General continued fractions*, with numerators other than unity.

IV   Use of the *Euclidean algorithm*.

[1,1,1,4,2]; so all we can be really sure about is that he did not, at that stage, use anything related to continued fraction techniques.

[15]   *Correspondence*, vol. 5, nos. 921 & 931, pp. 305 & 318. We do not have Newton's reply to the first letter.

[16]   For references, see A. Weil, *Number Theory, an Approach through History* (Boston, etc.: Birkhäuser, 1984), p. 183.

V   Some procedures for *evaluating convergents*, equivalent to

$$p_{k+1} = n_{k+1}p_k + m_{k+1}p_{k-1} \quad \& \quad q_{k+1} = n_{k+1}q_k + m_{k+1}q_{k-1}.$$

(For simple continued fractions, all $m_k = 1$.)

VI   Some procedure for evaluating *intermediate convergents*.

VII   *Approximation* by convergents, alternately *lesser* and *greater*.

VIII   *Best approximation* from above and below.

IX   Best *absolute approximation*.

X   Any *analytic properties*, such as expansions for π or e.

Authors

1   Rafael Bombelli (1526–1573), in *L'Algebra parte maggiore dell'
aritmetica divisa in tre libri* (Bologna, 1572), reprinted with different
title page and dedicatory letter in *L'Algebra Opera* (Bologna, 1579).
In order to extract square roots (he gives the examples of $\sqrt{13}$ and
$\sqrt{8}$), Bombelli describes, on pp. 35–7, a procedure tantamount to
evaluating, one by one, the successive convergents of the general
continued fraction (III)

$$\sqrt{(a^s+r)} = a + \cfrac{r}{2a + \cfrac{r}{2a\ldots} +},$$

in which r may be a negative. But he gives none of the features under
headings I–X above, nor does he offer any justification of his method.

2   Pietro Antonio Cataldi (1548–1626), *Trattato del modo brevissimo di
trovare la radice quadra delli numeri, et regole da approssimarsi di
continuo al vero nelle radici de' numeri non quadrati, con le cause, &
inventioni loro...* (Bologna, 1613). Cataldi describes different ways of
generating approximations to $\sqrt{(a^2+r)}$, and then, on pp. 70ff, gives
a procedure, very similar to that of Bombelli (to whom he makes no
reference), for expressing it as a general continued fraction expansion.
He goes on to describe a layout (I) of the general continued fraction
(III),

$$\sqrt{18} = 4. \; \& \tfrac{2}{8}.$$
$$\& \tfrac{2}{8}.$$
$$\& \tfrac{2}{8}.$$

and a more compact version: 'Note that it is not possible to show
fractions conveniently in print, nor fractions of fractions, as they
should appear, that is thus: $4.\tfrac{2}{8}.\&\tfrac{2}{8}.\&\tfrac{2}{8}$ as we have made ourselves do
in this example; from now on we shall show them all like this: 4. $\&\tfrac{2}{8}$.
$\&\tfrac{2}{8}$. $\&\tfrac{2}{8}$., letting a period by the 8 in the denominator of each fraction
indicate that the following fraction is a fraction of this denomi-

Table 1

233		1	1 0
177	1	0	1
56	3	1	1
9	6	1	4
2	4	19	25
1	2	79	104
0	0	177	233

nator',[17] He observes that the approximations are alternately less and greater (VII).[18]

3 Daniel Schwenter, in *Delicæ Physico-Mathematicæ*, Nuremberg, 1636, and *Geometria Practica Nova*, Nuremberg, in its third edition of 1641.[19] In the brief treatment of approximation to fractions, in Chapter 87, pp. 111–13, of *Deliciæ Physico-Mathematicæ*, Schwenter takes the example of 177/233 and explains how to build up the numbers in Table 1 which we recognise as the computation of the terms of the expansion using the Euclidean algorithm (IV), and the evaluation of the convergents (V).

4a John Wallis, in his *Arithmetica Infinitorum* (1656), Proposition CXCCI and its Scholium describes Lord Brouncker's general continued

fraction (III) of $\dfrac{4}{\pi}$ (X), set out (I) as

$$\square = 1\tfrac{1}{2}\ \tfrac{9}{2}\ \tfrac{25}{2}\ \tfrac{82}{2}\ \&c.$$

He describes how to evaluate the convergents (V) of a general continued fraction, and how they are alternately lesser and greater (VII). But the convergents of this fraction converge very slowly and

[17] The Italian, with the fractions omitted, reads, 'Notisi, che nó si potendo cómodamétte nella stámpa formare i rotti, & rotti di rotti come andariano … come ci siamo sforzati di fare in questo, noi da quì inázi gli formaremo tutti à questa similitudine … facendo un punto all'8. denominatore di ciascun rotto, à significare, che il seguente rotto è rotto d'esso denominatore.'

[18] For discussions of Bombelli, Cataldi, and some later episodes in the history of continued fractions, see S. Maracchia, *Da Cardano a Galois, Momenti di storia dell'algebra* (Milan: Feltrinelli, 1979).

[19] I have so far been unable to consult this book. I know of no details of Schwenter's life, nor of any discussion of his work, apart from that in M. Cantor, *Vorlesungen über Geschichte der Mathematik*, 4 vols. (Leipzig: Teubner, 1900–8), in vol. 2 (1200–1668), pp. 763–5, and those subsequent accounts that clearly draw on this source.

he does not here enter the discussion of best absolute approximation, though he does give a tantalising hint of Brouncker's remarkable convergence-accelerating procedure. This material is then summarised very briefly in Chapter LXXXIV of the *Treatise of Algebra* (1685).

4b John Wallis, first as an appendix to the second edition of Jeremiah Horrocks, *Opera Posthuma* (1678), then in his *Treatise of Algebra*, Chapters X & XI, describes how, starting from the decimal expansion of a given number, to construct what we now describe as its intermediate convergents (VI) and he observes, without proof, that these give the best possible approximations from above and below (VIII); hence, implicitly, his calculation contains all of the information needed to deduce the simple continued fraction expansion (II). But I have argued (see note 12) that he shows no awareness of the connection between his procedure here and the continued fraction process that he described in his *Arithmetica Infinitorum*.

4c John Wallis and Lord Brouncker, in *Commercium Epistolicum* (1685), describe how to solve Pell's equation $x^2 - ay^2 = 1$ (a a non-square integer) in integers; but again they show no awareness of the connection between their method and the continued fraction process. This connection will not be made explicit until Leonard Euler, 'De usu novi algorithmi in problemate Pelliano solvendo'.[20]

5 Christiaan Huygens, in his *Descriptio Automati Planetarii* (written sometime after 1691, possibly several years later; published posthumously in 1703) gives a fluent and complete treatment of simple continued fractions with sketches of proofs, covering explicitly topics I, II, IV, VII, VIII, and IX but, curiously, neither V nor VI. Elsewhere, in his notes, we find details of earlier explorations of simple continued fractions, starting in 1686 or 1687.[21] He surely knows the recursive way of calculating the convergents and intermediate convergents, but he does not appear to describe it.

6 Roger Cotes, in 'Logometria' (1714), reprinted in his *Harmonia Mensurarum* (1722), described here, gives procedures that complement those of Huygens for simple continued fractions.

[20]   Written in 1759; published in *Novi Commentarii Academiae Scientiarum Petropolitanae*, 11(1765):26–66; and rept. in *Leonardi Euleri Opera Omnia*, 71 vols. to date (Leipzig: Teubner, 1911–), Series I, vol. 3, pp. 73–111.

[21]   See Christiaan Huygens, *Œuvres Complètes*, 22 vols. (The Hague, 1888–1950); *Descriptio Automati Planetarii* is reprinted in vol. 21, pp. 579–652, and a discussion of and references to Huygens' other notes on continued fractions is given in my 'Approximation technique', (note 12, above).

Table 2

	I	II	III	IV	V	VI	VII	VIII	IX	X
Bombelli			×							
Cataldi	✓		✓				✓			
Schwenter				✓	✓					
Wallis & Brouncker	✓	×	✓		✓	×	✓	×		✓
Huygens	✓	✓		✓			✓	✓	✓	
Cotes				✓	✓	✓	✓			✓

Recall the headings of columns: I: Layout; II: Simple continued fractions; III: General continued fractions; IV: Euclidean algorithm; V: Evaluation of convergents; VI: Intermediate convergents; VII: Alternating approximations; VIII: Best approximations; IX: Absolute approximations; X: Analytic properties.

We can summarise this sketch in Table 2, where a tick indicates an explicit occurrence of the feature in question, while a cross indicates that the author is deploying some procedure that we recognise as equivalent, though he shows no indication of this connection.

— IV —

Newton and eighteenth-century mathematics and physics

14   A study of spirals: Cotes and Varignon

RONALD GOWING

Acknowledgement

Professor D. T. Whiteside kindly arranged for me to have access to the Cotes papers in Trinity College and Clare College, Cambridge, and in the Cambridge University Library. He also encouraged me in my close study of the work of Pierre Varignon (1654–1722). I would like to acknowledge his friendly help and advice, always readily given.

Roger Cotes was the first Plumian Professor of Astronomy and Experimental Philosophy, in the University of Cambridge. His generally accepted importance as the editor of the second edition of Newton's *Philosophiae Naturalis Principia Mathematica* (Cambridge, 1713) (*the Principia* of subsequent history as Professor Rupert Hall has called it) has tended to divert attention from his own mathematical work. He published but one paper in his lifetime, the 'Logometria'[1] and as Professor Whiteside has observed, this might very well not have reached the public until after Cotes' early death in 1716. In this paper Cotes developed the relationship between logarithmic and trigonometric quantities, which he said 'have a truly remarkable affinity', summarised in the formula $\ln(\cos\theta + i\sin\theta) = i\theta$,[2] (a result restated by Euler more than thirty years later, and sometimes known as Euler's formula). This insight enabled him to extend substantially the range of algebraic forms which he could integrate, and these he collected in eighteen tables of integrals, published posthumously as 'Logometria part 2' in his collected works.[3]

Cotes begins a long scholium generale in the 'Logometria' by rectifying two pairs of curves which he says are related to each other in a similar way. These are the Archimedean spiral and the parabola (of Apollonius); and the reciprocal spiral (the name is due to Cotes) and the logarithmic curve. Why these particular curves and how did Cotes rectify them? Cotes does not reveal his methods, nor mention the integrals, but simply states the lengths of the curves in geometrical terms. The first pair of curves will serve as an illustration (Figure 1). The arc length AP of the parabola, vertex A, ordinate PQ, focus F, L the mid-point of PQ is given as $AL + LM$ where

$$LM = AF \cdot \ln[(LA + AQ)/QL].$$

[1]   Roger Cotes, 'Logometria' in *Harmonia Mensurarum*, ed. R. Smith (Cambridge, 1722), pp. 4–41.   [2]   Roger Cotes, *ibid*, p. 28.

[3]   Roger Cotes, *Harmonia Mensurarum*, pp. 43ff.

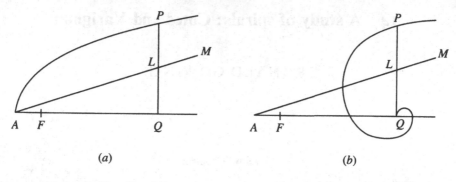

(a)   (b)

Arc length $AP = AL + LM$ Arc length $QP = AL + AM$

Figure 1. From R. Cotes, *Harmonia Mensurarum*, Cambridge University Press, 1722.

The arc length QP of the Archimedean spiral, pole Q, L the mid-point of QP, LA parallel to the tangent at P, F a point such that QL is a mean proportional between AF and AQ is also $AL + LM$ where LM is as defined above, but with reference to the second figure.

Such are Cotes' correct solutions stated with no indication of method. If we take the equation of the parabola as $y^2 = ax$, the arc length AP is

$$(1/a)\int_0^y (a^2 + 4y^2)^{\frac{1}{2}}\,dy,$$

an integral which Cotes can clearly do as

$$(y/2a)(a^2 + 4y^2)^{\frac{1}{2}} + (a/4)\ln\{[2y + (a^2 + 4y^2)^{\frac{1}{2}}]/a\}$$
$$= AL + LM \quad \text{as given by Cotes.}$$

Taking the equation of the Archimedean spiral as $r = \frac{1}{2}a\theta$, calculation of the arc length QP leads to the same integral, once again expressible as $AL + LM$ as defined above.

In the second pair of curves (Figure 2), the arc length of the reciprocal spiral $r\theta = a$ is

$$\int_{r_1}^{r_2} (1/r)(a^2 + r^2)^{\frac{1}{2}}\,dr$$

and that of the logarithmic curve $x = a\ln y$ is

$$\int_{y_1}^{y_2} (1/y)(a^2 + y^2)^{\frac{1}{2}}\,dy$$

and Cotes gives similar geometric expressions for the lengths in both cases, details of which need not detain us here.

The answer to the question 'How did Cotes rectify the curves?' is clearly, by means of his tables of integrals, which he refers to at one point as 'Certain

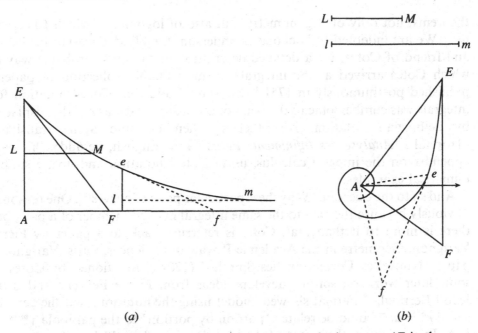

Figure 2. Arc length $Ee = (EF - LM) - (ef - lm)$ in both cases. AF is the subtangent, constant in both cases. In (a) $AL = EF - AF$. In (b) LM is computed as $AF \ln[AE(EF - AF)]$. From R. Cotes, *Harmonia Mensurarum*, Cambridge University Press, 1722.

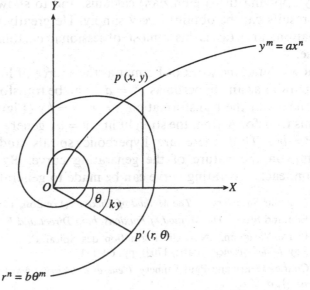

Figure 3. $P(x,y)$ on $y^m = ax^n$ transforms to $P(r,\theta)$ on $r^n = b\theta^m$.

theorems, not only of trigonometry, but also of logometry, which I keep by me.' We are indebted to Nicholas Saunderson, the blind Lucasian Professor and friend of Cotes, for a detailed description of the most probable way in which Cotes arrived at the integrals, in his valuable collection of papers, published posthumously in 1751.[4] One should add that Cotes' notation for integrals was cumbersome to the point of absurdity, and was rapidly displaced by Leibnizian notation. Nevertheless, when Edmund Stone translated l'Hôpital's '*Analyse des Infiniments Petits*'[5] into English, he added his own appendix on the integral calculus, using Cotes' notation and giving simple numerical examples.

And so to the question 'Why these particular pairs of curves?'. One reason is obviously because they led to the same integral in each member of a pair; but there is more to it than that. Cotes is referring back to a paper by Pierre Varignon, Géomètre in the Académie Royale des Sciences, Paris. Varignon's paper 'Nouvelles Formation des Spirales' (1704),[6] mentioned by Cotes in some later work on spirals, develops ideas from Pierre Fermat[7] and from Jean I Bernoulli.[8] Fermat showed, modernising the notation, that the general spiral $r^n = a\theta^m$ could be related 'portion by portion' to the parabola $y^{m+n} = bx^m$. In particular, the spiral of Archimedes $r = a\theta$ could be related to the parabola $y^2 = bx$. Varignon, in his *mémoire*, and Fontenelle (in the *Histoire*[9]) comment that investigation of the properties of spirals by 'the methods of the ancients' had always been long and difficult. Varignon proposes to simplify the work by applying the (Leibnizian) calculus, and to show how all previously known results can be obtained very simply. He greatly widens the range of investigation, and (as is his almost obsessional custom) seeks for general formulae.

Acknowledging the work of Fermat as the source of his idea, he points out that all parabolas and hyperbolas $y^m = ax^n$ can be transformed to spirals $r^n = b\theta^m$ by means of the transformation $x = r$; $\theta = ky$ (Figure 3). In particular, under this transformation, the straight line $y = ax$ generates the Archimedean spiral $r = b\theta$. Thus there are hyperbolic spirals and parabolic spirals, depending on the nature of the generating curve. By varying the transformation, each generating curve can be made to generate many spirals, and

[4] Nicholas Saunderson, *The Method of Fluxions* (London, 1751).

[5] Edmund Stone, *The Method of Fluxions, both Direct and Inverse* (London, 1730).

[6] Pierre Varignon, 'Nouvelle Formation des Spirales', in *Mémoires de l'Académie Royale des Sciences* (Paris, 1704), pp. 69–131.

[7] Charles Henry and Paul Tannery, *Oeuvres de Fermat*, 4 volumes (Paris, 1891–1912), vol. 1, p. 207.

[8] Jean I Bernoulli, *Acta Eruditorum* (Leipzig, 1691), pp. 16, 17.

[9] Bernard Fontenelle, 'Sur les Spirales à l'Infini', in *Histoire de l'Académie Royale des Sciences* (Paris, 1704), pp. 47–57.

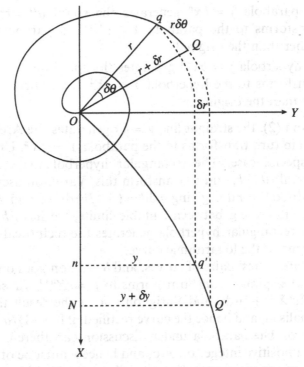

Figure 4. The spiral OqQ is transformed to the curve $Oq'Q'$. The distance nN is chosen so that $\text{Lim}_{\delta\theta\to0} nN = \text{Lim}_{\delta\theta\to0} qQ = r\,d\theta$.

indeed any curve could be used as generator. This leads Fontenelle to make one of his charming philosophical asides, 'Thus, there are more spirals than curves, although spirals are themselves curves, which is a paradox'. If the origin of the spiral is at the centre of a generating hyperbola, we have a *spirale hyperbolique asymptotique cocentrale*, of which, more later.

To investigate the properties of spirals, Varignon uses a device proposed by Jacques Bernoulli and reported by his brother Jean.[10] It is reminiscent of Fermat's work. The spiral developed from the parabola or hyperbola is subjected to a further transformation as follows (Figure 4). Neighbouring points q, Q on the spiral, coordinates (r, θ), are transformed to neighbouring points q', Q' on the new curve with coordinates (X, Y) such that $X = \int r\,d\theta$, $Y = r$.

There are three points of particular note about this transformation.

(1) $[(\mathrm{d}r)^2 + (r\,\mathrm{d}\theta)^2]^{\frac{1}{2}} = [(\mathrm{d}X)^2 + (\mathrm{d}Y)^2]^{\frac{1}{2}}$

  i.e. arc length is preserved.

---

[10] Jean I Bernoulli, in *op. cit.* note (8), refers to work by J. Wallis, G. P. de Roberval and T. Hobbes.

(2)   The parabola $y = bx^n$ generates the spiral $a\theta = r^n$ which in turn transforms to the parabola $Y = cX^{n+1}$, a parabola of degree one higher than the original.

(3)   The hyperbola $y = bx^{-n}$ generates the spiral $a\theta = r^{-n}$, which in turn transforms to the hyperbola $Y = cX^{1-n}$, a hyperbola of degree one less than the original.

Note that from (2), the straight line $y = kx$ generates the Archimedean spiral $r = b\theta$ which in turn transforms to the parabola $Y = cX^2$. Under (3), $n = -1$ presents a special case. The rectangular hyperbola $xy = a$ generates the reciprocal spiral $r\theta = b$, and to transform this, Varignon uses the differential form $dX = dr$, $dY = r\,d\theta$, giving $r\,d\theta = (-1/b)\,dr/r$; and $b\,dY = -dX/X$; from which, integrating between suitable limits, we have $bY = \ln X$. Thus, from (3), the rectangular hyperbola generates the reciprocal spiral, which in turn transforms to the logarithmic curve.

Here then are Cotes' pairs of curves, and Varignon goes on to consider the arc lengths. The spiral $r = a\theta^n$ transforms to $y = cX^{n+1}$ whose element of arc ds is $[(n+1)^2 c^2 X^{2n} + 1]^{\frac{1}{2}} dx$ and Varignon quotes the result that $ey^p(f + gy^q)^r$ can be rationalised, and hence the curve rectified, if $[(p+1)/q - 1]$ is a positive integer or zero. The parabola under discussion can therefore be rectified if $(1/2n) - 1$ is a positive integer or zero, and hence n must be of the form $1/2m$, where m is a positive integer. Varignon does not investigate which curves, according to his result, can or cannot be rectified; and this is surprising since the object of the transformation is to deduce properties of spirals from properties of the transformed curves. (For example, areas are preserved, up to a factor of 2, the constant polar sub-normal of the Archimedean spiral goes over into the constant sub-normal of the parabola of Apollonius, and so on.) Possibly he doubted, quite rightly, the completeness of his result $[(p+1)/q - 1]$. Fontenelle has no such inhibitions, and declares firmly that the case $n = 1$ does not fulfil the required condition $n = 1/2m$, where m is a positive integer, and hence the parabola of Apollonius is not rectifiable and so neither is the Archimedean spiral. Similar arguments would apply to the logarithmic curve and the reciprocal spiral. Thus for the curve $Y = \ln X$, $ds = x^{-1}(1 + x^2)^{\frac{1}{2}} dX$ and this does not meet Varignon's requirement for rationalisation and hence rectification of the curve.

The curves therefore presented a challenge to Cotes, who is easily able to do the integrals, and to find interesting logarithmic properties in the results.

By the time Varignon came to write his *mémoire* on spirals, the properties of the logarithmic spiral were well-known. Christaan Huygens had investigated it thoroughly at least by 1673[11] and read a paper before the Académie Royale

---

[11]   Christiaan Huygens, *Horologium Oscillatorium* (Paris, 1673), in *Oeuvres Complètes*, vol. 18, p. 40.

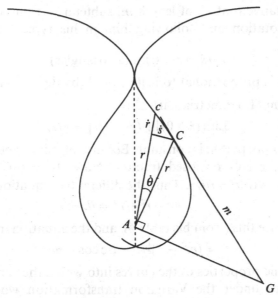

Figure 5. The spiral or enclosed tractrix. Arc length is proportional to the logarithm of the ratio of the bounding radii.

des Sciences, Paris on 3 September 1678, showing such a spiral to be self evolute. Newton used its geometrical properties to show that it was the curve described by a gravitating body under the inverse cube law of attraction,[12] and under the inverse square law in a suitably resistant medium.[13] Cotes developed the work of Edmond Halley (reported in the *Philosophical Transactions* for 1686) showing the logarithmic spiral to be the stereographic projection of a loxodrome (as a contribution to the longitude problem) and included it in his own investigation of the five spiral orbits obtainable under the inverse cube law – a paper which enabled him to display his prowess in integration in a bravura performance.[14] It was then quite natural for Cotes to pick up Varignon's comment that since in a spiral we deal with three elements, arc length, radius and angle, and these can be permuted in pairs in six ways, there are six logarithmic spirals, five of them new. Cotes investigates one of the five, and adds a seventh.

Of Varignon's five new spirals, Cotes finds *The spiral whose arc length is proportional to the logarithm of the ratio of the bounding radii.* This is problem 4 of 'Logometria part 3'[15] which Cotes was preparing for publication.

With reference to the diagram (Figure 5), *C,c* are neighbouring points on the

[12]   Isaac Newton, *The Mathematical Principles of Natural Philosophy*, trans. Andrew Motte, 2 vol. (University of California Press, 1966). Book 1, proposition 9, p. 52.

[13]   Isaac Newton, *ibid.*, Book 2, propositions 15, 16, pp. 282–8.

[14]   Roger Cotes, *Harmonia Mensurarum*, pp. 30–5.   [15]   Roger Cotes, *ibid.*

curve. *CG*, the tangent at *C*, of length *m*, subtends a right angle at the pole. Using Cotes' notation and following him in his typical semi-geometrical analysis,

$$m/\dot{s} = r/\dot{r} \text{ (similar triangles)} \tag{1}$$

$$\dot{s} \text{ is proportional to } \ln\left[(r+\dot{r})/r\right] \text{ (by definition).} \tag{2}$$

A key result from 'Logometria part 1' is

$$\text{Lim} \, (\dot{r} \to 0) \ln\left[(r+\dot{r})/r\right] = \dot{r}/r. \tag{3}$$

(1), (2) and (3) imply that *m* is constant. Because of this property, Cotes called the curve the spiral or enclosed tractrix. Now $AG = (m^2-r^2)^{\frac{1}{2}}$, and from similar triangles, $AG/r = r\dot{\theta}/\dot{r}$. Thus the differential equation of the curve is

$$\dot{r}(m^2-r^2)^{\frac{1}{2}}/r^2 = \dot{\theta}.$$

Cotes can integrate this (from his table 4), and the equation in modern form is

$$\theta = (m^2-r^2)^{\frac{1}{2}}/r - \arccos r/m.$$

The logarithmic properties of the curves into which the various logarithmic spirals transform under the Varignon transformation would be of great interest to Cotes. The transform of the spiral tractrix would be difficult and he does not seem to have attempted it. However the train of thought is clear. The logarithmic property of the rectangular hyperbola was well known – i.e. the *area* under the curve is proportional to the logarithm of the ratio of the bounding abscissae. Therefore this must be the transform of the spiral in which the sectorial *area* is proportional to the logarithm of the bounding radii. In Varignon's scheme, the hyperbola $x^2y = a$ generates the spiral $r^2\theta = b$ which transforms to the hyperbola $XY = c$. The required spiral is then $r^2\theta = b$. Cotes refers to this as Varignon's hyperbolic spiral for $m = 2$, (it is one of Varignon's *spirales hyperboliques asymptotique cocentrales*), and from the defining equation obtains the logarithmic property in the form $A = k\ln(r/r')$, i.e. sectorial area is proportional to the logarithm of the bounding radii. To Varignon's three elements, arc, angle and radius, Cotes has added a fourth, sectorial area. This would give twelve possible permutations in pairs, but he does not comment on this. The spiral under discussion (Figure 6) is the lituus, named by Cotes. *Lituus* is an Etruscan word meaning a magician's staff, the origin according to some authorities of the bishops' staff. In a final flourish, Cotes declares 'This curve is related to the hyperbola in the same way that the Archimedean spiral is related to the parabola and the reciprocal spiral to the logarithmic curve.'

One cannot fairly leave this essay without referring to the Cotes spiral. This is the path followed by a gravitating particle projected under a centrifugal force obeying the inverse cube law. It forms a sort of codicil to the earlier work on spiral orbits described under the inverse cube law of attraction. The work was done in connection with Cotes' able and industrious editing of the

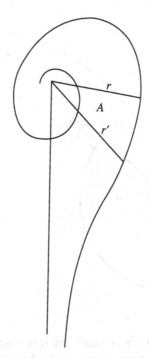

Figure 6. The lituus. The sectorial area is proportional to the logarithm of the ratio of the bounding radii.

Principia but also in my opinion owes something to Varignon's paper of 1700 'Du Mouvement en General'.[16] By 1700 Varignon had thoroughly studied the *Principia*. Indeed, when thanking Newton for his gift of a copy of the second edition in 1713, he wrote that he was particularly pleased to have it, since his own first edition had been so thoroughly annotated as to be almost illegible. Varignon said he had been thinking about spirals for some years, and in the 1700 paper 'Du Mouvement en General' he turns to the problem of finding the law of force under which a gravitating body will describe certain spirals. (His treatment of the inverse problem, given the law, find the orbit, has been very interestingly discussed by Eric Aiton.[17]) Varignon shows, as Newton had done, that the logarithmic spiral would be described under the inverse cube law of attraction, and then turns to seeking a general law (as is his custom) which would apply to all parabolic and hyperbolic spirals. Varignon had become adept in the use of the Leibnizian differential calculus, learned initially from

[16]   Pierre Varignon, 'Du Mouvement en General, par toutes sortes de Courbes, & des Forces Centrales', in *Mémoires de l'Academie Royale des Sciences* (Paris 1700), pp. 83–102.

[17]   Eric Aiton, 'The inverse problem of central forces', in *Annals of Science*, 20(1) (1964):81–99.

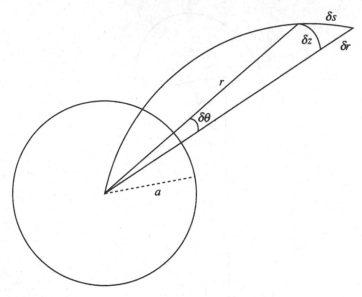

Figure 7

Jean Bernoulli during the latter's youthful visit to Paris in 1692, and developed in a sustained and fruitful correspondence which lasted until Varignon's death in 1722. Varignon's procedure can be summarised as follows.

From the general spiral $r^m = a\theta$, where a is the radius of the generating circle (Figure 7), we have

$$mr^{m-1}\,\mathrm{d}r = a\,\mathrm{d}\theta$$

also $\mathrm{d}z = r\,\mathrm{d}\theta$ (geometry)
and $r\,\mathrm{d}z = \mathrm{d}t$ (constancy of areal velocity).
From these we obtain

$$\mathrm{d}r = a\,\mathrm{d}z/mr^m.$$

Combining these results with $\mathrm{d}s^2 = \mathrm{d}r^2 + \mathrm{d}z^2$ we obtain

$$\text{the velocity}^2 = (\mathrm{d}s/\mathrm{d}t)^2 = (a^2/m^2r^{2m+2}) + 1/r^2.$$

Hence $v\,\mathrm{d}v/\mathrm{d}r$, and therefore the central force, is proportional to

$$[(m+1)a^2 + m^2r^{2m}]/m^2r^{2m+3}$$

and this is the general law for all parabolic or hyperbolic spirals of the form $r^m = a\theta$. In the case of the reciprocal or first hyperbolic spiral, $r\theta = a$, we have $m = -1$ and thus $v\,\mathrm{d}v/\mathrm{d}r$ is proportional to $1/r^3$, i.e. the spiral would be described under the inverse cube law. Varignon was possibly disappointed to find only one such spiral obeying such a simple law, but contents himself with observing that the spiral of Archimedes, in which $m = 1$, would be described under a force proportional to $(2a^2 + r^2)/r^5$, i.e. a combination of inverse fifth and inverse cube laws.

Cotes, in his investigation of the five spiral orbits described under the inverse cube laws finds the two (the logarithmic spiral and the reciprocal spiral), and three others, not of the parabolic/hyperbolic form; and shows clearly how the particular spiral described depends on the initial conditions of projection. It is unlikely that Cotes had not read Varignon's *mémoire*, but his work stems more directly from Newton's fairly brief discussion in *Principia*, 'Book 1, section viii, propositions XL, XLI, XLII' – on which see also Eric Aiton's paper referred to above; and also, D. T. Whiteside, *The Mathematical Papers of Isaac Newton* (Cambridge, 1967–81), vol. 6, pp. 435–8.

If Cotes had corresponded with Varignon, might we have known something?

15 The fragmentation of the European mathematical community

LENORE FEIGENBAUM

Acknowledgements

Archival research for this essay was made possible by the National Science Foundation, under Grant #SES – 8709820, whose support I gratefully acknowledge. Special thanks are also due to the College Council of St John's College, Cambridge, to the Library of the Royal Society of London, and to the Universitätsbibliothek and the Bernoulli Edition of Basel for permission to publish excerpts from the letters in their possession. Finally, my deepest appreciation goes to Tom Whiteside, for enriching our profession in unsurpassable ways with his exceptional scholarship, his daunting energy, and his unfailing generosity and encouragement.

When asked once in 1701 to give his opinion of Isaac Newton, Gottfried Wilhelm Leibniz replied that

> taking Mathematicks from the beginning of the world to the time of S^r I.
> What he had done was much the better half...[1]

Newton received a similar compliment a decade later from a select committee of the Royal Society, who used the occasion to insinuate that Leibniz had plagiarized Newton's calculus. Confronted with an earlier such charge by John Keill,[2] Leibniz had naively sought redress from the Royal Society, only to be told 'officially' that

> we reckon Mr Newton the first Inventor; and are of opinion that Mr Keill
> in asserting the same has been noways injurious to Mr Leibnitz.[3]

We know now that the committee responsible for this judgment not only relied on Newton for its direction, but produced a report composed almost

---

[1]   Quoted in Richard S. Westfall, *Never at Rest* (Cambridge University Press, 1983), p. 721. According to an eyewitness, Sir A. Fontaine, this was Leibniz' answer to the Queen of Prussia during dinner in the royal palace in Berlin.

[2]   Inserted in a paper on the laws of centripetal force in the *Philosophical Transactions* 26(1708):185, which appeared in 1710.

[3]   From the Royal Society report of April 24, 1712, reprinted in *The Correspondence of Isaac Newton*, vol. 5, ed. A. Rupert Hall and Laura Tilling (Cambridge University Press, 1975), p. xxvi

entirely by Newton.[4] As for the controversy itself, historians have sifted through enough of the unpublished papers to reach a conclusive verdict. Unlike the twentieth-century habit of destroying incriminating evidence, the early eighteenth-century penchant for saving compromising documents has enabled us to identify the crucial authors of anonymously published articles so that no doubt remains: Newton and Leibniz are considered the independent and relatively simultaneous inventors of the calculus.

Nor is there any doubt that some of the greatest mathematicians of the age that witnessed the adolescence of the calculus engaged in behavior that would cause even adolescents to blush. One can easily fall victim to historical voyeurism, watching with fascination and embarrassment as grown men, brilliant and powerful, betrayed their friends, lied shamelessly to their enemies, uttered hateful chauvinistic slurs, and impugned each other's characters, each convinced that the justness of his cause and the accumulation of enemy injustices could excuse any improper behavior.

D. T. Whiteside has aptly likened the priority dispute to a

> long-festering boil ... which came slowly to a head over many years ... broke suddenly open early in 1712 and ... polluted the whole European world for a decade afterwards with the corruption of its discharging pus.[5]

As this graphic description suggests, my emphasis will be epidemiological in part. Although the details of the quarrel and the personal and mathematical activities of the principal characters have been well documented,[6] much less has been done to chart the spread of the disease throughout the rest of the community. This is mainly due to the lack of evidence, hitherto unavailable or at least unexamined, and to the notoriety of the central priority dispute which has succeeded in capturing most of the historical headlines. If the controversy failed to corrupt, it managed to touch at least every European mathematician in some way. How that in turn affected mathematical activity, the transmission of scientific news and results, the production of textbooks, the actions and attitudes of lesser known and amateur mathematicians, personal relationships, and national conduct are all of interest if we seek a more comprehensive picture of the mathematical community during these years.

The focus here will be on a subgroup of this community, on three mathematicians whose professional lives were intertwined in a special way: the

---

[4] See *The Mathematical Papers of Isaac Newton*, vol. 3, ed. D. T. Whiteside (Cambridge University Press, 1969), p. 20, note 1.

[5] *Math. Papers*, vol. 8, ed. Whiteside (1981), p. 469.

[6] The most comprehensive account of the priority dispute, with extensive references to the sources and to other modern studies, is A. Rupert Hall's *Philosophers at War: The Quarrel Between Newton and Leibniz* (Cambridge University Press, 1980).

Englishman Brook Taylor (1685–1731), his chief mathematical opponent Johann I Bernoulli (1667–1748) of Basel, and the intermediary in their own secondary controversy, the French probabilist Pierre Rémond de Monmort (1678–1719). Since a sizable portion of their correspondence exists,[7] we are able to witness through contemporary eyes the distressing psychological, social, and professional consequences of the priority dispute on the larger community and to understand more fully the fragmentation of this group during those years.

It is not my intent, however, to focus exclusively on the events that divided mathematicians in the early eighteenth century. Behind the glamor of the Newton–Leibniz controversy an ultimately more powerful unifying development was taking place, namely the gradual formation of a professional group of mathematicians, the precursor of our modern mathematical community. From the latter half of the seventeenth century, European mathematicians began to view themselves as a working group with institutionalized connections. In addition to the growing number of university chairs and the traditional contact through private correspondence and public challenge problems, mathematicians of the late seventeenth and early eighteenth centuries communicated through the newly established journals, through scientific societies with their own corresponding secretaries, and through an active book trade, which included textbooks published in the vernacular by mathematically inclined amateurs.

Despite the intense antagonisms that accompanied the birth of the calculus, the remarkable progress it engendered and the universally shared confidence in its potency and fecundity all contributed to solidifying this emergent professional community. These were the first generations to apply the new calculus to a prodigious number of scientific and mathematical problems and to appreciate the extent of its applicability.

Leibniz' role in this was justly hailed in the first textbook to appear on the calculus, L'Hôpital's *Analyse des infiniment petits* of 1696:

> His Calculus has carried him into Countries hitherto unknown; and he has made Discoveries by it astonishing to the greatest Mathematicians of Europe. The Messieurs Bernouli were the first who perceived the Beauty of

---

7   Most of Taylor's letters, including those exchanged with Monmort, are preserved in the Library of the Royal Society of London (hereafter RS) and in the Library of St John's College, Cambridge (hereafter SJC). For a list of the published sources containing letters or extracts written to or by Taylor see Lenore Feigenbaum, 'Brook Taylor and the method of increments', *Archive for History of Exact Sciences* 34(1985):1–140, especially pp. 130–40. I am currently at work on an edition of Taylor's collected correspondence. The Universitätsbibliothek Basel (hereafter UB) contains the correspondence of the Bernoulli family, to be published as part of the monumental Bernoulli Edition.

the Method; and have carried it such a length, as by its means to surmount Difficulties that were before thought insuperable.[8]

Similar praise of Newton was echoed across the Channel several years later by one of the first English textbook writers, Charles Hayes:

What incredible advances natural Philosophy has received since it began to flourish is apparent to those that are but a little conversant in the profound Writings of the Great Mr. Newton, whose Immortal Genius will be a lasting Ornament to the English Nation; and the application of Fluxions in Astronomy, Mechanicks, Dioptricks, Catoptricks, Gunnery, Navigation &c. are numberless...[9]

The calculus provided the analytical tools for solving problems that had troubled or confounded the ancients. Calling Newton 'the Second Archimedes', Humphry Ditton told the readers of his text on fluxions what all of us try, in vain, to impress upon our calculus students:

this Method helps a Man to arrive at Truth, by a most easie, cheap, and compendious Way.[10]

Whether Newtonian or Leibnizian, these early practitioners all shared the optimism of explorers who believed they had discovered the key to unlocking the secrets of the natural world.

Thus within this broader context – of a professional group torn by personal and national hostilities and simultaneously brought closer by a common program and spirit – I want to examine more closely some aspects of the mathematical community as revealed in the correspondence of Taylor, Bernoulli, and Monmort.

No letters ever passed directly between Taylor and Bernoulli, however each corresponded with Monmort who had proclaimed himself neutral in the battle between the English and the Germans. As such Monmort provided not only an unusual link between the two adversaries, but an important connection between the Newtonian and Leibnizian camps in general.

And both camps used and abused it to full advantage. The steady communication back and forth across the Channel allowed Taylor to send challenge problems to Monmort for transmission to the mathematicians of Germany, Italy, and Switzerland; permitted Bernoulli to lie to Newton about his authorship of a stinging article; allowed both sides to learn about the latest scientific news, political and religious intrigues, and embarrassing personal gossip; gave Monmort the occasional opportunity to reveal a confidence he

[8] From the translation by Edmund Stone in *The Method of Fluxions Both Direct and Inverse* (London, 1730), p. viii.
[9] Charles Hayes, *A Treatise of Fluxions* (London, 1704), Preface.
[10] Humphry Ditton, *An Institution of Fluxions*, 2nd. edn. (London, 1726), Dedication.

had promised to keep; provided for the exchange of books and journals; facilitated the communication of threats, insults, and libel; and last but not least, made it possible for gifts of French wine and English cider to fall victim to the summer heat, rough seas, and more often than not, the greedy hands of customs agents.

What circumstances first brought Taylor, Monmort, and Bernoulli together?

We know that while Newton was in Cambridge, absorbed in composing the *Principia*, Taylor was born into an affluent family that owned extensive property near Canterbury. After graduating from Cambridge with a doctorate in civil law, Taylor moved to London – under no apparent pressure to secure employment – where he joined the group of scientists active in the Royal Society under Newton's presidency and became one of Newton's staunchest partisans. Only two weeks after his election as a Fellow in early 1712 he was chosen to serve on the committee charged with adjudicating the priority dispute. Appointed Secretary of the Royal Society in 1714, he resigned from the office in 1718 for reasons I shall return to later.

On a visit to Paris in 1715 he met Monmort, whose first impressions of the young man were reported to Nikolaus I Bernoulli:

> Mr. Brook Taylor, one of the Secretaries of the Royal Society is here and I see him often... This Mr. Taylor is a very polite young man who will splendidly sustain the reputation the English have acquired through their extensive knowledge of the profound geometry. Like all other Englishmen he is terribly biased in favor of Newton's philosophy.[11]

Upon Taylor's return to England the two maintained a close friendship – chiefly through correspondence but also through Taylor's occasional visits to France – that lasted until Monmort's death in 1719.

Monmort had studied the sciences and mathematics in Paris in the early 1700s as a disciple of the Oratorian Nicholas Malebranche, who actively promoted Cartesian physics and the Leibnizian calculus. A competent mathematician who was known for his contributions to the theory of probability, Monmort was nonetheless aware of his limitations, as he lamented to Taylor:

> I am very upset that my ignorance reduces me to the simple rank of spectator. I would really like the pleasure of being an actor in these intellectual battles in which so much honor is to be gained. It is disgraceful for France that we have no one capable of entering the lists with the English and the Germans.[12]

[11] Monmort to Nikolaus I Bernoulli, 14 November 1715, UB.
[12] Monmort to Taylor, 5 August 1718, SJC.

Whatever intellectual dissatisfactions troubled Monmort, these were more than compensated for by his social and financial status. A rich man with Court connections, a residence in Paris, and a chateau in Champagne, he, like Taylor, had no employment concerns, although this never prevented either from complaining about the lack of time to do mathematics. As Monmort would later remark to Taylor, at a time when Johann Bernoulli was overburdened with students and university duties as Professor of Mathematics in Basel:

> What you say, Sir, about the advantages that Mr. Bernoulli has over you is very true. We are geometers for our pleasure. These men do so for pleasure and by profession. They always have their heads full of figures and theorems, whereas a man of the world like you is distracted by business, pleasures, the obligations of the Royal Society, from which I conclude that there has to be a lot more talent and genius in a person like you than in a professor in order to make great progress in the sciences.[13]

Like Mersenne before him, Monmort could list the foremost mathematicians of the day as his correspondents. In 1703 he initiated a correspondence with Johann Bernoulli while the latter was in Groningen. Bernoulli had already impressed the Parisian mathematicians years earlier with his exceptional talent, having visited the savants gathered at Malebranche's home in 1691. As he later commented on the experience to Monmort:

> From a conversation I had there with the Marquis [de L'Hôpital], I soon realized he was a good geometer for the common geometry, but he knew nothing at all of the differential calculus, whose name he hardly knew, and still less... of the integral calculus.[14]

L'Hôpital in turn found it difficult to believe that the young man of twenty-four had succeeded in solving the problem of the catenary which had eluded Galileo. When Bernoulli demonstrated further that it was 'mere child's play',[15] as he called it, to compute the radius of curvature of a curve, L'Hôpital was charmed by the new calculus and delighted when Bernoulli agreed to instruct him in its secrets. After a period of informal lessons, L'Hôpital offered Bernoulli one of the most surprising arrangements in the history of mathematics, an arrangement made even more astonishing by Bernoulli's rapid acceptance of it:

> I will be happy to give you a retainer of 300 pounds... I promise shortly to increase this... which I know is very modest, as soon as my affairs are somewhat straightened out... I am not so unreasonable as to demand in return all of your time, but I will ask you to give me at intervals some hours of your time to work on what I request and also to communicate to me

[13]   Monmort to Taylor, 25 January 1718, SJC.
[14]   Johann Bernoulli to Monmort, 21 May 1718, UB.   [15]   *Ibid.*

your discoveries, at the same time asking you not to disclose any of them to others. I ask you even not to send here to Mr. Varignon or to others any copies of the writings you have left with me; if they are published, I will not be at all pleased.[16]

The subsequent events are not too surprising: in 1696 L'Hôpital published the first textbook on the differential calculus, based entirely on Bernoulli's discoveries but with only the following short acknowledgement in the preface:

> I must own my self very much obliged to the Labours of Messieurs Bernoulli, but particularly to those of the present Professor at Groenengen, as having made free with their Discoveries as well as those of Mr. Leibnitz: So that whatever they please to claim as their own, I frankly return them.[17]

Of course, when Bernoulli later tried to claim the book as his own, there were very few who would believe him.

Bernoulli's troubles with Taylor began many years later, but over an issue not unrelated to his grievance against L'Hôpital. In an article published anonymously in the *Acta Eruditorum* for 1716, Bernoulli came close to charging Taylor with plagiarism:

> It is nothing new with certain Englishmen, who think they alone are allowed with impunity to usurp the discoveries of others as if they were their own...There are examples like...Hayes, Taylor, and others who everywhere have used what the Bernoullis and others have discovered, either with no mention made of the author, or only naming him ambiguously in the preface, so that what properly belongs to him is not apparent from the context. This is especially noticeable in Hayes, who certainly stole the greater part of his book from the Bernoullis...[18]

In naming Taylor, Bernoulli was referring to the treatise Taylor had published the previous year entitled the *Methodus incrementorum*. An excessively concise and abstruse work, it contained not only his famous series but an impressive list of results in the theory of finite differences, differential equations, and their applications to the solution of several problems in mechanics. Except for an occasional reference to Newton, Taylor had failed to cite anyone else, although some of his problems had already been treated by continental authors, in particular by Johann Bernoulli and his older brother Jacob. The evidence shows that Taylor never committed plagiarism;[19] however he was grossly negligent in failing to acknowledge his predecessors and refused

---

16   L'Hôpital to Johann Bernoulli, 17 March 1694, translated in C. Truesdell, 'The New Bernoulli Edition', *Isis*, 49(1958):54–62, p. 61.

17   Translated by Stone in *Method*, p. xi.

18   'Epistola pro Eminente Mathematico', *Acta Eruditorum* (July 1716): 296–314, pp. 307–8. See footnote 26.

19   See Feigenbaum, 'Brook Taylor', especially pp. 63–72.

to take responsibility for his oversights. In defending himself against Bernoulli's charges in the *Philosophical Transactions*, Taylor offered only the following weak excuse:

> I have indeed treated of many things in common with others; but I have by no means used other men's inventions as my own...There may have been some little negligence in the matter, that being wholly intent on things of importance, I omitted some little historical matters.[20]

While Bernoulli was justified in upbraiding Taylor, he had much less reason to be upset with Charles Hayes, who had borne the brunt of his attack in the *Acta Eruditorum*. Unlike Taylor whose work was original, Hayes had written a textbook on fluxions for a much wider audience. As he explained in his preface:

> such a Treatise was wanting in the English Tongue, as should contain a full and plain account of the best Methods, the most celebrated Geometers of our Age have made use of in their wonderfull Discoveries...[21]

Disavowing any pretense to originality, Hayes had generously attributed the contents to a long list of luminaries:

> Dr. Wallis, Dr. Barrow, Mr. Newton, Mr. Libnitz, the Marquess De l'Hospital, Mess[rs.] Bernouilli, Mr. Craig, Dr. Cheyne, Dr. Gregory, Mr. Tchirnhaus, M. De Moivre, Mr. Fatio, Mr. Varignion, Mr. Newintiit, Mr. Carre...And tho' in the ensuing Treatise, the Author has made no scruple to borrow from any of those excellent Persons as occasion requires, yet he acknowledges himself more particularly indebted to Mr. Newton, Mr. Libnitz, the Mess[rs.] Bernouilli, the Marquess De l'Hospital and Mr. Craig...[22]

In light of the extraordinary tutoring deal made between Bernoulli and L'Hôpital, it is understandable that Bernoulli would find it difficult to distinguish between the obligations of a textbook writer and a creative mathematician in expressing indebtedness to one's predecessors. Even for us today the issue of attribution is not a simple one.

Our modern calculus textbooks, for example, present a host of techniques – partial fractions, the chain rule, areas inside polar curves, integration by parts, and so on – all without mention of the inventors, and all followed by problems handed down anonymously to each new generation of students, often from the time of Newton, Leibniz, and the Bernoullis. There is a certain validity in Taylor's argument to Monmort that

---

[20] Brook Taylor, 'Apologia D. Brook Taylor...contra...V. C. J. Bernoullium...', *Philosophical Transactions* 30(1719):955–63, pp. 957–8, translated in Charles Hutton, George Shaw, and Richard Pearson, *The Philosophical Transactions of the Royal Society, Abridged*, vol. 6 (London, 1809), p. 397.

[21] Hayes, *Treatise*, Preface.   [22] *Ibid.*

If greater niceness was necessary in all those who write after others, it would be a very great discouragement to any one to pretend to meddle wth a Subject that had been considered before...[23]

However who decides whether techniques or results require attribution or not?

Of course we should not forget the ultimate irony in this issue: the few results we do manage to attribute in today's calculus texts are often incorrectly cited. The Maclaurin Series, for example, is really no more than a special case of the Taylor Series, a fact duly noted by Maclaurin although not by his successors.[24] And L'Hôpital's Rule we know today is the invention of Johann Bernoulli.[25]

In view of the vehemence with which mathematicians of the early eighteenth century fought over priority and sought to claim authorship of unpublished works, it also seems ironic that a substantial amount of time and energy was devoted to publishing anonymous articles and to denying one's authorship. Given the small number of mathematicians capable of producing such papers, it took little effort to identify the authors, or to limit the possibilities to at most two or three.

Among the most controversial of the anonymous pieces was the 'Letter on Behalf of an Eminent Mathematician', which contained the accusations against Taylor and Charles Hayes mentioned above. Although the evidence points conclusively to Johann Bernoulli,[26] he passionately repudiated his role, admitting only to supplying some of the facts used by the anonymous author. Bernoulli's protestations notwithstanding, the English had no trouble determining the author on the basis of a small but telltale mistake: whereas Bernoulli had been referred to throughout in the third person, on the last page the editor had accidentally let slip the phrase 'meam formulam' which exposed him immediately.

If we recall Newton's part in the Royal Society report condemning Leibniz, Bernoulli would certainly seem to be Newton's match in deception. Nor does the similarity end there. As the foremost mathematicians of Europe after the death of Leibniz, Newton and Bernoulli were both confident in their talents, fiercely protective of their accomplishments, insistent upon unquestioning allegiance from their supporters, and frustrated and resentful at having been deprived of the credit each felt he justly deserved.

[23]   Draft of a letter from Taylor to Monmort, 18 March 1719, SJC.

[24]   Colin Maclaurin in *A Treatise of Fluxions* (Edinburgh, 1742), vol. 2, p. 611 found the Taylor Series about zero and then stated 'This theorem was given by Dr. TAYLOR *method. increm.*'

[25]   The rule was communicated in a letter of Johann Bernoulli to L'Hôpital, 22 July 1694, published in Johann Bernoulli, *Briefwechsel*, ed. O. Spiess (Birkhäuser, Basel, 1955), vol. 1, pp. 235–6. See D. J. Struik, *Mathematics Teacher* 56(1963): 257–60.

[26]   See *The Correspondence*, vol. 6, ed. Hall and Tilling (1976), pp. 303–4, note 1.

Neither suffered inferior intellects gladly. Just as Newton disdained 'little Smatterers in Mathematicks',[27] Bernoulli complained to Monmort about one of the early critics of the calculus:

> Mr. Rolle is a wretched intellect and he meddles in things that he understands no better than my cat; he attacked the integral calculus as ridiculously as a blind man trying to criticize a beautiful painting.[28]

Although Newton and Leibniz were the two contending for priority, it was Newton and Bernoulli who played parallel roles in the feud. While Newton secretly directed his own defense through his spokesperson John Keill, Bernoulli plied Leibniz with arguments, documents, and challenge problems to expose the supposed weaknesses in the fluxional calculus.

Each lived into his eighties, outlasting many friends and rivals. Newton survived Leibniz, while Taylor died well before Bernoulli at the age of 46. When he learned of Taylor's death, Bernoulli wrote to Maupertuis about the loss of his most prominent mathematical adversary in England:

> I did not know about the death of Mr. Taylor, but I suspected it when I did not see his name on the list of academicians that my nephew brought from England. He must already have been dead more than a year... Be that as it may, Taylor *is* dead. It is a kind of fate that my antagonists died before me, all younger than I. He is the sixth one of them to die in the last 15 years... Parent, Keil, Hartsoaker, Clark, Taylor, and de Louville; perhaps there are others whose death has not reached me. All these men attacked and harassed me, some more than others, though I did them no wrong. It seems that heaven would avenge the wrong they have done me.[29]

While the conflict between Taylor and Bernoulli is well known to historians of the period, the papers of Taylor, Monmort, and Bernoulli reveal an unexpected antagonism within the Newtonian camp itself, between Newton and his own supporter Taylor.

In 1718 Taylor resigned from his post as Secretary of the Royal Society, explaining to Halley that

> my circumstances for the future will be such that I shall not be in London long enough to discharge the duty of that Office so compleatly and so punctually as I ought to do...[30]

---

[27] Quoted in Westfall, *Never at Rest*, p. 459. Newton's remark was reported by William Derham to John Conduitt, 8 July 1733.

[28] Johann Bernoulli to Monmort, 25 November 1704, UB.

[29] Johann Bernoulli to Pierre Louis Moreau de Maupertuis, 27 November 1732, UB.

[30] Taylor to Edmund Halley, 21 October 1718, RS. The letter was printed in William Young, *Contemplatio Philosophica* (London, 1793), pp. 103–4.

We know now that Taylor was in fact pressured to resign so that Newton could install a more pliant ally, John Machin, in his place.[31]

As Newton once admitted in connection with the Mint, 'I am safest in people that are afraid of me.'[32] It would appear that Taylor was not sufficiently afraid of Newton, which caused the inner circle around the President of the Royal Society to consider him a threat. We learn this from Monmort who confided in Taylor:

> I was not aware that your merit...made many people jealous of you in England...You have to let it run its course...I heard a long time ago...about a kind of cabal against you; it was nothing, but I caught a glimpse of Mr. Newton...beginning to feel his age.[33]

A month later Monmort discussed the issue at greater length:

> I already had the honor of informing you that I suspect Mr. Moivre of putting me in a bad light with Mr. Newton. When I learned of...the cabal...against you concerning the secretariat of your academy, Mr. Moivre was accused of the same thing with regard to you, that is of having tried to put you in a bad light with Mr. Newton...inspiring him with suspicions and even a kind of jealously...You would do well to go no longer to Mr. Newton's house. It seems to me that you could not do so honorably...The Abbé Conty...is as surprised as I that Mr. Newton after all the attachment you have shown him...has failed you so absolutely.[34]

Despite the absence of additional evidence, one might surmise some of the reasons for Newton's resentment toward one of his most vocal adherents. Unlike Keill, who allowed himself to be Newton's sycophant, Taylor functioned more independently, promoting the English cause but conducting his own battles as he generally saw fit. Likewise in his mathematical work he demonstrated a large degree of independence from his master. Although the language and notation of Taylor's book were Newtonian, not only did he propose what he considered to be a more sound foundation for the fluxional calculus based on his finite increments, but the subject matter, physical applications, and the analysis he employed were all more characteristic of continental mathematics. As Monmort told Nikolaus I Bernoulli upon receiving a copy of Taylor's treatise from the author himself:

[31] On 8 June 1719, UB, Monmort wrote about Machin to Nikolaus I: 'Mr. Newton praises him, protects him, and has made him secretary of the Royal Society in place of Mr. Taylor, who is very displeased about it...' Nikolaus replied on 26 August 1719, UB: 'I know that this Mr. Machin is well protected by Mr. Newton and that...the latter is not too happy with Mr. Taylor because of the *Methodus Incrementorum*.'

[32] Quoted in Westfall, *Never at Rest*, p. 845. From the Mint Papers 19.2, f. 442.

[33] Monmort to Taylor, 30 January 1719, RS.

[34] Monmort to Taylor, 27 February 1719, SJC.

He wanted to entitle his book *Methodus differentialis* and this title would have been good, since he considers the change in quantities by means of their differences, whether finite or infinitely small. But this title would have displeased the English.[35]

We should remember that Taylor was in his early thirties and at the height of his creativity when Newton was in his mid seventies. Thus Newton may have been jealous of Taylor's success in solving a challenge problem, the so-called orthogonal trajectory problem, devised by Johann Bernoulli and proposed by Leibniz in 1716 in order 'to feel ... the pulse of the English'.[36] Although Newton had published his own solution anonymously, according to Whiteside

> he was for the first time in his life at a loss to choose an effective mode of approach, and instead he was put feebly and inadequately to uttering loose generalities ...[37]

Thus it fell upon Taylor to uphold the mathematical reputation of the English.

Whether Newton felt threatened by Taylor's abilities at that time is not a question I can answer. However we know that jealousy is not the most rational of reactions.

Indeed one begins to crave a little rationality in the midst of the ugly behavior and incessant tensions characteristic of the decade. One of the few who brought a refreshingly calm and balanced perspective to a community contaminated by prejudice, hostility, and deceit was the neutral Frenchman Pierre Rémond de Monmort.

As in the case of the orthogonal trajectory problem, the solutions to challenge problems were often sent to Monmort for safekeeping prior to publication. Having proven himself an impartial friend and – probably to his great dismay – a threat to no one, he had won the trust and respect of both sides. Unwilling to behave like his preeminent colleagues, he attempted to reconcile them privately and when that failed, to soften if possible their public exchanges. Caught in the middle between his two friends, Bernoulli and Taylor, he was unable to prevent a rupture between them but succeeded in being frank with both while alienating neither.

Although Taylor and Monmort were intimate friends, their association was not without difficulty owing to their strong disagreements about natural

---

[35] Monmort to Nikolaus I Bernoulli, 8 June 1715, UB.

[36] Leibniz to Johann Bernoulli, [December 1715], translated in *The Correspondence*, vol. 6, ed. Hall and Tilling (1976), p. 262.

[37] *Math. Papers*, vol. 8, ed. Whiteside (1981), p. 65. See Whiteside's discussion of Newton's response to the challenge problem in the same volume, pp. 425–34. Ironically, Bernoulli assumed initially that the anonymous solution came from Taylor, as he wrote in his last letter to Leibniz. See *Leibnizens Mathematische Schriften*, vol. 3, ed. C. I. Gerhardt (Halle, 1856), p. 972.

philosophy. Concurrent with the calculus dispute, the debate over Newtonian and Cartesian physical principles further divided the European mathematical community. However, unlike the mathematical conflict in which both sides acknowledged essential agreement in theory but not in form, the clash was between two mutually exclusive philosophies. While fluxions and differentials could easily coexist and be translated – too freely, I might add – one into the other, vortices and attractions were incompatible and further served to separate England from the Continent. Referring to the Newtonian theory of attractions, Monmort once characterized it as

> a national malady that no remedy can cure; thank God it has not crossed the sea...[38]

Hoping to convert Taylor, Monmort sent a long memoir defending Cartesian principles. Although Taylor remained unconvinced, it is a testament to the strength of their friendship that Monmort could later admit:

> I shall love you without loving your attractions, and you shall love me without loving our little vortices...[39]

Lasting friendships were unfortunately the exceptions during this period. As we saw in the case of the two allies Newton and Taylor, very few personal relationships escaped unscathed. Perhaps the most harmful psychological consequence of the belligerent climate was the extent to which it poisoned the conduct of ordinarily just and honest men.

As we learn from Monmort, Jacob Hermann, who was well known for his politeness, composure, and fairness, wrote an uncharacteristically harsh critique of Taylor's work on the vibrating string. In response to Monmort's displeasure with Hermann's tone and approach, Hermann excused himself by arguing:

> The liberty that Mr. Keill has taken with regard to me has made me treat Mr. Taylor in a similar way.[40]

In a subsequent letter to Taylor, Monmort commented on Hermann's personality:

> If Mr. Keil and Mr. Clark have acted badly toward him is it fair that you suffer?...I know from several people who know Mr. Hermann well that he is the sweetest and most pleasant man in the world. My friend Nikolaus Bernoulli told me a very curious thing about him...namely that from the moment he first came into this world he has never gotten angry...[41]

[38] Monmort to Nikolaus I Bernoulli, 31 December 1718, UB.
[39] Monmort to Taylor, 5 November 1718, SJC.
[40] Jakob Hermann quoted in Monmort to Taylor, [5] August 1718, SJC.
[41] Monmort to Taylor, [28] September 1718, SJC.

At least in the case of Hermann and Taylor, Monmort was successful in effecting a reconciliation.

If the one form of communication desperately needed during these troubled years was dispassionate debate, then it is a tribute to Monmort's character that he accomplished it so often with each side. To be sure, he was naive in assuming that Taylor and Newton could be converted to Cartesian physics; however his assessment of the calculus priority controversy, which he discussed at length with Taylor and the Bernoullis, was remarkably astute, and what is more, historically accurate. As he frankly explained to Taylor:

> I have always resolutely maintained that Mr. Newton was master of the differential and integral calculus before any other geometer and that from the year 1677 he knew all that the works of Mr. Leibniz and the Bernoullis have discovered since... It is true that Mr. Leibniz published the rules of our calculus first and it is an advantage he has over Mr. Newton. In my opinion the Bernoullis also have some rights to the title of inventors. They have filled the journals of Leipzig with a thousand lovely things. These journals have trained geometers well, who would find nothing comparable anywhere else...[42]

With the Leibnizian camp Monmort was no less honest. To Nikolaus I Bernoulli he remarked:

> They would do best to recognize each other as coinventors... Mr. Leibnitz published first... I also think that from the year 1676 Mr. Newton possessed all that we have now of this calculus. I have seen incontestable evidence in the old books whose dates cannot be held suspect.[43]

In keeping with his commitment to do justice to all those who had contributed to the development of the calculus, Monmort had undertaken to write a 'History of Geometry', a project never completed because of his untimely death, but one for which he made the following promise:

> I will not try to please interested parties, nor fear displeasing them; I will speak of the living as I would if they were dead, and in particular, in speaking of a man, I will forget his country.[44]

Since one of the greatest challenges for the historian of early eighteenth-century mathematics is to uncover instances of kind and decent behavior, it is especially pleasing to find an example in Monmort's correspondence. The incident described took place between a Newtonian and a Leibnizian in northern Italy – and it is perhaps in a similar spirit of friendship and respect that we toast the man we honor here with these essays. Nikolaus I Bernoulli,

[42] Monmort to Taylor, 22 June 1717, SJC.
[43] Monmort to Nikolaus I Bernoulli, 15 March 1716, UB.
[44] Monmort to Johann Bernoulli, 14 September 1717, UB.

the nephew of Johann Bernoulli and a professor at Padua, and the Scotsman James Stirling visited one another at the latter's residence in Venice. As Nikolaus later told Monmort:

> I drank with Mr. Sterling to the health of Mr. Newton and to that of my uncle, and we emptied a glass of English beer together for peace between the English and the Germans...[45]

[45] Nikolaus I Bernoulli to Monmort, 15 October 1718, UB.

16   Euler on action-at-a-distance and fundamental equations in continuum mechanics

CURTIS WILSON

Euler's contributions to astronomy based on the inverse-square law are numerous and important. The method of trigonometric series that he introduced in a prize essay of 1748,[1] and the method of variation of orbital constants that he published in an appendix to his lunar theory of 1753,[2] became the mainstays of celestial mechanics; without them Lagrange's and Laplace's achievements in astronomy would be unimaginable.

Other Eulerian innovations include: discovery of induced eccentricity in a perturbed orbit, 1752;[3] derivation of the secular variation in the obliquity of the ecliptic, 1755;[4] first determination of the planetary perturbations of the Earth, 1756.[5] His final lunar theory,[6] completed with the aid of assistants after he had become totally blind, introduced the rotating coordinates used a century later by G. W. Hill in the theories of the outer planets.

These innovations were premissed on the inverse-square law of gravitation. During the 1740s Euler had doubted its exactness – doubted whether all anomalies, and in particular, the whole 40° of annual motion of the Moon's apse, were derivable from it. By April 1751, with Clairaut leading the way, he had become persuaded that the apsidal motion was so derivable.[7] Henceforth, he took the inverse-square law as working hypothesis.

Yet Euler disbelieved in action-at-a-distance. The full import of his disbelief

---

[1]   *Piece qui a remporté le prix de l'Académie Royale des Sciences en 1748 sur les inégalités du mouvement de Saturne et de Jupiter* (Paris, 1749); printed in Euler, *Opera Omnia* II *25*, pp. 45–157. The *Opera Omnia* will hereafter be cited as *OO*.

[2]   *Theoria motus lunae* (1753), Additamentum, in *OO* II *23*, p. 283ff.

[3]   *Recherches sur les irrégularités du mouvement de Jupiter et de Saturne*, printed in *Recueil des pièces qui ont remporté les prix de l'Académie des Sciences*, vol. VII (Paris, 1769).

[4]   'De la variation de la latitude des étoiles fixes et de l'obliquité de l'écliptique', read at the Berlin Academy on 24 April 1755, and published in *Mémoires de l'Académie des Sciences de Berlin 10* (1754), 1756, 296–336.

[5]   *Investigatio perturbationum quibus planetarum motus ob actionem eorum mutuam afficiuntur*, in *Recueil des pièces qui ont remporté les prix*, vol. VIII (Paris, 1771).

[6]   *Theoria motuum lunae...* (Petersburg, 1772); *OO* II *22*.

[7]   Euler to Clairaut, 10 April 1751, in *OO* IVA, vol. 5, pp. 206–7.

has sometimes been missed.[8] According to Euler, all actions over distance, whether interplanetary or submicroscopic, must be the result of forces of contact. Gravitation must be due to the action of a subtle matter that does not itself gravitate.

But, you may object, wasn't the eighteenth century the era for subtle fluids – phlogiston, latent heat, caloric, electric and magnetic fluids, and so on? In invoking such a fluid, wasn't Euler merely following the fashion of the day?

Among subtle fluids we had better distinguish. When Newton in the second English edition of his *Opticks* (1717) tentatively proposed an aethereal cause for gravity, his intent was not to avoid action-at-a-distance, but to show that he did not 'take Gravity to be an essential Property of Bodies'; the particles of his aether mutually repelled one another.[9] From the 1740s onwards it was a rare thing for an aethereal mechanism to be proposed with the intent of avoiding action-at-a-distance; more commonly the mechanism explicitly or tacitly presupposed such action. The Newtonian pattern had set the style.

For Euler, early and late, action-at-a-distance was to be rejected. During the 1730s he believed this rejection to be in accord with the views of 'all physicists following reason and experience'.[10] During the 1740s, he came to realize that his was an embattled view. During the 1750s and 1760s he continued to defend it. It does not appear that he ever changed his mind. He was the last great partisan of the doctrine of action-by-contact-alone, in the direct line that begins with Descartes and continues through Huygens and Leibniz to the Bernoullis. The attempt, by William Thomson and Maxwell during the 1860s and 1870s, to derive all physical effects from contiguous action through a universal aether, would have warmed his heart.

To more than one contemporary, this stand seemed odd. Daniel Bernoulli, writing Euler in February 1744, confessed:

> I cannot hide from you that on this point I am a complete Newtonian, and I marvel that you so long adhere to the Cartesian principles; perhaps some passion is mixed with it. If God could create a soul whose nature is incomprehensible to us, so could he impress a universal attraction on matter, even if such an attraction is beyond our comprehension, for the

8 B. L. van der Waerden in 'Eulers Herleitung des Drehimpulssatzes', *Leonhard Euler, 1707–1783: Beiträge zu Leben und Werk* (Basel: Birkhauser Verlag, 1983), 271–81, hypothesizes that Euler, like Poisson later, thought of the law of moment of momentum as deriving logically from the law of linear momentum. But Poisson had to assume that the parts of a body are attracted to one another by central forces, whereas Euler denies the existence of such forces.

9 Newton, *Opticks* (Dover, cop. 1952), Advertisement II and Query 21.

10 Euler, *Dissertatio de igne*, para. 2; printed in *Recueil des pièces qui ont remporté les prix*, vol. IV (Paris, 1752).

Cartesian principles, too, always involve something beyond our com-
prehension.[11]

In June 1744 Bernoulli explained more fully why he thought attraction
should be granted. Any system of bodies in motion would, because of mutual
impacts, disperse itself in infinite space, were there not some principle to
counteract the dispersal.[12] Euler appeared to concede the importance of this
argument.[13] But in fact his opinion did not change.

During the later 1740s and early 1750s, in a series of memoirs, Euler
elaborated a metaphysical defence of his position. This involved explications
of the nature of body, force, space and time, and a strict dichotomy between
body and soul. His opinions on these matters were reiterated in his *Letters to
a German Princess*, published in Petersburg in 1768. To his contemporaries, his
continuing commitment to these opinions could seem stubbornly naive.

D'Alembert writing to Lagrange in June 1769 mentions wanting to see
Euler's *Letters to a German Princess*. 'According to what you say, it is his
Commentary on the Apocalypse [the allusion is to Newton's work on the
Apocalypse]. Our friend Euler is a great analyst, but a very bad philosopher.'[14]
Lagrange, replying in August, remarks that the work has no other merit than
that of coming from the pen of a great geometer.[15] D'Alembert, responding a
few days later, agrees:

> You are right to say that for the sake of his honor he should not have had
> this work printed. It is unbelievable that so great a genius as he in geometry
> and analysis should be in metaphysics so inferior to the least schoolboy,
> not to say so dull and absurd, and it is a case of the saying that the gods do
> not shower all gifts on the same man.[16]

Whence the condescension of these savants? The *Letters to a German
Princess* met with enormous success; for seventy years it was one of the most
popular of popularizations of science. By 1800 it had been translated from the
original French into eight other European languages; by 1840 it had gone to
over forty editions.[17] In it Euler had clearly indicated the tenets of his simple
Christian faith – love of God and man, belief in salvation and a future life as
guaranteed by biblical revelation; and no doubt d'Alembert the Encyclopedist

---

[11] P. H. Fuss, *Correspondance mathématique et physique de quelques célèbres géométres
du XVIIIème siècle* (Petersburg, 1843), II, pp. 550–51 (Lettre XXXII).

[12] *Ibid.*, pp. 557–58 (Lettre XXXIV).   [13] *Ibid.*, p. 563 (Lettre XXXV).

[14] *Oeuvres de Lagrange* (ed. J.-A. Serret), vol. 13 (Paris, 1882), p. 135.

[15] *Ibid.*, p. 145.   [16] *Ibid.*, p. 147.

[17] Ronald Calinger, 'Euler's "Letters to a Princess of Germany" as an expression of his
mature scientific outlook', *Archive for History of Exact Sciences* 15 (1975–76), p. 215.

and friend of Voltaire shared none of this, but there is no sign that he wished or dared to ridicule Euler's religious beliefs. D'Alembert's objections had to do with metaphysics. And Euler's disquisitions on metaphysics in the *Letters* had as their central concern, if I am not mistaken, the foundations of mechanics: the nature of matter, force, space, and time.

D'Alembert had written on these topics in the *Encyclopédie*. His opinions on them were different from Euler's. Of the two sets of opinions, I suspect we shall find ourselves more in tune with d'Alembert's. In the article on 'Attraction' in Vol. I, which appeared in 1751, d'Alembert says:

> When we see that two separated bodies approach one another, we should not be pressed to conclude that they are pushed toward one another by the action of an invisible fluid or other body, until experience has demonstrated it, as happened with the phenomena that the ancients attributed to the horror of a vacuum, and which we now recognize as the effect of air pressure....
>
> As we find weight in all bodies we can observe, we are justified in concluding by one of the accepted rules of philosophy that it is found in all others; moreover, as we find it to be proportional to the quantity of matter in each body, it must exist in each part of every body; and hence it is a law of nature that each particle of matter tends toward every other.... How wrong are those modern philosophers who proudly declare themselves opposed to the principle of attraction without giving any other reason than that they cannot conceive how one body can act on another which is distant from it.... Nothing is wiser and more in agreement with the true philosophy than to suspend our judgment on the nature of the force that produces these effects.[18]

But d'Alembert's positivism is not above reproach. In his *Traité de Dynamique* of 1743 he had claimed to found mechanics on principles 'necessarily true and evident in themselves',[19] and he had further claimed to eliminate all 'motive causes', drawing the principles from the idea of movement alone.[20] But his attempted derivation of mechanics is logically flawed[21] – a fact of which he remained blithely oblivious. If we, too, reject Euler's metaphysics, we had best have our own basis for doing so.

In the next two sections I shall review, first, Euler's attempts in the 1730s and early 1740s to construct physical theories for fire, light and color, magnetism, and gravity, and secondly, his efforts in the late 1740s and early 1750s to articulate and defend a metaphysical foundation for mechanics. In the final

[18] D'Alembert, art. 'Attraction', *Encyclopédie* I (Paris, 1751), pp. 847, 849, 850.

[19] D'Alembert, *Traité de Dynamique* (Paris, 1743), Preface, p. i.   [20] *Ibid.*, p. xvi.

[21] See the analysis given by Thomas L. Hankins in his *Jean d'Alembert, Science and the Enlightenment* (Oxford: the Clarendon Press, 1970), p. 152ff., esp. p. 165.

section I shall be concerned with the relation, heuristic rather than logical, between Euler's metaphysics and the fundamental equations of continuum mechanics that he was the first to formulate.

Euler on fire, light and color, magnetism, and gravity

In 1738 Euler received one-third of the prize in the Paris Academy's contest on the nature and propagation of fire.[22] At the start of his essay he announces that all effects of nature are determined by matter and motion; on this principle all physicists following reason and experience agree. To discover the causes of particular natural phenomena, hypotheses are necessary, and here the difficulty arises that more than one hypothesis may satisfy the phenomena. To reach truth in the surest way, Euler advises, we should try to select a phenomenon that can be explained only in a single way. Such a phenomenon is the propagation of fire.

This phenomenon appears contrary to the laws of nature and motion, for an effect is always proportional to its cause, yet the least spark can ignite the greatest fire. The only possible resolution of the paradox, Euler asserts, is that the particles of combustible materials contain, in confinement, vast quantities of a highly compressed elastic (that is, expansive) fluid. When just a few of these particles are shattered, the release of flying bits of matter moving at high speed causes other particles to shatter as well, and so fire spreads.

Euler goes on to explain the flame, heat, and light produced by fire, and argues that the matter of fire is distinct from the aether which serves to convey light. The crux of his theory lies in his insistence that wherever there appears to be a multiplication of force, it is only apparent; the forces are but hidden. And, says Euler, if we are not to flee to occult qualities, these hidden forces can only be elastic forces, which, he further claims, derive their origin from centrifugal force. He concludes that the hypothesized structure agrees with the laws of nature, satisfies the phenomena, and is indeed the only one possible.

In a letter to Euler of 20 January 1742, Daniel Bernoulli writes disapprovingly of Euler's use here of 'the method of exclusion'.[23] And in the same letter he objects on the same grounds to another of Euler's essays, an essay on the cause of the tides for which Euler received one-third of the prize in the Paris contest of 1740.[24] The main body of the essay is mathematical, and for this Bernoulli has the highest praise. But when Euler pronounces concerning the physical cause of gravitation, Bernoulli takes exception.

[22]   *Dissertatio de igne.*   [23]   P. H. Fuss, *Correspondance*, pp. 481–2 (Lettre XXII).

[24]   Euler, *Inquisitio physica in causam fluxus ac refluxus maris*, as printed in Le Seur et Jacquier, *Philosophiae naturalis principia mathematica; auctore Isaaco Newtono...* (Grenoble, 1740), III, pp. 283–374.

Euler had begun by asserting that 'occult qualities and imaginary causes are now completely rejected',[25] then gone on to insist that only two possible explanations of gravity are admissible: either there is a continual flux of subtle matter into the central sun or planet, bearing down with it all bodies it passes through, or there is a vortex about the sun or planet, the centrifugal force of which causes gross bodies to accelerate towards the center. The first hypothesis entails a force moving aether to the center, and then a mechanism for evacuating it from the center, and so is implausible. Therefore, 'no other cause of these forces is left except one consisting in centrifugal force.'[26] The vortices must be so disposed as to yield an inverse-square force toward the center, but Euler does not enter into this problem, claiming it has been sufficiently explained by others. He perhaps has in mind the essay by his former teacher, Jean Bernoulli, 'New Thoughts on the System of Descartes', which won the prize of the Paris Academy for 1730, and which argues that a vortex can give rise to an inverse-square field.[27]

To these assertions Daniel Bernoulli, son of Jean Bernoulli, reacted sharply: 'I was astonished that you spoke with such great praise of the vortices, indeed claimed that they had been demonstrated, and that the truth could not possibly be otherwise...I believe I have gone as far as anyone else in the examination of the hypothesis of vortices, and they still appear to me quite apocryphal...'. Bernoulli goes on to point out that the centrifugal force if varying inversely as the square of the distance would be greater at the center, so that the vortex would be unstable.[28]

We do not have Euler's responses, but from Bernoulli's next letters[29] we know that a discussion of the objections to vortices ensued. Euler soon abandoned the vortex theory of gravitation, as becomes apparent from his *Nova theoria magnetis*, which won part of the prize of the Paris Academy for 1744.[30]

'I have never doubted', Euler begins, 'that all effects of nature derive from mechanical causes, and I always preferred to confess my ignorance openly

[25]   *Ibid.*, p. 283.   [26]   *Ibid.*, p. 290.

[27]   Jean Bernoulli, 'Nouvelles Pensées sur le Système de M. Descartes', printed in Johannis Bernoulli, *Opera Omnia* (Lausanne and Geneva, 1742), III, pp. 131–73. According to Bernoulli, Newton's attempted refutation of vortices in Props. 52 and 53 of the *Principia* rests on incorrect analyses of both friction and rotational motion. Bernoulli's notion of frictional forces is wrong, although consistent with his denial of forces of attraction. He is right in insisting that a moment-arm enters into the correct analysis; see Horace Lamb, *Hydrodynamics*, 6th edn (Dover, 1945), pp. 587–9.

[28]   P. H. Fuss, *Correspondance*, pp. 481–2 (Lettre XXII, 20 Jan. 1742). Bernoulli had made the same point in his *Hydrodynamics*, Ch. XI, para. 6.

[29]   P. H. Fuss, *Correspondance*, pp. 485–7, 490–2 (Lettres XXIII and XXIV).

[30]   Euler, *Nova theoria magnetis...*, in *Opuscula*, III (Berlin, 1751).

rather than to say that anything happens without a cause.'[31] 'There is no doubt that all effects of the magnet arise from a corporeal or mechanical cause.'[32] As Euler acknowledges, his theory has features in common with the theory given by Descartes a hundred years before.[33] The body of any magnet, including the Earth, is imagined to contain parallel channels of uniform cross-section; Euler equips them with valves or fibrils so oriented as to allow a fluid to pass through only in one direction. The fluid that is able to enter and traverse the channels is an especially fine-grained subspecies of aether.

According to Euler, aether consists of tiny vortices of different sizes, the smaller ones filling in the angular spaces between the larger ones. The idea of composing the aether of vortices stems from Malebranche, who introduced it in 1699 to supply the lack in Descartes's system of a believable explanation of solidity. The vortices, being elastic or expansive because of the centrifugal force due to their rotations, compress gross bodies and so give them their hardness.[34] Euler retains this explanation of solidity. He nevertheless assumes all gross bodies to be permeable to all varieties of aether, one reason being that ordinary gross bodies do not form magnetic shields.

But the special channels in magnets can be traversed only by the smaller aethereal vortices. These smaller vortices, under the general pressure of the aether, are forced into the channels and traverse them at high speed; in emerging at the exits, they do not at once mix with the general aether, but are deflected to the sides, moving round the magnet but more slowly, so that the stream widens, and there is some mixing here with the general aether. Farther on, the stream narrows once again as it approaches the entrances to the channels, where it is once again pressed in at high speed. Euler sees the trace of the flow in the pattern that iron filings form round a magnet.

But whence the force causing this vehement aethereal circulation? It comes, Euler replies, from the elastic force of the general aether; this force must then be weakened in the neighborhood of magnets, including the Earth. Here, Euler decides, is the explanation of gravity. Let the elasticity or pressure in the general aether be E; a body Pp is placed at the distance OP from the center of the Earth O, its thickness Pp being negligible with respect to the distance OP. (See the figure.) It seems probable enough, Euler says, that the diminution of

---

[31]   *Ibid.*, p. 2.   [32]   *Ibid.*, p. 4.

[33]   See Descartes, *Principles of Philosophy* (tr. V. R. Miller and R. P. Miller; Dordrecht: D. Reidel, 1984), pp. 242–72.

[34]   See Paul Mouy, *Le développement de la physique Cartésienne, 1646–1712* (Paris, 1934), pp. 289–90. Whether Euler read any edition of Malebranche's *Recherche de la vérité* is unclear, but he undoubtedly read Jean II Bernoulli's 'Recherches physiques et géométriques sur la question: comment se fait la propagation de la lumière', in *Recueil des pièces qui ont remporté les prix*, vol. III (Paris, 1736), in which Malebranche's theory of the constitution of the aether is described.

the aether's elastic force should be inversely proportional to the distance from the center of the Earth. Then the pressure downward on the top of the body at P will be $E - A/OP$, where A is a constant, and the pressure upwards on the bottom of the body at p will be $E - A/Op$, which is smaller, so that the net pressure downwards will be given by $A/Op - A/OP = A \cdot Pp/Op \cdot OP$. Thus this downwards pressure is inversely proportional to the square of the distance from the center. The other planets and the Sun, Euler proposes, are similarly endowed with magnetic vortices. But the aethereal motion around the outside of the celestial magnets is so slow as not to develop detectable centrifugal effects; hence the theory is free from the objections to which the vortex theory of planetary motion is subject.

For this theory Euler claims so high a degree of probability that it amounts nearly, he says, to certitude.[35]

After reading Euler's essay, Daniel Bernoulli wrote:

> Your essay is deficient in a single point, for you suppose no force, or rather a false force, to cause the subtle magnetic matter to circulate through the magnetic channels with great rapidity. You would not suppose it possible to construct a hollow cylindrical tube of any figure, length, and thickness, equipped in whatever way with valves, through which elastic air would circulate spontaneously. You understand this as well as I.... At many points you use expressions implying that you view this theory as fully demonstrated. Since, however, I know very well your deep insight, I ascribe this to pure politics.[36]

By the end of 1751 Euler was admitting that his explanation of gravity was doubtful, because the tails of comets, in pointing directly away from the Sun, failed to indicate motion in the aether.[37] But in his faith that gravity was due to *some* aethereal mechanism, he remained undaunted.

One of the reasons Euler remained confident that an aether existed and filled cosmic space, was that only thus, in his opinion, could the transmission of light be plausibly explained. An early formulation of his ideas on this subject, 'Pensées sur la lumière et les couleurs', was presented to the Berlin Academy on 6 February 1744; his *Nova theoria lucis et colorum*, differing in some respects from the earlier paper, was published in 1746.[38] Euler opposes

---

[35] Euler, *Nova theoria magnetis*, p. 8.

[36] P. H. Fuss, *Correspondance* p. 644 (Lettre LIV, without date, from early 1749).

[37] Eric G. Forbes, *The Euler–Mayer Correspondence (1751–1755)* (New York: American Elsevier, 1971), p. 44.

[38] The *Nova theoria lucis et colorum* was published in *Opuscula varii argumenti*, I (Berlin, 1746), pp. 169–244. On the 'Pensées' of 1744, and its difference from the later work, see R. W. Home, 'Leonhard Euler's "Anti-Newtonian" theory of light', *Annals of Science 15* (1988): 521–33, especially p. 529.

Newton's emission theory of light. The continuous emission of light corpuscles from the Sun ought soon to exhaust the Sun's substance. The filling of interplanetary space with such corpuscles, moving at high speeds, ought to interfere with planetary motion as much as or more than the presence of a rare elastic medium. The transparency of transparent solids would require that they be riddled in every direction by open, rectilinear channels, making solidity impossible. The violent collision of corpuscles ought to interfere with the rectilinear crossing of rays or their being brought to a focus.

During the 1690s Malebranche had developed a theory in which light consists of regular trains of waves propagated through the aether, with different wave-lengths corresponding to different colors.[39] Euler defends and elaborates this feature of Malebranche's theory,[40] developing the mathematics and mechanics of the propagation of sinusoidal pulses. The colors of bodies arise, he argues, not from differential reflection of the incident rays, but from re-transmission of the incident motion, by oscillators within the body, each having its own characteristic frequency. He assigns to red a higher frequency than to violet, because it is less refracted in the denser medium, so that its speed is less diminished, owing, he believes, to the successive pulses being so close together as to impinge on one another and so increase the speed. But on this question he will reverse his position in 1753, having by then examined the succession of colors in Newton's rings.[41] He does not appear to recognize polarization as a difficulty, although Newton had clearly described it and argued that it was compatible only with an emission theory.

From the approximately known velocity of light, and the formula Euler derived for that velocity in terms of the density and elasticity of the aether,[42] it followed that the aether must be highly elastic and very rare. Yet however rare the aether, it must, because of its inertia, resist the motion of bodies through it. Euler devoted a special memoir to this subject, showing that the resistance must cause both the periods and eccentricities of the planets to diminish.[43] For a time he thought he had evidence for a secular diminution in the Earth's annual motion, until Tobias Mayer in 1754 pointed out that, with the exception of the questionable equinoxes reported by Ptolemy, ancient

[39] See Mouy, *Le développement de la physique Cartésienne*, p. 305ff.

[40] Euler's acquaintance with a theory of light associating colors with regular vibrations of given frequency may have come from Jean II Bernoulli's essay cited in note 34.

[41] 'Essai d'un explication physique des couleurs engendrées sur des surfaces extrêmement minces' (E.209), read at the Berlin Academy on 13 April 1753; printed in *Mémoires de Berlin 8* (1752), 1754, pp. 262–82, and in *OO* III *5*, pp. 156–71.

[42] Euler's formula is the same as Newton's: the speed is proportional to $(E/d)^{\frac{1}{2}}$, where E = elasticity and d = density.

[43] Euler, 'De relaxatione motus planetarum', in *Opuscula varii argumenti*, I (Berlin, 1746), pp. 245–76.

observations compared with modern indicated a constant year. Euler continued to regard aethereal resistance as the most eligible explanation for the secular acceleration of the Moon.

In a letter to Euler of 25 December 1743,[44] Daniel Bernoulli argued that the resistance exercised by the aether could be nil. The action of a vein of fluid on a body, he explained, involved changes in both speed and direction of the fluid particles. The aether in flowing through bodies is forced to twist and turn, but likely without change in speed, and the effects of all the twistings could destroy one another. This analysis pleased Euler,[45] and he incorporated it in his commentary on Robins's *New Principles of Gunnery* (1745), where it stands as the first published statement of the so-called d'Alembert's paradox. For resistance to be absent, Euler thinks, the fluid material must be perfectly fluid and perfectly elastic; but such may be the properties of the celestial matter in which the planets and comets move.[46]

Euler on the metaphysical foundations of mechanics

On 18 June 1744 Euler read to the Berlin Academy a memoir entitled *Physical Investigations Concerning the Least Parts of Matter*.[47] This is the first of several memoirs in which Euler concerns himself with the nature of matter, force, space and time, while challenging the metaphysical speculations of his Leibnizian and Wolffian colleagues. The memoir begins with a reference to the Leibnizian principle of the identity of indiscernibles, which was taken by the philosophers to imply that the least particles of matter must all be different in quality, in order to be plural in number. Euler asserts: 'I shall show so clearly that no one can any longer doubt it that all the least molecules which constitute the bodies surrounding us are equally heavy, or that they all have the same specific gravity.'[48]

Euler's argument is identical with one given by Daniel Bernoulli in Chap. XI of his *Hydrodynamics* (1738), which treats hypothetically of the possibility of an aethereal explanation for gravity. The least particles of bodies must be without pores and hence impervious to the subtle fluid causing gravity, and the force with which they are pressed must, as in hydrostatic effects generally, be proportional to the volume of subtle matter they displace. But Newton has shown that all terrestrial bodies fall toward the Earth with the same

44   P. H. Fuss, *Correspondance*, pp. 545–6 (Lettre XXXI, 25 Dec. 1743).

45   *Ibid.*, p. 550 (Lettre XXXII, 4 Feb. 1744).

46   *Neue Grundsatze der Artillerie aus dem Englischen des Herrn Benjamin Robins übersetzt und mit vielen Anmerkungen versehen* (Berlin, 1745); *OO* II *14*, pp. 268–9.

47   'Recherches physiques sur la nature des moindres parties de la matiere', *Opuscula varii argumenti* I (Berlin, 1746), pp. 287–300; *OO* III *1*, pp. 6–15.

48   *OO* III *1*, p. 9.

acceleration. It follows that their true volumes – the volumes actually occupied by the ultimate molecules – are as their masses. To Euler, who unlike Bernoulli took an aethereal explanation to be necessary, the conclusion was inescapable.

An aethereal theory of gravity still allows gravitation to be 'universal' – Euler in the *Letters to a German Princess* condones this way of speaking[49] – in the sense of applying to all gross bodies. Since the gravitational force to which these bodies are subject depends only on the volume of aether their solid parts displace, Euler can endorse Newton's conclusion that 'the weights of bodies do not depend on their forms and textures.'[50] According to Newton, aethereal theories of gravity in which aether is assumed to be transformable into ordinary bodies and *vice versa* – including Descartes's theory and one that Newton had himself proposed in 1675 – were incompatible with the experimental demonstration that mass and weight are proportional;[51] but the refutation does not touch Euler's theory provided that aether is not assumed to mix or combine with gross matter in such a way as to change the density of the latter.[52]

As for the conclusion that the ultimate solid particles of bodies are all of the same density, it was in agreement with atomist and Cartesian thought. Newton himself was inclined to assume such sameness of density, although for him it was conjectural.[53] But Euler's argument for the conclusion would have surprised him.

Indeed, for Newton the experiment whereby he demonstrated the proportionality of mass and weight, and which he carried out in the winter of 1684–85, had played an important role in enabling him to dismiss the aethereal theories of gravity he had earlier entertained, and so to espouse universal gravitation. Euler, by contrast, with the rest of the eighteenth century and popularizers ever since, regarded Newton's discovery of universal gravitation as a remarkably simple inductive generalization. The story of the apple, told by Newton in his last years and publicized by Voltaire, led to the supposition that the discovery had come already in 1666. Euler in his *Letters to a German Princess* emphasized its accidental character, and introduced a note of jocularity by making the apple hit Newton on the head.[54] Neither he nor any of his contemporaries realized that the gravity Newton allegedly thought of

---

[49] *Lettres à une princesse d'Allemagne*, Lettre 53, 5 Septembre 1760, in *OO* II *11*, p. 118.

[50] Newton, *Principia*, all edns, Book III, Prop. 6, Cor. 1.

[51] *Ibid.*, Cor. 2. The first edition differs from the later editions in an interesting way.

[52] In his *Anleitung zur Naturlehre*, written in the mid-1750s, Euler explains the flexibility of flexible bodies by their containing aether in enclosed pores; but he says that the amount of aether so contained is negligibly small. *OO* III *1*, pp. 142–6 (para. 134–7).

[53] Newton, *Principia*, 2d and 3d edns, Book III, Prop. 6, Cor. 4.

[54] *OO* III *11*, p. 116 (Lettre 52 for 3 Sept. 1760). The story as told by Newton and reported by Conduitt and Voltaire contained no such collision.

extending to the Moon in 1666 was undoubtedly a gravity mediated by aethereal action; or that the idea of universal gravitation emerged only much later.[55]

One of the ironies of the situation is that Euler, in using Newton's experimental result to support the sameness of density of the ultimate parts of gross matter, understands himself to be opposing not only Leibnizians and Wolffians, but also the English 'attractionists', followers of Newton. Both groups assume secret, 'active' forces or powers within the least parts of matter; thereby, as Euler sees it, undermining mechanics.

In 1746 there was published in German an anonymous memoir bearing the title *Thoughts on the Elements of Bodies, in which the System of Simple Things and Monads is Examined, and the true Essence of Bodies Discovered.*[56] The author is unquestionably Euler. The memoir is a volley in a metaphysical battle that Euler had much to do with initiating.

The Berlin Academy as reorganized in 1744 was the first European academy to have a philosophical class or division.[57] Every fourth year, as a consequence, the question proposed for the annual prize contest was of a philosophical or metaphysical character. For the contest of 1747, at Euler's urging, the topic proposed was the doctrine of monads. The contestants, who could not be members of the Academy, were asked either to refute this doctrine, or to establish it and derive from it an intelligible explication of the principal phenomena of the universe, and in particular of the origin and movement of bodies.[58] Before any of the competing essays were submitted, Euler's *Gedancken von den Elementen der Körper* appeared, opposing the doctrine of monads.

The premises of this doctrine as commonly formulated, Euler tells us, are two. First, given that bodies are extended and therefore divisible, it is claimed that by successive subdivisions one must arrive at the least parts of a body. Secondly, given the continuous change observable in the world, it is claimed

---

[55]   On the lateness of Newton's discovery of universal gravitation, see D. T. Whiteside, 'Before the *Principia*: the maturing of Newton's thoughts on dynamical astronomy, 1664–1684', *Journal for the History of Astronomy* 1(1970):5–19; Richard S. Westfall, *Force in Newton's Physics* (American Elsevier, 1971), pp. 456–66; Curtis Wilson, 'From Kepler's Laws, so-called, to universal gravitation: empirical factors', *Archive for History of Exact Sciences* 6(1970):136–70. The thesis will receive impressive new corroboration in B. J. T. Dobbs's forthcoming book, *The Janus Faces of Genius: the Role of Alchemy in Newton's Thought* (Cambridge University Press). Professor Dobbs kindly allowed me to read a typescript of Chaps. 5 and 6.

[56]   *Gedancken von den Elementen der Körper, in welchem das Lehr-Gebäude von den einfachen Dingen und Monaden geprüfet, und das wahre Wesen der Cörper entdecket wird* (Berlin, 1746); *OO* III 2, p. 347ff.

[57]   Adolf von Harnack, *Geschichte der Königlich Preussischen Akademie der Wissenschaften zu Berlin* (Berlin, 1900), I, 309–11.   [58]   *Ibid.*, II, p. 305.

that bodies and indeed the least parts of bodies must be endowed with living force (*lebendige Kraft*, *vis viva*), understood as a power working to cause change. To each least part or monad is thus ascribed a continuous striving to change its condition.

According to Euler, Leibniz and Wolff differed as to whether the monads making up a finite body were infinitely many (Leibniz) or finite in number (Wolff). Neither position, Euler says, is tenable: infinite divisibility implies that there are no least parts, and if the monads are indivisible yet finite in number they do not constitute the body.

As for the doctrine that the monads are endowed with forces by which they strive continuously to change their state or condition, its authors have failed to understand the law of inertia. This law is proved by both reason and experience, the rational ground being the principle of sufficient reason, according to which nothing happens without a cause. If a body at rest begins to move, or a body in uniform rectilinear motion changes its speed or direction, there must be an agency causing the change.[59] That agency cannot be in the body, for inertia, which belongs to the essence of body, is the body's tendency to remain in its state either of rest or uniform rectilinear motion, and nothing in the body can contradict its essence. Thus the Leibnizian and Wolffian explanation for the continuous change in the world is wrong; the true cause is that bodies, each of which individually seeks to maintain its inertial state, interfere with one another and so interact.

A further conclusion to be drawn, Euler claims, is that there is an infinite difference between the elements of bodies and the essence of souls or spirits. Whereas bodies are endowed with inertia, souls or spirits are endowed with the power to represent the world to themselves and to change their state. Leibniz was thus wrong to attempt to make the distinction between soul and body one of degree.

This absolute dualism is characteristic for Euler. He had articulated it earlier in an essay printed in his *Opuscula* of 1746.[60] Concerning this piece, and Euler's new involvement in metaphysical disputation, Daniel Bernoulli complained in a letter of 29 April 1747:

> Herr Ramspeck wrote my father that you are involved publicly in several metaphysical controversies. You should not meddle with such subjects; for from you we expect only sublime things, and it is not possible in these [metaphysical] things to excel. I fear that your fifth question in the *Opuscula* will find little approbation among the metaphysicians. I think all

[59] In his *Theoria motus corporum solidorum*, para. 87 (*OO* II 3, p. 50), Euler appears to allow that some reasoning on experience is required in order to recognize that, in the absence of interfering forces, uniform rectilinear motion would persevere.

[60] 'Enodatio quaestionis: Utrum materiae facultas cogitandi tribui possit nec ne?', *Opuscula varii argumenti*, I (Berlin, 1746), pp. 277–86.

cogitations must have an impulsive force, and so your argument, derived from inertia, fails.... Forgive my free manner of speaking, which I use only with my true friends.[61]

Again on 16 August 1747, after informing Euler that 'Herr Wolf in a letter to my father complained much of you', Bernoulli expresses his worry: 'I fear that you will still bring down on yourself much vexation from the number of your antagonists, and it is certainly not worth the trouble.'[62]

Euler's notes on twenty-five of the pieces submitted in the Berlin prize contest of 1747 have survived.[63] Here, in commenting on Samuel König's essay, he recognizes that the system of monads would be sustainable on an idealist basis, the monads being assumed to be neither in place nor capable of spatial movement, but the representations they have of one another giving rise to space and time. The essayists who defended monads as constituents of the spatio-temporal world fell inevitably. Euler believed, into inconsistencies. It is worth noting that König made attraction (or 'English attraction', as Euler calls it) a necessary consequence of Leibniz's monads – a conflation that would have surprised Leibniz, but which Euler did not find incongruous.

Neither Daniel Bernoulli's warnings nor the passions unleashed in the debate kept Euler from persisting in his critique of Leibnizian and Wolffian notions. On 1 February 1748 he presented to the Academy his 'Reflections on Space and Time'.[64] In the usual formulations of the Leibnizians and Wolffians, space was the order of coexistence, and time the order of succession, of things as they appear in the representations that a monad forms of the world. But, argued Euler, space and time are not imaginary or unreal, or merely relational. The law of inertia, which is solidly established, presupposes the notions of *same direction* and *equal time interval*; these ideas, in turn, presuppose absolute space and time, independent of relations among bodies. However, Euler adds, 'I address myself here to the Metaphysicians who recognize some reality in bodies and motion; as for those who deny absolutely this reality, I do not flatter myself that these reflections will make the least impression...'

On 1 October 1750 Euler read to the Academy his 'Investigation Concerning the Origin of Forces'.[65] In this memoir he reaches his final formulation of the essence of matter. It confirmed him in his opposition to action-at-a-distance. Once more he starts from inertia as a *sine-qua-non* property of body. The principle of sufficient reason implies that a body will remain in its inertial state

---

[61] P. H. Fuss, *Correspondance*, p. 621 (Lettre XLVII).

[62] *Ibid.*, pp. 624–5 (Lettre XLVIII).   [63] *OO* III 2, pp. 416–29.

[64] 'Reflexions sur l'espace et le tems', *Mémoires de Berlin* (1748), 1750, pp. 324–44; *OO* III 2, pp. 376–83.

[65] 'Recherches sur l'origine des forces', *Mémoires de Berlin* (1750), 1752, pp. 419–447; *OO* II 5, pp. 109–31.

of rest or uniform rectilinear motion unless caused to do otherwise – unless acted upon by a force. The memoir seeks to answer the question: whence force?

The answer, according to Euler, lies in the property of impenetrability, a property as essential to body, he claims, as extension or inertia. Bodies are such that two of them cannot occupy the same place at the same time, and all things having this property are bodies. In two later treatments of the properties of bodies, Euler will urge that impenetrability implies both extension and mobility, while mobility implies inertia.[66] Almost all philosophers, he will note in his *Theoria motus corporum solidorum* of 1765, assert the essence of bodies to be completely unknown to us.[67] Euler believes he has discovered it in the property of impenetrability.

Now impenetrability, in the sense in which Euler uses the word, does not admit of degrees; body *qua* body does not allow itself to be penetrated. Hence impenetrability requires that there be forces sufficient to prevent penetration. It follows that impenetrability is a source of force. The essentialist course of Euler's analysis leads him to suppose that it is the only source of force exerted by bodies on bodies.

The forces Euler has discovered, unlike the forces of which the Leibnizians and Wolffians speak, are not *in* bodies. Forces come into existence only when required to prevent two bodies from penetrating each other. The magnitude and direction of the force are not determined in advance, but in each case are just such as to prevent penetration, the force always being the least required for the purpose. All changes in the world are thus in accordance with the principle of least action.

Philosophers who introduce imaginary forces (Euler means the Leibnizians and Wolffians) throw into confusion the first principles of physics; it is probable that the forces of attraction and adhesion are no better founded. As for the forces by which the heavenly bodies are solicited,

> it appears rather probable that all these bodies, being surrounded unquestionably by a subtle matter, are also put in movement by it, although we do not know the manner of it. Yet that two bodies separated by a space entirely empty should attract each other mutually by some force, seems as strange to reason, as it is unproved by an experience. With the exception therefore of the forces by which spirits are perhaps capable of acting on bodies, which are undoubtedly of a quite different nature, I conclude that there are no other forces in the world than those which derive their origin from the impenetrability of bodies.[68]

---

[66]   *Anleitung zur Naturlehre*, para. 38, in *OO* III *1*, p. 48. *Theoria motus corporum solidorum*, para. 122, in *OO* II *3*, pp. 61–2.

[67]   *Theoria motus corporum solidorum*, para. 129, in *OO* II *3*, pp. 63–4.

[68]   'Recherches sur l'origine des forces', para. 59, *OO* II *5*, p. 131.

Euler's metaphysics and the fundamental equations of continuum mechanics

During the 1750s Euler developed, by stages, the fundamental equations for the motion of rigid bodies and fluids. Mechanics as Euler defines it is the science of motion. Hence it is not surprising that, in Euler's mind, the equations of motion for fluids and rigid bodies are closely associated with the metaphysics in which he believes mechanics to be grounded.

The first of Euler's memoirs specifically devoted to the rotation of rigid bodies was the 'Discovery of a New Principle of Mechanics',[69] presented to the Berlin Academy on 3 September 1750. His 'Investigation Concerning the Origin of Forces' was read to the academy just a month later, on 1 October. The principle described as new in the first memoir is the principle of moment of momentum, but Euler gives it no name. He asserts that, like all principles of mechanics, it is grounded in the principle of linear momentum, $F = ma$. This does not mean that it is logically derivable from $F = ma$. Euler says rather that 'in order to know the state of [the] forces, it is necessary to take account of their moments...'. To know how forces act in certain circumstances, you must know their moments. As in the case of Euler's metaphysics, it appears inappropriate to seek here for formal logical rigor. Somehow the principle of moment of momentum is grounded in the principle of linear momentum, and, as the second memoir shows, the latter is grounded in Euler's metaphysics of body or impenetrability.

The importance for Euler of a metaphysical grounding of the equations of motion is corroborated if we follow the further development of his mechanics of rigid bodies. The memoir of September 1750 culminated in a differential equation that was not integrable. The single coordinate system used was fixed in space with respect to the center of mass of the rotating body, so that the material elements of the body were not identifiable by their coordinates. To remedy this lack, Euler, in October 1751, introduced a second coordinate system with the same origin as the first but fixed in the body.[70] To specify the relations between the two systems, he introduced the angles now called 'Eulerian'. But for most applications, the equations remained intractable.

The solution emerged in two memoirs of 1758. The first of these bears the title, 'Investigation of the Mechanical Knowledge of Bodies';[71] it is concerned principally with calculating moments of inertia, and establishes that in every body there are (at least) three mutually perpendicular axes about which the body can rotate freely without wobbling. The second memoir is entitled 'On

---

[69] 'Découverte d'un nouveau principe de mécanique', *Mémoires de Berlin* (1750), 1752, pp. 185–217; *OO* II 5, pp. 81–108.

[70] 'Du mouvement d'un corps solide quelconque lorsqu'il tourne autour d'un axe mobile', *OO* II 8, pp. 313–56.

[71] 'Recherches sur la connoissance mécanique des corps', *OO* II 8, pp. 178–99.

the Movement of Rotation of Solid Bodies about a Variable Axis'.[72] In it Euler employs the principal axes for the coordinate axes fixed in the body, and so finds that he can obtain integral solutions to problems he had previously believed 'to surpass the powers of the calculus'. These include determining the continuation of motion of a body initially rotating about an axis other than a principal axis.

Between 1758 and 1760 Euler then produced his great treatise on the motion of solid bodies. The full title is worth attending to: *Theory of the Motion of Rigid or Solid Bodies, Established on the First Principles of Our Cognition, and Accommodated to All Motions which can Occur in such Bodies.*[73] Here in an introductory section Euler sets forth once more the metaphysical foundations of mechanics: absolute space and time, impenetrability as the essence of body, force arising from the necessity that penetration be avoided. And he states his opposition to action-at-a-distance: 'It seems very probable that bodies exert on each other no other forces than those by which penetration is avoided.'[74]

In the case of fluid mechanics, the fundamental equations in the spatial description emerge in the period 1750–1755.[75] They are two in number: the general equation of continuity, and the equation specifying acceleration in terms of forces per unit mass. Both are essentially the equations recognized today, except that the second lacks the term for viscous forces, which Euler's ontology did not permit him to recognize.

At some point in the 1750s, after arriving at these equations, Euler drafted an extended treatise entitled *Anleitung zur Naturlehre, Introduction to Natural Science.*[76] In its final paragraphs Euler derives the equations just referred to. They contain, he asserts, all possible motions that fluids may undergo; but analysis has not yet been able to supply a complete, integral solution.

In the earlier portions of the treatise, Euler presents both the metaphysics and the hypothetical aethereal physics that we have come to recognize as Eulerian. At the start he tells us that natural science has to do with changes occurring in bodies, and no change occurs without a cause. Impenetrability, we learn once again, is the essence of body, and it implies the additional

---

[72] 'Du mouvement de rotation des corps solides autour d'un axe variable', *OO* II *8*, pp. 200–35.

[73] *Theoria motus corporum solidorum seu rigidorum ex primis nostra cognitionis principiis stabilita et ad omnes motus, qui in huiusmodi corpora cadere possunt, accommodata* (Rostochii et Gryphiswaldiae, 1765), in *OO* II *3*.   [74] *Ibid.*, p. 67.

[75] For an account of this emergence, the reader is referred to Clifford Truesdell, 'Rational fluid mechanics, 1687–1765', in *OO* II *12*, esp. pp. LVIII–C.

[76] *Anleitung zur Naturlehre Worin die Grunde zu Erklarung aller in der Natur sich Ereignenden Begebenheiten und Veranderungen festgesetzt werden, OO* III *2*, pp. 16–178. This treatise was probably not composed before about 1755, when Euler published the two basic equations of fluid dynamics in another memoir (E.226; *OO* II *12*, pp. 54–91).

properties of extension, mobility, and inertia; from it all forces derive except those caused by the direct action of spirits on bodies. A subtle matter or aether serves to carry light waves. It is highly elastic, its elasticity appearing as an unexplained primitive property involving compression; it is beyond our powers, Euler avers, to know how or why the aether should be compressed. Solid bodies are made solid by its pressure; gross bodies are elastic in the measure in which they contain tiny amounts of this aether in closed pores. Once more, from the proportionality of mass and weight, Euler infers that the solid parts of all gross bodies are of the same density. Gravity, he once more claims, is due to gradients of pressure within the aether in the neighborhood of gross bodies, caused by motions in the aether; if we knew the nature and source of these motions, Euler opines, we would have the complete explanation of gravity.

Thus the equations of fluid mechanics appear as the basic equations for understanding the natural world as Euler believes it to be. The *Anleitung* was never put in final form or published during his lifetime; we do not know why. Its formulations are even more dogmatic than usual. Can Euler have had doubts?

He ought to have had, of course; his metaphysics is impossible. Impenetrability does not imply extension or mobility; mobility, even with the principle of sufficient reason thrown in as an added premiss, does not imply inertia as understood in modern physics. Nor can forces be derived from impenetrability; if any derivation is possible here, it will be of impenetrability in a restricted sense from forces of a certain kind. Euler, a rationalist believing in innate ideas, has failed to examine the strict logical relations among the ideas he takes as fundamental. And, like other rationalists before him, he has confused logical implication, or something pretending to be such, with causality.

Nevertheless, I wish to argue that we owe a debt of gratitude to Euler's mistaken metaphysics – a debt quite other than the one that German philosophy owes Euler for persuading Kant that philosophy must take account of the premises found necessary (or successful?) in natural science, including the ideas of space and time, which, as Euler had pointed out, are neither purely empirical nor purely ideal.[77]

If it is asked whether, during the eighteenth century, anyone else but Euler might have originated the fundamental equations of rigid-body and fluid mechanics, the answer must surely be 'No'. It is Euler who has the magnificent sense for mathematical form, and who by persistently returning to the same problems achieves true clarity and generality. If there could be a contemporary to rival him in continuum mechanics, it would have to be d'Alembert. In fact,

[77] See Andreas Speiser, *Leonhard Euler und die deutsche Philosophie* (Zurich, 1934).

continuum mechanics owes a debt to d'Alembert as well as to Euler, for in a memoir of 1749 and another of 1750 d'Alembert provided stimulus and clues that set Euler on the path to finding his fundamental equations. But it is quite unlikely that d'Alembert himself would have found these equations. And it is not merely a matter of the execrable form in which he left his mathematical writings. It is a matter, as well, of ideas.

Consider first d'Alembert's *Researches on the Precession of the Equinoxes*, published in 1749; it was from this treatise, if I am not mistaken, that Euler extracted the principle of moment of momentum.[78] In order to determine the motions of the Earth's axis due to the attractions of the Sun and Moon on the equatorial bulge, d'Alembert first formed the product of each mass-element of the Earth by the reversed component of its acceleration in each coordinate direction, then the moment of this product about each of the coordinate axes, then the integrals of these moments about the three axes. Invoking 'd'Alembert's principle', he next asserted these integrals to be in equilibrium with the torques due to the external forces. To express this equilibrium, he set the integrals equal to the torques, thus committing an error in sign. Fortunately he committed a compensating error of sign in evaluating the external torques.

Euler wrote d'Alembert that he was unable to follow him in his preliminary propositions, but that, 'having been assisted by some insights in your work by which I was little by little enlightened, I have come to be able to judge your excellent conclusions.'[79] To d'Alembert, the principles of equilibrium were self-evident, and dynamics was to be reduced to statics by treating the reversed accelerations, multiplied by the masses accelerated, as forces balanced by other forces. Euler, by contrast, thought in terms of causes producing effects, forces producing accelerations, torques producing angular accelerations. Thus, in his *Scientia Navalis*, completed by 1740 but not published till 1749, he had already introduced the notion and name of moment of inertia, and constructed the equation torque = moment of inertia × angular acceleration in analogy to $F = ma$. In his 'Discovery of a New Principle of Mechanics' he forms the same integrals as d'Alembert but with the direct rather than the reversed accelerations, and sets them equal to the corresponding external torques. He gives no explicit justification of the resulting equation. Are we to take it as a statement of the equality of cause and effect?

Clarification comes a decade later, in Euler's *Theory of the Motion of Solid Bodies*. Here he defines the *elementary forces* as forces which, applied to the single elements of a body separately, would produce in them the same change of state as they undergo in the actual motion of the body. They are thus ideal

---

[78] See my 'D'Alembert *versus* Euler on the precession of the equinoxes and the mechanics of rigid bodies', *Archive for History of Exact Sciences* 37(1987):233–73.

[79] *OO* IV A 5, p. 306.

rather than actual.[80] The actual forces are, first, the forces externally applied and, second, the forces, whatever they may be, that come into play to maintain the rigidity of the body. The elementary forces are given by the difference between the external forces and the forces maintaining rigidity. It is because the latter forces, on the assumption of rigidity, must be in equilibrium or add to zero, that d'Alembert's principle holds in the present case, and that Euler's use of the principle of moment of momentum is justified. It is indeed a case, Euler tells us, of the metaphysical principle that the cause equals its full effect; but he warns that, in the general fog of things metaphysical, he prefers to rely not on the metaphysical principle but on physical demonstration. He means that the situation needs to be analyzed in terms of all and only the actual forces in play.[81]

In their views as to what mechanics or dynamics is or should be, Euler and d'Alembert differed profoundly. In his *Traité de Dynamique* (1743) d'Alembert sought to develop a science as certain as geometry. To do this, he believed it necessary to reduce dynamics to kinematics, to make it as like geometry as possible, and so to eliminate from it consideration of motive causes, with which the world might indeed be replete, but of which we know nothing. As axioms he adopted the laws of inertia, equilibrium, and the parallelogram of forces. Of the flaws in the logical structure of his dynamics – for instance the fact that he uses the notion of mass without defining it – he remained unaware. But the point I wish to make is that he regarded his dynamics as a branch of analysis, complete as to principle, and applicable in limited ways to a world full of forces that are unknown and perhaps largely unknowable.

Euler, by contrast, had since the time of his *Mechanica* of 1736 envisaged the project of developing a mechanics dealing with all possible kinds of body. The *Mechanica* itself was to deal with infinitely small bodies, to which the laws of motion directly apply. The remaining subdivisions of the subject would deal with (1) rigid bodies, (2) flexible bodies, (3) compressible bodies, (4) bodies variously connected in constraint systems, (5) fluids. In all these finite bodies the motion followed is composed of the tendencies of the individual parts, and up to now, Euler says, this cannot be defined 'because of insufficiency of principles'.[82] The principles he has in mind, we may be sure, involve pushes and pressures.

This project remained a persistent concern of Euler's. Thus during the 1740s he mentions more than once, as an unsolved problem, the determination of the motion of a freely rotating rigid body.[83] And he mentions it again in his letter to d'Alembert of 7 March 1750, for he sees that the solution must somehow be

---

[80]   *OO* II *3*, p. 137.   [81]   *Ibid.*, pp. 139–40.   [82]   *OO* II *1*, pp. 38–9.

[83]   In the *Scientia Navalis*, *OO* II *18*, p. 82; P. H. Fuss, *Correspondance*, pp. 583, 590 (Lettres XXXIX, XL).

contained in d'Alembert's memoir on the precession.[84] For d'Alembert, on the contrary, the only project in view was that of deriving the precession and nutation from the Newtonian law of gravitation.

The difference in Euler's and d'Alembert's views of dynamics appears in another way. In treating the motions of solids and fluids, Euler separates the kinematic from the dynamic; that is, the determination of what is geometrically possible, from the determination of the effects of forces. In the case of rigid-body mechanics, by dealing with the kinematic determinations first – his expertise in algebra and spherical trigonometry assisting – he wonderfully simplifies and clarifies the formulation of the problem. One of the reasons d'Alembert's presentation is so helter-skelter is that, intending as he does to reduce dynamics to kinematics, he is not philosophically prepared to make the distinction between the two. To be sure, there are other reasons, including his lack of command of spherical trigonometry and his repeated reliance on crabbed diagrams.

In fluid mechanics the story is similar. Once more a memoir by d'Alembert plays a role, namely his *Essay on a New Theory of the Resistance of Fluids*, which was submitted in the Berlin Academy's competition of 1750, then, when it failed to receive the prize, published independently.

As Truesdell has pointed out, d'Alembert's *Essay* is a turning-point in mathematical physics: 'For the first time, a theory is put (however obscurely) in terms of a *field* satisfying partial differential equations.'[85] For the special case of axially-symmetric, steady flow past a submerged body, d'Alembert derives the equation of continuity for both incompressible and compressible fluids, and obtains certain conditions on the velocity components, with a solution in terms of complex variables. At the same time the unclarities in the essay go deep. D'Alembert's principle here leads d'Alembert into error, and he falls into logical fallacy in attempting to reconcile two supposedly equivalent results.[86] But this essay was what triggered Euler's endeavor, over the next few years, to lay general foundations for fluid mechanics.

Truesdell gives a detailed account of what Euler achieved.[87] I have already mentioned the most fundamental equations, both of which Euler is the first to obtain: the general equation of continuity, limiting kinematic possibility, and the equation specifying acceleration in terms of forces per unit mass. D'Alembert had not sought the first equation, because he did not have the aim of a general fluid mechanics. He could not have written the second, since he did not have a measure or symbol for pressure. In fairness it should perhaps be noted that d'Alembert frequently speaks of the 'tenacity' of fluids: he was free

[84] *OO* IV A 5, p. 306.

[85] Clifford Truesdell, 'Rational fluid mechanics', p. LVII.   [86] *Ibid.*, p. LIV.

[87] *Ibid.*, pp. LVIII–C.

to recognize viscosity as Euler was not. But the idea of formulating an equation analogous to $F = ma$, in which all the terms of the Stokes derivative of the velocity appear on one side, and all the forces per unit mass on the other, had not occurred to him.

Without Euler, the fundamental equations of the mechanics of fluids and rigid bodies would not have been written in the 1750s, or indeed until another century. Much is owing here to Euler's persistent drive toward clarification in terms of pushes and pressures.

And the naive metaphysics – what kept him tied to the top of that flagpole? Opposition, for one thing, to the metaphysics of his Berlin colleagues, so contrary, as he found it, to clarity in mechanics. Opposition, again, to the Enlightenment figures that Frederick II brought to Berlin, like de la Mettrie and Voltaire, who left no place for biblical revelation or the spiritual.

The development of mathematical physics in the nineteenth century was to carry it beyond the limits that Euler had set for it. In his day analysis sought its justification in physical interpretation; not till decades later would the foundational work of Bolzano, Cauchy, and others begin to give it the independence to stand on its own feet. It would be still later when Maxwell, following up on ideas of William Thomson, relying on the new principle of energy and on the Lagrangian formulation of dynamics (itself dependent on d'Alembert's principle), would appeal to analogy in coping with the unknown, without insisting upon the reality of particular detailed mechanisms. If there were steps in the direction of such sophistication that Euler could have taken, they held no attraction for him. But his stubbornly naive metaphysics brought its own rewards: a clear analysis in terms of forces considered as causes, and the first adequate formulation of fundamental equations in continuum mechanics.

17   St Peter and the rotation of the earth: the problem of fall around 1800

DOMENICO BERTOLONI MELI

Acknowledgements

Part of the research for this paper was carried out in Berlin (West) with a generous grant from the Fritz Thyssen Stiftung. I thank the editors of this volume for their observations on a previous draft of my essay. I also thank the staff of the Biblioteca Universitaria, Biblioteca dell'Archiginnasio, and Archivio dell'Istituto delle Scienze di Bologna, for having allowed me to study several manuscripts in their possession. The illustrations are reproduced by permission of the Syndics of the University Library, Cambridge.

Introduction

In the seventeenth century the problem of fall occupied a central position in the literary world. The magisterial works by Alexandre Koyré, Paolo Galluzzi, Tom Whiteside and others have shown that debates on the trajectory of falling bodies lay at the intersection of a number of themes ranging from Copernicanism to the new science of motion, from biblical hermeneutics and religious censorship to the role of experiments. Several leading mathematicians and philosophers of the period, such as Galileo Galilei, Johannes Kepler, Pierre Fermat, Pierre Gassendi, Gianbattista Riccioli, Giovanni Alfonso Borelli, Robert Hooke, and, of course, Isaac Newton, dwelt on this problem.[1]

Following some isolated publications, the issue emerged again in a different context between the end of the eighteenth and the beginning of the nineteenth century. Although the rotation of the earth was no longer a controversial matter,[2] mathematicians and astronomers took part in an investigation

---

[1]   A. Koyré, 'A documentary history of the problem of fall from Kepler to Newton', *Transactions of the American Philosophical Society*, new series, vol. 45, 4(1955):329–96, and *Newtonian Studies* (Cambridge, Mass.: Harvard University Press, 1965), ch. 5; P. Galluzzi, 'Galileo contro Copernico', *Annali dell'Istituto e Museo di Storia della Scienza* II, 2(1977):87–148; J. Lohne, 'Hooke versus Newton', *Centaurus* 7(1960):6–52; D. T. Whiteside, 'Newton's early thoughts on planetary motion', *British Journal for the History of Science* 2(1964):117–37, on pp. 131–6.

[2]   P. Varignon, 'Incompatibilité géométrique de l'hypothêse du tournoyement de la terre sur son centre, avec celle de Galilée touchant la pesanteur', *Mémoires de l'Académie des Sciences de Paris* (1707):12–17; J. Hermann, 'De nova accelerationis lege', *Acta Eruditorum* Sept. (1709):404–411; G. Poleni, 'Considerazioni intorno alle varie

involving long and difficult experimental trials as well as sophisticated calculations. The reasons behind this renewed interest varied from case to case: the young Abbé Gianbattista Guglielmini, for example, was eager to publish on an issue where he thought that the Catholic Church had lost considerable ground and needed to regain the initiative. Despite his initial impatience to make his ideas known before having started his trials, he later claimed that an experiment could settle the matter once and for all. However, some aspects of the problem became controversial, and other protagonists, such as the astronomer Jérôme Lalande, Pierre Simon Laplace, the political pamphleteer and astronomer Johann Friedrich Benzenberg, the astronomer and physician Wilhelm Olbers, and Carl Friedrich Gauss, stimulated each other to tackle the issue.

Around 1800 theoretical investigations imposed strict quantitative constraints on experimental techniques. The problem was no longer to drop balls from the masts of moving ships or from high towers – as, for example, Gassendi and Riccioli had done – in order to see where they fell, but to confront with the greatest accuracy theoretical predictions with experimental results. Expected deviations were of the order of a few centimetres for a height of fall of almost one hundred metres. As a consequence, experimenters had to develop sophisticated techniques involving microscopes, engage in night sessions in order to avoid traffic, and employ data analysis based on a primitive error theory. On the other hand, theoreticians were calculating deviations using the principle of virtual velocities, rotating reference frames, series expansions and other approximation techniques. Thus many of the resources employed to tackle the problem of fall were typical of the late eighteenth century, and even if the problem was not new, its investigation provides us with a lively picture of experimental and mathematical techniques at work around 1800.

The issue involved an East as well as a South deviation. Everybody agreed on the former, although initially its magnitude varied by a factor of 2/3. Concerning the latter, there seemed to be no reason to doubt it until Laplace made it known that according to his theory the South deviation was too small

accelerazioni, o ritardazioni, che i corpi gravi nelle loro cadute patirebbero se la terra girasse per l'orbe annuo', *Giornale de' Letterati d'Italia* 8(1711):199–215; J. d'Alembert, 'Sur le mouvement des corps pesans, en ayant égard a la rotation de la terre autour de son axe', *Opuscules Mathématiques* 7, 1780, pp. 314–71, and 'Réflexions sur le mouvement des corps pesans, en ayant égard a la rotation de la terre autour de son axe', *Mémoires de l'Académie des Sciences de Paris* (1771):10–20. On the doubts about the rotation of the earth compare J. F. Benzenberg, *Briefe geschrieben auf einer Reise nach Paris*, 2 vols. (Dortmund, 1805–6; hereafter *Briefe*), vol. 1, p. 228: 'Ich fragte ihn: woher es denn komme, dass so beruehmte Astronomen und Geometer, als La Lande und La Place, das copernikanische System glauben? – "Das finde ich sehr natuerlich, sagte Mercier, das traegt ihnen jaehrlich 40,000 Liv. ein."'

to be observed. First Lalande, then Guglielmini, Gauss and Olbers became convinced that there was virtually none; Guglielmini and eventually Benzenberg either reinterpreted their experiments or performed new ones which showed no South deviation. At the end everybody was convinced, but persuasion followed the most unpredictable paths. In the case of Benzenberg, for example, who was incompetent in mathematics and sceptical about theoretical predictions, a meeting with Laplace apparently exerted a greater impact than any mathematical theory. Despite the general agreement, however, Laplace, Olbers and probably Gauss were not happy with the way experiments were carried out. They judged the accuracy of Guglielmini's and Benzenberg's results to be insufficient even if their own theory was apparently confirmed. Benzenberg's lively and contradictory account in which he reported with comparable emphasis factual and redundant information, success and failure, certainly affected his credibility.[3] Setting aside Benzenberg's catastrophic presentation, however, I wish to argue that although at that time error theory and data analysis were not well established, experimental accuracy was questioned even when theory and experiment appeared to be in agreement.

In the following section I examine Guglielmini's theory and experiment, which indicated an East and a slightly smaller South deviation. Although he had originally planned to perform the experiment from the cupola of St Peter's in Rome, 'next to the grave of the holy Peter', eventually he had to settle for the rather less prestigious Asinelli Tower in Bologna. In the third section we shall see how his findings were publicized in France and Germany by Lalande. By the end of the 1790s it was known that Laplace had some objections. On becoming aware of this, Guglielmini swiftly changed his mind, reinterpreted his previous results, and performed new experiments in agreement with the new predictions. Laplace's theory is outlined at the end of the section. The fourth section is devoted to the early reception of the problem in Germany and Benzenberg's first trials in Hamburg. Olbers, who acted as Benzenberg's theoretical adviser, followed Guglielmini in predicting an East and a slightly smaller South deviation. Whilst Guglielmini aimed at a few highly accurate and consistent results, Benzenberg was forced by circumstances to develop a primitive error theory based on a high number of trials. Once again, experimental results agreed with theory. In the fifth section I examine the theory by Gauss, the debate between him and Olbers, and the way Olbers was persuaded. On developing his own theory based on a rotating reference frame, Gauss realized that the predictions by Olbers were inaccurate. The latter, while accepting his friend's results, could not see where he had been mistaken. Their correspondence provides us with a wonderful example of what counted as

---

[3]   J. B. Benzenberg, *Versuche ueber das Gesetz des Falls, ueber den Widerstand der Luft, und ueber die Umdrehung der Erde, nebst der Geschichte aller frueheren Versuche von Galilei bis auf Guglielmini* (Dortmund, 1804; hereafter *Versuche*).

convincing evidence. Olbers wanted to perceive through the mathematical formalism what was going on in nature: only when Gauss provided a new mathematical theory together with an intuitive explanation allowing Olbers to understand how gravity and air resistance affected motion was Olbers satisfied. Lastly, in the sixth section I discuss Benzenberg's second trials in a mine-shaft near Schlebusch in October 1804. They were performed after Benzenberg's return from Paris, where he had had an extraordinary series of meetings, including those with Lalande and Laplace himself. Benzenberg published one of the most interesting and lively accounts of Parisian life under Napoleon. Back in Germany the new series of trials showed no South deviation.

In this paper I focus on the interaction between theories and experiments. If experiments were set up to test theoretical predictions, these were calculated having recourse to several problematic approximations aimed at simplifying the problem and producing measurable values. This interdependence emphasizes the need to study these two aspects together. More generally I show that institutions and individuals, expectations and authority, mathematics and experiment interact and all contribute – admittedly in different degrees – to the course of this case-study. By neglecting any one of these elements, the picture I draw would lose one of its dimensions.[4]

Gianbattista Guglielmini's theory and trials

At the end of the 1780s the Bolognese Gianbattista Guglielmini, benefiting from the atmosphere of tolerance following the abolition of the anti-Copernican decree by Benedict XIV in 1757,[5] began considering the experiment of dropping balls from high edifices to see where they fell with respect to the foot of the plumb line. A brief preliminary work, *Riflessioni sopra un nuovo esperimento in Prova del Moto Diurno della Terra*, was published in Rome in 1789 and dedicated to Cardinal Ignazio Boncompagni, 'protector of the sciences' and Secretary of State to Pius VI. In the sixteenth century, under Pope Gregorio XIII Boncompagni, the annual period of the

[4] In the recent literature on experiments I have found the following texts particularly stimulating: H. M. Collins, *Changing Order* (London and Beverly Hills: Sage, 1985); S. Shapin and S. Schaffer, *Leviathan and the Air Pump* (Princeton University Press, 1985); P. Galison, *How Experiments End* (Chicago University Press, 1987); D. Gooding, T. Pinch, S. Schaffer, eds., *The Uses of Experiment* (Cambridge University Press, 1989).

[5] For a general survey on the philosophical activities in the Papal States and in particular at Bologna compare U. Baldini, 'L'Attivitá Scientifica del Primo Settecento', *Storia d'Italia. Annali, 3*, ed G. Micheli (Torino: Einaudi, 1980), especially pp. 513–45; P. Redondi, 'Aspetti della Cultura Scientifica negli Stati Pontifici', *ibid.*, especially pp. 782–9; R. L. Rosen, *The Academy of Sciences of the Institute of Bologna, 1690–1804*, Ph.D. Thesis, Case Western Reserve University, 1971 (hereafter *Academy of Bologna*), pp. 136–62 and the useful indices.

earth was determined in connection with the reform of the calendar; therefore Guglielmini thought it very appropriate that the determination of the earth's daily motion should be associated with a prominent member of the same family. The *Riflessioni* were followed by *De Diurno Terrae Motu Experimentis Physico-Mathematicis Confirmato Opusculum*, published in Bologna in 1792.[6]

By the late eighteenth century agreement in favour of the rotation of the earth was virtually universal and was based on phenomena such as the shape of the earth and the delay of the pendulum at the equator, both mentioned by Guglielmini. What prompted him to provide a direct proof of something hardly in question can be inferred from some passages in the *Riflessioni*, *Opusculum*, and correspondence with the Secretary of the Bolognese Academy Sebastiano Canterzani. Guglielmini, from 1788 to 1790 a student of mathematics in Rome, was eager to publish on a topic he deemed appealing and important. Fearing that he may be preceded by others, he wished it to appear in print before having performed the experiment. Canterzani, however, insisted that the experiment be performed first, reassured Guglielmini about his priority, and improved some preliminary drafts composed by his young protégé. Eventually Guglielmini published first the *Riflessioni* – to be considered as an announcement – and then, after having performed several trials, the *Opusculum*, where he stated that an experiment was the best way to settle the matter without a dangerous dispute [*lis periculosis*]. In the *Riflessioni* he reproached Galileo for not having thought that the cupola of Santa Maria del Fiore in Florence would have been a most suitable place to perform the experiment. Indeed, whilst his 'sincere accusers' were talking of the effects of the earth's motion on the spires and on the *cupolino* of St Peter's in Rome, by chance they were 'innocently' setting in front of his eyes a good proof to settle his case, namely the experiment of dropping balls. Hence the bold proposal of the Abbé Guglielmini to perform the experiment in St Peter's, since the height of the *cupolino* and the 'very appropriate place' were good auspices of success. In order not to disturb the pilgrims, he proposed to perform the experiment at night.[7] It is possible to interpret this proposal by a member of the Catholic Church as an attempt both to emerge with an ingenious publication on an appealing topic, and to regain the initiative on a matter in which the prestige

[6] G. Guglielmini (1763–1817) held the chair of mathematics at the University of Bologna between 1794 and 1817, and of astronomy in 1801. He officiated as Rector of the University in 1814–15. Compare S. Mazzetti, *Repertorio di Tutti i Professori della Famosa Università, e del Celebre Istituto delle Scienze di Bologna* (Bologna, 1843). In his letter to Benzenberg of 23 March 1803 he says of himself: 'In der Religion bin ich Katholick, aber im gesellschaftlichen Leben ein guter Stoiker – ein Freund meiner Freunde, – und gleichgueltig gegen die uebrigen'. *Versuche*, p. 387.

[7] Biblioteca Universitaria, Bologna, MS 2096, Busta IV, f. 85–112, esp. f. 97, 100, 108 (11 July 1789–16 Jan. 1790). On Canterzani compare *Dizionario Biografico degli Italiani* (hereafter *DBI*). *Riflessioni*, p. 7–8. *Opusculum*, p. 4.

of the Church had been badly shaken. By providing direct evidence for the earth's rotation in the very centre of Catholicism, Guglielmini was hoping to make St Peter's and Rome the centre of scientific truth as well. As Benzenberg was to put it a few years later, 'It would be indeed a strong proof of the progress of the times if the trial for a system which the Church had banned were arranged in the holiest city of the entire Christendom, next to the grave of the holy Peter.'[8]

Unfortunately the project could not be carried further in Rome, probably because of the political troubles surrounding Cardinal Boncompagni.[9] In summer 1790 Guglielmini went back to Bologna, where he wasted no time in setting up the experiment from the Asinelli tower, the same used by Riccioli in the seventeenth century. Before considering Gugielmini's trials, however, we need to know what he expected to find.

Guglielmini approximated the trajectory of the falling body with a parabola resulting from gravity and a rectilinear uniform motion which is the difference between the speed of the top and the foot of the tower. He predicted an East deviation equal to uat/r, where t is the time, a the height of fall, and u/r is the angular speed of the earth times the cosine of the latitude. He also predicted a South deviation due to air resistance. In a vacuum, falling bodies and a plumb line would be equally deflected to the South because of centrifugal force, at least in the Northern hemisphere. In the case of air resistance, however, falling bodies would be deflected to the South by a factor proportional to $(K+t)^2$, where K is the time of fall in a vacuum and t is the delay due to the air, which Guglielmini tried to determine experimentally. Since the deflection of the plumb line is only proportional to K^2, a falling body should experience a South displacement with respect to the plumb line, because centrifugal force would act for a time longer than K, namely $K+t$. The proof was based on the assumption of a spherical earth, but later Guglielmini tried to show that even if the earth is a spheroid the result is the same.[10]

After having obtained permission by the Bolognese authorities to use the Asinelli Tower at his leisure, and even financial support, he started an experimental programme which lasted until February 1792. By that time Guglielmini had discovered that dropping balls from high towers was no easy matter. He had to overcome several problems and spend many sleepless nights before finding consistent results.

8   *Versuche*, pp. 432f. J. F. Benzenberg (1777–1846) studied theology, astronomy and physics. He published several pamphlets on politics and economics.

9   The worsening of the political position of Cardinal Boncompagni (1743–1790) forced him to resign in September 1789 (compare *DBI*). On 18 September Guglielmini wrote to Canterzani that his *Riflessioni* were being printed (Bibl. Universitaria, Bologna, Ms 2096, Busta IV, f. 102). Guglielmini's initiative could not have fallen at a less propitious time.   10   *Opusculum*, pp. 23–6 and 58–74.

The Asinelli tower measures 300 Parisian feet. Cutting some concentric holes of approximately a square foot through some intermediate plates inside the tower, Guglielmini was able to free a height of 240 feet. Unfortunately the tower had some holes – due to the scaffolding employed to build it – allowing the wind to blow through, and this imposed a strict restriction on the times when the experiment could be performed. Lead balls were used, of one inch diameter; once they looked perfectly immobile, Guglielmini burnt the silk, linen or metal thread from which they were hanging and let them fall. Although the plumb line from the initial position passed through the centre of the holes he had cut, the first few balls did not even reach the wax plate at the bottom, but were stopped by the intermediate plates. Several trials over a few days led to similar results. Having thought more carefully about this major inconvenience, Guglielmini found by means of a microscope that the balls were still oscillating for a long time – some 25 minutes – after he thought they had stopped. The discovery of microscopic oscillations, while providing a convincing explanation for his previous failures, brought a new range of problems including traffic and even perturbations due to the flame of the candle used to burn the thread.

Between 12 and 14 September 1790 Guglielmini, with the assistance of Luigi Zanotti, and the cooperation of the astronomers Petronio Matteucci [*vir summae authoritatis*] and Francesco Sacchetti [*vir doctissimus*], set up six preliminary trials in the Astronomical Observatory in Bologna. With a height of 90 feet, they found an East deviation from the plumb line of two lines (twelfths of an inch) and no significant South deviation.[11]

The perturbations due to the candle could be overcome, he alleged, by means of a suitable protective device, and by paying special attention; traffic, however, forced Guglielmini and his team including the young Petronio Colliva – who unfortunately died in 1791 –, Luigi Tagliavini, Senator Alemanno Isolani, and Alfonso Bonfioli [*regnantis Pii Papae Sexti Praelatus domesticus*] to work in the Asinelli tower at night.[12] Even then problems were not over: since the balls seemed to rotate, it became necessary to make sure that they were perfectly concentric. It also happened that when, after long preparations, everything was set to drop the ball, a carriage passed down the square and vibrations started again, forcing the team to yet another long pause. As a result of all inconveniences, they managed to drop only a couple of balls during favourable nights. Even so, balls kept falling quite far from one another for no apparent reason. The cold Bolognese winter nights, frustration

---

[11] The hole used by Guglielmini to perform the experiment can still be seen from the stairs in the Bologna Observatory. See also A. Braccesi, 'Un dimenticato experimentum crucis', *Giornale di Astronomia* (1983):319–32, esp. pp. 327–8.

[12] Information about Guglielmini's team is in *Academy of Bologna*, indices. It is worth noticing Guglielmini's emphasis on the social and academic status of his collaborators.

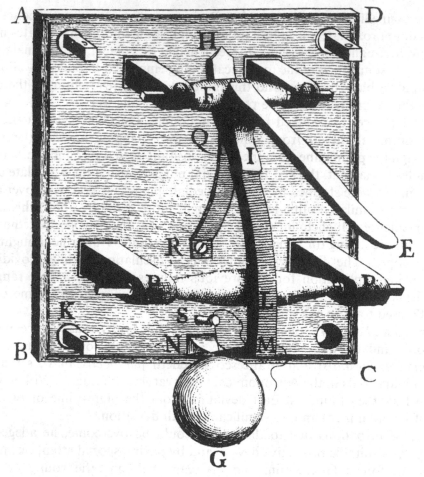

Figure 1. The suspension device built by Francesco Comelli; from Gianbattista Guglielmini, *Opusculum*.

and failure, and even the loss of the young Colliva, who was soon to die, all conspired to the same end: in January 1791 Guglielmini abandoned experiments.

Letters from very distinguished people in Rome, Turin and elsewhere, however, induced him to resume his trials. This time he used a new suspension device built by Francesco Comelli, which allowed him to drop the balls by simple pressure on a lever (see Figure 1). In this way he could eliminate the flame of the candle, which may well have been a cause of vibrations.[13] Eventually, between June and September 1791, sixteen balls were dropped

[13] Some of the correspondence of this period can be found in the Biblioteca dell'Archiginnasio, Bologna, classmarks B4021 to B4023. *Opusculum*, pp. 17–18 and 75–7.

Anno 1791. Experimentorum institutorum dies.	Deiectorum globorum numerus; atque Atmosphæræ status tempore lapsus.	Distantia centri cuiusque fossæ a meridiano TV, per pollices lineas, & decimales partes.	Distantia centri eiusdem fossæ a parallelo TZ, antecedenti menaura putata.
Experimentum I. Tertio nonas Iunii. Ab hora noctis prima in tertiam usqu.	I. quietus. Aere tranquillo.	Pol. 6: 11,25	Pol. 3: 6,83
	II. pene quie. Flante ven to.	Pol. 7: 6,50	Pol. 3: 10,67
Exper. II. Pridie nonas-iusdem. Ab hora noctis prima in tertiam.	I. quietus. Cælo tranquillo.	Pol. 7: 3,00	Pol. 3: 10,67
	II. quietus. Aere leviter flante.	Pol. 7: 0,00	Pol. 3: 11,17
Exper. III. Decimotertio kalendas Augusti. Ab hora noctis prima in tertiam.	I. quietus. Aere tranquillo.	Pol. 7: 2,00	Pol. 4: 0,33
	II. quietus. Aere tranquillo.	Pol. 7: 2,25	Pol. 4: 0,00
Exper. IV. Tertio nonas-eiusdem. Ab hora noctis prima in secundam cum horæ dimidio.	I. quietus. Aere pene tranquillo.	Pol. 7: 0,00	Pol. 3: 11,50
	II. quietus. Cælo tranquillo.	Pol. 7: 3,00	Pol. 4: 0,00
Exper. V. Postridie nonas eiusdem. Ab hora noctis prima in tertiam cum dimidio.	I. quietus. Aere tranquillo.	Pol. 7: 2,00	Pol. 3: 11,50
	II. quietus. Aere pene tranquillo.	Pol. 7: 1,25	Pol. 4: 0,00
Exper. VI. Postridie idus eiusdem. Ab hora noctis prima in quartam.	I. quietus. Aere pene tranquillo.	Pol. 6: 11,25	Pol. 3: 10,67
	II. pene quie. Vento valide flante.	Pol. 8: 5,00	Pol. 4: 4,17
Exper. VII. Tertio nonas Septembris. Ab hora noctis prima in quintam.	I. quietus. Aere tranquillo.	Pol. 7: 2,00	Pol. 4: 0,00
	II. quietus. Aere tranquillo.	Pol. 7: 1,25	Pol. 4: 0,00
	III. quietus. Aere tranquillo.	Pol. 7: 0,00	Pol. 3: 11,17
	IV. quietus. Aere pene tranquillo.	Pol. 6: 11,25	Pol. 4: 1,00

Distantiarum a meridiano TV summa - - - - - - - - Pol. 115: 2,00
quæ per globorum numerum sexdecim divisa dat distantiam centrorum fossarum mediam - - - - - Pol. 7: 2,375

Distantiarum a parallelo TZ summa - - - - - - Pol. 63: 5,68
quæ per globorum numerum divisa dat mediam distantiam - - - - - - Pol. 3: 11,605

Figure 2. Results of Guglielmini's trials from the Asinelli tower, Bologna.

over seven nights. The time of fall was determined experimentally by means of a pendulum and a light at the bottom of the tower which was extinguished by the impact of the ball: it was $4\frac{1}{5}$ seconds. Lastly Guglielmini had to determine with great accuracy the foot of the plumb line in order to measure the deviations, which were previously determined with respect to a suitable conventional point. Although the experiment ended at the beginning of September 1791, it was not until February 1792 that a favourable night with absolutely no wind occurred. Guglielmini placed some candles in the holes of the tower at different heights and waited until the flames had ceased to tremble. He found an average East deviation of 8.375 lines, and an average South deviation of 5.272 lines. The theoretical deviations were 7.581 and 6.163 respectively, in excellent agreement with experimental results. Consistency among experimental data and between theory and observations were such that Guglielmini proceeded with publication (see Figure 2).[14]

Reception in France. The theory of Laplace

In 1797 Jérôme Lalande, who had devoted a few pages to the problem of fall in his *Astronomie*, published a very favourable review of Guglielmini's work. Guglielmini was by then professor of mathematics at Bologna. Events were soon precipitated, however, when Laplace made known that according to his own calculations the East deviation was $\frac{2}{3}$ of the value indicated by Guglielmini, and the South deviation was negligible.[15]

Laplace's statement proved as effective as Napoleon's army:[16] Lalande immediately changed sides, and Guglielmini soon carried out new experiments in agreement with the new predictions and reinterpreted his previous results. Much of this story is known through Lalande's reports and the documents

[14] Gugielmini's publications stimulated debates and controversies in Italy. His most prominent opponent was the Professor of Hydraulics at Ferrara, Teodoro Bonati (1724–1820), who published *Di uno esperimento proposto per iscoprire, se realmente la terra sia quieta, oppure si muova* (without year and place of publication: the preface, though, is dated 'Ferrera, 28 Aprile 1791'). Bonati claimed that the South deviation ought to be six times greater than the East deviation. The Roman astronomer Giuseppe Calandrelli (1749–1827) performed allegedly very accurate experiments in agreement with Bonati's predictions. While criticising Bonati's theory, Guglielmini objected that Calendrelli had not used microscopes in his trials. *Opusculum*, pp. 18–22 and 37–74. On Bonati and Calandrelli compare *DBI*.

[15] J. Lalande, *Astronomie* (Paris, 1771²), vol. 1, book 5, pp. 517–20; J. Lalande, 'Sur une expérience qui prouve les mouvemens de la terre', *Magasin Encyclopédique*, vol. 14 (1797):433–5 and vol. 16 (1797):452. Compare also *Magazin fuer den neuen Zustand der Naturkunde* I, 3, Jena (1798):45–8, reporting Lalande's review with Laplace's objections in Germany. Regarding the South deviation see below.

[16] French armies entered Bologna on 16 June 1796.

later published by Benzenberg. In a letter to Benzenberg of March 1803, for example, Guglielmini wrote that he still did not know Laplace's demonstration.[17] Further experiments by Guglielmini in analogous circumstances gave an East deviation of approximately 7 lines and a South deviation of 3 lines. He now attributed the South deviation to the air 'which softly breezes through the holes of the tower even if outside it is completely still.' Interestingly, he also reinterpreted his six trials in the Astronomical Observatory, which were admittedly preliminary, as evidence that in a closed space where wind cannot interfere the South deviation disappears.[18] Guglielmini changed his theory too and presented an essay on the topic to the Bolognese Academy, where it can still be read. By means of simple proportions, he tried to show that air resistance opposed vertical descent as well as the motion towards the South. In conclusion, air resistance would produce two compensating effects so that not even a virtually undetectable South deviation would occur with respect to the plumb line.[19] Other reinterpretations concerned different aspects. Guglielmini had completed his trials at the beginning of September, but he could not determine the foot of the plumb line before the following February. The considerable differences in temperature between a very hot and a very cold month were referred to by him and Benzenberg as a probable cause for a different inclination of the tower which may have affected the result. Moreover, it was claimed that the measure of time was not accurate enough. As Benzenberg put it rather ungenerously, at the beginning Guglielmini had the support of the authorities and their financial help, as well as the prestige deriving from a theory which was confirmed by experiments; and at the end it became clear that not only was the theory wrong, but also the experiments.[20]

Laplace's theory, which had occasioned this debacle, appeared in print in

[17] J. Lalande, *Magasin Encyclopédique* vol. 41 (1801): 160–1 and vol. 52 (1803): 358, and *Bibliographie Astronomique* (Paris, 1803), p. 789 and 856; *Versuche*, pp. 290–5 and 384–6. Compare also p. 401, where it is reported that in 1797 Guglielmini had agreed that there was no South deviation and had already performed new experiments.

[18] *Versuche*, pp. 294–7. Benzenberg objected that he did not believe that the wind, at different times and in varying conditions, could produce a constant deviation in one direction. Moreover, in his opinion the trials in the Observatory were too few and from too small a height to be significant. He also questioned the new suspension device, which opened always on one side; *ibid.*, p. 335.

[19] Archivio dell'Accademia della Scienze di Bologna. The manuscript consists of six not numbered bound folii (*Academy of Bologna*, p. 307): on 18 November 1802 Guglielmini presented the paper 'On the Northern deviation of falling bodies'. Another paper in defence of his book was read on 20 November 1794: *ibid.*, p. 296. This paper is no longer to be found in the Archivio.

[20] *Versuche*, pp. 291–2. Compare also pp. 385–6, Guglielmini to Benzenberg, 23 March 1803.

the *Bulletin de la Société Philomathique* for 1803, and again, with some variations, in 1805 in the fourth volume of the *Mécanique Céleste*.[21] In the *Bulletin* paper Laplace made clear that despite the general agreement about the East deviation, great uncertainty surrounded the South deviation and its dependence on air resistance. This uncertainty, continued Laplace, stimulated him to provide an exact solution to the problem of fall taking into account air resistance in order to compare theory and observation.

Naming X, Y, Z the three rectangular coordinates of the point from which the body falls; x, y, z the coordinates of the body; r, θ, w its spherical coordinates; n the angular motion of the earth, he set:

$$X = r\cos\theta; \qquad\qquad x = (r - \alpha s)\cos(\theta + \alpha u);$$

$$Y = r\sin\theta\cos(nt + w); \qquad y = (r - \alpha s)\sin(\theta + \alpha u)\cos(nt + w + \alpha v);$$

$$Z = r\sin\theta\sin(nt + w); \qquad z = (r - \alpha s)\sin(\theta + \alpha u)\sin(nt + w + \alpha v);$$

where αs, αu and αv – expressing the change of coordinates during the fall – are so small that their squares and mixed products can be neglected. By taking into account the attractive forces of all the molecules of the earth, considered to be a spheroid, and air resistance, Laplace found by means of the principle of virtual velocities and after having neglected higher order terms, the following equations:

$$0 = \alpha\frac{\mathrm{dd}s}{\mathrm{d}t^2} + \alpha S\frac{\mathrm{d}s}{\mathrm{d}t} - g;$$

$$0 = \alpha\frac{\mathrm{dd}u}{\mathrm{d}t^2} + \alpha S\frac{\mathrm{d}u}{\mathrm{d}t} - g\left(\frac{\mathrm{d}y}{\mathrm{d}\theta}\right);$$

$$0 = \alpha\frac{\mathrm{dd}v}{\mathrm{d}t^2}\sin\theta - 2\alpha n\frac{\mathrm{d}s}{\mathrm{d}t}\sin\theta + \alpha S\frac{\mathrm{d}v}{\mathrm{d}t}\sin\theta - \frac{g}{\sin\theta}\left(\frac{\mathrm{d}y}{\mathrm{d}w}\right);$$

here the radius is set equal to 1; $S = m\alpha\,\mathrm{d}s/\mathrm{d}t$ represents air resistance; m is a coefficient which can be supposed to be constant, and g is local gravity. The first equation regards vertical descent; the second and third equations give respectively the South and East deviations of the falling body.[22] Laplace's reasoning about the South deviation at this point is purely formal, being based on his own mathematical model rather than on a spatial perception of the path

21 P. S. Laplace, *Mécanique Céleste*, vol. 4 (Paris, 1805), pp. 294–305; the *Bulletin* version is reprinted in Laplace, *Oeuvres*, vol. 14 (Paris, 1912), pp. 266–77, and in *Versuche* together with all theoretical investigations.

22 It is worth specifying that y is no longer one of the rectangular coordinates, but results from the equation $r = a + y$, where r is the radius and a is a constant for a given level surface, where pressure is constant. The notion of 'level surface' had been introduced by Colin Maclaurin in *Treatise of Fluxions* (Edinburgh, 1742), article 640, and was later used in studies on the shape of the earth by Clairaut and others.

of the falling body. The first equation gives αs as a function of time. Since the plumb line is perpendicular to a level surface, which in general is not perpendicular to the radius, a plumb line with length αs will deviate Southwards of the radius – in the Northern hemisphere – by the amount $\alpha s (dy/d\theta)$. From the second equation, since g and $dy/d\theta$ can be considered to be constant during the time of fall, we find that a possible solution is $\alpha u = \alpha s(dy/d\theta)$. Indeed, adds Laplace, this is the only solution appropriate to the present question, because of the initial conditions $u = 0$; $du/dt = 0$; $s = 0$; $ds/dt = 0$. Thus the solution to the second equation tells us that there is no South deviation of the falling body with respect to the plumb line.

With regard to the third equation, Laplace claimed that the East deviation of the plumb line is

$$\frac{\alpha s}{\sin \theta}\left(\frac{dy}{dw}\right),$$

whereas that of the body is $\alpha v \sin \theta$. Hence the relative East deviation is

$$\alpha v' = \alpha v \sin \theta - \frac{\alpha s}{\sin \theta}\left(\frac{dy}{dw}\right).$$

Rewriting the differential equation with respect to $\alpha v'$ and taking the integral he found:

$$\alpha v' = \frac{2n}{m} t \sin \theta - \frac{4n \sin \theta}{m\sqrt{(mg)}} ang \, tang \left\{ \frac{c^{\frac{t}{2}\sqrt{(mg)}} - c^{-\frac{t}{2}\sqrt{(mg)}}}{c^{\frac{t}{2}\sqrt{(mg)}} + c^{-\frac{t}{2}\sqrt{(mg)}}} \right\};$$

The first term of the series expansion of this solution gives, neglecting air resistance, a value $\frac{2}{3}$ of that found by Guglielmini:

$$\alpha v' = \frac{ngt^3 \sin \theta}{3}.$$

As one can easily judge even from this extremely summarized account, Laplace's reasoning was very difficult to follow in mathematical terms – this alone was far above Guglielmini and Benzenberg – and with respect to a spatial visualization of what was going on. Mathematical formalism had over-shadowed intuition of the physical causes and their effects.

Early developments in Germany. Olbers and Benzenberg's first trials

On 12 November 1802 Wilhelm Olbers wrote to Gauss that Benzenberg had finished his experiments on the deviation of falling bodies. The experiments had been carried out in the tower of St Michael's in Hamburg from a height of 235 feet and had given an average East deviation of 4 lines and a South deviation of $1\frac{1}{2}$ lines, compared to theoretical predictions of 5.89 and 1.58 lines

	Often		Süden	
1	5,25	Lin.	0,50	Lin.
2	12,50		4,34	
3	9,00		4,34	
4	6,00		4,84	
5	8,00		6,00	
6	8,25		5,67	
7	6,00		5,17	
8	9,00		5,67	
9	8,00		5,17	
10	7,25		5,67	
11	5,25		4,34	
12	23,00		9,84	
13	8,00		5,67	
14	7,25		5,67	
15	6,00		4,84	
16	5,25		6,67	
Mitt. o. Nro. 12	7,4 Lin.		5,02 Lin.	

Figure 3. Guglielmini's results presented in a different fashion by Johann Friedrich Benzenberg.

respectively. The theoretical calculations had been carried out by Olbers, whose help had been sought by Benzenberg. The only reason why Olbers was not completely convinced was that Laplace had given an assurance that there was no South deviation. Hence he submitted the problem with his own solution to the young and brilliant mathematician Gauss, who had recently astonished the scientific world with his computation of the orbit of Ceres, a minor planet between Mars and Jupiter which had been discovered by Giuseppe Piazzi in Palermo, then lost, and eventually rediscovered thanks to Gauss's work.[23]

[23] *Versuche*, p. 346; *Wilhelm Olbers. Sein Leben und Seine Werke*, ed. C. Schilling, 3 vols., vol. 2 (Berlin, 1900; hereafter *Olbers* II), pp. 107–11. For useful background information compare: W. K. Buehler, *Gauss* (Berlin: Springer Verlag, 1981); *Carl*

The theory by Olbers was based on the same principles as Guglielmini's. The Bremen astronomer predicted the same East deviation as the Bolognese mathematician. Further, on the assumption that the earth is a sphere, he found a South deviation of the falling body from the plumb line dependent on air resistance and proportional to $t^2 - a/g'$, where t and a are time and height of fall, and g' is the space traversed by a body falling on a rotating earth in one second, at the required latitude. His initial reply to Olbers shows that Gauss was having some difficulties in thinking about the matter. His speculations about the lack of any South deviation did not convince Olbers, who asked his friend to consider 'that both Guglielmini's and Benzenberg's trials really show this South deviation'. The reply to this observation is very revealing: while yielding to Olbers's pressure, Gauss declared himself more convinced by his friend's authority [Ihre Autoritaet] than by experiments, since the East deviation found by Benzenberg differed from the theoretical value by the same amount as the South deviation differed from zero. As we are about to see, this observation questioning the accuracy of the experimental results was soon followed by further blows to Benzenberg's credibility.[24]

Benzenberg's great faith in experiments was accompanied by a mistrust of theory. In his book he claimed that as a result of his own experiments, even Ptolemy, Tycho Brahe, and Riccioli would have had to believe in the rotation of the earth. He also maintained that for over a hundred years not only had theory been in disagreement with experiments, but also theoreticians had almost always held different ideas even among themselves, and closed with the sentence: 'On the basis of theory there is nothing more to find.' His views are of particular interest because, like many experimentalists today, Benzenberg was unable to understand the theory he was testing, and had to rely on the results and prestige of mathematicians.[25]

Initially he was trying to replicate Guglielmini's experiment. However, Benzenberg objected to the way data were presented in the *Opusculum*: by taking the deviation from the plumb line rather than from conventional points – as Guglielmini had done – he showed that the results appeared in a different

Friedrich Gauss (1777–1855), ed. I. Schneider (Muenchen: Minerva-Publikation, 1981).

[24] Compare the correspondence of November and December 1802, *Olbers* II, p. 112–17.

[25] *Versuche*, p. VI and 402. In a letter to Gauss of March 1803, while expressing some scepticism about the accuracy of Benzenberg's trials, Olbers claimed that 'Benzenberg ist eine trefflicher Kopf und recht zu solchen Versuchen geboren.' *Olbers* II, p. 131. A few years later Gauss expressed his views rather ruthlessly: 'Hr. Benzenberg scheint oft zu glauben, man habe nicht eher gemerkt, dass die Sonne scheine, bevor er es ausgerufen.' *Olbers* II, p. 660, 2 August 1817. Compare also J. F. Benzenberg, *Briefe geschrieben im Paris im Jahr 1815* (Dortmund, 1816), pp. 99–134 (hereafter *Briefe 1815*). On Gauss and Benzenberg see M. Kuessner, *Karl Friedrich Gauss und seine Welt der Buecher* (Goettingen: Musterschmidt, 1979), pp. 41 and 86.

light (see Figure 3).[26] His trials, beginning with a preliminary attempt on 7 October with six gun bullets which fell everywhere [*ueberall*] including six feet to the West, were run in autumn 1802. Despite the tower of St Michael's having no holes and being allegedly not affected by wind, Benzenberg had considerable difficulties. For fear of fires he was not given permission to use artificial light and therefore could not work at night or use microscopes. He had to acknowledge that traffic might affect his results, and that the clock in the church tower produced vibrations too. The solution to these at first sight insurmountable difficulties was brilliant: to run many trials on the assumption that the errors would cancel out under suitable conditions.[27] The longer the series, the more accurate the average; the accuracy of the measure, namely the difference between the calculated average value and the true one, was optimistically defined as the *Fehlergrenze* divided by the number of trials. Benzenberg spelt out the three guiding criteria of his theory:

(1) to eliminate all constant errors which make a long series of trials less accurate than a short one, because errors of equal sign are added together;

(2) when calculating the average, to select only those trials which have been run correctly, and to neglect only those deviating so much, that a major external action must have interfered;

(3) to form series from the observations, such that the greatest accuracy can be attained with the smallest number of trials. This rule applies when observations are affected by an uncertain value, such as time in his fall experiments. In other words, it would be misleading to evaluate the accuracy of the result whilst neglecting the uncertainty over the measure of time.[28]

For the Hamburg experiments time of fall had been determined by means of a clock and an extraordinarily complicated procedure in some previous trials on air resistance: they had given almost exactly four seconds for a height of fall of 235 feet.[29]

Benzenberg was bold enough to tell Gauss about his estimate of accuracy, claiming that in over 31 trials the *Fehlergrenze* was 18 lines. Gauss could not understand what *Fehlergrenze* meant. After consultations with Olbers, it became clear that Benzenberg meant the greatest deviation from the average,

---

[26]   *Versuche*, p. 290. Further, Benzenberg gave the result in lines rather than feet. Curiously, Guglielmini reported East and South deviations with a precision of 1/4 and 1/1000 of line respectively.

[27]   Benzenberg reports that according to some the average of an odd number of trials is better than the average of an even number. Others are of the opposite opinion. He thought this was irrelevant. *Versuche*, p. 339.   [28]   *Versuche*, p. 325ff.

[29]   *Versuche*, pp. 178–80 and 382–3.

1802	Tag.		Nord.	Süd.	Oſt.	Weſt.
Julius	23	1	5	—	—	5
	—	2	—	5	8,5	—
	—	3	—	—	—	—
	—	4	1	—	—	—
	—	5	—	3	21,5	—
	—	6	—	—	—	6
Auguſt	14	7	—	—	16	—
	—	8	—	2,8	1,8	—
	—	9	6,8	—	5,2	—
Auguſt	16	10	—	1,7	—	10
	—	11	2,2	—	12	—
	—	12	—	0,8	—	1,5
	—	13	2	—	—	1
	17	14	4	—	—	8
	—	15	8	—	7	—
	—	16	—	9	1,5	—
Octob.	14	17	5,7	—	0,5	—
	—	18	—	17	12	—
Octob.	15	19	—	10,5	17,3	—
	—	20	—	2	7	—
	—	21	—	8	0,8	—
	—	22	—	2	17,5	—
	—	23	0,2	—	14,7	—
	—	24	—	16,7	0,2	—
Octob.	23	25	—	—	—	14,8
	—	26	—	3	8	—
	—	27	—	5,6	—	4,2
Octob.	26	28	1	—	6	—
	—	29	—	5	13,5	—
	—	30	10,5	—	1,8	—
	—	31	—	0,5	1,7	—
Summe			46,4	92,6	174,5	50,5

Süden 92, 6   Oſten 174,5

Norden 46, 4   Weſten 50,5

46, 2   124, 0

Figure 4. Results of Benzenberg's first series of trials from the tower of St Michael's in Hamburg.

namely 18 lines. Although 8 out of 31 balls had actually fallen to the West, Benzenberg claimed that the uncertainty in his attempts was 18/31, or approximately 0.6 lines (see Figure 4). As Olbers wrote to Gauss, and as Benzenberg himself made clear, his speculations on error theory were influenced by Johann Heinrich Lambert's work.[30] At that time Gauss was very interested in these issues, since some of his astronomical computations were based on probability. In this case he explained that, whatever the meaning of *Fehlergrenze*, the accuracy of Benzenberg's experiment was such that the average could not be determined with the precision of single lines. Despite the shortcomings of his estimate of accuracy, Benzenberg had a relatively innovative attitude to error theory.[31]

At the end of 1802 the situation in Germany was very uncertain: Olbers believed in a South deviation, but he was worried about Laplace's objections, which were not yet in print; details about Guglielmini, who had originally predicted and observed a South deviation, and now seemed to have changed his mind, were not known; Benzenberg claimed that he had observed a South deviation, but Gauss was not convinced by his experiments. Olbers had plenty of reasons to seek the theoretical advice of the most brilliant German mathematician.

The theory of Gauss and its reception

At the beginning of 1803, stimulated further by the correspondence with Benzenberg, Gauss developed the equations of relative motion [*scheinbare oder relative Bewegung*] for a body moving with respect to the surface of the earth. If at the beginning he had relied on the authority of the Bremen astronomer, now Gauss found with astonishment, on the basis of his own

[30] *Versuche* pp. 290, 330–1 and 341ff; *Olbers* II, p. 126. Attention to Lambert's pioneering work on error theory has been drawn by O. B. Sheynin, *Nature* 211(1966): 1003–4; see also the more extensive 'J. H. Lambert's work on probability', *Archive for the History of the Exact Sciences* 7(1970): 244–56, on pp. 249–55; on p. 249 it is reported that according to Galle and Gauss himself, Gauss arrived at the idea of least squares reading Lambert. See also P. Dupont, 'Lambert et le Calcul de Probabilités', in *Colloque International et Interdisciplinaire Jean-Henri Lambert* (Paris: Ophrys, 1979), pp. 261–71; J. H. Lambert, *Beitraege zum Gebrauche der Mathematik und deren Anwendungen* (Berlin, 1792²), pp. 215–313 and 424–88; compare also *Photometria* (Augsburg, 1760), paragraphs 271–306. Lambert estimates the precision of a series of observations as the difference between the arithmetic mean of all observations, and the arithmetic mean of all but the most deviating observation: *Photometria*, par. 294; *Beitraege*, pp. 426–7. Compare *Versuche*, p. 290.

[31] Useful background information can be found in: S. M. Stigler, *The History of Statistics. The Measurement of Uncertainty before 1900* (Cambridge, Mass.: Harvard University Press, 1986); L. Daston, *Classical Probability in the Enlightenment* (Princeton University Press, 1988); *The Probabilistic Revolution*, vol. 1, *Ideas in History*, ed. L. Krueger, L. Daston, M. Heidelberger (Cambridge, Mass.: The MIT Press, 1987); I. Hacking, *The Taming of Chance* (Cambridge University Press, 1990).

calculations, that the South deviation was so small as to be practically undetectable, and that the East deviation was only two thirds of the value predicted by Guglielmini and Olbers. The new results were in perfect agreement with Laplace's, although the mathematics used was very different indeed, being based on rotating reference frames.[32]

At the outset of his calculations Gauss chose two terns of perpendicular planes, where one tern is fixed and the other rotates with the earth. Starting from the coordinate transformations between the two frames, after taking twice the derivative with respect to time, Gauss found the general expression for relative acceleration on the rotating earth. He distinguished between two kinds of terms: those acting on all bodies alike, and those acting only on bodies in motion. Only the latter produce an observable effect. Taking the earth to be a spheroid, and assuming that the combination g of gravity and centrifugal force is constant and acts along the perpendicular Gauss set:

$$0 = \frac{\mathrm{d}\mathrm{d}x}{\mathrm{d}t^2} - 2n\sin\phi\frac{\mathrm{d}y}{\mathrm{d}t} + Mu\frac{\mathrm{d}x}{\mathrm{d}t};$$

$$0 = \frac{\mathrm{d}\mathrm{d}y}{\mathrm{d}t^2} + 2n\sin\phi\frac{\mathrm{d}x}{\mathrm{d}t} + 2n\cos\phi\frac{\mathrm{d}z}{\mathrm{d}t} + Mu\frac{\mathrm{d}y}{\mathrm{d}t};$$

$$0 = \frac{\mathrm{d}\mathrm{d}z}{\mathrm{d}t^2} - 2n\cos\phi\frac{\mathrm{d}y}{\mathrm{d}t} + g + Mu\frac{\mathrm{d}z}{\mathrm{d}t};$$

where x, y, z are the coordinates along moving axes in the directions South–North, East–West and vertical respectively; n is the angular velocity of the earth; ϕ the latitude; M a factor dependent on air resistance; and u the velocity in the moving frame.[33] The approximate solution – neglecting air

[32] *Olbers* II, pp. 126–7, 1 March 1803; Gauss, *Werke*, 5, pp. 495–503, Gauss to Benzenberg, 2 February and 8 March 1803. On rotating axes and reference frames compare D. Bertoloni Meli, 'The relativization of centrifugal force', *Isis* 81(1990):23–43; 'The emergence of reference frames and the transformation of mechanics in the Enlightenment', forthcoming in *Historical Studies in the Physical and Biological Sciences*. C. Wilson has drawn attention to Leonhard Euler's usage of a tern of rotating axes in his works on perturbation theory in 'Perturbations and solar tables from Lacaille to Delambre. The rapprochement of observations and theory, Part II', *Archive for the History of Exact Sciences* 22(1980):189–304, on p. 193, and in the present volume.

[33] The expression 'centrifugal force' is not referred to by Gauss. Notice the terms representing the fictitious acceleration later named after Coriolis. Compare A. Sommerfeld, *Mechanik* (Leipzig: Akademische Verlagsgesellschaft, 1948[4]), pp. 160–3. The same equations were later used, with minimal changes in the conventions, by S. D. Poisson, 'Mémoire sur le mouvement des projectiles dans l'air, en ayant égard à la

resistance – to the second of these equations, representing the East deviation, is:

$$y = -\tfrac{2}{3}z\cos\phi\,nt;$$

where z is the height of fall. The South deviation is proportional to n^2 and is therefore very small.

With regard to these experiments, together with the criticisms we have seen in the previous section, Gauss seemed to object in principle to their precision. He surmised that even if the air appears to be at rest, the motion of the falling ball in the tower may create a wind producing systematic deviations. Thus all trials would be affected by a constant factor dependent on the structure of the tower.[34]

Olbers, while sharing his friend's doubts about the accuracy of the experiments,[35] was not entirely happy with the mathematical theory. In asking where he had been mistaken, he was hoping to find an intuitive cause related to a three-dimensional visualization of the relative trajectory, rather than a refined reasoning based on differential equations.

In order to convince his friend, Gauss tackled the problem again, this time without rotating reference frames. Using a more elementary formulation, he was able to find the same equation with the factor 2/3 he had found before by considering that gravity cannot be assumed to act along parallel lines, since they all converge to the centre of force. Now Olbers was convinced and elated: he could see that even over very small distances, the more refined approximation produced a non-negligible effect. As far as he was concerned, the theoretical problem of the East deviation had been solved.[36]

With regard to the South deviation, the calculations were slightly more complicated, but the intuitive principle explaining the effect was also easy to grasp. Gauss was able to show that, in the reasoning followed by Olbers, air resistance was not treated symmetrically: while considering vertical descent, Olbers had assumed that air resistance prolongs the time t of fall and opposes gravity; when calculating the South deviation, however, Olbers had assumed that centrifugal force acts for the longer time t without being opposed by air resistance. As it happens, the additional South deviation due to centrifugal force acting for the longer time t is virtually exactly compensated by a North

rotation de la Terre', *Journal de l'École Polytechnique* 16(1838):1–68; J. Binet, 'Suite de la note sur le mouvement du pendule simple en ayant égard à la revolution diurne de la terre', *Comptes Rendu de l'Académie des Sciences de Paris* 32(1851):197–205.

34 *Olbers* II, p. 127, 1 March 1803.

35 *Olbers* II, p. 131: 'In general, I consider the determination of a quantity by means of trials to be very uncertain if the possible errors of the trials are 4–5 times larger than the very quantity one wants to determine.' 4 March 1803.

36 *Olbers* II, pp. 139–40 and 143–4.

deviation due to air resistance opposing centrifugal force – neglecting terms in M^2. Therefore the falling body does not deviate appreciably from the plumb line.[37]

The elegant reasoning by Gauss convinced Olbers. Once again, he was able to perceive through the formalism the interplay of actions in nature. The mathematical reasoning accompanied and represented physical phenomena at each step. Indeed, approximations were not obvious and the role of many effects was not clear a priori. Looking for deviations of the order of a few lines, it was by no means certain whether the earth could be approximated as a sphere; whether gravity could be taken to be acting along lines parallel among themselves; whether air resistance had a sizeable effect on the result, and whether it affected the East and South deviations in the same way. Exact solutions could not be calculated and, in order to find a measurable value, approximations had to be introduced at some stage. While looking innocuous at first sight, however, they could well have affected – and in fact they did affect – theoretical predictions.

Paris and Schlebusch. Laplace and Benzenberg's second trials

Despite the advice of theoreticians, Benzenberg was still dubious. As he wrote in *Versuche*, by 1803 Laplace, Gauss and Olbers had agreed that there was no South deviation. Yet not only his own allegedly very accurate experiments, but also those by Hooke and Guglielmini had given a South deviation. However, Benzenberg was isolated and in order to settle the matter he had to rely on what he believed was the most powerful argument: experiment. The decision to test the theory again and to delay publication was taken after a large portion of his book had already been printed, and after Olbers had assured him that there was no South deviation. All previous experiments had been carried out from towers. Perhaps this may have had something to do with the result, surmised Benzenberg, since the side of the South is hotter than the others?[38] These speculations led him to perform a new series of trials in a mine-shaft over 260 feet deep near Schlebusch, not far from Cologne.

---

[37] *Olbers* II, pp. 140–1 and 143–4. See also *Olbers* I, pp. 611–17, Olbers to Benzenberg, 5 October 1803. Centrifugal force is referred to by Olbers, not by Gauss. More directly, one could consider that the earth is not a sphere – as Olbers had assumed – but a spheroid where the combination of gravity and centrifugal force is always perpendicular to the surface. Therefore, since the plumb line is always perpendicular to the earth, air resistance cannot induce a South deviation.

[38] *Versuche*, pp. 357–8 and 405–6. In *Versuche ueber die Umdrehung der Erde aufs neue Berechnet* (Duesseldorf, 1845; hereafter *Versuche aufs neue Berechnet*), p. 27, Benzenberg tells us that he had the first 403 pages of the book printed before the second trials.

Figure 5. Portrait of Laplace from von Zach's *Ephemeriden* (Weimar, 1799).

In the spring the shaft was too wet. In the summer Benzenberg was in Paris and it was not until October 1804 that the experiment could be carried out. The locals assured him that the shaft was as dry as it would ever get.[39] Before examining the new trials, however, we need to consider some events which left a lasting impression on Benzenberg.

In the French capital he encountered a remarkable series of people. His

[39]   *Versuche*, p. 406.

Briefe contain one of the most vivid and entertaining accounts of Napoleonic Paris. As a good German, he was to write eleven years later, he visited everything systematically: the Louvre and the Museum of Natural History, the Collège de France and the Institut National, the Astronomical Observatory and the libraries.[40] Joseph-Louis Lagrange, we are told, was getting old and frail. Charles Augustin Coulomb, though also old, was extremely lively and spoke so fast that one could barely catch a tenth of what he was saying. Benzenberg found it hard to understand how someone so restless could have made such accurate observations on attraction and magnetism with the torsion balance.[41] Lalande was somewhat disappointing. On the basis of the letters he had received from the French astronomer, Benzenberg expected him to look active and full of energy. Quite on the contrary, age had turned him into a quiet and calm man. He repeatedly said: 'I am weak, I am very weak'.[42] His lectures at the Collège de France were not exactly a success: on one occasion Benzenberg counted only six people in the audience. At least there were no soldiers, in contrast to the popular lectures by Antoine Fourcroy and René-Just Haüy, where Benzenberg counted 112 and 100 people respectively. As a German, he remarked, he found the presence of soldiers unusual and unpleasant, since at lectures in Goettingen and Jena there were three times as many people and yet no soldier was to be seen.[43]

Encounters with Laplace greatly impressed the young German. When he first saw Laplace at the Institut, Benzenberg thought he was not quite as remarkable as Lazare Carnot. Benzenberg found Laplace's portrait in Franz von Zach's *Ephemeriden* very life-like (see Figure 5). From the way he looked, his posture and the way he talked, it was easy to discern the self-assurance of a man who believed his superiority was acknowledged.[44]

At a more private meeting in the Observatory Laplace and Benzenberg discussed experiments on the rotation of the earth, and Laplace seemed to be surprised that the shaft near Schlebusch was wet. He probably thought that shafts are as dry as the tower of his observatory, commented the German. After Benzenberg promised to send the results of his new trials as soon as they would be available, Laplace offered to invite him to the Institute. Although Benzenberg had already an invitation by Lalande, he was most impressed by

---

[40] *Briefe 1815*, p. 35. Standard sources on French science under Napoleon are: M. P. Crosland, *The Society of Arcueil. A View of French Science at the Time of Napoleon I* (London: Heinemann, 1967); *Science in France in the Revolutionary Era*, ed. M. P. Crosland (Cambridge, Mass.: Harvard University Press, 1969), which refers to Benzenberg; R. Fox, 'The rise and fall of Laplacian physics', *Historical Studies in the Physical Sciences* 4(1974):89–136.

[41] *Briefe*, I, p. 193. Compare C. S. Gillmor, *Coulomb and the Evolution of Physics and Engineering in Eighteenth-Century France* (Princeton University Press, 1971).

[42] *Briefe*, I, p. 182.   [43] *Briefe*, I, pp. 224–5.   [44] *Briefe*, I, pp. 191–2.

this courtesy from someone of Laplace's rank, the Chancellor of the Senate and former Minister of the Interior. The world thought of Laplace as someone very aristocratic and proud, but from the little he had seen, added Benzenberg, he thought otherwise.[45] If the first impression was not very favourable, the second marked a notable improvement, and the third was to end in deification.

The session at the first class of the Institut turned out to be rather boring [*recht langweilig*]. Benzenberg slipped out in the middle of it and went to the library to leaf through *De Revolutionibus Orbium Coelestium* by Copernicus. In the library he found someone who had left before him: Laplace. Soon they were joined by a third man who had found the session equally tedious: it was Lagrange. He and Laplace started talking about the motion of light, but Lagrange looked tired and had no pertinent comments to offer. However Laplace, who had a very pleasant voice, spoke with extraordinary clarity and precision about what was known, without concealing what was uncertain or unknown. Indeed, we are told, Laplace is an exceptional man: if a meeting is boring he leaves, but if it is interesting, he is very attentive regardless of the subject. Claude Louis Berthollet, Georges Cuvier and Charles Augustin Coulomb, on the other hand, pay attention only to what pertains to chemistry, anatomy and physics respectively, because only few minds like Laplace's embrace 'the entire triangulation of human knowledge'. In the motion of planets he discovered secular perturbations returning after centuries, he applied the same laws to the 'cohesion of bodies, the elective affinities of analytical chemistry and the motions of the humours of organic nature'. Benzenberg concluded that Laplace was indeed the Newton of his time.[46]

Back in Schlebusch life must have looked almost unbearable to Benzenberg, also because of his unenviable accommodation.[47] Laplace, however, was awaiting the results and everything had to be completed before the autumnal rains. Time of fall was found to be 4.16 seconds. Unfortunately these trials were also plagued by several problems, including air currents and dropping water. At the beginning Benzenberg tried to use microscopes, but on realizing that after 57 minutes one ball was still oscillating, he decided to abolish them. Neglecting some results, allegedly affected by drops of water or by the rotation of the ball, Benzenberg was left with 29 suitable trials ranging from a 19 line North to a 15 line South deviation, and from a 20 line East to a 10 line West deviation; 8 balls had fallen to the West (see Figure 6). On the basis of the usual computations, he found a very small North deviation and a 5.1 line East deviation, compared to a predicted value of 4.6 lines. At the end of his report of the experiment Benzenberg noted that setting aside the most divergent trial – number 21, giving a 19 line North deviation – his results would be in even

[45] *Briefe*, I, pp. 210–11.   [46] *Briefe*, I, pp. 236–41.

[47] *Versuche*, I, pp. 411–2.

1804	Okto-ber.	Num-mer.	Nord.	Süd.	Oſt.	Weſt.
	7	1	11,5	—	—	3
	—	2	14,5	—	12	—
	—	3	5	—	3	—
	8	4	4	—	13	—
	—	5	1,5	—	20	—
	—	6	—	2	—	2
	—	7	—	9	11,5	—
	—	8	—	0,5	—	4
	—	9	—	15	2	—
	—	10	8	—	2	—
	—	11	—	5	12	—
	—	12	13,5	—	7	—
	—	13	3,5	—	13,5	—
	—	14	—	0,5	11	—
	9	15	13	—	9	—
	—	16	—	4	—	8
	—	17	8	—	8	—
	—	18	10	—	10	—
	—	19	—	13	7	—
	—	20	—	11	7,5	—
	—	21	19	—	6	—
	—	22	—	8,5	—	2
	—	23	—	0,5	11	—
	10	24	—	15	—	4
	—	25	—	6	—	9
	—	26	—	7	—	10
	—	27	8	—	8,5	—
	—	28	—	6	10	—
	—	29	4,5	—	5,5	—
Summa			124	103	189,5	42

Figure 6. Results of Benzenberg's second series of trials in a shaft near Schlebusch.

better agreement with theory. In his opinion the new series of trials proved conclusively that theoretical predictions were correct and that there was no South deviation. Publication followed at the end of 1804.

The reception Benzenberg encountered was mixed. Gauss had already expressed his doubts and acknowledged receipt of the book without further comments. Olbers, while observing that this time no South deviation had been found and that the cause of the South deviation in Hamburg could not have been the rotation of the earth, added 'as long as trials giving such different results are at all reliable with $1\frac{1}{2}$ lines from the average.'[48] Laplace claimed that the experiments performed in Italy and Germany were in accordance with his theory, but added that they had to be replicated with greater accuracy. According to Benzenberg, however, who met him again during a second journey to Paris in 1815, Laplace had investigated his second trials with the calculus of probability and found that he could bet 8000 to 1 that the earth rotates, and this only on the basis of Benzenberg's trials.[49]

Debates and experiments on the problem of fall continued for many decades. Although the scope of this paper extends only to the first few years of the nineteenth century, it is worth mentioning that in 1831 a further experiment by Ferdinand Reich in a shaft near Freiberg gave a South deviation and fuelled additional speculations. Hans Christian Oersted, John Herschel and William Grove discussed the possibility that the South deviation, if confirmed, could result from magnetic or electrical phenomena. The existence of a South deviation was still being tested in 1903 with a 948 trial experiment at the Jefferson Laboratory of Harvard University. Guglielmini and Benzenberg have been alternately praised and ridiculed until our century. An article published in the 1980s still defends Guglielmini's 1792 theory and experiments, and claims that with regard to the South deviation Guglielmini saw further than Laplace.[50]

In this essay I have shown that the resources employed to tackle the problem of fall involved sophisticated and comparatively novel techniques including a rigorous quantification of experiments, data analysis and error theory, and rotating reference frames. Thus the problem I have investigated intersects some crucial themes at around 1800 exactly as it had done in the seventeenth century. Although the refined theoretical studies by Laplace and Gauss had a crucial role, they were not accepted by the other protagonists simply on the

48   *Olbers* II, pp. 244–5 and 249.

49   P. S. Laplace, *Mécanique Céleste*, vol. 4, p. 302; *Briefe 1815*, pp. 97–8; *Versuche aufs neue Berechnet*, p. 46; on p. 28 Benzenberg attributed the South deviation in Hamburg to temperature differences between the North and the South side of the tower.

50   F. Reich, *Fallversuche ueber die Umdrehung der Erde* (Freiberg, 1832); H. C. Oersted, 'On the deviation of falling bodies from the perpendicular', *The American Journal of Science and Art* (1847):138–40, and *Report of the Sixteenth Meeting of the British Association for the Advancement of Science* (London, 1847), Miscellaneous Communications, pp. 2–3; L. P. Gilbert, 'Les Preuves Mécaniques de la Rotation de la Terre', *Bulletin des Sciences Mathématiques et Astronomiques*, second series, 6(1882):189–223. The Harvard experiment gave virtually no South deviation: E. H. Hall, 'Do falling bodies move south?', *Physical Review* 17(1903):179–90 and 245–54; J. G. Hagen, *La Rotation de la Terre. Ses preuves mécaniques anciennes et nouvelles* (Rome, 1911), esp. pp. 23–29; A. Armitage, 'The deviation of falling bodies', *Annals of Science* 5(1947):342–51.

basis of their contents: persuasion involved prestige and authority, agreement between theoretical predictions and observations, accuracy and internal consistency of experimental results, and a theory allowing physical actions to be mirrored – and not obscured – by the mathematical formalism. Only the interaction between these elements led to the provisional settling of the debate.

— V —

After Newton: optics and dynamics

18 Why Stokes never wrote a treatise on optics

JED Z. BUCHWALD

With Stokes, mathematics was the servant and assistant, not the master. His guiding star was natural philosophy. Sound, light, radiant heat, chemistry, were his fields of labour, which he cultivated by studying properties of matter, with the aid of experimental and mathematical investigation. [From Kelvin's posthumous tribute to Stokes in *Nature*.]

The indiscretion of plausible conjecture

In the fall of 1878 George Gabriel Stokes had been secretary to the Royal Society of London for nearly a quarter-century and Lucasian Professor of Mathematics at Cambridge for five years beyond that. Well-known in Britain and on the Continent for his papers on mathematics, hydrodynamics and optics, as well as for his experimental investigations of fluorescence, at the age of fifty-nine Stokes was nevertheless something of a disappointment to his British colleagues. Older than William Thomson (later Lord Kelvin) by four years, and than James Clerk Maxwell by fourteen, Stokes had failed to produce a major text. Thomson (together with Tait), had created the immensely influential *Treatise on Natural Philosophy* in 1867.[1] Maxwell's 1873 *Treatise on Electricity and Magnetism* already informed the understanding of a generation, and Rayleigh's 1877 *Theory of Sound* treated that subject in ways that often paralleled the two *Treatises*. The acknowledged master of hydrodynamics and optics, Stokes had produced nothing similar, despite the fact that during the 1860s 'the scientific world expected from him a systematic treatise on Light, and indeed a book was actually advertised as in preparation'.[2]

Not only did Stokes fail to write the awaited *Treatise*, by the late 1870s his overall scientific productivity was not what it had been a decade before. Rayleigh, no doubt reflecting a rather common opinion, attributed the absence of the book on light to 'pressure of work, and perhaps a growing habit of procrastination'. There is ample evidence in his *Memoir and Scientific Correspondence* that Stokes did often put things off, and he was certainly

---

[1] 2 vols. (Cambridge University Press).

[2] Lord Rayleigh (J. W. Strutt), 'Obituary notice of Sir George Gabriel Stokes, Bart. 1819–1903', *Roy. Soc. Proc.* 75(1904):199–216, on p. 211.

burdened by his long secretaryship of the Royal Society, which he took quite seriously. However he could at any time after the early 1860s easily have produced a text in optics that presented a unified view of the subject based in major part on two important memoirs that he had published in 1849 and 1862. But he never did, and in Britain there was no one of comparable stature and knowledge to produce the missing text. Consequently for nearly forty years British students lacked a modern treatment of optics whereas they were well-supplied in the books of Maxwell, Thomson and Tait and of Rayleigh with elaborate, highly-developed discussions of electromagnetism, mechanics and sound.

Stokes's failure to put pen to paper does not reflect the absence of a coherent view of the subject on his part. On the contrary, his published articles present a unified scheme for optics, one which remained surprisingly constant. Moreover Stokes lectured on the subject at Cambridge from 1849 for nearly half a century. And for at least half of this time many future British mathematicians and physicists heard him speak, gleaning a substantial part of their understanding of optical principles from him.[3] Yet Stokes's career ceased being on the forefront of research sometime between the mid 1850s and the early 1860s. From that time on Stokes began increasingly to sit on the sidelines, his influence bracketed on one end by his often detailed critiques of papers submitted to the Royal Society and on the other by the extensive correspondence that he maintained.

Rayleigh perceptively captured the character traits that, in part, pulled Stokes away from novel experimental and mathematical research after his fortieth birthday:

> Perhaps [Stokes] would have been the better for a little more wholesome desire for reputation. As happened in the case of Cavendish, too great an indifference in this respect, especially if combined with a morbid dread of mistakes, may easily lead to the withholding of valuable ideas and even to the suppression of elaborate experimental work, which it is often a labour to prepare for publication.[4]

Lack of competitiveness, fear of error, and a certain laziness – all of these traits were remarked by Stokes's friends and family or even by himself, and they do go far in explaining why he did not fulfil his early promise, and even why he

---

[3] David B. Wilson, *Kelvin and Stokes. A Comparative Study in Victorian Physics* (Bristol: Adam Hilger, 1987), pp. 42–53 nicely discusses Stokes's lecturing career, and points out that 'Stokes's course was one for the best students until the mid-1870s, with about 80 per cent of the top ten wranglers during that time enroled' (p. 45). After that time enrolment increasingly declined, until by the early 1890s Stokes's course had apparently become almost completely irrelevant.

[4] Rayleigh, 'Obituary notice', p. 210.

never produced a text in optics or in hydrodynamics. But in the case of optics at least he had all the material ready to hand by the early 1860s, and it would not have been an immense labor to write the book.

Why, then, did he never write the text? In the absence of definite evidence one can of course only conjecture, but it seems probable that Stokes never produced the book because he feared that he did not have a proper subject to write about, and that to attempt to create the subject might lead him into what he most feared: public error. A hint of his attitude can perhaps be gleaned from his much later (1887) Burnett Lectures, which were published in the *Nature* series of science popularizations.[5] The lectures were divided into three 'courses', the courses being given in separate years (1883, 1884 and 1885). The first one, entitled 'On the nature of light', contained the substance of physical optics; the second concerned 'investigations carried on by the aid of light' (including absorption and emission, as well as spectral phenomena); the third dealt with light's 'beneficial effects'. In the third lecture of the second course Stokes made the following remark, which reveals something of the difficulty he had had for over two decades in producing a mathematical account of contemporary optics as a unified science:

> In all the phenomena which I have brought before you in my last lecture and in this, and indeed in all that I shall have occasion to mention in this year's course, there is a very intimate relation between molecular grouping and the optical features observed. We touch here on the boundaries of our present physical knowledge. That light consists in the vibrations of a subtle medium or *ether*, that self-luminous bodies, including phosphorescent bodies, which are for the time being self-luminous, are in a state of molecular agitation which they are capable of communicating to the ether, that consequently in the phenomenon of absorption molecular disturbance is excited in bodies at the expense of etherial vibrations – all this is so well established as to leave no reasonable room for doubt. But what may be the mode of connexion by which the vibrations of the ether agitate the

[5] George G. Stokes, *Burnett Lectures on Light. In Three Courses* (Nature Series. London: Macmillan and Co., 1887). Like other texts in the series Stokes's lacked mathematics but was nevertheless written at a very high level – sufficiently high, in fact, that one doubts that most of his auditors could have gleaned much from it. Stokes was the first Burnett lecturer, having no doubt been chosen both for his eminence and (even more so) for his overwhelming (if rather idiosyncratic) interest in revelation, since the terms of the lectures included that '...the lecturer [shall have] regard, in treating of the special subject prescribed, to the illustration afforded by it of the theme proposed by the testator', which was 'That there is a Being, all-powerful, wise, and good, by whom everything exists...' and so on in like vein. The Burnett trust was originally established in 1784 for 'various charitable and pious objects', in Stokes's words, including essays on this theme. Its terms were revised to form the lecture series that began in 1878.

molecules, or the molecules in their turn are able to agitate the ether, what may be the cause of the diminished velocity of propagation in refracting media, what may be the mechanical cause of the difference of velocity of propagation of right and left-handed circularly polarized light in media like sirup of sugar, which is manifested by a rotation of the plane of polarization of plane-polarized light through bodies – all these are questions concerning the true answers to which we can affirm nothing, though plausible conjectures may in many cases be framed.[6]

'Plausible conjecture' – *that* was Stokes's intellectual problem, because he refused to commit himself in print to anything that he felt to be merely 'plausible',[7] and yet he was also unwilling to treat a subject less than thoroughly. For him the *causes* (in the sense of the quotation) of phenomena as fundamental as refraction would necessarily have formed an important part of the subject's textual treatment, and in their absence a fully-fledged subject simply did not exist – and so no reasonable text could be written for it. Moreover in the case of optics the need for the kind of text that Stokes might have written – one without causal conjecture – was not so pressing as it was, e.g. in electricity and magnetism, since the fundamental physical and analytical structure of optics had been reasonably well fixed since the late 1820s, whereas in electricity and magnetism the structure had undergone wholesale revision in the hands of Faraday, Thomson and Maxwell over a forty year period that ended with the production of the requisite text (Maxwell's *Treatise*).

By the late 1870s Stokes was no longer contributing in a substantial way to contemporary optics – to the extent that Richard Tetley Glazebrook, instead of Stokes, wrote the major report on the subject for the British Association in 1885. The previous report in 1862 had indeed been written by Stokes, and it had dealt fully with double refraction.[8] But, Glazebrook revealingly noted a quarter-century later, 'unfortunately [Stokes] confined himself to that one

---

[6] Stokes, *On Light*, pp. 166–7.

[7] This could go quite far. During 1882 he and Rayleigh corresponded about viscous effects at the boundary between a solid and a fluid. Stokes felt that eddies would necessarily form in the boundary layer which would expel dust particles from the region. This affects Rayleigh's analysis of 'the Dark Plane which is formed over a heated wire in dusty air', which he had sent to Stokes for publication in the *Proceedings of the Royal Society*. Stokes wanted to add what would have been an interesting but innocuous note on the point, but he could not bring himself to do so: 'I have abstained from putting in a note about this', he wrote to Rayleigh, 'because the existence of such a narrow stratum of eddies is at present only conjectural'. See George G. Stokes, *Memoir and Scientific Correspondence*, selected and arranged by Joseph Larmor, 2 vols. (Cambridge University Press, 1907. Reprinted New York: Johnson Reprint Corporation, 1971), vol. 2, p. 109.

[8] George G. Stokes, 'Report on double refraction', *British Association Reports* (1862), pp. 253–82.

branch of the subject'. What had Stokes *not* treated in 1862 that he might have? He could not very well have dealt with anomalous dispersion (as Glazebrook did in 1885), since it was not discovered until 1870, but there were two areas he certainly could have treated in some detail but did not: reflection and dispersion. Instead he chose to discuss only double refraction – though the previous British Association *Report*, written by Humphrey Lloyd in 1834 on the very eve of the wave theory's wide dissemination, had discussed nearly every possible topic that the subject might embrace. In 1862 Stokes had limited himself severely indeed, and this I believe was already a symptom of his growing unwillingness, or temperamental inability, to stay on the cutting edge of research. To do that he would *in print* – and not just in correspondence – have had to say much that was merely plausible.[9] And that he refused to do.

Analysis and experiment

If you gave Stokes the Sun there was no experiment he could not do for two-pence.[10]

Despite the fact that Stokes chose in 1862 to write a *Report* solely on double refraction his own reputation in optics at the time had little to do with the subject, though he had certainly thought much about it. Stokes had instead worked primarily on three other, distinct topics, two of which were closely bound to laboratory investigation. The earliest (1845), which had little to do with contemporary experiment, involved the conditions that should be imposed on the ether's motion in the vicinity of the earth in order to yield a correct account of stellar aberration.[11] After 1849 he rarely discussed this topic, but it was important early in his career in that, through an extended and public controversy with Challis to which it led, Stokes had a lesson in how dangerous premature assertion can be.[12]

[9]   Stokes began the *Report* by 'regretting' that 'in consequence of other occupations the materials for a complete report on Physical Optics, which the British Association have requested me to prepare, are not yet collected and digested'. He offered a report on the 'single branch' of double refraction alone, which we shall see could be treated analytically with results that gave nothing away to mere plausibility.

[10] Saying round Cambridge: Stokes, *Memoir*, vol. 1, p. 19.

[11] On which see Wilson, *Kelvin and Stokes*, chap. 6 for details. Stokes required that the ether have no slip relative to the earth at its surface, that its velocity must lack divergence, and that it possess a velocity potential. In 1886 Lorentz demonstrated that Stokes's conditions are inconsistent (on the effects of which see Jed Z. Buchwald, 'The Michelson experiment in the light of electromagnetic theory before 1900', *AIP Conference Proceedings* 179(1988): 55–70).

[12] During their controversy Stokes was forced to admit on one critical point that Challis had the better of him (Wilson, *Kelvin and Stokes*, p. 141). Then, in 1849, Stokes became Lucasian professor at Cambridge, where Challis lectured on all of physics. On

Shortly before Stokes became Lucasian professor at Cambridge he finished his first major analytical work in optics, which was entitled *On the dynamical theory of diffraction*.[13] In it Stokes penetrated very far into the mathematical core of the contemporary wave theory – farther than anyone had since Fresnel – and he also derived an important new result that he immediately sought to confirm in the laboratory. It is worth spending a moment on Stokes's theory of diffraction, both because he did *not* apparently discuss the subject in detail in his optics lectures,[14] and because it illustrates the way in which, when convinced that he was on the firmest ground, Stokes could nicely draw physical conclusions with important and direct empirical significance from intricate analysis.

Stokes's purpose was to uncover the function that governs the dependence of the amplitude of secondary waves on the direction with respect to the normal to the front which they form.[15] To do so he began at once with the general differential equation of motion for an isotropic elastic solid:[16]

$$\partial^2 \mathbf{u}/\partial t^2 = b^2 \nabla^2 \mathbf{u} + (a^2 - b^2) \nabla(\nabla \cdot \mathbf{u})$$

where $\mathbf{u}$ is the displacement, and a, b are elastic constants. He then separated the equation by defining 'for shortness' δ as the negative compression $\nabla \cdot \mathbf{u}$ (or 'dilatation' as he called it), and ω as the rotation (or, again in Stokes's terminology, the 'distortion') $(\frac{1}{2})\nabla \times \mathbf{u}$:

$$\partial^2 \delta/\partial t^2 = a^2 \nabla^2 \delta,$$

$$\partial^2 \omega/\partial t^2 = b^2 \nabla \omega.$$

The single equation for the compression, and the three for the components of the rotation, all have precisely the same form, and Stokes at once wrote down the following solution for them, which he obtained from Poisson:

$$U = (t/4\pi)\iint F(at)\,d\sigma + (1/4\pi)(d/dt)\{t\iint f(at)\,d\sigma\}.$$

Stokes's arrival Challis climbed to the lofty perch of astronomy, handing fluid mechanics and optics over to him (Wilson, *ibid.*, p. 44).

[13]   According to Rayleigh – who himself had a rather jaundiced view of Stokes's conclusions (see Stokes, *Memoir*, vol. 2, pp. 110–11) – this work guaranteed him the Lucasian [Rayleigh, 'Obituary notice', p. 204].

[14]   As is apparent from John Ambrose Fleming's notes of the lectures given in 1879: Papers of John Ambrose Fleming, 1879, MS ADD 122/36. The Library. University College, London.

[15]   This was an important question that had much vexed Fresnel over thirty years before, on which see Jed Z. Buchwald, *The Rise of the Wave Theory of Light* (The University of Chicago Press, 1989), pp. 194–6.

[16]   As developed in this form by Stokes himself in 'On the theories of the internal friction of fluids in motion, and of the equilibrium and motion of elastic solids', *Cambridge Philosophical Society Transactions* 8(1845):287–319.

Here U is the solution at some point P, t is the time, and f, F are respectively the initial values of the function and its time derivative. The integrals are taken over a surface of radius at (or bt) that surrounds the field point P.[17]

It is particularly ironic that Stokes took this from Poisson, because he at once used it to argue that the amplitude of the secondary waves varies with direction in a fashion that Poisson himself would probably not have accepted.[18] The solution expresses the value of the function at a given place and time in terms of integrals over the region that contained the initial disturbance.[19] Through clever manipulation Stokes was able to conclude from it that the integral for the pulse of distortion contains the factor $1 + \cos \theta$, where θ is the angle between the normal to the pulse at the element of integration and the line from there to the field point. Significant though this was – since it justified in retrospect Fresnel's assumptions – it was not the centerpiece of Stokes's analysis.[20]

[17] Here $d\sigma$ is $\sin \theta \, d\theta \, d\phi$ where θ, ϕ are the angular spherical coordinates for the surface surrounding the field point P. The integrals therefore correspond to the *mean values* of the functions over the surfaces: see B. B. Baker and E. T. Copson, *The Mathematical Theory of Huygens' Principle* (Oxford: Clarendon Press, 1939), chap. 1 for details. Poisson's solution has the peculiarity of representing the effect at P in terms of a time-dependent radius that is drawn from P. Instead, that is, of following a pulse as it expands outwards, with Poisson's solution we start at a given point and cut space with surfaces drawn about it until we find surfaces that pass through the regions which contain the initial disturbance. Poisson, as it were, held fixed the initial disturbance and went looking for it from the field point.

[18] Buchwald, *Wave Theory*, p. 192.

[19] Baker and Copson, *Huygens' Principle*, pp. 12–15 discuss Poisson's solution, which is difficult to formulate in a rigorous manner. Note that the solution does not require periodicity but applies to any disturbance that begins at some moment, i.e. to pulses. Stokes justified the extension to infinitely long wave trains from pulses in the following way:

> In the investigation it has been supposed that the force [disturbance] began to act at the time 0, before which the fluid was at rest, so that $f(t) = 0$ when t is negative. But it is evident that exactly the same reasoning would have applied had the force begun to act at any past epoch, so that we are not obliged to suppose $f(t)$ equal to zero when t is negative, and we may even suppose $f(t)$ periodic, so as to have finite values from $t = -\infty$ to $t = +\infty$.

Stokes is a bit disingenuous here, since not only his investigation, but Poisson's solution, requires the limitation. Stokes's quick attempt to extend the class of allowable functions to cover those which are not temporally delimited requires a great deal more justification than this.

[20] Neither Rayleigh (*Theory of Sound*, 2 vols. (Cambridge University Press, 1894. Reprinted New York: Dover Publications, Inc., 1945), vol. 2, chap. 14) nor Todhunter (*A History of the Theory of Elasticity and of the Strength of Materials from Galilei to Lord Kelvin*, edited and completed by K. Pearson, 3 vols. (Cambridge University Press, 1886–93. Reprinted New York: Dover Publications, Inc., 1960), vol. 1, secs. 1263–75)

In Stokes's eyes and in the eyes of his contemporaries the core result of his investigation – the one that had great importance for ongoing controversies – concerned a relation that he obtained between diffraction and the plane of polarization.[21] If α_i, α_d are respectively the angles between the plane of *oscillation* and the plane normal to that of diffraction for the incident and diffracted rays then with θ the angle of diffraction Stokes found:[22]

$$\tan \alpha_d = \cos \theta \tan \alpha_i.$$

This could be examined in the laboratory. Stokes performed a series of difficult experiments in his Cambridge rooms, followed by complicated estimations of error, which – though not without a great deal of discussion concerning disturbing effects – supported the oscillation's being normal to, rather than in, the plane of polarization. The conclusion did not achieve immediate assent, in part because the experiment was extraordinarily difficult to replicate, but also because Stokes's analysis was itself hard to penetrate and also because it seemed critically to depend upon the propagation equation that

mentions Stokes's discovery of the inclination factor, though both point out his use of Poisson's solution. Baker and Copson (*Huygens' Principle*, sec. 4.6) approximate in Helmholtz's diffraction integral for a *periodic* disturbance (to small wavelengths) and obtain thereby the inclination factor. They write that this solution is 'due to Stokes', referring to his 'On the dynamical theory of diffraction', *Transactions of the Cambridge Philosophical Society*, 9(1849).

Stokes certainly did obtain the inclination factor, and he did also deploy Poisson's solution. But it is extraordinarily difficult to see how he obtained the former from the latter when the *only* way that it has been obtained elsewhere is through Helmholtz's integral for *periodic* disturbances. Stokes's solution was apparently much more general, embracing any retarded function at all. These are difficulties here which badly need clarification, but this will require a thorough study of the subsequent history of diffraction theory, as well as a detailed analysis of Stokes's own *Dynamical theory*.

[21] The best, and most comprehensive, treatment of this and related issues concerning ether dynamics remains E. T. Whittaker's *A History of the Theories of Aether and Electricity*. 2 vols. (New York: Humanities Press, 1974), vol. 1, chap. 5. As always with Whittaker, however, the discussion reads very much like a retrospective British Association *Report* on a newly-deceased issue. Great technical insight can be gleaned from such things, but equally great care must be taken to maintain a critical, historical distance from the argument. Whittaker's *History*, as I have previously remarked, must itself be treated as a kind of primary material.

[22] The plane of polarization was conventionally specified by its angle with respect to the plane of diffraction. Consequently if the direction of oscillation is *perpendicular* to the plane of polarization then the angles α in Stokes's equation correspond directly to that plane, but if the oscillation is *in* the plane of polarization then the α are the complements of the polarization angles. Stokes's α can therefore always be treated as conventional polarization angles if, in his formula, $\cos \theta$ is used for oscillations perpendicular to the plane of polarization, and $\sec \theta$ is used for oscillations in that plane.

he had chosen.[23] Indeed, during the next forty years a great deal of experimental and theoretical attention was devoted to this topic, one that nicely mixed physical and mathematical complexities with laboratory intricacy.[24]

Stokes devoted nearly half of his paper to an elaborate discussion of the experiments, for here he hoped to present a new 'discovery' and not merely a new result of analysis. Despite Stokes's present (and indubitably deserved) reputation as a master of analysis, he himself was most strongly attracted to the unearthing of novel experimental lore. This is hardly surprising given the temper of the era, which placed an extremely high premium on discovery, but his desire for great success in the optical laboratory was only partially fulfilled, and in any case not by these experiments on polarization and diffraction. They remained controversial, both on theoretical and on experimental grounds, so that Stokes had no hope of claiming here a universally-recognized discovery.

This changed dramatically in 1852, for in that year Stokes discovered something that rapidly extended his reputation from mathematics into natural philosophy. In his own words, written at the time:

[23] In a note appended to the version printed in his *Memoir* Stokes noted that Holtzmann in 1856 obtained results that led to the opposite conclusion from his own, but that Lorenz in 1860 – while rejecting Stokes's analysis – obtained the same empirical result that Stokes had. Clearly the boundaries of contemporary optical technique were stressed by the polarization of diffracted light. Glazebrook, e.g., remarked in 1885 that the 'experiments are troublesome, and the comparison of the results with theory is complicated by the fact that the refraction through the glass plate on which the [diffraction] grating is ruled also produces a change in the position of the plane of polarisation' (R. T. Glazebrook, 'Report on optical theories', *British Association Reports* 1885:157–261, on p. 203). M. E. Mascart, *Traité d'Optique*, 3 vols. (Paris: Gauthier-Villars et Fils, 1893), vol. 3 still regarded the experiments as doubtful, remarking 'It does not seem to me that the agreement of the measurements [with theory] can authorise [Stokes's conclusion]'. The question was eventually settled when the electromagnetic theory of light achieved widespread acceptance. According to it the Fresnel sine law for reflection applies to the electric field vector when the latter is *normal to* the plane of reflection. Since, empirically, that law holds for light polarized *in* this plane, then it follows that the direction of oscillation must be normal to the plane of polarization *if* the electric field is the optical vector. It seems that the latter question was reasonably well-settled by Otto Wiener in 1890, who produced standing waves in photographic emulsions that were only fractions of a wavelength in thickness (see M. Born and E. Wolf, *Principles of Optics*, Fifth edn (Oxford: Pergamon Press, 1975), pp. 279–80 for a brief discussion of Wiener's experiments). Of course once the electromagnetic theory prevailed this issue was no longer particularly interesting since the relationship of the optically-effective vector to the plane of polarization no longer had any importance for distinguishing between equally viable, alternative theories.

[24] See the lengthy discussion in Glazebrook's 'Report', chap. 6 for details. Some of the conceptual problems that arose concerned the applicability of Stokes's solution to wave trains, as well as problems in understanding how to use Huygens's principle.

> I discovered on Monday, April 28th, 1852, that in the phenomenon of interior dispersion a ray of light actually *changes its refrangibility*. In sulphate of quinine … the violet rays of a certain refrangibility produce the interior dispersion noticed by Sir D. Brewster, while the invisible, or at any rate barely visible, rays beyond the extreme violet produce the narrow band of light described by Sir J. Herschel …[25]

The following September Stokes presented his discovery of fluorescence at the British Association meeting in Belfast, where, following Stokes's talk, the Association's president, Col. Edward Sabine, effusively remarked:[26]

> many would look back with delight to their presence there that evening, as they watched the onward progress of him whose present discovery was but a first step, of him who, if God is pleased to spare his life, promises to be one of the first scientific men of his age or of any other; that his countrymen have good reason to be proud of him …[27]

John Herschel, in his report for the Royal Society on the formal paper that Stokes submitted, considered it 'to be one of the most remarkable and important contributions to physical optics which have appeared for a long time'.[28]

Stokes many years later described his discovery as having thrown open 'a new field of research',[29] and he had great hopes for continuing in the same vein, as his reply to a question concerning his 'favorite occupation' from his prospective wife, Mary Robinson, nicely shows: '8. Occupation. Scientific investigations, especially when they lead to discoveries.'[30] Stokes's preoccupied

[25] Stokes, *Memoir*, vol. 1, p. 9. Herschel called his observation of what, after Stokes, was understood to be fluorescence, 'epipolic dispersion', and it had not occurred to him that a substance could actually change the wavelength of light. Stokes noted in his Burnett Lectures that he had arrived at his discovery by 'reflecting' on Herschel's epipolic dispersion.

[26] According to the recollection of Stokes's sister Elizabeth.

[27] Stokes, *Memoir*, vol. 1, p. 10.

[28] Stokes, *Memoir*, vol. 1, p. 128. Stokes's discovery stimulated a great deal of interest indeed, including a request from the Prussian Prince of Salm-Horstmar for an appropriate bit of glass to carry out the experiments (p. 136). The correspondence between Stokes and Salm-Horstmar went on for a decade.

[29] Stokes, *Memoir*, vol. 1, p. 9.

[30] Stokes, *Memoir*, vol. 1, p. 62. His daughter draws an evocative image of Stokes at experiment:

> As a child I loved to watch him working at experiments in his study; I can still see the Rembrandt effect of the strong light and shade cast upon his face, when he opened the shutter from time to time to alter the position of the things resting on the bracket, and the absorbed and delighted expression of his countenance. (*Ibid.*, p. 19)

abstraction, no doubt enhanced by his long bachelor years and his recent success, nearly repelled the young Miss Robinson, who was looking for romance and close companionship. He, at least, found family life extremely congenial, but after his marriage in 1857 he produced nothing quite so original as his *Dynamical theory* nor did he ever again generate something new in the laboratory.[31] His career at the cutting edge of physical research was over by the mid-1850s, even though it had come into full bloom only in 1852.

Psychological and social factors were unquestionably instrumental in collapsing Stokes's career into a comparatively small compass by 1860. Marriage and family life blunted the edge of his none-too-sharp ambition; he immersed himself in Society business to the detriment of research. Most important of all, Stokes mightily feared public commitment, and not only in science. These factors operated in conjunction with his overpowering conviction that partial theories were not enough, that there was little point in developing accounts that had to stop just where they became most interesting. His daughter reminisced that Stokes 'could not bear "scientific romancing", as he called it'.[32] This was already apparent in his work on fluorescence, which, in print, he did not pursue far beyond the discovery of the change in wavelength itself. We can see an oblique reflection of his objection to 'romancing' in remarks he made many years later in his Burnett Lectures:

> in speaking of a change of refrangibility I would guard against being misunderstood. All that is intended is that light of one refrangibility being incident on the substance, light of a different refrangibility is emitted so long as the first remains in action. It is not to be supposed, according to a view which has erroneously been attributed to me by more than one writer, *but which I never for a moment entertained, much less published*, that the refrangibility is changed in the act of reflection from the molecules. The view which I have all along maintained is that the incident vibrations caused an agitation among the ultimate molecules of the body, and that these acted as centres of disturbance to the surrounding ether, the disturbance lasting for a time which, whether it was long enough to be rendered sensible in observation or not, was at any rate very long compared with the time of a single luminous vibration.[33] [emphasis added]

The distinction here between views entertained and views published corresponds to Stokes's lifelong refusal to commit himself in print to anything that seemed to be even slightly speculative, or that lacked analytical and

---

[31] Of course he was in any case nearly forty when he married and already deeply involved as a secretary in the Royal Society.   [32] Stokes, *Memoir*, vol. 1, p. 33.

[33] Stokes, *On Light*, p. 150.

conceptual rigor.[34] Correcting a claim erroneously attributed to him, Stokes explains his own 'view', one that he never developed in any detail at all.

With an attitude like this, Stokes could hardly have written anything like Maxwell's *Treatise*, nor would he ever have indulged in public talks such as Kelvin's 1884 *Baltimore Lectures* (much less allow them to see print). Maxwell's *Treatise*, which Stokes scarcely ever mentioned, disavowed any claim to explicating through a model the concept of electric charge, and reconstructed electromagnetics in radical fashion with little contemporary support from the laboratory.[35] Kelvin's *Baltimore Lectures*, conversely, went into exactly the kind of speculative detail that Stokes abhorred; he surely regarded many of the lectures as 'scientific romances'. Stokes balanced uncomfortably between these two extremes, unable or unwilling to strike out in either direction. Of course in the early 1860s the extreme *Maxwellianism* of the later *Treatise* was scarcely developed, and Maxwell's reputation was high but not overpowering. Moreover Maxwell had engaged in careful, detailed model-making at just this time in discussing the electromagnetic field. And yet when Stokes produced his own *Report* for the British Association in 1862 he failed to include in it the model that he had himself invented and that, as far as he then knew, was apparently as good as the alternatives that he *did* discuss. Though he excluded the model itself, he nevertheless *did* include its implication for the wave surface: even slightly doubtful models were never to be discussed; statements with direct laboratory consequences could be, albeit tentatively. This episode nicely illustrates Stokes's extreme reluctance to delve publicly into something about which he was not thoroughly certain.

Stokes's *Report* rather neatly divides his career. His seminal work in hydrodynamics, elasticity and diffraction, as well as his discovery of

[34] Stokes passionately insisted on the absolute distinction between private and public views. An apposite example concerns his verbal suggestion to William Thomson concerning the link between the absorption and emission D line for sodium. 'On the strength of this conversation', Stokes wrote to John Lubbock in 1881,

> and of his having mentioned the thing is his lectures to his class, he tried to make out that I was the first to point out the existence of sodium in the sun. I think he was quite wrong; *for if a man's private conversations with his friends are to be brought in, there is an end to all evidence that such a man suggested or pointed out in such a thing.* [emphasis added. Stokes, *Memoir*, vol. 2, p. 76]

The words of disclaimer in this letter are *almost verbatim* the same ones that he had used twenty years before in a letter to Henry Roscoe (viz. 'if a man's conversations with his friends are to enter into the history of a subject there is pretty nearly an end of attaching any mention or discovery to any individual' [p. 83]).

[35] See Jed Z. Buchwald, *From Maxwell to Microphysics* (The University of Chicago Press, 1985) for details. On the mathematical structure of the *Treatise* see Peter Harman, 'Mathematics and reality in Maxwell's dynamical physics', in *Kelvin's Baltimore Lectures and Modern Theoretical Physics*, eds. Robert Kargon and Peter Achinstein (Cambridge, Mass.: The MIT Press, 1987), pp. 267–97.

fluorescence, all precede it. Afterwards, though he wrote many influential papers on limited subjects, he never again achieved the breadth of his previous analytical work nor did he again feel the exhilaration of experimental discovery. The *Report* bears the traces of Stokes's growing recognition that he could not himself see how to unify optics, a recognition that almost certainly acted to dampen his enthusiasm – always well contained in any case – for the kind of 'scientific romancing' that might integrate the subject's disparate branches.[36]

The *Report on double refraction*

Stokes's *Report* was probably the most widely-read article of his career. It dissected a generation of work at the frontiers of optical theory, pointing out where the several theories either failed or where they were, in his view, unsatisfactory. It is worthwhile examining Stokes's remarks here, both for what they tell us about his own outlook in the early 1860s (which froze solidly in place at just about that time), and for what they have to say about research concerning the problem that is so often taken to encapsulate the entire nineteenth century – the structure of the ether.

Stokes quickly rehearsed Fresnel's failure adequately to have derived his several surfaces for birefringence from his ether structure – from the supposition that the ether consists of particles that are mutually repulsive.[37] Cauchy, Stokes continued, remedied this defect in Fresnel's analysis by properly constructing the equation of motion for such a medium.[38] Assuming that the ether's particles are distributed symmetrically with respect to three orthogonal planes, Stokes continued, Cauchy obtained a differential equation with nine disposable constants – the constants being extremely complicated functions of the forces and the spacing of the particles in equilibrium. Three of the constants represent the equilibrium pressures, and Cauchy at first assumed that they vanish. This led to a formula that determines the wave speeds as a function of the direction of the wave normals, or for what one may call (using modern terminology) Cauchy's version of the 'normal surface'. The formula still contained six disposable constants, whereas Fresnel's normal surface contained three, and so Cauchy constrained his expression by requiring a priori that it must have the same sections as Fresnel's surface in the three coordinate planes.

Though Cauchy's result was *not* Fresnel's surface it differed from Fresnel's by amounts that were vastly too small to be detected in the contemporary

---

[36] And that, according to Glazebrook, *did* nearly succeed in so doing at the hands of Kelvin, a point we will shortly return to.

[37] For details of Fresnel's difficulties see Buchwald, *Wave Theory*, chap. 11.

[38] For details of Cauchy's theory see Jed Z. Buchwald, 'Optics and the theory of the punctiform ether', *Arch. Hist. Exact Sci.* 21(1980):245–78.

laboratory.[39] Despite this apparent empirical success Stokes felt that the theory was unsatisfactory in two respects. First, the relations that Cauchy established by requiring his surface and Fresnel's to have the same sections in the coordinate planes are 'forced', by which he meant that they have no external warrant beyond the necessity of experiment. Second, as in all molecular theories, incompressibility cannot be imposed on Cauchy's structure because its reaction to compression is ineluctably tied to its reaction to shear. Consequently pressure (or longitudinal) waves with finite speeds must necessarily exist in such a medium, whereas optics has to do only with transverse waves.

Stokes offers four reasons that such waves – at least in Cauchy's form – must not occur. First, strictly speaking Cauchy's theory does not imply the existence of *either* transverse (torsion) or longitudinal (pressure) waves, because according to it in every wave the oscillation must have components both in and normal to the front. The waves however divide into two types, for one of which the oscillation is nearly in the front, for the other of which it is nearly perpendicular to the front. Although the former's obliquity is hardly insignificant (amounting to ten degrees for propagation in the principal section of Iceland spar) in fact no empirical test could possibly detect it because observations are made in air, not within the crystal.[40] Stokes nevertheless regarded the implication as objectionable. He felt that a mere turning of the oscillation (from being nearly in to being nearly normal to the front) should not transform it from something that could be detected optically to something that could not be.[41]

This objection to Cauchy's theory was hardly water-tight, but Stokes had other, stronger ones, based on what occurs during crystalline reflection, which necessarily involves both types of waves (the nearly-normal and the nearly-transverse). The boundary conditions that govern this particularly difficult problem were quite uncertain, but Stokes was nonetheless able to provide a rough estimate of the amount of energy that would be carried off by the nearly-normal wave within the crystal.[42] Using Cauchy's implication that the nearly-

---

[39] Stokes found, e.g., that in Arragonite the difference in velocity for waves travelling at equal angles to the principal directions appears only in the 10th decimal place, unity being the velocity in air. 'Such a difference as this', he remarked, 'would of course be utterly insensible in experiment'.

[40] So that, e.g., an interference experiment *within a crystal* could not easily be conceived that might detect the obliquity.

[41] In his words: 'We can hardly suppose that a mere change of inclination in the direction of vibration of from 10° to 80° with the wave-front makes all the difference whether the wave belongs to a long-known and evident phenomenon, no other than the ordinary refraction in Iceland spar, or not to any visible phenomenon at all.'

[42] To do so he chose a plane of incidence such that the oscillation (which, remember, is *in* the plane of polarization) is inclined at 10 and 80 degrees respectively to the nearly-

normal wave has a speed equal to $\sqrt{3}$ times that of the nearly-transverse wave, Stokes deduced from this that the former's intensity must be $\frac{1}{26}$ that of the latter which, he continued, 'is by no means insignificant, and therefore it is a very serious objection to the theory that no corresponding phenomenon should have been discovered'. Even if the phenomenon itself had not been observed the loss of energy involved in its existence would unquestionably have shown itself in many different kinds of experiments.[43]

Clearly any pressure wave that would carry off, or generate, a detectable amount of energy was empirically unacceptable. However the greater the speed of the normal wave the *less* energy it carries off, so that if it were possible to make the speed sufficiently large then the empirical difficulty could be avoided. This is precisely what can be done, it turns out, if in Cauchy's equations the equilibrium pressures do not vanish, and the oscillation is in consequence normal to the plane of polarization. Here again Fresnel's normal surface does not emerge exactly, but the difference is once more undetectable. This theory – which can grant the normal wave high speed in virtue of an extra disposable constant – 'to a certain extent' alleviates the difficulty, Stokes admitted, though it still suffers from arbitrariness.

Only George Green's analysis came close to satisfying Stokes's rigorous strictures on what a satisfactory theory for double refraction would be, and even it failed. Green took his stand on what Stokes termed 'the method of Lagrange': he developed a potential energy density for a completely general strain and applied what Green termed d'Alembert's principle' to it to obtain differential equations of motion and boundary conditions (the latter through partial integration). The general density contains twenty-one coefficients among which relations must be established to obtain the normal surface. This would seem to make it even *more* arbitrary than Cauchy's equations, but Stokes thought not. On the contrary, he felt that Green's potential suffered from a great deal less arbitrariness because it could be constrained in a manner much more closely tied to the basic requirements of the wave theory than Cauchy's equations could be.

The relations required by Cauchy are arbitrary in that their *sole* justification is the necessity to reach Fresnel's normal surface as closely as possible, and even then the result only approximates Fresnel's surface, at the further cost of incompletely satisfying transversality within the crystal. Green's method for

transverse and the nearly normal oscillations that occur within the crystal. Stokes then takes refracted amplitudes to be approximately proportional to the projections of the incident amplitude onto their directions. This amounts to a tacit boundary condition, albeit an admittedly approximate one.

[43]   Stokes also brought up Green's old objection to normal waves – that, if they do exist, then one would expect them to generate nearly-transverse waves on reflection, and yet no unaccounted-for waves had ever been observed.

constraining his coefficients is thoroughly different from Cauchy's, because he did not have to refer in any way at all *ab initio* to Fresnel's normal surface. Instead, Green required that, of the three possible waves which his equations permitted in general, two must have their oscillations restricted entirely to the wave front. In other words Green took exact transversality within the crystal as the constraint, whereas Cauchy had abandoned transversality and used instead the sections of Fresnel's normal surface. In this way Green was able *exactly* to obtain Fresnel's surface, but only at the cost of having the oscillation *in* the plane of polarization and not normal to it, supposing the equilibrium pressures to vanish.

Stokes felt that Green's method approached perfection, but that there were unfortunately other arguments against the particular theory that he had developed:

> Were it not that other phenomena of light lead us rather to the conclusion that the vibrations are perpendicular, than that they are parallel to the plane of polarization, this theory would seem to leave us nothing to desire, except to prove that we had a right to neglect the *direct* action of the ponderable molecules, and to treat the ether within a crystal as a single elastic medium, of which the elasticity was different in different directions.

Green, like Cauchy, also essayed a theory in which the equilibrium pressures do not vanish – and so in which the oscillation is *normal* to the plane of polarization. He was again able to obtain Fresnel's surface exactly, but this time only at the cost of introducing a certain arbitrariness (in Stokes's sense) into the constraints that governed his coefficients, for he had to introduce a condition that derived from his goal – Fresnel's surface of elasticity.[44]

The only other theory that Stokes considered had been developed by the Irish mathematician James MacCullagh. It became extremely influential twenty years later, when its analytical structure proved to be similar to that required by Maxwell's electromagnetic field theory.[45] At the time, however, Stokes thought it to be unacceptable because, though it did give Fresnel's normal surface exactly, and though it did (like Green's) utilize a potential in the fashion that Stokes approved, nevertheless it was not 'mechanically' acceptable.[46] Unlike the later Maxwellians, Stokes always felt that dynamical

---

[44] This surface is such that the semi-axes of its sections by a plane give the directions of polarization and the wave speeds for fronts that are parallel to the section. The normal surface follows easily from it. See Buchwald, *Wave Theory*, appendix 5 for details.

[45] See Buchwald, *From Maxwell*, appendices 2 and 4 for details.

[46] The difficulty amounted to this: MacCullagh had not employed Green's potential, but had rather developed one that seemed to be completely different from it (though in fact MacCullagh's potential can be obtained from Green's by dropping certain terms from the latter). The resulting energy density is proportional to the square of the medium's absolute rotation at a point $(\nabla \times \mathbf{u})$. This, Stokes could easily show, meant that the

structure in itself – that is to say, workable potential and kinetic energy densities – did not suffice, that the structure had to be 'mechanically' realizable. He accordingly remarked that MacCullagh's 'methods have been characterized as a sort of mathematical induction', and, though useful, they were not in themselves sufficient for theory construction.

So even here, in the highly limited area of double refraction, Stokes was unsatisfied. And yet one might easily wonder *why* he seems to have been so convinced that there was little purpose in pursuing the subject. For even in 1862 he himself suspected that there might be a satisfactory route to a theory of double refraction. Indeed, he more than suspected it, he even knew what it was and how to formulate it. We shall now see that, had he pursued the issue, he might in the early 1860s have discovered what twenty-six years later (when the point had in any case become essentially moot) *did* prove to be a satisfying unification of optics on mechanical (and not merely dynamical) principles.

Stokes's failure to seize the moment

Although Stokes had carefully limited his report to double refraction, he was well aware that reflection theory was, if anything, an even more difficult subject. In double refraction, as we have seen, there were ways to generate the results that experiment demanded. Reflection theory failed even in this minimally-necessary respect.[47] Yet Stokes had to hand an idea that could be applied successfully to *both* double refraction and to reflection, an idea whose consequences for double refraction he mentioned in 1862. That theory does not lead to Fresnel's surface of elasticity but to one that is its reciprocal.[48] 'In the present state of the theory of double refraction', Stokes insisted, 'it appears to be of especial importance to attend to a rigorous comparison of its laws with actual observation' – that is, to find out whether the replacement works.

Yet Stokes did not undertake the investigation for five years, despite the fact that he loved experimental work, that he had conceived the theory behind the new surface nearly *twenty years before* (in 1843), and that Rankine had independently published much the same thing shortly after Stokes's *Report on*

medium as it stood violated conservation of angular momentum (since the corresponding 'stress' tensor is not symmetric). This was of course unacceptable, yet there was an escape that Stokes himself recognized. If the continuum itself produces a torque in reaction then conservation can be maintained. But Stokes did not see how this could occur, and he was not willing simply to assume that it did.

47 For a brief discussion see Buchwald, *From Maxwell*, appendix 2. Whittaker, *History*, vol. 1, chap. 5 provides a great deal of insight into the many difficult questions of the period, though it often elides problems that arose decades apart.

48 Fresnel's surface has the form $a^2x^2 + b^2y^2 + c^2z^2 = 1$; Stokes's suggested replacement for it is $x^2/a^2 + y^2/b^2 + z^2/c^2 = 1$.

double refraction was printed. In fact, the idea was sufficiently obvious that the young Rayleigh, unaware of either Rankine's *or* Stokes's work on the point, himself re-invented the theory in 1871. In March of that year Stokes wrote him the following letter, which is worth quoting at length since it is the only detailed evidence that he ever provided concerning these developments:

> Prof. Maxwell called my attention a day or two ago to your paper on double refraction ... I have just been reading it.
>
> In a paper of mine on some cases of fluid motion [...about 1843 ...] I obtained an expression for the equivalent inertia of an incompressible fluid moving relatively to a solid ... I saw at the time this would lead to a theory of double refraction, differing from Fresnel's in having reciprocals of velocities in place of velocities themselves. But having calculated the velocity at 45° to the axis in Iceland spar, I found it to differ from that given by Huyghens' construction by a quantity large enough to deter me from publishing the result without a careful scrutiny of the observations of Wollaston and Malus to see whether such an error could be tolerated. I had always a hankering after this theory, and developed it for myself much as you have done, and even investigated the form of the wave surface. After my experiments on diffraction came out Rankine published in the *Phil. Mag.* a similar theory. About four years ago I carried out the suggestion in my Report for examination by prismatic refraction on a crystal of Iceland spar ... The result was perfectly decisive. The *difference of inertia theory must be rejected, and Huyghens' construction adhered to.* The difference between the results of the two theories is something like 100 times the probable errors of observation. I ought to have published the results before this.[49]

So we see that by the time of his *Report on double refraction* Stokes had had a theory in hand for nearly twenty years that *might* have worked (he only suspected that it would not), that he discussed its consequences in his *Report*, that he there urged new empirical investigations, that he did not himself undertake them for five years, and that he did not publish his results until Rayleigh revived the theory a few years later.

This 'difference of inertia' theory, as Stokes called it, differs fundamentally from every contemporary alternative because it does not at all alter the ether's coefficients of elasticity. Instead, it transforms (in effect) the ether's density into a symmetric tensor, which is to say that it makes the ether's inertial reaction depend upon direction.[50] But why was he so attracted by this theory?

[49] Stokes, *Memoir*, vol. 2, pp. 99–100. See J. W. Strutt, 'On double refraction', *Phil. Mag.* 48(1871):369–81 for his independent discovery of the Stokes–Rankine inertia theory.

[50] The difference between the inertia and elasticity theories can be formulated in the following way. Suppose that we require (as Stokes did) that every theory must be derivable from variational principles ('the method of Lagrange') applied to the ether's

The answer is, I believe, quite simple and can be divined from remarks that Stokes made in his refraction report. After criticizing the elasticity theories, but before presenting the results (not the substance) of the inertia alternative to them, he concluded:

> The various theories which have just been reviewed have this one feature in common, that in all, the direct action of the ponderable molecules is neglected, and the ether treated as a single vibrating medium. It was, doubtless, the extreme difficulty of determining the motion of one of two mutually penetrating media that led mathematicians to adopt this, at first sight, unnatural supposition; but the conviction seems by some to have been entertained from the first, and to have forced itself upon the minds of others, that the ponderable molecules must be taken into account in a far more direct manner.
>
> ... In concluding this part of the subject, I may perhaps be permitted to express my own belief that the true dynamical theory of double refraction has yet to be found.

That missing 'true dynamical theory', Stokes evidently hoped, lay in the idea of the ether's anisotropic inertia because that was how material particles could most naturally act upon the ether.

What most troubled Stokes about the elasticity theories had little to do with their empirical adequacy since all of them seemed to yield reasonably accurate results in double refraction. Nor was he profoundly troubled by the difficulty of melding them with an adequate account of reflection, since the latter was in any case missing because (Stokes felt) of difficulties in forming appropriate boundary conditions. Rather, in his eyes their inadequacy derived primarily from what he felt to be their arbitrary assimilation of matter's affect on ether to a change in the internal forces that govern the latter.

Stokes was hardly alone in this sentiment, and it is not at all surprising that he first strongly felt it as early as 1843. In 1842 Matthew O'Brien, whom Stokes probable knew quite well,[51] had proposed an intricate scheme, modeled on Cauchy's, in which two interpenetrating systems of particles – the one representing ether, the other matter – act upon one another. The resulting equations were extremely complicated, and O'Brien was not able to obtain much from it.[52] Stokes knew also that Cauchy had attempted to develop such

---

energy density. The elasticity theories begin with a hypothetical expression for the density of the potential energy, but they leave the kinetic density alone. The inertia theory does the reverse: it introduces a new form for the kinetic density, but it leaves the potential energy untouched.

[51] Stokes thanked 'his friend' O'Brien in the 1849 *Dynamical theory* for providing him with certain instruments.

[52] Though it sufficed to immerse him in controversy with Kelland and Earnshaw over precisely how to represent the effect of the material on the ethereal particles. Stokes never mentioned the controversy, which he would certainly have found distasteful,

a theory (in ultimate response to an extremely damaging critique developed by MacCullagh), and he was no doubt thinking of all such theories when he remarked in the *Report* that Cauchy 'does not seem to have advanced beyond a few barren generalities, towards a theory of double refraction founded on a calculation of the vibrations of one of two mutually penetrating media.'[53]

Instead of these flamboyant schemes, which invented forces and spread particles about according to whatever hypothetical symmetries seemed to be necessary, Stokes's inertial theory made the simplest possible assumption. Namely, that whatever else they may do to the ether, *material particles act to load its motions* since both they and the ether posses inertia, much as a grape embedded in a jelly would load it: the vibrating ether (jelly) must carry the material particle (grape) along. In crystals the loading would be anisotropic (due to some sort of anisotropy in *material* – not ethereal – structure), and elaborate, arbitrary mathematics are not necessary to represent the possibility.

While O'Brien and others in the early 1840s were avidly publishing their blueprints for complicated ether–matter lattices of point particles, Stokes remained silent. He had an alternative, though one that might not work empirically. Yet he did not examine it in the laboratory, nor did he even publish it. Twenty years later he mentioned the possibility in print, called for experiments, and then he did not perform them. When he did finally carry them out he remained silent until someone else rediscovered the theory.

Stokes's initial reticence fits his lifelong pattern of refusing even to mention in print something about which he was at all doubtful. His subsequent failure to test the hypothesis shows something else as well. It shows that Stokes would generally not carry through an experimental investigation unless he felt that it would produce novel results, new discoveries. Both his examination of polarization in diffraction and his discovery of fluorescence five to ten years later *did* produce something new. At best, he knew, an examination of double refraction would indicate either that the inertia theory was unacceptable or else that contemporary technique could not tell the difference between its requirements and Fresnel's normal surface.[54] There was, consequently, little

and in any case he felt that the entire scheme was thoroughly misguided. For details see Buchwald, 'Punctiform ether' and 'The quantitative ether in the first half of the nineteenth century', in Cantor and Hodge (eds.), *Conceptions of Ether* (Cambridge University Press, 1981), chap. 7. Ironically, considering Stokes's critical remarks about his work, MacCullagh was motivated by very much the same distaste for that kind of theorizing.

[53] This was not entirely fair, since Cauchy *had* developed an extremely elaborate scheme, although it was based rather on a new form of differential equation (one with periodic coefficients) than it was on a specific mechanical structure. See Buchwald, 'Punctiform ether' for details.

[54] Fresnel's surface worked extremely well at the accuracies of Wollaston's and Malus's old experiments, which were about 1% or so (see Buchwald, *Wave Theory*, chaps. 1–2

incentive for Stokes, who was in any case so prudent, to spend his time pursuing the issue, particularly since any publication about it in the early 1840s was certain to land him in the midst of controversy.

The situation had however changed by the early 1860s. The danger of angry controversy was past, in part because the old issues were dead, and in part because Stokes's reputation could easily suppress it. Yet he still refused to publish more than a snippet or, for five years, to try it in the laboratory. No doubt he continued to feel that the likelihood of success in such an investigation was small. This – together with Stokes's acknowledged tendency to procrastinate and his immersion in Society business – again put the issue aside until he found the time to look into it, which evidently occurred sometime during 1867. And then the results were just what he had long anticipated – the inertia surface failed abysmally.

We might end the story there – what after all is there to say after such a definitive result from the laboratory? – were it not for one thing. Stokes was wrong. Not about the failure of the *surface* that he had deduced: it certainly was frustrated by experiment. Rather, the *theory* he had developed did not have to be abandoned. It could have been evolved into something extraordinarily successful had Stokes persevered with it – if he had had the confidence, and perhaps the competitive desire, to examine precisely *why* the theory led to the empirically-unacceptable surface.

The belated success of mechanical optics

A quarter-century after Stokes closed the book on contemporary optical theory his old friend, William Thomson (by then Lord Kelvin), re-opened it in a remarkable article that was published in the *Philosophical Magazine*. Thomson remarked:

> Having...after a great variety of previous efforts which had been commenced in connexion with preparations for my Baltimore Lectures of this time four years ago, seemingly exhausted possibilities in respect to *incompressible* elastic solid [for generating a theory of reflection], without losing faith either in light or in dynamics, and knowing that the condensational-rarefactional wave disqualifies any elastic solid of *positive*

for details). Any new surface had to be at least this accurate, so either it would rapidly prove inadequate, or else (in these antique experiments) no better than Fresnel's. Given no difference between them in these experiments, new ones would have to be performed, and these would not likely uphold the inertia surface *over* Fresnel's since the latter had been successfully used in a wide variety of experiments. At best, then, more accurate experiments would show only that the new surface was tenable – and so that the difference between it and Fresnel lay at the boundary of contemporary technique. In other words the most that could be hoped for from experiment would be a demonstration of *possibility*.

compressibility, I saw that nothing was left but a solid of such negative compressibility as should make the velocity of the condensational-rarefactional wave zero. So I tried it ...[55]

The first point to remark here is that Thomson in the 1880s, unlike Stokes in the early 1860s, was particularly concerned to retrieve Fresnel's reflection laws from ether structure. Second, Thomson's own lectures in Baltimore had stimulated him to probe deeply *why* only one of the two laws could be deduced in this way.[56] Third, Thomson was well prepared to formulate a new attack because he was convinced throughout his life that the ether must be able to sustain pressure waves.[57]

Since the time of Green, and particularly with the advent of Maxwellian electrodynamics, British physicists had tacitly assumed that the ether is incompressible. The first stimulus to this belief had been Green's apparent demonstration that otherwise it would be unstable, that a slight disturbance would cause it to collapse. Thomson, convinced for decades for a complex of reasons that the ether should be compressible, began in the mid 1880s to wonder about Green's demonstration of instability. He discovered a hidden flaw, or, rather, that it contained a hidden supposition. If, he demonstrated, we take Green's potential and perform a partial integration, with a boundary at infinity, then Green's conclusion – that an ether without compressibility must be unstable – fails provided that we ignore the resulting surface integral.[58] This permitted Thomson to assume the precise opposite of Green: namely, that, far from being incompressible, the ether is infinitely compressible (i.e. that it stores no energy in compression). This 'labile' structure, he went on to show, successfully reproduces both of Fresnel's reflection laws provided, significantly, one assumes that the elasticities remain the same across media boundaries but that the densities alter.

But what of double refraction? Thomson did not treat the subject, but later that same year the British Maxwellian and optical specialist, Richard Tetley Glazebrook did. Glazebrook was particularly well-prepared to take up the subject since his first major published work concerned experiments on birefringence that were directly motivated by Stokes's 1862 *Report*. That work

55 W. Thomson (Lord Kelvin), 'On the reflexion and refraction of light', *Phil. Mag.* 26(1888), pp. 414–25, on p. 414.

56 See W. Thomson, *Baltimore Lectures on Molecular Dynamics and the Wave Theory of Light* (1884). Stenographically reported by A. S. Hathaway. In particular, it was quite simple to obtain Fresnel's sine law for light polarized in the plane of reflection, but no one had succeeded in deducing his tangent law for light polarized in the perpendicular plane, at least not without introducing controversial principles that went beyond the normal constraints imposed by elasticity. See Buchwald, *From Maxwell*, Appendix 2 for details.

57 See, e.g., Norton Wise, 'Mediating machines' in *Science in Context* 2(1988):77–113, on p. 107.   58 Which means that there is no displacement at infinity.

was done at the Cavendish Laboratory, where he was a demonstrator, and was communicated to the Royal Society in 1878 by Maxwell himself. At the time Glazebrook was a fellow of Trinity College. He knew Stokes (who guided his work) and had attended his lectures on optics while an undergraduate. It seems that a major purpose of Glazebrook's work was to see just how badly the old Stokes–Rayleigh theory for double refraction, which required the density to become anisotropic, failed. He remarked after presenting an elaborate series of data:

> though some of the apparent difference [between the Rayleigh–Stokes equation and experiment] may be due to the error made in assuming the principal plane of the prism to coincide with one of the crystal, that cannot account for the whole; for we have seen that in Fresnel's surface the error made by the same assumption appears only in the fourth place of decimals, in the value of the refractive index, while the differences between Lord Rayleigh's theory and experiment show themselves in the third place, and tend to increase [with the incidence].
>
> Thus it seems that Lord Rayleigh's theory will not account for the phenomena of double refraction in arragonite. This result agrees with that arrived at by Professor Stokes for Iceland spar.[59]

But in 1888 Glazebrook, thoroughly familiar with the structure of the Rayleigh–Stokes theory, now saw that Thomson's referral of optical processes to changes in the effective ether density could be used to rescue it from empirical disaster. He easily demonstrated that Fresnel's original surface will emerge *exactly* if, as Thomson required, the ether has no resistance to compression.[60] This would have been a stunning result in the early 1860s, for it meant that one could generate a consistent theory for reflection and double refraction on the basis of Green's potential – on the basis, that is, of an elastic solid – provided densities, but not elasticities, are manipulated. However by the late 1880s this was hardly enough. By then anomalous dispersion, absorption and metallic reflection were at the center of many optical physicists' attention. Glazebrook, who had written the century's third *Report on optics* for the British Association in 1885, knew in detail how continental physicists, particularly in Germany but also elsewhere, had created structures for optics that embraced the new phenomena. They had done so by building two equations of motion. One, for the ether, never changed its form, but it contained a term that linked it to the second equation, which governed matter.

[59]   R. T. Glazebrook, 'An experimental determination of the values of the velocities of normal propagation of plane waves in different directions in a biaxal crystal, and a comparison of the results with theory', *Phil. Trans.* (1879):287–377, on p. 318.

[60]   R. T. Glazebrook, 'On the application of Sir William Thomson's theory of a contractile aether to double refraction, dispersion, metallic reflection, and other optical problems', *Phil. Mag.* 26(1888):521–40.

That equation could be manipulated according to necessity, and German physicists in particular had built a cottage industry on doing so.[61] Glazebrook now sought to *adapt* their methods to Thomson's labile ether. He remarked:

> Refraction occurs because the optical density of the aether is different in different media; double refraction, because in a crystal the optical density is different in different directions.
>
> It remains now to consider what is meant by the optical density of the aether, and how it can vary in different media, or in different directions in the same medium. The phenomena of aberration and the other optical effects produced by the motion of transparent bodies are more easily explicable if we suppose the actual density of the aether as well as its rigidity to be the same in all bodies.[62] Let us make this supposition for the present. Now the motion of the aether within a transparent body is not free; in addition to the forces arising from its own rigidity there must be others arising from the action of the transparent matter; and though we are ignorant of the nature of this action we can show, remembering that light-waves travel through the medium with a velocity which is independent of the amplitude, that the forces resolve themselves into two sets. One of these makes its appearance in such a way as to be equivalent to an increase in the density of the aether, while the other is equivalent to an increase in its rigidity.[63]

Where German physicists had taken their stand on manipulating the ether–matter link, and the material equation, in a way that changed the ether's effective *elasticity*, and with only partial success, Glazebrook now demonstrated that modifying the effective *density* leads to a thoroughly comprehensive optics.

Glazebrook's demonstration, as well as Thomson's original paper, generated very little contemporary reaction.[64] It is as though British physicists

---

[61] For a brief discussion of the origins of this 'twin equation' structure, which derives from Helmholtz's response to the discovery of anomalous dispersion, see Buchwald, *From Maxwell*, chap. 27.

[62] Referring implicitly here to Stokes's theory for aberration, which required the earth completely to drag the ether along. That theory was already in trouble, though Glazebrook was not aware of the fact, and in any case he was not interested in aberration *per se*; he wanted only to provide a nice, extrinsic reason for accepting what he was about to base an entire new optics on. In general the problem of optics for moving bodies held very little interest for most physicists, in Britain or on the Continent, until well into the 1890s: see Buchwald, 'Michelson experiment'.

[63] Glazebrook, 'On the application', pp. 530–1.

[64] One major exception to the silence was Gibbs, who regarded the labile ether as a major accomplishment, who wrote that 'A REMARKABLE [sic] paper by Sir William Thomson ... has opened a new vista in the possibilities of the theory of an elastic ether'. (J. W. Gibbs, 'A comparison of the electric theory of light and Sir William Thomson's theory of a quasi-labile ether', *American Journal of Science* 37(1889): 139–144). Gibbs

regarded what was apparently a long-sought unification with collective apathy. The reason was that, as Glazebrook himself pointed out, one could – analytically at least – do just as well with Maxwell's electromagnetic field theory.[65] Interesting but anti-climactic, the Thomson–Glazebrook unification of optics was simply not at the forefront of contemporary research. Stokes made no public remarks about it at all, though he was hardly a convinced Maxwellian.

For our present purposes three aspects of the unification are particularly striking. First, it derives ultimately from an *analytical* perception on Thomson's part which was based on a prior belief that the ether must be compressible. Second, the account of double refraction to which it naturally leads amounts to the one that Stokes had himself quietly developed nearly a half-century before. Third, Glazebrook's extension of the scheme was founded on the concept that ether and matter interact without altering one another's inherent structure, that their mutual affects are to be sought in a single mathematical link that expresses their dynamical connection. Of these three, the first two were available to Stokes in the early 1860s, and he was himself instrumental in suggesting the third.[66] Stokes's enduring belief that the ether is

went on to demonstrate that Thomson's labile ether satisfies the same equations as the electromagnetic field in non-conducing media. This is perhaps obvious in retrospect because the labile ether's potential function is essentially the same as MacCullagh's, and the latter governs the electromagnetic field (see Buchwald, *From Maxwell*, Appendix 2 for further details). There is however a difference between the two subjects' energy functions which involves the divergence of the displacement. Gibbs pointed out that crystalline refraction could be used to probe the disagreement, but that recent experiments on birefringence in Iceland spar 'do not encourage us to look in this direction for the decision of the question'. Gibbs felt in the end that the 'electrical theory' remained superior to the mechanical because 'it is not obliged to invent hypotheses, but only to apply the laws furnished by the science of electricity'. (I thank Martin Klein for emphasizing to me the importance of Gibbs's paper in this

[65] Specifically, Glazebrook noted with some exaggeration:

There seems ... to be no reason – as has been pointed out by Professor Fitzgerald – against applying to the oscillations of the electro-magnetic field the methods and reasoning [of the twin-equation system]. Almost the whole of the work can be translated into the language of the electro-magnetic theory at once. Periodic electric displacement in the ether will produce periodic electric displacement in the matter, and the relations between the two will depend on the ratio of the period of the ether vibrations to the possible free periods of the electric oscillations in the matter molecules; and it is not difficult to see how the action between the two might depend on the relative electrical displacements and their differential coefficients. (Glazebrook, *Report*, p. 256)

[66] Since it was Stokes who had insisted on the necessity of considering the ether–matter connection in optics, and who had himself devoted much attention to it in his discussion of fluorescence, which he suggested might involve an anharmonic material restoring force.

incompressible goes far to explain why he did not see what Thomson, who thought otherwise, found many years later. Nevertheless physicists do often play with their mathematics and concepts in restricted ways when faced with empirical recalcitrance, particularly when they are convinced that their overall approach requires preservation. Stokes did not play; he did not attempt to preserve. He made no effort *at all* to manipulate his equations. I believe that the primary reason for Stokes's apparent refusal to carry on, to probe further was his nearly palpable distrust of 'scientific romance'. As late as 1883, when he undoubtedly knew that there were excellent reasons for thinking that paths to a comprehensive optics had opened, he still refused to move very far from certainty, remarking in his Burnett Lectures:

> It may readily be imagined, as more probable than the contrary, that the presence of the ponderable molecules interspersed through the ether ... may have the effect of altering the velocity of propagation of the ethereal disturbances ... and very probably diminish it. But what may be the precise mechanism by which this result is brought about we do not know. It is easy to frame plausible hypotheses which would account for the result, but it is quite another matter to establish a theory which will admit of, and which will sustain, cross-questioning in such a variety of ways that we become convinced of its truth.[67]

The foundation of Stokes's variable-density formulation for ether dynamics was this very belief that matter does not affect the structure of the ether, but rather that it merely adds to the mass that moves *with* the ether. That much Stokes always admitted to be 'probable', but he never pursued it intensely at any time, and when the experiments that he finally performed did not favor it, he let the theory die. Stokes stimulated new physics but he rarely produced it after the mid-1850s.

[67]   Stokes, *On Light*, p. 81.

19 Maxwell and Saturn's rings: problems of stability and calculability

P. M. HARMAN

Acknowledgements

This essay draws on work carried out in the preparation of my edition of *The Scientific Letters and Papers of James Clerk Maxwell*, which is in course of publication by Cambridge University Press. I am grateful to Lord Rayleigh, to the Syndics of Cambridge University Library, to the Cambridge University Observatory, to the Keeper of Western MSS, Bodleian Library, Oxford, and to the Librarian of Glasgow University Library, for kind permission to reproduce documents. I am indebted to the Council of the Royal Society, the Leverhulme Trust, and the National Science Foundation for generous financial support of my work on the edition; and to Clare Hall, Cambridge and the Department of the History of Science, Harvard University for providing facilities for this work. I thank Jed Buchwald for helpful comments on the draft of this paper.

Introduction

In his essay on 'Newton the mathematician' D. T. Whiteside observed that Newton's choice of scientific models in his mathematical physics was shaped by 'the art of the formulable and the workable'.[1] These words aptly characterise James Clerk Maxwell's investigation of the stability of the rings of Saturn in his memoir *On the Stability of the Motion of Saturn's Rings* (1859), the essay which won him the University of Cambridge Adams Prize for 1857. Finally concluding that the rings are due to waves in a satellite system, he made it very clear that he believed that tracing the pattern of the collisions of the particles forming the rings would be incalculable. This was 'a subject above my powers at present',[2] for 'we can no longer trace the mathematical laws of their

---

[1] D. T. Whiteside, 'Newton the mathematician', in *Contemporary Newtonian Research*, ed. Zev Bechler (Dordrecht/Boston/London: Reidel, 1982), pp. 109–27, on pp. 118–19.

[2] Maxwell to William Thomson, 14 November 1857, in *The Scientific Letters and Papers of James Clerk Maxwell. Volume I: 1846–1862*, ed. P. M. Harman (Cambridge University Press, 1990), p. 555; henceforth cited as Maxwell, *Letters and Papers*, vol. 1.

motion with any distinctness'.[3] Thus the investigation of formulable and intelligible models of the ring system could not encompass such problems.

In the introduction to his memoir he explained that

> when we contemplate the Rings from a purely scientific point of view, they become the most remarkable bodies in the heavens, except, perhaps, those still less *useful* bodies – the spiral nebulæ. When we have actually seen that great arch swung over the equator of the planet without any visible connexion, we cannot bring our minds to rest. We cannot simply admit that such is the case, and describe it as one of the observed facts in nature, not admitting or requiring explanation. We must either explain its motion on the principles of mechanics, or admit that, in the Saturnian realms, there can be motion regulated by laws which we are unable to explain.[4]

Maxwell's examination of the various models which could provide workable representations of the motion and stability of the rings of Saturn, occupies an important – if little studied[5] – place in his science.

A component of Maxwell's mathematical physics that commanded his sustained and energetic effort for two years, at a critical period in his intellectual development, is clearly of considerable intrinsic interest. I will therefore commence with an account of the circumstances which led to the selection of the problem of the stability of the rings of Saturn as the Adams Prize topic in 1855. This account provides the basis for a study of the development of Maxwell's work on the Saturn's rings problem, which forms the second part of this paper.[6] I will conclude with a discussion of the place of the problem of the stability of the rings of Saturn within Maxwell's broader

---

[3] James Clerk Maxwell, *On the Stability of the Motion of Saturn's Rings* (Cambridge, 1859), in *The Scientific Papers of James Clerk Maxwell*, ed. W. D. Niven, 2 vols. (Cambridge University Press, 1890), vol. 1, p. 354; henceforth cited as Maxwell, *Scientific Papers*.

[4] Maxwell, *Scientific Papers*, vol. 1, p. 291.

[5] There have however been two recent studies of Maxwell's memoir on *Saturn's Rings*: Stephen G. Brush, C. W. F. Everitt and Elizabeth Garber eds., *Maxwell on Saturn's Rings* (Cambridge, Mass./London: The MIT Press, 1983); and A. T. Fuller, 'James Clerk Maxwell's Cambridge manuscripts: extracts relating to control and stability – II and III', *International Journal of Control* 36(1982):547–74; *ibid.*, 37(1983):1197–238.

[6] The complete body of Maxwell manuscripts relating to his work on the Saturn's rings problem will be published in my edition of Maxwell's *Letters and Papers*; documents relating to the Adams Prize essay itself are included in the first volume. While Brush, Everitt and Garber's *Maxwell on Saturn's Rings* includes only a few of the Maxwell manuscript letters and papers bearing on the topic, their introduction prints the texts of four pertinent letters (of 28 February, 14 March and 23 March 1855, and of 31 December 1856) from James Challis to William Thomson; see *Maxwell on Saturn's Rings*, pp. 6–10, 12.

mathematical physics and natural philosophy. It has long been argued that Maxwell's work on Saturn's rings, in which he finally came to the conclusion that the rings should be considered as a collection of particles, led him in 1859 to his kinetic theory of gases.[7] This first study of gases was undertaken, as he made clear at the time, for 'mathematical work' and as 'an exercise in mechanics'.[8] In this work, his first discussion of gas theory, he was concerned to formulate mechanical theorems expressing the regularity of the motions of molecules,[9] introducing a statistical formula for the distribution of velocities among gas molecules. Both the problem of Saturn's rings and that of the gas system were construed by Maxwell as problems in particle mechanics; but the link between them can be strengthened if emphasis is placed on the question of the calculability of a complex system of particles in motion.

He had first confronted the problem of calculability in bringing his work on the memoir on *Saturn's Rings* to a conclusion. He noted that he was unable to compute the trajectories of the individual particles constituting the rings of Saturn 'with any distinctness'.[10] The introduction of a probabilistic argument brought a degree of calculability to such complex problems in particle mechanics. However, the problem of calculating the trajectories of particles in the case of the rings of Saturn is distinct from Maxwell's method in his kinetic theory of gases, where the distribution of velocities is computed by a statistical formula. The statistical method of the theory of gases does not compute the spatio-temporal trajectories of gas particles, but does provide a means of describing the complex pattern of the motion of these particles in terms of statistical regularities. The problem of stability (of Saturn's rings) is transformed into a problem of molecular regularity. Maxwell himself subsequently made the link between the two problems. In 1864 he attempted to apply his statistical method for the distribution of velocities among particles to the case of the particles forming Saturn's rings, 'to throw some light on the theory of a confused assemblage of jostling masses whirling round a large central body'.[11] The statistical method of Maxwell's kinetic theory of gases was possibly developed as a partial resolution of the problem of calculability confronted in the memoir on *Saturn's Rings*.

The problems of stability and calculability engendered by the Saturn's rings

[7]   See William Garnett's comment in L. Campbell and W. Garnett, *The Life of James Clerk Maxwell* (London, 1882), p. 562; and C. W. F. Everitt, *James Clerk Maxwell: Physicist and Natural Philosopher* (New York: Scribner's, 1975), p. 131.

[8]   Maxwell to George Gabriel Stokes, 30 May 1859, in Maxwell, *Letters and Papers*, vol. 1, pp. 610–11.

[9]   Theodore M. Porter, 'A statistical survey of gases: Maxwell's social physics', *Historical Studies in the Physical Sciences* 12(1981):77–116, esp. pp. 79, 97.

[10]   Maxwell, *Scientific Papers*, vol. 1, p. 354.

[11]   University Library, Cambridge, Add. MSS 7655, a/7.

problem thus have an important bearing on major themes in Maxwell's physics. As will become clear, there are further ramifications of the argument, encompassing his investigation of the stability problems of governors; his discussion of the spontaneous fluctuations of molecules; and his intriguing remarks on the inherent incalculability of a mechanical system subject to instabilities at points of singularity. The present essay makes no attempt to deal systematically and in detail with this complex of problems in mathematical physics and natural philosophy. My primary aim is to establish the significance of the memoir on *Saturn's Rings* within the Maxwellian corpus; and to emphasise the value of considering Maxwell's physics from a vantage point other than that of electromagnetism or of gas theory and statistical mechanics. Tracing the themes of stability and calculability offers a new perspective on the scope of Maxwell's physics.

The Adams Prize problem

The Adams Prize had been established in 1848 by members of St John's College in honour of John Couch Adams' prediction of the existence of the planet Neptune. The subjects for the prize on the first three occasions it had been set were on celestial mechanics; but the prize had attracted few candidates and had only been awarded once.[12] For the Adams Prize for 1857, James Challis, as Plumian Professor of Astronomy at Cambridge, was determined to excite greater interest among Cambridge graduates who alone were eligible to compete. Writing to William Thomson, Professor of Natural Philosophy at Glasgow, on 28 February 1855, Challis informed Thomson of the latter's appointment to serve as an examiner for the prize,[13] and began the discussion of possible subjects for the prize.

> The subjects hitherto proposed have been astronomical, – but they may be in 'other branches of Natural Philosophy' & in 'Pure Mathematics'. We have not hitherto been successful in inducing competitors to come forward.... I fear that Cambridge mathematicians have no taste for investigations that require long mathematical calculations. I should be

---

[12] R. Peirson, 'The theory of the long inequality of Uranus and Neptune, depending on the near commensurability of their mean motions: an essay which obtained the Adams Prize for the year 1850, in the University of Cambridge', *Transactions of the Cambridge Philosophical Society*, 9(1853):i–lxvii.

[13] The third examiner was Stephen Parkinson of St John's College; see the formal 'Advertisement' for the Adams Prize for 1857 printed as a preliminary to Maxwell's essay (Maxwell, *Scientific Papers*, vol. 1, p. 288).

glad if you can suggest some subject that will be more likely to attract candidates.[14]

Challis enclosed a list of four 'Subjects suggested for the Adams Prize'.[15] Two of these were problems in celestial mechanics, and one on the aberration of light (this last being a topic of special interest to Challis himself). The third topic, 'An investigation of the perturbations of the forms of Saturn's Rings, supposing them to be fluid', was the one to which Challis gave preference. He told Thomson that this topic 'acquires an interest on account of the singular conclusions Otto Struve has recently come to respecting the approach of the inner Ring to the ball of Saturn'.

As Challis here indicated, the problem of the rings of Saturn had excited considerable recent interest among astronomers. A dark 'obscure ring' interior to the two bright rings had been first observed by the Harvard astronomer George Phillips Bond in 1850,[16] and on a visit to Europe the following year he had discussed his discovery with his colleagues. Visiting the Pulkovo Observatory (near St Petersburg) in August 1851 he took the opportunity of observing Saturn in the company of Otto Struve. His Journal records that Bond perceived that Struve 'was seeing the new ring for the first time & with entire certainty – I suspected so before he spoke'. He recorded a discussion the following day with Struve and his father Wilhelm Struve, where the view was advanced that the ring system 'is in process of change'.[17] Bond's discovery of an 'obscure ring' prompted Otto Struve to a comprehensive series of observations and an exhaustive literature search, reported in a substantial memoir on the 'Dimensions des anneaux de Saturne' (1853), where he maintained that the newly discovered ring was in fact a 'recent formation'. Reviewing two centuries of observations of the ring system he concluded that

the inner edge of the interior bright ring is gradually approaching the body of the planet while at the same time the total breadth of the two bright rings is constantly increasing...[and] that during the interval which elapsed between the observations of J. D. Cassini and those of Sir William

14 James Challis to William Thomson, 28 February 1855, University Library, Cambridge, Add. MSS 7342, C 76A; printed in Brush, Everitt and Garber, *Maxwell on Saturn's Rings*, p. 6.

15 University Library, Cambridge, Add. MSS 7342, C 76Aa; in *Maxwell on Saturn's Rings*, p. 7.

16 See 'Inner ring of Saturn', *Monthly Notices of the Royal Astronomical Society* 11(1851):20–7.

17 Edward S. Holden, *Memorials of William Cranch Bond...and of his Son George Phillips Bond* (San Francisco/New York, 1897), pp. 102–3.

Herschel, the breadth of the inner ring had increased in a more rapid ratio than that of the outer ring.[18]

If it were the case that the rings were changing in form over time, then this would support the hypothesis (as Challis implied in stating his suggested Adams Prize problem), that the rings were of fluid rather than solid construction.[19]

Bond had already made the acquaintance of the Cambridge scientific luminaries William Whewell, George Gabriel Stokes and Challis himself in July 1851;[20] and on his return to England the following October he paid a visit to Challis at the Cambridge Observatory, also meeting Adams and Whewell.[21] While Bond did not record the nature of his discussions at Cambridge, it would be surprising if the problem of the form of Saturn's rings passed without comment. Certainly by February 1855, having read Otto Struve's memoir, Challis was keen to set an Adams Prize subject on the topic.

Responding to Challis' request for suggestions for possible subjects, Thomson came up with various ideas. In his reply of 14 March 1855 Challis looked with favour on Thomson's suggestion of '"a careful examination of the demonstrations given by Laplace of the stability of the Solar System"', but considered the 'Elasticity of solids', a problem of interest to Thomson himself at this time, to be 'a subject somewhat removed from the general tenor of Cambridge mathematics'. Challis expressed a desire for 'the subject on this occasion not to be wholly unconnected with Astronomy'. He continued to express a preference for 'the question about Saturn's Rings'; and as Thomson 'seemed to approve' he enclosed 'a scheme...of instructions to the competitors, supposing that subject should be fixed on'.[22] Challis' draft on 'The mechanical stability of Saturn's rings', with Thomson's emendations, is reproduced in the appendix below. These emendations, with one trivial exception,[23] were duly incorporated into the published advertisement for the Adams Prize for 1857 on 'The Motions of Saturn's Rings', which Challis

---

[18]   Otto Struve, 'Sur les dimensions des anneaux de Saturne', *Mémoires de l'Académie Impériale des Sciences de Saint-Petersbourg* ser. 6, 5(1853):439–75, on pp. 444, 473; passage quoted from the abstract of Struve's paper published in the *Monthly Notices of the Royal Astronomical Society* 13(1852):22–4.

[19]   A conclusion drawn by Bond in his Journal; see Holden, *Memorials*, p. 103.

[20]   See Holden, *Memorials*, pp. 93–4.   [21]   See Holden, *Memorials*, pp. 112–17.

[22]   James Challis to William Thomson, 14 March 1855, University Library, Cambridge, Add. MSS 7342, C 76B; printed in Brush, Everitt and Garber, *Maxwell on Saturn's Rings*, pp. 8–9.

[23]   See James Challis to William Thomson, 23 March 1855, University Library Cambridge, Add. MSS 7342, C 76C; printed in *Maxwell on Saturn's Rings*, pp. 9–10.

dated 23 March 1855 and forwarded to Thomson for signature on that day.[24] Thomson had suggested the emendation in the title of the subject 'because it may perhaps be found that the Rings do not possess mechanical stability',[25] a possibility raised by Otto Struve's argument.

In his letter to Thomson of 14 March 1855 Challis explained that

> I quite think that a definite result respecting the stability of the forms of the Rings may be arrived at by supposing no other forces to be concerned with the mutual action of the parts than that of gravity. As soon as the friction of the parts, or a resisting medium enters into the account, we have a cause operating to produce permanent change. In the instructions to the candidates I have separated the part of the Problem which may admit of a definite answer, from that which can hardly be answered except upon gratuitous hypotheses. The latter part may give rise to speculation and conjecture, which it may not be useless to encourage.

The discussion of such hypotheses did indeed play a role in Maxwell's attempt to answer Struve's claim of a change in form of the rings over time. The consideration of friction as a characteristic of a particle system was to lead him to questions of gas viscosity, and ultimately to investigate the dynamical theory of gases.

The problem of Saturn's rings was, as Challis had stated to Thomson, consonant with 'the general tenor of Cambridge mathematics'. Whewell had set a Smith's Prize examination question in February 1854 (the year Maxwell was, together with E. J. Routh, awarded the prize) requiring that candidates 'Shew that a fluid may revolve in a permanent annulus, like Saturn's ring. How does it appear that Saturn's ring is not a rigid body?'.[26] The second part of the question involved reference to Chapter 6 of Book III of Laplace's *Traité de Mécanique Céleste*, where Laplace had established that the motions of a uniform solid ring were dynamically unstable; he concluded that the rings could be irregular solid bodies whose centres of gravity did not coincide with their geometrical centres.[27] The first part of Whewell's question, on the rotation of a fluid ring, was probably prompted by Joseph Plateau's suggestion that the appearance of the rings of Saturn was analogous to the effect of rotation on a sphere of oil immersed in a mixture of alcohol and water, where the sphere is transformed into a ring. This experiment, moreover, served as an

---

[24] The formal advertisement was enclosed with the letter of 23 March 1855 cited in note (23).

[25] See Thomson's emendations to Challis' draft reproduced in the appendix to this paper.

[26] See *The Cambridge University Calendar for the Year 1854* (Cambridge 1854), p. 413.

[27] P. S. de Laplace, *Traité de Mécanique Céleste*, 5 vols. (Paris, 1799–1825), vol. 2, pp. 155–66.

illustration of Laplace's 'nebular' cosmogony, where planets were supposed to have been formed by the condensation of gaseous matter surrounding the primeval sun. Laplace himself had suggested that the satellites and rings of Saturn were formed by the condensation of gaseous matter in the planet's atmosphere.[28] Plateau had found that 'under the influence of a sufficient centrifugal force' a sphere of oil 'is transformed into a perfectly regular ring'. He noted that 'the heavens exhibit to us also a body of a form analogous to our liquid ring... [this was] Saturn's ring'; and he observed that the rupturing of the fluid ring presented 'an image in miniature of the formation of the planets, according to the hypothesis of Laplace, by the rupture of the cosmical rings attributable to the condensation of the solar atmosphere'.[29]

Challis himself had assisted in the translation of Plateau's memoir which was published in Taylor's *Scientific Memoirs*.[30] Familiar with Laplace's argument, he considered the 'hypothesis of rigidity' of the ring to be 'inconsistent with the condition of stability'; though 'the proof that it is so is not without interest, and may give scope for clever handling'. Favouring the hypothesis of a fluid ring, he noted that this case 'virtually reduces the Problem to one in Hydrostatistics' (as he informed Thomson).[31] Challis himself had strong research interests in the mechanics of fluids; and he had clear grounds for considering the subject of the stability of Saturn's rings, involving celestial mechanics, rigid body dynamics, and the mechanics of fluids, to be consonant with the 'general tenor of Cambridge mathematics' as pursued in the Mathematical Tripos of the 1850s.[32]

It is not clear when precisely Maxwell took up the Adams Prize problem. At the time of the announcement of the subject of the prize, in March 1855, he was a bachelor-scholar of Trinity, becoming a Fellow the following October, and heavily engaged in work leading to his first paper on electromagnetic field theory, 'On Faraday's lines of force' (1856). This paper is based on theorems expressing the flow of an incompressible fluid in tubes formed by lines of force, an argument first formulated as a purely hydrodynamical study, in May 1855, of the equations of fluid motion, a treatment of the conditions of stability of

---

[28]   *Ibid.*, vol. 5, p. 291.

[29]   Joseph Plateau, 'Sur les phénomènes qui présente une masse liquide libre et soustraite à l'action de la pesanteur', *Mémoires de l'Académie Royale des Sciences de Bruxelles* 16(1843), (trans. with the assistance of James Challis) as 'On the phenomena presented by a free liquid mass withdrawn from the action of gravity', *Scientific Memoirs*, ed. R. Taylor, vol. 4 (London, 1846), pp. 16–43, on pp. 27–8, 35–6.

[30]   See note (29).   [31]   Challis to Thomson, 14 March 1855; see note (22).

[32]   See P. M. Harman, 'Newton to Maxwell: the *Principia* and British physics', *Notes and Records of the Royal Society* 42(1988): 75–96, esp. pp. 80–2, 85; and my 'Introduction' to Maxwell, *Letters and Papers*, vol. 1, pp. 9–11.

stream lines in an incompressible fluid.[33] In the spring of 1856 he was working on his 'dynamical top', an instrument which demonstrated the dynamics of rotation and the maintenance of dynamical stability.[34] There is the suggestion here of the process that Maxwell later termed the 'cross-fertilization of the sciences',[35] a thematic link between different aspects of his physics (fluid flow, the motion of Saturn's rings, and the 'dynamical top'), here grounded on the common theme of stability problems.

Quite apart from the intrinsic interest of the problem there may well have been more practical considerations which led Maxwell to take up the subject. The cash value of the prize, 'of about £130',[36] was a not inconsiderable sum, comparing favourably with Stokes' current salary of £155 p.a. as Lucasian Professor.[37] He may well have wished to establish his reputation with a memoir on mathematical astronomy in the Cambridge style and which bore directly on Laplace's classical work on celestial mechanics. But it was surely the intrinsic interest of the subject, which was not wholly unrelated thematically to his current work, which will probably have weighed most heavily with him. As he remarked in the introduction to his essay, 'our curiosity with respect to these questions is rather stimulated than appeased by the investigations of Laplace'.[38] His first extant reference to the topic is in a letter to his college friend Richard Buckley Litchfield of 4 July 1856. He there remarked that he had 'been giving a portion of time to Saturn's Rings which I find a stiff subject but curious, especially the case of the motion of a fluid ring', and was able to report one of his major conclusions on the motion of a fluid ring: 'The very forces which would tend to divide the ring into great drops or satellites are made by the motion to keep the fluid in a uniform ring'.[39] In a letter to another Trinity friend Cecil James Monro of 14 October 1856, he states 'the equation I am come to in Saturns rings'.[40] This is a

---

[33] These documents are printed as texts 63 and 66 in Maxwell, *Letters and Papers*, vol. 1, pp. 295–9, 309–13.

[34] See Maxwell's references to 'a form of top' in letters to his father and to Lewis Campbell in February and April 1856; Maxwell, *Letters and Papers*, vol. 1, pp. 384, 405.

[35] Maxwell, *Scientific Papers*, vol. 2, p. 744. For correspondences between the 'dynamical top' and the essay on Saturn's rings, see his references in 1857 to a paper by James Elliott, with reference both to precession and to the rotation of Saturn's ring round the planet; see Maxwell, *Letters and Papers*, vol. 1, pp. 500, 536.

[36] As stated in the Adams Prize notice on 'The Motions of Saturn's Rings' (Cambridge Observatory Archive).

[37] See *The Cambridge University Calendar for the Year 1855* (Cambridge, 1855), p. 147.

[38] Maxwell, *Letters and Papers*, vol. 1, p. 440.

[39] Maxwell, *Letters and Papers*, vol. 1, p. 411.

[40] Maxwell, *Letters and Papers*, vol. 1, p. 417.

biquadratic equation for the angular velocity (relative to the rotation of the rings) with which a system of waves travels round the rotating ring. The solution of this equation forms the core of the argument of the second part of his Adams Prize essay, concerned with the conditions of stability and motion of a fluid ring.

It is therefore clear that by the summer of 1856 Maxwell was well advanced in the work on his essay, which he completed and submitted to the examiners (according to the terms of the competition) by 16 December 1856.[41] It is likely that he had been engaged on the work for at least some months prior to July 1856. A letter to Thomson of 25 April 1856 shows that he was still struggling to formulate the analytical theorems of the second part of his paper 'On Faraday's lines of force' in their finished form.[42] It would seem likely that he commenced sustained and concentrated work on the essay around May 1856.[43]

Maxwell's Adams Prize essay

The argument of Maxwell's essay 'On the Stability of the Motion of Saturn's Rings' followed the terms of reference suggested by the examiners in framing the formal notice for the prize. The essay is divided into two parts, the first being concerned with the motion of a rigid ring, the second with the motion of a fluid ring or a ring formed of disconnected particles. The mathematical argument rests on potential theory, Taylor's theorem, and Fourier analysis, all methods familiar to a Cambridge wrangler. He begins with the classic work of Laplace, seeking to determine the conditions under which the rotation of a solid ring would be stable. The basic mathematical technique derives from Laplace's treatment of potential theory in the *Mécanique Céleste*. He sought to determine the potential at the planet due to the ring: 'We have next to determine the forces which act between the ring and the sphere, and this we shall do by means of the *potential*, V, due to the ring.'[44] Obtaining the equations of motion for the rotation of the ring about its centre of gravity, he derived conditions under which a uniform motion of the ring would be possible.

[41] As stated in the Adams Prize notice (see note (36)); and see Challis to Thomson, 31 December 1856, University Library, Cambridge, Add. MSS 7342, C 76D; in Brush, Everitt and Garber, *Maxwell on Saturn's Rings*, p. 12.

[42] Maxwell, *Letters and Papers*, vol. 1, pp. 406–9.

[43] William Thomson was Maxwell's leading scientific correspondent in this period, but correspondence with him on the subject of Saturn's rings would have been impossible, as Thomson was one of the examiners for the Adams Prize. Absence of letters on the subject prior to July 1856 does not therefore firmly establish that Maxwell did no work on Saturn's rings prior to the late spring or early summer of 1856.

[44] Maxwell, *Letters and Papers*, vol. 1, p. 446.

He then proceeded to consider the effect of disturbances on the motion of the ring, seeking to establish whether the effect would be merely a periodic disturbance or would be sufficiently small as to maintain dynamical stability, or such as to 'produce a displacement which would increase indefinitely and derange the system altogether'. Expanding the potential to first order by Taylor's theorem he obtains 'Equations of the motion when slightly disturbed'. To investigate these deviations from a steady state he approximates and linearises the differential equations of the motion of the rings, and reduces the equations of motion to a biquadratic equation

$$An^4 + Bn^2 + C = 0$$

where n stands for the operation d/dt.[45] The coefficients A, B, and C establish whether the motion of the ring would be stable or unstable.

Maxwell sets out five conditions for values of n.

(1)   If n were positive, this would 'indicate a displacement which must increase indefinitely so as to destroy the arrangement of the system'.

(2)   If n 'be negative the disturbance which it indicates will die away'.

(3)   If n 'be a pure impossible quantity…this would indicate a periodic variation'.

(4)   If n 'be the sum of two terms, of which one is positive and the other impossible…which indicates a periodic disturbance, whose amplitude continually increases till it destroys the system'.

(5)   If n 'be the sum of a negative quantity and an impossible one…which indicates a periodic disturbance, continually dying away'.

It is clear that cases (1) and (4) are 'inconsistent with the permanent motion of the system'. But because the equation contains only even powers of n every root coming under cases (2) and (5) implies the existence of another root belonging to cases (1) and (4). Maxwell therefore concludes 'that to ensure stability all the roots must be of the third kind, that is, pure impossible [imaginary] quantities'.

> That this may be the case both values of n^2 must be negative, and the condition of this is
> 1st   That A, B & C should be of the same sign
> 2nd   that $B^2 > 4AC$.
> When these conditions are fulfilled a periodic disturbance is possible. When they are not both fulfilled the motion cannot be permanent.[46]

[45]   Maxwell, *Letters and Papers*, vol. 1, pp. 448–50.

[46]   Maxwell, *Letters and Papers*, vol. 1, p. 451.

Thus, if all real roots and all real parts of the complex roots were negative, the rings would be stable; and Maxwell establishes that only case (3) would correspond to stability. In his review of the published memoir on *Saturn's Rings* George Biddell Airy, the Astronomer Royal, 'commend[ed] these propositions to the study of the reader, as an interesting example of a beautiful method, applied with great skill to the solution of the difficult problems which follow'.[47] Maxwell was to employ the same method in establishing stability conditions in his 1868 paper 'On governors'.[48]

He goes on to apply these stability conditions to determine the circumstances under which the rotation of a solid ring would be stable, finding that if a solid ring were supposed to be lopsided, conditions of stability would be attained, though 'the necessary amount of this inequality…must be so enormous as to be quite inconsistent with the observed appearance of the rings'.[49] Nevertheless, his discussion of the stability conditions was not free from error, as Challis noticed and tried (without success)[50] to remedy. Maxwell had erred in establishing the equations for the gravitational potential of the ring; and this entailed that a uniform solid ring would be stable, contrary to Laplace's demonstration. Maxwell, however, failed to draw this conclusion as the result of a slip; the consequence of this compensating error was that he inferred that 'the motion is unstable' in the case of a uniform solid ring: 'The result of the theory of a rigid ring shows not only that a perfectly uniform ring cannot revolve permanently about the planet, but that the irregularity of a permanently revolving ring must be a very observable quantity.'[51] This was therefore in accordance with his earlier observation that '[Laplace] proves most distinctly (Liv III Chap VI) that a solid uniform ring cannot possibly revolve about a central body in a permanent manner'.[52] Only when he reworked the argument the following August did he perceive the source of his error, and conclude, on valid grounds, that a uniform solid ring would be unstable.[53]

In the second part of the essay he turned to the case of a fluid ring. Here 'every particle of the ring is now to be regarded as a satellite of Saturn'. The different parts of the ring are now considered as being capable of independent motion; hence 'we must take account of the attraction of each portion of the

[47] G. B. Airy, 'On the stability of the motion of Saturn's rings', *Monthly Notices of the Royal Astronomical Society* 19(1859):297–304, on p. 300.

[48] See Otto Mayr, 'Maxwell and the origins of cybernetics', *Isis* 62(1971):425–44, esp. pp. 428–9.   [49] Maxwell, *Letters and Papers*, vol. 1, p. 442.

[50] Maxwell, *Letters and Papers*, vol. 1, p. 455n.

[51] Maxwell, *Letters and Papers*, vol. 1, p. 455.

[52] Maxwell, *Letters and Papers*, vol. 1, p. 442.

[53] Maxwell to William Thomson, 24 August 1857, in Maxwell, *Letters and Papers*, vol. 1, p. 533.

ring as affected by the irregularities of the other parts'.[54] To handle this complicated dynamical problem, to determine the forces acting on the ring in terms of the instantaneous state of the ring, he first considers the ring to be at rest. Assuming the ring initially to be uniform but subject to small disturbances, he considers radial and tangential forces to act on an element of it. He finds that if the ring were assumed to be at rest 'the whole ring would collapse into satellites'; but that 'when we treat the question dynamically ... we are able to understand the possibility of the stable motion of a fluid ring'. A fluid ring would only be stable when itself in motion about the rotating planet. The effect of the motion of the ring is to compensate the tangential force produced by the action between the parts of the ring; any disturbance in the ring produces four sets of waves in the plane of the ring, as well as two sets of waves oscillating normal to the plane. He therefore concludes that: 'the same tangential force, which would, of itself, break up the ring by increasing its irregularities, becomes converted by the motion of the ring into a cause which actually tends to diminish the irregularities and to reduce the ring to uniformity'.[55]

As in the first part of the essay, he linearises the equations of motion and finds the characteristic roots of a biquadratic equation for the angular velocity (relative to the rotation of the ring) with which the system of waves travels round the ring. The solution of this equation forms the core of the argument of the second part of the essay. On discussing the question (raised by Struve) of the possible change in form of the rings over time, he considers the effect of disturbing causes on the stability of the rings. These are the friction of the rings, and an external force due to the irregularities of the planet, the mutual attraction of the satellites, or the effect of the irregularities on neighbouring rings.[56]

In calculating the effect of long-continued disturbances on a system of rings he takes 'advantage of two general principles in Dynamics'. The first of these is 'the principle of the Conservation of Angular Momenta'.[57] The principle of conservation of angular momentum was not generally presented as a fundamental dynamical axiom at this time; Cambridge texts emphasised the fundamental status of Newton's laws of motion.[58] Maxwell's argument here suggests cross-reference to his paper on the 'dynamical top', on which he was working at the same time, where he makes important 'use of the mechanical conception of Angular Momentum' in a contemporary Cambridge paper by

[54]   Maxwell, *Letters and Papers*, vol. 1, pp. 443, 456.
[55]   Maxwell, *Letters and Papers*, vol. 1, p. 444.
[56]   Maxwell, *Letters and Papers*, vol. 1, pp. 468–73.
[57]   Maxwell, *Letters and Papers*, vol. 1, p. 472.
[58]   See Harman, 'Newton to Maxwell', pp. 85–6.

Robert Baldwin Hayward.[59] He there stated 'the permanence of the *original angular momentum* [of a rotating body] in direction and magnitude', and made it clear that this principle had important dynamical status by declaring that it formed part of 'our stock of appropriate ideas and methods',[60] an allusion to Whewell's notion of 'fundamental ideas', concepts which are necessary in the sense that their negation could not be conceived.[61]

The second general dynamical principle employed is that of the conservation of energy, where again his usage and terminology reflects contemporary practice.[62] He expresses the principle in terms of 'half the vis viva of the system [V]' and 'the potential energy of the system due to Saturn's attraction [P]'. Noting that if the motion of the system was an 'angular rotation, and if no loss of power took place on account of internal friction we should have $V + P =$ const. But if there be loss of power by internal friction then $V + P$ will continually diminish'. On considering the rotation of the ring system he finds that 'the more the rings spread out from one another...the less will be the value of $(V + P)$'. Hence 'the ultimate effect of internal friction is to make the outer rings extend farther from the planet and the inner rings come nearer to it'.[63] The dynamical argument thus implied that Struve's conclusion, that the ring system had changed in form over time, could be correct.

> The result of a long-continued series of disturbances among the rings [would be]...that the exterior rings would recede from the planet and the interior ones approach towards his surface. This perhaps is the only one of our results which has been observed, or believed to have been observed.

He concluded the essay by looking to future astronomical observations using the actual telescopes used by the old astronomers, to resolve the question as to whether the changes apparently observed in the rings were due to 'this

[59] R. B. Hayward, 'On a direct method of estimating velocities, accelerations, and all similar quantities with respect to axes, moveable in any manner in space, with applications', *Transactions of the Cambridge Philosophical Society* 10(1856):1–20, esp. pp. 7–10.

[60] J. Clerk Maxwell, 'On a dynamical top, for exhibiting the phenomena of the motion of a system of invariable form about a fixed point', *Transactions of the Royal Society of Edinburgh* 21(1857):559–70, esp. 560–2; see Maxwell, *Scientific Papers*, vol. 1, pp. 249–51.

[61] William Whewell, *Philosophy of the Inductive Sciences, founded upon their History*, 2 vols., 2nd edn (London, 1847), vol. 1, p. 66. For Maxwell's use of the term see Maxwell, *Letters and Papers*, vol. 1, pp. 378, 421.

[62] See W. J. M. Rankine, 'On the general law of the transformation of energy', *Philosophical Magazine* ser. 4, 5(1853):106–17, esp. 106; and see also P. M. Harman, *Energy, Force, and Matter. The Conceptual Development of Nineteenth-century Physics* (Cambridge University Press, 1982), pp. 58–64.

[63] Maxwell, *Letters and Papers*, vol. 1, pp. 472–3.

continual source of decay' arising from the friction of the rings or to 'the improvement of telescopes'.[64]

Challis was the first examiner to read the essay, which he received on 17 December 1856,[65] appending his comments. He acknowledged its receipt back from Thomson, who in turn had added his own annotations, on 28 April 1857.[66] In his 'Book of Minutes relating to the Adams Prize' Challis recorded: '1857 May 30. The Adams Prize to be awarded in 1857 was this day adjudged to James Clerk Maxwell B.A. of Trinity College.'[67] Maxwell visited Cambridge early in July 1857 to take his MA degree, shortly after the award of the Adams Prize.[68] As he reported to William Thomson (with whom correspondence on the subject was now at last possible) in a letter of 1 August 1857

> I have been brewing at Saturns Rings with infusion of your letters for a month during most of which I have been on the move but I hope to explain myself now. I have had talks with Challis and hunts in the [Cambridge] University Library and sights of diagrams.[69]

He outlined some contemporary work on the Saturn's rings problem, and drew attention to Otto Struve's memoir which he described as the 'best historical treatise' on the subject. He assured Thomson, who had queried his discussion of the issue,[70] that 'the encroachment of the inner ring seems very certain and not due to the improvement of telescopes'.[71]

During the summer and autumn of 1857 Maxwell revised the argument of his Adams Prize essay. The course of these revisions can be followed from his letters to Thomson, Challis, Tait and Campbell.[72] The first major point to be clarified was the error in his treatment of a uniform solid ring. He soon spotted his mistake,[73] restructuring the argument in much the form as it appears in the published memoir *Saturn's Rings*.[74] He now establishes unambiguously that a uniform solid ring would be unstable; and that a ring thicker at one side than the other would also be unstable. Only in the case of a solid ring loaded with a heavy particle where the ratio of the mass of the particle to the mass of the ring is as '82 to 18' would the ring be stable. In this case the attraction of the

[64] Maxwell, *Letters and Papers*, vol. 1, pp. 476–7.

[65] Challis to Thomson, 31 December 1856 (see note (41)).

[66] Challis to Thomson, 28 April 1857, University Library, Cambridge, Add. MSS 7342, C 76E. [67] Cambridge Observatory Archive.

[68] See Maxwell, *Letters and Papers*, vol. 1, pp. 506, 509.

[69] Maxwell, *Letters and Papers*, vol. 1, p. 527.

[70] In his annotations to the text of Maxwell's Adams Prize essay: see Maxwell, *Letters and Papers*, vol. 1, p. 473n. [71] Maxwell, *Letters and Papers*, vol. 1, p. 531.

[72] These (and other) letters are reproduced in Maxwell, *Letters and Papers*, vol. 1, pp. 527–41, 553–67, 573–84. [73] See note (53).

[74] Maxwell, *Scientific Papers*, vol. 1, pp. 307–10.

planet acts at a point outside the ring; a couple acts on the ring, and the stability of the system depends on 'numerical considerations', on the adjustment of the load on the ring.[75]

Turning to the problem of a fluid ring, he reported to Thomson on 14 November 1857 that he had 'abolished my off hand theory of the attractions of a thin fluid filament affected by waves, *long* compared with the diameter of the filament, and this because the *short* waves are the only dangerous ones'. Calculating the forces between the parts of a ring, he found equations for the radial and tangential displacements of elements of a ring in terms of the angular velocity (n) of a system of m waves in the ring, giving 'a biquadratic for n whose four values are possible, provided Saturn be large enough'. He found that for stability the tangential force should not exceed $\frac{1}{14}\omega^2$, where ω is the angular velocity of the ring. This had the consequence that 'the density of Saturn must be at least 42.5 times that of the ring'. However, 'that the outer and inner parts should have the same angular velocity Laplace shows that Saturn must be not more than 1.3 times as dense as the ring'. Hence 'the liquid continuous ring is doomed'.[76]

He then turns to a discussion of the conditions in which the parts of broken ring could form a permanent ring of satellites orbiting the planet. He considers the case of a 'cloud of aerolites' or 'brickbats',[77] finding that the condition of stability is that 'the mean density of the cloud must be less than 1/330 of that of Saturn', as he reported to Challis on 24 November. 'But such a ring could not revolve with uniform angular velocity (See Laplace) so we are driven to a plurality of rings with independent angular velocities'.[78] This then became the conclusion of the essay on *Saturn's Rings*: the ring system consists of concentric rings of satellites.

To facilitate understanding of the operation of systems of waves in a ring of satellites he devised a model 'for the edification of sensible image worshippers',[79] which he described in a letter to Thomson of 30 January 1858. By adjusting the position of each satellite in the model, 'the ring of satellites may be thrown into waves of any length which travel round the ring'.[80] This model provided a visual representation of Maxwell's explanation of the structure and stability of the rings of Saturn.

[75] See Maxwell, *Letters and Papers*, vol. 1, pp. 533–5.

[76] Maxwell, *Letters and Papers*, vol. 1, pp. 553–4.

[77] Maxwell, *Letters and Papers*, vol. 1, pp. 554, 566.

[78] Maxwell, *Letters and Papers*, vol. 1, p. 566.

[79] Maxwell, *Letters and Papers*, vol. 1, p. 576.

[80] Maxwell, *Letters and Papers*, vol. 1, p. 579.

Problems of stability and calculability

While engaged in revising the essay on Saturn's rings, Maxwell remarked to Thomson that the 'general case of a fortuitous concourse of atoms each having its own orbit & excentricity is a subject above my powers at present',[81] a point he amplified in the published memoir: 'When we come to deal with collisions among bodies of unknown number, size, and shape, we can no longer trace the mathematical laws of their motion with any distinctness.' In the memoir on *Saturn's Rings* this remark served as a disclaimer:

> All we can now do is to collect the results of our investigations and to make the best use we can of them in forming an opinion as to the constitution of the actual rings of Saturn which are still in existence and apparently in steady motion, whatever catastrophes may be indicated by the various theories we have attempted.[82]

These statements on calculability have important bearing on themes in Maxwell's mature physics.

The most immediate implication was in the initiation of his work on the kinetic theory of gases. Some time between February and May 1859 he happened to notice the translation of a paper by Rudolf Clausius (published in the *Philosophical Magazine* of February 1859) on the kinetic theory of gases.[83] Clausius' paper led Maxwell to a study of the collisions of particles as a means of establishing the properties of gases. In an important letter to George Gabriel Stokes of 30 May 1859 he gave a full account of the genesis of his interest in the subject. He emphasised that he had 'taken to the subject for mathematical work' and as 'an exercise in mechanics', but subsequently looked for confirmation of his argument in work on molecular science where he hoped to be 'snubbed a little by experiments'.[84] In answer to the problem of calculability he was led to formulate mechanical theorems expressing the regularity of molecular motions. The investigation of the Saturn's rings problem, which had alerted him to the problem of calculability, led him to introduce a probabilistic argument to describe the complex motions of particles.

While Maxwell had already been concerned with the problem of collisions among particles, in their effect on the stability of the rings of Saturn, two special features of Clausius' work may have attracted his attention. The first was in the physical properties of gases. In concluding his work on Saturn's

81   See note (2).   82   Maxwell, *Scientific Papers*, vol. 1, p. 354.

83   Rudolf Clausius, 'On the mean length of the paths described by the separate molecules of gaseous bodies on the occurrence of molecular motion: together with some other remarks upon the mechanical theory of heat', *Philosophical Magazine* ser. 4, 17(1859):81–91.   84   Maxwell, *Letters and Papers*, vol. 1, pp. 610–11.

rings he had drawn on data on gas viscosity, relevant to establishing the effect of friction in disturbing the stability of the rings, in a paper by Stokes on the damping of pendulums.[85] The viscosity of gases became a central feature of Maxwell's own work on kinetic theory. Related to the problem of gas viscosity was Clausius' own explanation of the slow diffusion of gas molecules. Arguing that the molecules would repeatedly collide, he had introduced a statistical argument to calculate the probability of a molecule travelling a given distance (termed the 'mean free path') without collision. Maxwell had been interested in probability theory as early as 1850:

> They say that Understanding ought to work by the rules of right reason. These rules are, or ought to be, contained in Logic; but the actual science of Logic is conversant at present only with things either certain, impossible or *entirely* doubtful, none of which (fortunately) we have to reason on. Therefore the true Logic for this world is the Calculus of Probabilities, which takes account of the magnitude of the probability (which is, or which ought to be in a reasonable man's mind).[86]

These remarks, in a letter to Lewis Campbell probably dating from July 1850, may have been aroused by an essay by Sir John Herschel in the *Edinburgh Review* of July 1850 on Adolphe Quetelet's *Theory of Probabilities*; and he would have encountered the review when it was reprinted in Herschel's collected *Essays* which he read in the winter of 1857–8.[87] There are similarities between Maxwell's derivation of the velocity distribution law in his paper 'Illustrations of the dynamical theory of gases' (1860) and Herschel's proof of the law of least squares, which suggests that Maxwell had studied Herschel's argument when reading his collected *Essays*.[88] A prior interest in the logic of probabilities as a method of calculation, and in problems of computing particle collisions (in part deriving from the work on Saturn's rings), encouraged his ready response to Clausius' paper.

[85] Maxwell to George Gabriel Stokes, 7 September 1858, in Maxwell, *Letters and Papers*, vol. 1, pp. 597–8; and see Maxwell, *Scientific Papers*, vol. 1, pp. 354–6. Stokes' discussion is in his paper 'On the effect of the internal friction of fluids on the motion of pendulums', *Transactions of the Cambridge Philosophical Society* 9 part 2 (1851):[8]–[106], on pp. 16–17, 65 and 81.

[86] Maxwell, *Letters and Papers*, vol. 1, p. 197.

[87] See Maxwell's letters to Lewis Campbell and R. B. Litchfield of 22 December 1857 and 7 February 1858, in Maxwell, *Letters and Papers*, vol. 1, pp. 576, 583. Herschel's review of 'Quetelet on probabilities', *Edinburgh Review* 92(July 1850):1–57, was reprinted in his *Essays from the Edinburgh and Quarterly Reviews, with Addresses and other Pieces* (London, 1857), pp. 365–465.

[88] See E. Garber, S. G. Brush and C. W. F. Everitt eds., *Maxwell on Molecules and Gases* (Cambridge, Mass./London: The MIT Press, 1986), p. 10. Maxwell's paper is cited in note (90).

In his letter to Stokes of 30 May 1859 Maxwell outlined the main features of his theory of gases, emphasising his advance on Clausius' mathematical method by introducing a statistical formula for the distribution of velocities among gas molecules, a distribution function which is identical in form to the distribution formula in the theory of errors. He initiated a new analytical procedure, a method introduced as a partial resolution of the problem of the calculability of particle collisions. In a draft dating from 1864[89] in review of his 1859 memoir on *Saturn's Rings* he noted that,

> I was then of the opinion that 'When we come to deal with collisions among bodies of unknown number size and shape we can no longer trace the mathematical laws of their motion with any distinctness' (§(32)). I propose now to take up the question at this point and to endeavour to throw some light on the theory of a confused assemblage of jostling masses whirling round a large central body...Collisions will occur between these bodies and after collision each body will be projected with a velocity which will carry it into some other part of the cloud of particles, where it will meet with other particles moving with a velocity different from its own....The principles by which problems of this kind can be treated were first discussed by Prof^r Clausius in a paper 'on the nature of the motion which we call heat' and were applied to several cases in gaseous physics by myself in a paper on the Motions and Collisions of Perfectly Elastic Spheres.[90]

While the attempt to apply the statistical method to compute the collisions of the particles constituting Saturn's rings proved abortive, this document illustrates the thematic relation between the two fields of physics, both concerned with formulating mechanical theorems which described particle motions and collisions and which expressed the regularity of these motions.

The discussion of dynamical stability in the Adams Prize essay had a direct bearing on the argument of Maxwell's paper 'On governors' (1868). The problem of stability is basic to his mode of reasoning in this paper, which is concerned with theoretical problems in dynamics rather than the technology of governors and the utility of governor design. His interest in governors had been aroused by his work in 1863–4 on the measurement of electrical standards: the constant speed of rotation of a circular coil was maintained by

---

[89] As Maxwell states in a letter to George Biddell Airy of 16 October 1872, Royal Greenwich Observatory Archive, University Library, Cambridge, Airy Papers, 6/259, 204R–V.

[90] See note (11). Maxwell is referring to Clausius' paper 'On the kind of motion which we call heat', *Philosophical Magazine* ser. 4, 14(1857):353–80, and to his own paper 'Illustrations of the dynamical theory of gases. Part I. On the motions and collisions of perfectly elastic spheres', *ibid.*, 19(1860):19–32 (in Maxwell, *Scientific Papers*, vol. 1, pp. 377–91).

a centrifugal governor designed by Fleeming Jenkin. Maxwell immediately began to investigate the dynamical conditions regulating the operation of Jenkin's governor, reporting his conclusions to William Thomson in a letter of 11 September 1863. He discussed the dynamical conditions corresponding to his own suggested modifications to Jenkin's governor, obtaining a polynomial of the fifth degree.

> The roots of this eq^n are in this case of the form
>
> $$a, \quad b \pm \sqrt{-1}\, c, \quad b' \pm \sqrt{-1}\, c'.$$
>
> If either $a\, b$ or b' is positive there will be destructive oscillations. Can you find the conditions of their being all negative?[91]

The mathematical technique in considering stability problems here follows directly from his method in the memoir on *Saturn's Rings*.

This procedure forms the basis of the argument of the paper 'On governors'. Noting that a governor is stable if the controlled speed returns to equilibrium, he points out that stability is maintained when disturbances 'continually diminish' or 'may be an oscillation of continually decreasing magnitude'. As in the case of the essay on Saturn's rings, he linearises the equations of motion, obtaining linear differential equations. He reduces these equations to a characteristic polynomial; the coefficients of this characteristic equation determine whether the governor is stable. He notes that the condition of stability 'is mathematically equivalent to the condition that all the possible roots and all the possible parts of the impossible roots, of a certain equation shall be negative'.[92] This is the condition for dynamical stability first introduced in the essay on Saturn's rings.

The problems of calculating the trajectories of particles play an important role in shaping Maxwell's discussion of some of the deeper problems of natural philosophy. The distinction between reversible mechanical laws and the essential irreversibility of natural processes, as described by the second law of thermodynamics, was seen by Maxwell to rest on understanding the pattern of molecular motions. His first discussion of perfect reversibility comes in a letter to Thomson of 24 November 1857, written during his revision of the Adams Prize essay. Here he responded to some of Thomson's current speculations on hydrodynamics, and to Thomson's reference to his 1854 paper 'On a particular case of the descent of a heavy body in a resisting medium'.[93] In that paper

---

[91] Maxwell to William Thomson, 11 September 1863, Kelvin Papers M 15, Glasgow University Library; see A. T. Fuller, 'James Clerk Maxwell's Glasgow manuscripts: extracts relating to control and stability', *International Journal of Control* 43(1986): 1593–612, on p. 1599.   [92] Maxwell, *Scientific Papers*, vol. 2, p. 106.

[93] Published in the *Cambridge and Dublin Mathematical Journal* 9(1854): 145–8; see Maxwell, *Scientific Papers*, vol. 1, pp. 115–18.

Maxwell had reported that the fluttering action of a slip of paper falling through the air is due to a rapid rotation about a horizontal axis caused by air resistance. In his letter to Thomson Maxwell noted that

> In May you thought that these effects would take place in an incompressible fluid without friction, and now you think that opinion a delusion, because if all motions at any instant were reversed all would go back as it came.
> Now I cannot see why, if you could gather up all the scattered motions in the fluid, and reverse them *accurately*, the strip should not fly up again. All that you need is to catch all the eddies, and reverse them not approximately, but accurately.
> If you pour a perfect fluid from any height into a perfectly hard or perfectly elastic basin its motion will break up into eddies innumerable forming on the whole one large eddy in the basin depending on the total moments of momenta for the mass.
> If after a given time say 1 hour you reverse every motion of every particle, the eddies will all unwind themselves till at the end of another hour there is a great commotion in the basin, and the water flies up in a fountain to the vessel above. But all this depends on the *exact* reversal....[94]

This was written following his treatment of the dynamics of rotation in the Adams Prize essay and in the paper on the 'dynamical top'; the conservation of energy and of angular momentum are the dynamical principles determining perfect reversibility.

Following his introduction of a statistical analysis of particle motions in his kinetic theory of gases, he was led to speculate on the relations between particle dynamics and thermodynamics. As early as May 1855 he had been aware of the directionality of heat flow stated by the second law of thermodynamics, asking Thomson 'do you profess to account for what becomes of the vis viva of heat when it passes thro' a conductor from hot to cold?'.[95] If the answer lay in a theory of molecular motions, then the 'demon' paradox ('demon' being Thomson's term[96]) exposed the ambiguities of grounding the second law of thermodynamics on a supposedly regular pattern of molecular motions.

Maxwell first introduced his 'finite being' whose purpose was to 'pick a hole' in the second law of thermodynamics in a letter to Peter Guthrie Tait of

---

[94]   Maxwell, *Letters and Papers*, vol. 1, pp. 561–2.

[95]   Maxwell to Thomson, 15 May 1855, in Maxwell, *Letters and Papers*, vol. 1, p. 307.

[96]   See William Thomson, 'The kinetic theory of the dissipation of energy', *Nature* 9(1874):441–4, on p. 442n, where he ascribes the term to Maxwell. The term did not however receive Maxwell's approbation. In an undated note to P. G. Tait (see note (98)) he suggested that Tait 'call him no more a demon but a valve'. See Harman, *Energy, Force, and Matter*, p. 140. Maxwell had published his argument in his *Theory of Heat* (London, 1871), pp. 308–9.

11 December 1867. Here he suggested a way in which 'if two things are in contact the hotter' could 'take heat from the colder without external agency'.[97] Because of the statistical distribution of molecular velocities in a gas at equilibrium, the process of heat flow from one body to a hotter one occurs spontaneously at the molecular level; but it would require the intervention of the 'finite being' to produce an observable transfer. Hence, as he subsequently made clear to Tait, it had been his intention to 'show that the 2nd law of Thermodynamics has only a statistical certainty'.[98] The statistical method of Maxwell's theory of gases enabled mechanical theorems to be formulated which would express the regularity of molecular motions. But these regularities were inherently statistical; and though calculations of molecular motions could be made, these were based on the 'regularity of averages' and lacked the 'absolute uniformity of sequence' of dynamical laws.[99] Spontaneous, unstable fluctuations could occur among individual molecules; the trajectories of individual molecules were not computed in Maxwell's statistical method.

In a letter to Mark Pattison of 7 April 1868 Maxwell first brought together his discussion of the perfect reversibility of particle motions with his understanding of the essential irreversibility of natural processes as described by the second law of thermodynamics.

> Now one thing in which the materialist (fortified with dynamical knowledge) believes is that if every motion great & small were accurately reversed, and the world left to itself again, everything would happen backwards the fresh water would collect out of the sea and run up the rivers and finally fly up to the clouds in drops which would extract heat from the air and evaporate and afterwards in condensing would shoot out rays of light to the sun and so on. Of course all living things would regrede from the grave to the cradle and we should have a memory of the future but not of the past.
>
> The reason why we do not expect anything of this kind to take place at any time is our experience of irreversible processes, all of one kind, and this leads to the doctrine of a beginning & an end instead of cyclical progression for ever.[100]

Without dwelling here on the contemporary reference to problems of materialism, the implications for physics are made clear: natural processes are irreversible.

[97] Maxwell to Tait, 11 December 1867, University Library, Cambridge, Add. MSS 7655, I, b/8; see Cargill Gilston Knott, *Life and Scientific Work of Peter Guthrie Tait* (Cambridge University Press, 1911), pp. 213–14.

[98] University Library, Cambridge, Add. MSS 7655, V, i/11a; and see Knott, *Life of Tait*, p. 215.   [99] University Library, Cambridge, Add. MSS 7655, V, f/11.

[100] Bodleian Library, Oxford, MS Pattison 56, fols. 438r–441v.

Maxwell subsequently conjoined his arguments on the perfect reversibility of particle motions, the irreversibility of natural processes, and the impossibility of calculating the spontaneous fluctuations of molecules. He develops the full implications of his argument in a letter to John William Strutt (later Lord Rayleigh) of 6 December 1870.

> If this world is a purely dynamical system and if you accurately reverse the motion of every particle of it at the same instant then all things will happen backwards till the beginning of things the rain drops will collect themselves from the ground and fly up to the clouds &c &c and men will see all their friends passing from the grave to the cradle till we ourselves become the reverse of born, whatever that is.... The possibility of executing this experiment is doubtful but I do not think that it requires such a feat to upset the 2nd law of Thermodynamics.[101]

Once again the 'finite being' is introduced, now described as 'a mere guiding agent (like a pointsman on a railway with perfectly acting switches who should send the express along one line and the goods along another)', as a 'self-acting' device which could select molecules and so illustrate the essentially statistical nature of the second law of thermodynamics. This law was a statistical, irreversible law, and hence Maxwell draws the '*Moral*. The 2nd law of Thermodynamics has the same degree of truth as the statement that if you throw a tumblerful of water into the sea you cannot get the same tumblerful of water out again.' The perfect reversibility allowed for by the laws of mechanics is constrained by the irreversibility of natural processes and the incalculable, spontaneous motions of individual molecules.

Maxwell's discussion of the 'demon' paradox and its implications for a perfectly mechanical, reversible universe, bore on problems of materialism and determinism that exercised his circle in the late 1860s and 1870s.[102] Without dwelling on these ramifications of the argument, it will suffice to note that his reflections on the causal nature of mechanical laws led him to important conclusions about the nature of dynamical stability and calculability. He expresses the issue with great clarity in his 1873 essay on science and free will:

> It is a metaphysical doctrine that from the same antecedents follow the same consequents. No one can gainsay this. But it is not of much use in a

---

[101]   In private possession; and see R. J. Strutt, *John William Strutt Third Baron Rayleigh* (London: Arnold, 1924), p. 47.

[102]   See Theodore M. Porter, *The Rise of Statistical Thinking 1820–1900* (Princeton University Press, 1986), pp. 194–208; and C. W. Smith and M. N. Wise, *Energy and Empire. A Biographical Study of Lord Kelvin* (Cambridge University Press, 1989), pp. 628–33.

world like this, in which the same antecedents never again concur, and nothing ever happens twice.[103]

The universe was fundamentally causal, yet not deterministic. He illustrates this by alluding to the swerve of Lucretian atoms 'which at quite uncertain times and places deviate in an uncertain manner from their course'. He expounds this in terms of the instability of a dynamical system at 'singular points', invoking the example of a 'watershed, where an imperceptible deviation is sufficient to determine into which of two valleys we shall descend'.[104] Thus, because of the 'bifurcation of path' (as he terms it, following Boussinesq, in a letter to Francis Galton in 1879)[105] of a mechanical system at a point of singularity, a system will 'go off along that one of the particular paths which happens to coincide with the actual condition of the system at that instant'.[106]

The imperceptible deviation of a system at a singular point, leading to a bifurcation of path, implied the inherent incalculability of a mechanical system subject to such instabilities. As he explained in his essay on science and free will:

> Much light may be thrown on some of these questions by the consideration of stability and instability. When the state of things is such that an infinitely small variation of the present state will alter only by an infinitely small quantity the state at some future time, the condition of the system, whether at rest or in motion is said to be stable; but when an infinitely small variation in the present state may bring about a finite difference in the state of the system in a finite time, the condition of the system is said to be unstable.

Such instabilities were not uncaused, but were incalculable. While there were 'certain classes of phenomena...in which a small error in the data only introduces a small error in the result', there are 'other classes of phenomena which are more complicated, and in which cases of instability may occur, the number of such cases increasing in an exceedingly rapid manner, as the number of variables increases'. There were therefore limits to the perfect predictability of the Laplacian deterministic universe. These bifurcations at points of singularity were the result of 'potential energy, which is capable of

---

[103]   Campbell and Garnett, *Life of Maxwell*, p. 442.   [104]   *Ibid.*, pp. 441, 443.

[105]   In a letter to Francis Galton of 26 February 1879 (published in extract in Porter, *Statistical Thinking*, pp. 205–6), Maxwell refers to Joseph Boussinesq, *Conciliation du véritable déterminisme mécanique avec l'existence de la vie et de la liberté morale* (Paris, 1878), where the phrase 'lieux de bifurcation' is used (on p. 50). See also M. J. Nye, 'The moral freedom of man and the determinism of nature', *British Journal for the History of Science* 9(1976):274–92, on p. 281.

[106]   Maxwell to Galton, 26 February 1879; see Porter, *Statistical Thinking*, p. 206.

being transformed into motion, but which cannot begin to be so transformed till the system has reached a certain configuration'.[107]

For Maxwell there were therefore limits to the calculability of nature. The 'demon' paradox exposed one such limit, and explained the kind of regularity described by the second law of thermodynamics. His exposition of the imperceptible deviation of a dynamical system at a point of singularity revealed another limit to calculability. His discussion of these issues may be seen to have its origin in the memoir on *Saturn's Rings*, which above all was concerned with 'the art of the formulable and the workable'.

Appendix: Challis' draft for the Adams Prize notice, with Thomson's amendments[108]

The Adams Prize subject
The mechanical stability[a] of Saturn's Rings

The Problem may be treated on the supposition that the system of Rings is exactly or very approximately concentric with Saturn and disposed about the plane of his Equator, and different hypotheses may be made respecting the physical constitution of the Rings. It may be supposed (1) that they are rigid: (2) that they are fluid or in part aeriform: (3) that they consist of[b] incoherent matter. The question will be considered to be answered by ascertaining on these hypotheses severally, whether the conditions of mechanical stability are satisfied by the mutual attractions[c] of the planet and the Rings. It is desirable that an attempt should be made to determine on which of the above hypotheses the[d] phenomena both of the bright Rings and of the recently discovered dark Ring may be most satisfactorily explained; and to indicate any causes to which a permanent[e] change of form, such as is supposed from a comparison of modern with the earlier observations to have taken place, may be attributed.

(a) {Thomson} ⟨mechanical stability⟩ ⌐Motions⌐. This change is suggested because it may perhaps be found that the Rings do not possess mechanical stability.

(b) {Thomson} ⟨incoherent⟩ ⌐masses of [matter] not mutually coherent⌐. 'Incoherent matter' might possibly be thought to mean some vague 'nebulous' matter or some other kind of substance of which we know nothing by terrestrial observation.

(c) {Thomson} ⌐and motions⌐.

(d) {Thomson} ⌐varying appearances?⌐

107   Campbell and Garnett, *Life of Maxwell*, pp. 440–3. The relation between Maxwell's remarks and the 'catastrophes' of modern dynamical systems theory is considered by M. A. B. Deakin, 'Nineteenth-century anticipations of modern theory of dynamical systems', *Archive for History of Exact Sciences* 39(1988):183–94.

108   Cambridge Observatory Archive. Thomson's annotations to Challis' text are signalled (a), (b).... Thomson's name enclosed by brackets {...} indicates his annotations. His suggested deletions are marked within angle brackets ⟨...⟩ which precede his proposed interpolations which are enclosed within corners ⌐...⌐.

(e) {Thomson} ⟨permanent⟩. This probably would be correctly understood without alteration; but would a change to a state which appears to be being left for another, or would a continual progression of changes be properly called 'a permanent change'?

Responding to Thomson's emendations Challis noted that[109]

You will perceive that we have adopted all your suggestions, with the exception of omitting the word 'varying' before 'appearances'. It seems desirable to direct the Candidates to a consideration of the actual appearances apart from any change to which they may be subject, and in a particular manner to call their attention to the change which appears to have the best support from observation, namely the change of form.

[109]   Challis to Thomson, 23 March 1855; see note (23).

20 Poincaré, topological dynamics, and the stability of the solar system

JEREMY GRAY

Acknowledgements

I would like to thank Jesper Lützen, Eberhard Knobloch, and Curtis Wilson for their helpful remarks on earlier versions of this paper, and the organisers of an Oberwolfach conference for encouraging me to get started on the topic.

Introduction

The work of Newton and Poincaré in mathematics and planetary astronomy displays several remarkable complementarities. Each man first created new mathematics capable of a wide variety of applications. However, when Newton first turned his attention to the study of planetary orbits he sought to discover why they were conic sections with one focus at the sun; he concluded that his theory of gravity was the only theory that could readily explain why this was. Later generations were to find the idea of gravity, however obscure, convincing, and it became fundamental to more detailed treatments of the orbits.[1] Poincaré in some sense worked in the opposite direction. He accepted the physical theory of gravitation, and sought on that basis to deduce what the orbits ought to be. The orbits he sought, however, were not those of a single planet orbiting the sun, but of planets in orbit around the sun and interacting gravitationally with each other. The conclusion he came to was that such a dynamical system was capable of producing a bewilderingly complicated variety of orbits. Whereas Newton had used mathematics to develop a novel physical theory based on the assumption of simple orbits, Poincaré used mathematics to show that there could be no solution to the problem of planetary orbits within the simple physical framework that Newton had proposed.

This paper describes how substantial were the results and ideas of contemporary mathematics (many of them his own discoveries) that Poincaré applied, and argues that his work formed a new kind of applied mathematics, whose very novelty illuminates its subsequent reception.

---

[1] These investigations were also initiated by Newton, most notably in his study of the motion of the moon.

Planetary astronomy in the 1880s

Earlier approaches to the theory of planetary motion are so numerous it would be impossible to describe them in any detail here.[2] For the three body problem, it was customary to write down the equations of motion in Hamilton–Jacobi form. Jacobi was the first to point out explicitly that the equations of motion can be reduced to a system of order 6. Such is the complexity of the problem that many authors published a variety of ways in which the equations can be obtained. A central problem is to try to find equations which have comprehensible solutions. The principal distinction is between variables which vary periodically and those (termed secular) which do not. Only the secular terms can affect the long-term development of, say, the solar system. However, it can be difficult to determine whether a term in an approximate solution, even one as seemingly secular as t, the time variable itself, indeed represents a secular term in an exact solution or is but the first term in the expansion of $\sin t$ and so is periodic. Thus qualitative understanding of the three body problem was hard to obtain.

For this reason several authors studied what has come to be called, following Poincaré, the *restricted* three body problem.[3] In this problem two massive bodies travel in circles around their mutual centre of gravity, while a third body (sometimes called the planetoid) orbits them both in their plane of motion. The planetoid is assumed to have zero mass, which means that it is affected gravitationally by but does not affect, the other two bodies. The problem is to determine the orbit of the planetoid. It was first studied in a special case by Euler (see Szebehely, *Theory of Orbits*, p. 37) and later by Jacobi. For the slightly less restricted three body problem in which the massive bodies are supposed to have elliptic orbits, Delaunay showed that the time need only enter the solution as the argument of periodic functions. In 1883 and 1884 Lindstedt investigated the power series solutions that result from this approach using a new approximative method of his own; his conclusions were rederived along the lines of Delaunay's theory by Tisserand in 1884–5. Another theory of Delaunay's type was developed in the years from 1881 by Gylden, the director of the Stockholm observatory. He solved the differential equations by a method of successive approximations, and found results which proved applicable to the motion of several asteroids.

[2] See the survey by E. T. Whittaker, 'Report on the Progress of the Solution of the Problem of Three Bodies', *Report of the British Association for the Advancement of Science* (1899), pp. 121–55, and V. Szebehely, *Theory of Orbits: the Restricted Problem of Three Bodies* (New York: Academic Press, 1967), together with the literature cited there. Full references for this section will be found in Whittaker's paper.

[3] See Poincaré, *Les Méthodes Nouvelles de la Mécanique Céleste*, vol. 1 (Paris, 1892).

None of these methods were entirely rigorous; those of Lindstedt and Gylden left open the question of the convergence of the power series solutions that were obtained. For this reason, the long-term stability of the solar system was left unresolved. An important step forward in that direction was taken by the American astronomer G. W. Hill in 1877, which was much praised by Poincaré and Whittaker (*Report*, p. 130), who went so far as to state that it 'may be regarded as the beginning of a new era in Dynamical Astronomy'.[4] Hill considered how the planetoid would oscillate in the neighbourhood of a periodic orbit, when its orbit would not therefore be periodic.[5] In his next paper Hill showed how to find the periodic orbit, and indeed showed how to find a family of periodic orbits depending on a parameter m that is determined by the ratio of the time for the periodic orbit to the time in which the two massive bodies complete one circle of their own orbit.[6]

This, in brief, was the state of the art by 1885, when Mittag-Leffler circulated in the mathematical world the news of a prize to be awarded on 21 January 1889 by King Oscar II of Sweden (the date is that of the King's sixtieth birthday). There was to be a panel of three judges: Weierstrass, Hermite, and Mittag-Leffler himself. Four topics were proposed, of which the first was

> A system being given of a number whatever of particles attracting one another mutually according to Newton's law, it is proposed, on the assumption that there never takes place an impact of two particles, to

[4]   G. W. Hill, *On the part of the Motion of the Lunar Perigee which is a Function of the Mean Motions of the Sun and Moon* (Cambridge, Mass., 1877, reprinted in *Acta Mathematica* 8(1886):1–36 and in *Collected Mathematical Works of G. W. Hill* (Washington DC: Carnegie Institute of Washington, 1905), vol. 1, pp. 243–70).

[5]   Hill obtained this second order linear differential equation (p. 246):

$$\frac{\mathrm{d}^2\omega}{\mathrm{d}t^2} + \theta\omega = 0$$

where θ is a periodic function of t. The equation spawned a rapidly growing literature, and is nowadays known as Hill's equation. Hill looked for a periodic solution in order to establish that the planetoid does oscillate around the periodic orbit. He found it by the bold method of substituting a power series with unknown coefficients as solution and regarding the system of equations for the coefficients that resulted as one of infinite dimension. The vanishing of an infinite-dimensional determinant gave him the periodic solution he sought. One of Poincaré's contributions was to check on the convergence of this determinant. For the subsequent history of this topic, and its development into the theory of linear operators and infinite-dimensional vector spaces, see M. Bernkopf, 'The development of function spaces', *Archive for History of Exact Sciences* 3(1966):1–96.

[6]   G. W. Hill, 'Researches in the lunar theory', *American Journal of Mathematics* 1(1878):5–27, 129–48, 245–61, reprinted in *Collected Mathematical Works of G. W. Hill*, vol. 1, pp. 284–335.

expand the coordinates of each particle in a series proceeding according to some known functions of time and converging uniformly for any space of time.[7]

The topic and the timing could hardly have been better to catch the interests of Poincaré. The other topics had to do with Fuchs's theory of differential equations, first-order nonlinear differential equations of a type considered by Briot and Bouquet, and the application of Poincaré's Fuchsian functions to the theory of algebraic equations. So strongly does this selection suggest the interests of Poincaré that one wonders if there was not some wish to award him a prize in recognition of the remarkable contributions he had already made to mathematics.[8] The topics were probably suggested by Weierstrass, who was deeply interested in all of these topics, not least the n-body problem in dynamics, which had occupied him intermittently since 1878.[9]

The early work of Poincaré

Poincaré began his mathematical career with several different studies of differential equations, real and complex, ordinary and partial. This work rapidly led to spectacular new results. Indeed, it is not too much to speak of new branches of mathematics being created, which continue to occupy the attention of mathematicians to this day.[10] The later papers and books on

[7] 'The higher mathematics', *Nature* 32(1885):302.

[8] Such, we know, was the case with Kovalevskaya and the *Prix Bordin* of 1888, as R. Cooke has described in *The Mathematics of Sonya Kovalevskaya* (New York: Springer Verlag, 1984).

[9] See G. Mittag-Leffler, 'Weierstrass and Sonja Kowalewsky'. *Acta Mathematica* 39(1923):133–98, especially pp. 169–70, and G. Mittag-Leffler, 'Zur Biographie von Weierstrass', *Acta Mathematica* 35(1923):29–65.

[10] Some aspects of this work have been considered historically. There are essays by J. Hadamard, 'L'Oeuvre mathématique de Poincaré', *Acta Mathematica* 38(1921):203–87 = *Oeuvres de Poincaré*, (Paris: Gauthier-Villars, 1956), vol. 11, 152–242; H. von Zeipel, 'L'Oeuvre astronomique de Poincaré', *Acta Mathematica* 38(1921):309–85 = *Oeuvres de Poincaré*, 11(1956):262–346; and Poincaré himself, 'Analyse des travaux scientifiques de Henri Poincaré faite par lui-même', *Acta Mathematica* 38(1921):1–135 = *Oeuvres de Poincaré, passim*. More recently there have appeared Chr. Gilain, 'La théorie géométrique des équations différentielles de Poincaré et l'histoire de l'analyse' (Thesis, Paris, 1977); A. Schlissel, 'The development of asymptotic solutions of linear ordinary differential equations, 1817–1920', *Archive for History of Exact Sciences* 16(1976/77):307–78; and J. J. Gray, *Linear Differential Equations and Group Theory from Riemann to Poincaré* (Boston and Basel: Birkhäuser, 1986), but many issues remain unexplored. Gilain's useful paper certainly deserves to be better known. He had just published 'La théorie qualitative de Poincaré et le problème de l'intégration des équations différentielles', in

celestial mechanics grew out of his work on ordinary differential equations, which began in 1880, when he announced preliminary results about the curves defined by a first order equation. As Gilain has argued ('La théorie géométrique', p. 33) and as indeed Poincaré repeatedly stressed, one of Poincaré's most important novel ideas was to think of the *global* behaviour of the solution curves to a differential equation, and indeed to concentrate on the solutions as curves not as functions – what he called the qualitative theory of differential equations. The emphasis on geometry and the preference for geometry over power series methods makes Poincaré special amongst mathematicians and unique amongst the astronomers of his day. This went with a renewed attention to the importance of the real as opposed to the complex case, reversing the priorities of a generation of pure mathematicians.

One indication of the global nature of his theory is his statement of what is today called the Poincaré index theorem for flows on a sphere [Poincaré (a)].[11] Details (to be given below) were published when the index theorem for the sphere was proved [Poincaré (b), see p. 29]. Another indication was provided in the same year when Poincaré stated the index theorem for the general case of a flow on a surface of arbitrary genus [Poincaré (c)]. This is a remarkable application of topological ideas to the theory of differential equations, and without precedent in the literature. When in the next year he published the

La France Mathématique, ed H. Gispert, Cahiers d'histoire et de philosophie des sciences 34(1991): 215–42.

[11] The papers and books by Poincaré that are of interest here are: (a) 'Sur les courbes définies par une équation différentielle' *Comptes Rendus* 90(1880): 673–5 = *Oeuvres*, 1, 1–3; (b) 'Mémoire sur les courbes définies par une équation différentielle', *Journal de Mathématiques* (3) 7(1881): 375–422 and 8(1882): 251–96 = *Oeuvres*, 1, 3–84; (c) 'Sur les courbes définies par les équations différentielles' *Comptes Rendus* 93(1881): 951–2 = *Oeuvres*, 1, 85–6; (d) 'Sur certaines solutions particulières du problème des trois corps' *Comptes Rendus* 97(1883): 251–2 = *Oeuvres*, 7, 251–2; (e) 'Sur certaines solutions particulières du problème des trois corps' *Bullétin Astronomique* 1(1884): 65–74 = *Oeuvres*, 7, 253–61; (f) 'Sur les courbes définies par les équations différentielles' *Comptes Rendus* 98(1884): 287–9 = *Oeuvres*, 1, 87–9; (g) 'Sur les courbes définies par les équations différentielles' *Journal de Mathématiques* (4) 1(1885): 167–244 = *Oeuvres*, 1, 90–158; (h) 'Sur les courbes définies par les équations différentielles' *Journal de Mathématiques* (4) 2(1886): 151–217 = *Oeuvres* 1, 167–221; (i) 'Sur l'equilibre d'une masse fluide animée d'un mouvement de rotation' *Acta Mathematica* 7(1885): 259–380 = *Oeuvres*, 7, 40–140; (j) 'Sur les intégrales irréguliers des équations linéaires' *Acta Mathematica* 8(1886): 296–344 = *Oeuvres*, 1, 290–333; (k) 'Sur le problème des trois corps et les équations de la Dynamique' (Mémoire couronné du Prix de S. M. le roi Oscar II de Suède) *Acta Mathematica* 13(1890): 1–270 = *Oeuvres*, 7, 262–479; (l) *Les Méthodes Nouvelles de la Mécanique Céleste*, 3 vols. (Paris, 1892, 1893, 1898); (m) Correspondance de Henri Poincaré et de Felix Klein, *Oeuvres de Poincaré*, 11, 26–65. For brevity, they will henceforth be referred to as [Poincaré (a)] and so on, or more simply as (a) when no confusion can arise. All page references in this paper are to the *Oeuvres*.

second part of his memoir on flows [Poincaré (b)] he further enriched the global theory by introducing two mathematical ideas that were to be important later, the theory of consequents (the first return map) and the idea of limit cycles. Only in 1883 [Poincaré (d)], and in more detail in 1884 [Poincaré (e)], did Poincaré turn to astronomy and write at length about the three body problem.

In a major paper of 1885 [Poincaré (g)], where Poincaré proved the general index theorem, he discussed the stability of orbits for the first time. He also discussed in detail the idea of flows on a torus, and gave an interesting application of the first return map which remains of significance in the theory of chaos. Finally in 1886 he wrote a paper [Poincaré (h)] in which second order differential equations are discussed, thus bringing the equations of celestial mechanics into his new framework. Poincaré illustrated the wide range of phenomena that these equations can describe. In a further paper published that year [Poincaré (j)] Poincaré presented his ideas on the representation of functions by asymptotic series. He had established earlier that some of the behaviour of solution curves to problems in celestial mechanics arises from the fact that they can be represented by asymptotic series. Poincaré then turned to publish on a variety of other topics, notably the shape of the Earth, before publishing, in 1890 his paper [Poincaré (k)] on the three body problem and the equations of dynamics, awarded the prize by the King of Sweden the year before. In this paper he announced the Poincaré recurrence theorem, and we find the origins of the theory of chaos.

There are many new ideas to be found in these pages: Poincaré's analysis of the singular points of a differential equation into the various kinds which are the raw ingredients for the index theorem; his introduction of the ideas of limit cycles, stable and unstable solutions; the first return map, and the idea of iterating it indefinitely; asymptotic solutions and the theory of asymptotic expansions; most interestingly there are doubly asymptotic solutions and the Poincaré recurrence theorem. The algebraic topological ideas of genus and the point-set topological ideas of everywhere dense and perfect sets, although not original with Poincaré, are put to novel uses. Poincaré seems to have come to some of these ideas very early on. In the first major paper [Poincaré (b)] he described the problem facing mathematicians as that of going beyond the study of the solutions to a differential equation in a neighbourhood of a point of the plane (for which several methods were known) to study the solutions in the whole plane. Invoking the analogy with the solution of algebraic equations, he divided the problem into a part he called qualitative or geometric, and a quantitative or numerical part. It was natural, he said, that the qualitative part should come first, indeed such an analysis was required for the resolution of problems like the three body problem and the long-term structure of the solar system.

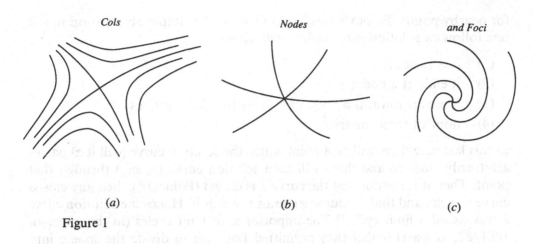

Cols   *Nodes*   *and Foci*

(a)   (b)   (c)

Figure 1

The qualitative part divides into a local and a global study. The local study concerns the nature of singular points for the flow, and the behaviour of solution curves that start off near to a given one. It is intimately connected to the global study, for, as Hadamard explained very lucidly, it is the idea of asymptotic solutions, coupled to a generalization of the idea of singular points to the situation where there are more variables, that led Poincaré to some of his deepest work in celestial mechanics. So it is instructive to see how Poincaré came to these ideas. Although the opening papers in this series deal with a differential equation too simple to be applicable to celestial mechanics, they led Poincaré to profound discoveries in due course. He analysed the singular points a system defined by this equation

$$dx/X = dy/Y$$

can have, and found that generically they were of the form shown in Figure 1.[12] The index theorem for a flow on a sphere was then formulated in this way by Poincaré: the total number of nodes and foci is equal to the number of cols plus 2.

As Poincaré remarked, a local analysis in terms of singular and non-singular points does not lead one to suspect the existence of one of the most important global features, the closed paths of the flow which have no singular points. These flow lines he called limit cycles; some are attractive and some repelling

[12]   Pictures of these and three other types of singular points for more complicated flows had been drawn earlier, by the Russian mathematician Zukovsky (*Math Sbornik* 8(1876): 163–238, see p. 180) in the course of an investigation of the kinematics of fluid bodies. We can be certain Poincaré did not know of this work, but some appreciation of the simplest types of singular point may well have been common amongst mathematicians working in this area.

for nearby points. Poincaré was led to them by the simple observation that if one follows a solution curve indefinitely either:

(1)   it is closed;

(2)   it ends at a node;

(3)   it moves towards a node or a focus but never arrives; or

(4)   none of these are true.

In this last case there will be a point which the solution curve (call it α) passes arbitrarily close to and there will be a solution curve (λ, say) through that point. Then it turns out that the curve λ is closed (Poincaré called any closed curve a cycle) and that the curve α spirals towards it. Hence the solution curve λ was called a limit cycle.[13] The importance of limit cycles (in the paper of 1881/82, at least) is that they permitted Poincaré to divide the sphere into regions having, none, one, or perhaps more than one limit cycle, and so to give a global description of the flow.

It is in 1885 [Poincaré (g)], that Poincaré first discussed the idea of stability of orbits and made a point of referring to the unresolved problem of the stability of the solar system. He regarded an orbit as stable if the moving point returned arbitrarily close to its initial position, a definition he later attributed to Poisson. He contrasted this with the asymptotic behaviour of solution curves approaching a limit cycle: the spiralling curves were unstable, the limit cycle itself was stable. Hence, he said (p. 95): 'Instability is therefore the rule and stability the exception'. In general, the flow would be stable if it was confined to a region having no limit cycles; limit cycles were a sign of instability.

Poincaré observed that it is the closed solutions, the periodic ones, that enable one to understand at least some other solution curves, those which must approach them or which are gradually driven away from them. These solutions Poincaré called asymptotic, and he distinguished between those which approach the closed orbit as $t \rightarrow \infty$, and those which approach it as $t \rightarrow -\infty$. Today these are called the forward and backward asymptotes. The realization of the existence and importance of asymptotic solutions is one of the first fruits of Poincaré's new point of view, and it seems to be original to him. It is missing, for example, in Hill's papers. However, as he pointed out, his own theory of differential equations was still not capable of dealing with

---

[13]   Made precise, this is the content of the Poincaré–Bendixson theorem, which asserts that if a solution curve is not closed, and if it does not tend towards a single point, it tends towards a closed orbit. See I. Bendixson, 'Sur les courbes définies par des équations différentielles', Ch. 1, Theorem 2, *Acta Mathematica*, 24(1901):1–88. Although it is in [Poincaré (b)] that Poincaré is at his most topological, it was Bendixson's great contribution to rework this theory in what is surely its intended setting, that of the continuously once-differentiable functions.

the equations of planetary motion. Nonetheless, he was able to provide the first substantial example of what was to develop into one of the major ideas that people were to draw from his prize-winning paper of 1889/90. This concerns the first return map, which he studied here for flows on the torus.

Poincaré regarded differential equations of the form $F(x, y, y') = 0$ where $y' = dy/dx$, as defining flows on the surface given by the equation $F(x, y, z) = 0$. As he rapidly saw, the qualitative nature of a flow is determined by the topology and specifically by the genus of this surface.[14] The general index theorem [Poincaré (c)] then asserted that if the genus of the surface was p, then the number of nodes plus the number of foci minus the number of cols is equal to $2 - 2p$. It follows from his general index theorem that the torus was the only surface upon which there could be a flow with no singularities, and partly for that reason, and partly because it is the first case where the genus is not zero, he devoted a long section of the paper to non-singular flows on the torus.

On the torus he took the angular variables ϕ and ω as coordinates of position, and wrote the differential equations as $d\omega/dt = \Omega$, $d\phi/dt = \Phi$, where Ω and Φ are functions of $\cos \phi$ and $\sin \phi$ and ω. The truly elementary example of what is today called the irrational flow on the torus, $d\omega/dt = a$, $d\phi/dt = b$ (a/b irrational), which has the solution curves $\omega/\phi = \text{constant} = a/b$, served him as an example of a case where there are no limit cycles. In accord with his general theory, the orbits are stable.

He then took the case where Φ and Ω are always positive and there are no singular points. He looked at the meridian $\phi = 0$, and since no solution curve is ever tangent to it he could look at the first return map on it. This is defined as follows. At each point on the meridian, say M, there is a flow line which is transverse to the meridian. Imagine that the point leaves the meridian and flows along this flow line. Eventually the flow line cuts the meridian again; where it does so for the first time is, say, at the point M'; M' is the image of M under the first return map.

Poincaré defined a set P as follows for an arbitrary point $M(0)$ on the meridian with $M(i)$ as its ith iterate: the set P is the set of all $M(i)$, $-\infty < i < +\infty$. Following Cantor, Poincaré defined the derived set of P, denoted P': a point is in the *derived* set if any of its neighbourhoods contains infinitely many points of the set P. Poincaré showed that P' is what Cantor called a perfect set (a set is *perfect* if it contains all its accumulation points and if every point of it is an accumulation point). For any choice of starting point, M, either the set P is a subset of the set P' or else they have no point in common; in the former case the orbit of M is stable, otherwise unstable. What Poincaré

---

14   The genus of a surface is, informally, the number of handles it has: a sphere has no handles, a torus one, a double torus two, and so on. If the surface has no holes, the genus is half the total number of curves on it with the property that no one can be deformed continuously into sums of the others.

found himself unable to determine, even after a long and delicate topological analysis, was whether some points could have stable orbits and others unstable ones.

He considered the case when there are no limit cycles, and asked in what order the points $M(i)$ occur along the meridian proceeding in a fixed direction from $M(0)$. He showed that the answer was determined entirely by an irrational number (today called the index of rotation of the flow) which he defined this way. Let $a(i)$ be the length of the interval $(M(i), M(i+1))$, and consider the quantity $(a(i) + \cdots + a(i+n))/n$. Poincaré showed that it has a limit as n tends to infinity which is finite and independent of i; he wrote the limit as $2\pi r/\mu$, where r is the radius of the meridian. He then demonstrated a simple rule involving μ which told you where on the meridian to place each successive $M(i)$.

Poincaré's prize-winning paper

The paper of 1889/90 was a show-case for Poincaré's ideas, in which they were brought together and applied to the question of the long-term stability of the solar system. It won the prize although it did not, strictly speaking, resolve the problem posed, because of the profundity of the new ideas it contained. Weierstrass[15] singled out the analysis of the stability of the solutions, the theory of invariant integrals,[16] the theory of periodic solutions, the contribution to the theory of partial differential equations, and the theory of asymptotic motion. We shall consider only two of these: the recurrence theorem (which followed from the theory of invariant integrals) and the existence of doubly asymptotic solutions. The existence of asymptotic solutions implies that some points P do not have stable orbits, but on the other hand there are cases of the three body problem where points do have stable orbits. He took the case where there are three degrees of freedom and the flow is volume-preserving, and assumed that there was a point P which moved in a bounded region of space. Then, he said, this point must return infinitely often arbitrarily close to its starting point. This is his celebrated recurrence theorem. The proof he gave it is very simple if somewhat informal: take any neighbourhood r_0 of the initial position of the point P. It has a certain volume. If the successive positions r_n of the neighbourhood r_0 never intersect, their

[15] See G. Mittag-Leffler, 'Zur Biographie von Weierstrass', p. 51. From the same article, p. 63, we learn that of Gylden's essay Weierstrass confided to Mittag-Leffler (p. 53) that 'I don't understand it, or rather, I remain uncertain whether the results are correct or not'.

[16] Invariant integrals are expressions of the form $\Sigma dx_i\, dy_i$, $dx_1\, dx_2 \ldots dx_n$ or more generally $\Sigma f(x, \ldots, x_n)\, dx_1\, dx_2 \ldots dx_n$. They are analogous to areas, and they arise for example when one is considering canonical changes of variable to a Hamiltonian system. More to the point, they are preserved by the flow of such a system of differential equations.

accumulating volume would eventually exceed the volume of the region bounding the orbit of P. So some r_n and r_m ($m > n$) intersect, whence r_0 and r_{m-n} intersect. The theorem followed by considering the successive consequents of this common region. A more careful consideration of the volumes involved showed that the probability that a point taken at random did not move in this way was zero, in a sense he made reasonably precise. (The first rigorous proof of this assertion required the techniques of measure theory, and was given by C. Caratheodory in 1919.)[17]

Towards the end of this long memoir Poincaré considered the question of what he called asymptotic surfaces. He narrowed his attention to the restricted three body problem, and let the position of the planetoid be described by two position and two angular variables, x_1, x_2, y_1, and y_2, respectively. The conservation of energy entails that these variables are connected by an equation $F(x_1, x_2, y_1, y_2) = 0$, so the orbit of P may be considered to lie in a space of three dimensions.[18]

Under certain conditions, which Poincaré discussed, the equations of motion can have asymptotic solutions, curves which either slowly approach a periodic solution or which slowly spiral away. Of particular interest to Poincaré was the case in which a given periodic solution was accompanied by asymptotic solutions of both kinds. Each of these families of curves fill out a surface and these surfaces meet along the curve which is the original periodic solution to the differential equations. Poincaré then investigated what could be said about these surfaces. He cut the three-space of all solutions by the surface $y_1 = 0$, which is transversal to the flow. The periodic solution is represented in space by a closed curve, so in the plane of section by a point, O'. The forward and backward asymptotic surfaces cut the plane in curves, which necessarily pass through the point O'. To investigate what this entails, the analogy with a col is useful. Most points do not behave asymptotically with respect to the periodic solution. Their images on the plane of section under successive iterates of the first return map may at first move in, only to be pushed away later. The points with asymptotic behaviour fill up two kinds of curve under iteration of the first return map, one moving in, one moving outwards away from the point where the periodic solution cuts the plane of section. So the iterates of points under the first return map on the plane of section make a picture that looks like a col. The defining lines of the col are the curves where the two asymptotic surfaces cut the plane of section.

Finally Poincaré considered the curve in the surface defined by the equation $y_2 = 0$. This meets the forward and backward asymptotic surfaces again, and

---

[17]   C. Caratheodory, 'Über den Wiederkehrsatz von Poincaré', *Sitzungsber. Preuss. Akad. Wiss. Berlin* (1919): 579–84.

[18]   Szebehely, *Theory of Orbits*, p. 12, points out that in the restricted three body problem energy is not in fact conserved.

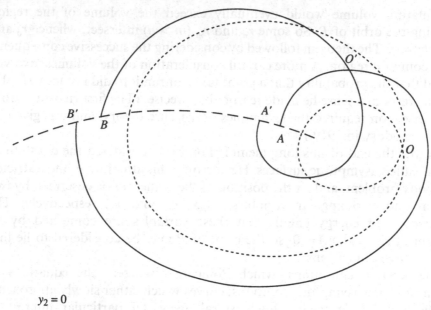

$y_2 = 0$

Figure 2. From J. H. Poincaré, *Oeuvres*, vol. 7, p. 438.

Poincaré drew the picture, Figure 2, where $AO'B'$ and $A'O'B$ are the curves to follow. One is part of the forward and the other part of the backward orbit of O'. (They are not necessarily closed.) When $O'B$ is extended it never crosses itself; the same is true of $O'B'$. But it is true that eventually $O'B$ and $O'B'$ cross. On relabelling (changing the origin in $y_2 = 0$), Poincaré said they cross at B.[19] So the two curves bound a closed region, which Poincaré called α. He now considered the effect of the flow on the region α. Eventually it returns to the plane $y_1 = 0$, and if the consequents of B and B' are B_1 and B_1' then $B_1 = B_1'$. So the region α overlaps itself, and so its boundary cuts the image of its boundary under the first return map in another point at least, N. It cannot be that the top half of the boundary cuts its iterate, so instead N is a point on the boundary through which passes a doubly asymptotic solution, that is a solution which is both a forward and a backward asymptote of the periodic solution. Iteration shows that there must be infinitely many of these. So each asymptotic surface cuts the surface $y_1 = 0$ in a curve through which pass infinitely many doubly asymptotic solutions. Poincaré rightly regarded this as an indication of how complicated the motion could be. Minkowski, in his admirable report on the paper for the *Jahrbuch über die Fortschritte der Mathematik* (1890, 1893, 22, 907–14, esp. pp. 912–13), said that this analysis

[19]   Points of this type were called homoclinic points by Poincaré in the third volume of his *Mécanique Céleste*. They have become widely regarded as indicators of chaotic behaviour in the present-day study of topological dynamics.

was a particularly difficult point. Later in the paper (Ch. 20) Poincaré showed that there are also infinitely many closed orbits.

Topology in the early work of Poincaré

Poincaré's work makes essential use of several then-novel concepts of pure mathematics. It is thoroughly immersed in the spirit of the nascent algebraic topology. As a preliminary to the global study of a flow given by a differential equation, Poincaré considered the topological character of the surface on which the flow took place. The idea that the best way to understand a surface given by an equation $F(x,y,z) = 0$ is to start by considering its genus, so as to determine its topological type, was not new. The concept of genus was introduced by Riemann in the 1850s, and after his death it was taken up by Clebsch. However, the context there was the study of analytic functions on an algebraic curve. Although, as Riemann emphasized, an algebraic curve can be seen as a real manifold of two dimensions (in short, as a real surface) the aims of those studying curves in this spirit were quite distinct from those studying real surfaces. Moreover, in the 1870s German algebraic geometers were trying to turn Riemann's way of thinking upside down, making the degree of the function defining the curve and the singularities of the curve the basic concepts (because they can be inferred from the equation). From this perspective the genus appears as a number defined algebraically and having analytical significance. The topological interpretation was found to be too obscure and was deliberately suppressed.[20]

Poincaré's first acquaintance with the concept of genus came in 1881, while he was developing his ideas about automorphic functions. He learned of it from Felix Klein, with whom he was corresponding.[21] He rapidly came to appreciate its importance for the study of automorphic functions and their associated Riemann surfaces. As Poincaré stressed, this approach to algebraic curves yields them as quotient spaces of real non-Euclidean two-dimensional space. The genus concept was further promoted in Klein's famous pamphlet about Riemann's ideas on algebraic functions and their integrals, which was published in 1882.[22] The date is significant, for Klein wrote his pamphlet as

[20] See J. J. Gray, 'Algebraic geometry in the late nineteenth century', in *The History of Modern Mathematics*, ed. J. McCleary and D. Rowe (Academic Press, 1990), pp. 361–88.

[21] The correspondence is reprinted in several places, for example Klein, *Gesammelte Mathematische Abhandlungen* 3(1925): 587–621, and [Poincaré (m)]. For a discussion, see J. J. Gray, *Linear Differential Equations*, Chapter 6.

[22] Klein, *Über Riemanns Theorie der algebraischen Funktionen und ihrer integrale* (Leipzig: Teubner, 1882) = *Abhandlungen*, 3, 499–573, tr. F. Hardcastle, as *On Riemann's Theory of Algebraic Functions and their Integrals* (Cambridge: Macmillan, 1893, Dover reprint, 1963).

therapy after the collapse of his health following his intense period of work on automorphic functions. Klein's approach, which stressed the intimate connection between harmonic and complex functions, likewise highlighted the real two-dimensional nature of a complex curve.

Nonetheless, these shifts of emphasis were not accompanied by any comparable development of the theory of real surfaces in their own right until Poincaré wrote his papers on flows. For this reason Hadamard (Poincaré, *Oeuvres*, *11*, 199) argued that it is largely due to Poincaré that mathematicians gave up emphasizing the degree over the genus, for he felt that it was these papers (rather than the ones on automorphic functions) that did the trick. Even if this is not correct, it indicates the effect he at least took Poincaré's work on celestial mechanics to have had.

Poincaré's work on automorphic functions grew out of his interest in differential equations in the complex domain, an interest originally focused on problems raised by some of the discoveries of Fuchs that were proposed for further investigation by the King of Sweden. It is in this connection that Poincaré first came across the work of Cantor on sets of points. At Hermite's instigation, Fuchs had tackled a problem in the theory of analytical continuation of certain functions and resolved it more-or-less correctly.[23] The difficulty concerns the distribution of singular points on the unit circle. Substantially the same problem presented itself in a more geometrical guise in Poincaré's work, when he had to consider the way the vertices of triangles and other polygons piled up on the boundary of a non-Euclidean tessellation. Although his own discussion was not without error, Poincaré found the ideas of Cantor that Fuchs had eschewed very helpful.[24] References to Cantor but to hardly anyone else except Kronecker can be found at many passages in Poincaré's papers of the 1880s, and one recalls that he was keen to have Cantor elected a member of the recently-founded Société Mathématique de France in 1884. So an appreciation of the unexpectedly delicate nature of point-set topology was another way in which Poincaré brought to the study of dynamics a real grasp of the latest discoveries in pure mathematics.

It is evident that Poincaré brought new pure mathematics to bear on the theory of dynamics. His work is also a new kind of applied mathematics: Poincaré changed the way one thought of answers to questions in differential equations and celestial mechanics, and greatly enlarged the kind of phenomena amenable to mathematical treatment. Poincaré turned decisively away from basing a theory of planetary motion on a theory of convergent series. He pointed out that the convergence of a series giving a solution function need not

---

[23]   L. I. Fuchs, 'Sur quelques propriétés des intégrales, etc', *Journal für die reine und angewandte Mathematik* 83(1877):13–37 = *Werke*, 2, ed. R. Fuchs (Berlin: L. Schlesinger, 1906), pp. 87–114.   [24]   Poincaré, *Oeuvres*, 2, *passim*.

have anything to do with the stability of the solution so described. What is crucial is the existence of limit cycles.

Poincaré put forward a theory in which novel geometric considerations were pre-eminent and ruled over the behaviour of series. He repeatedly stressed that a qualitative theory of differential equations had to precede any quantitative theory. As with his theory of automorphic functions, what Poincaré was able to do was to make a geometric insight sufficiently precise to suggest what range of behaviour was possible and to couple it to the rigorous methods of analysis.

The central idea was that of the asymptotic behaviour of solutions. It is this idea that enabled him to refine the idea of stability until it yields the result we know as the Poincaré recurrence theorem. It also enabled him to put forward his analysis of the doubly asymptotic solutions and to discover their remarkably intricate behaviour. In this way, and for the first time, the subject of chaos was broached in dynamics. No one before him had had any idea of how complicated the behaviour of simple-looking differential equations could be. Poincaré's ideas began to indicate what could actually be said.

The early reception of Poincaré's dynamics

By the time of his death in 1912, no one disputed the mathematical riches contained in these papers by Poincaré, although their impact was by then merged with his later contributions, notably his three-volume *Les Méthodes Nouvelles de la Mécanique Céleste*. Several mathematicians had added to the discoveries Poincaré had made. As far as the theory of ordinary differential equations was concerned, Bendixson had reworked Poincaré's ideas about singularities and cycles into a more acceptable and natural form. The nature of flows on surfaces of negative curvature (a class that includes all surfaces of genus greater than one) had been studied by Hadamard, and Birkhoff had begun to work on the general theory of topological dynamics. The study of invariant integrals was soon to be energetically taken up by Elie Cartan.

The restricted three body problem had continued to attract a rich literature since Poincaré's memoir of 1889. Bruns had published an important theorem on the non-existence of algebraic solutions to the problem, which was refined and corrected by Poincaré. Painlevé investigated the conditions that were necessary and sufficient to ensure that no collisions occurred. Perhaps most significantly, Liapunov had also given his account.[25] If we look at his work, we can obtain a more precise sense of the lasting novelty and significance of Poincaré's approach.

[25]   M. A. Liapunov, 'Problème général de la stabilité du mouvement, (tr. A. Davaux), *Annales de la Faculté des Sciences de Toulouse* (2) 9(1907):203–474, reprinted with the same pagination (Princeton University Press, 1947).

The stability of a path in a dynamical system is a topic with a long history. While developing his general ideas about dynamical systems, Poincaré also devoted a lengthy paper to the stability of the shape of the Earth, regarded as a rotating mass of fluid. This brought him into contact with the young Russian mathematician Liapunov, and since Liapunov's name is often invoked in modern studies of dynamics, it seems worthwhile to investigate this connection.[26]

The shape of the Earth is an old topic, which has its roots in the analysis that Newton so famously provided in his *Principia* (Book III, Prop. XVIII, Theorem XVI). In 1740 MacLaurin showed that there are two ellipsoids of revolution for any given angular velocity less than a certain amount. Almost 100 years later Jacobi showed that there is also a rotating ellipsoid having any given angular velocity (less than a second, smaller amount) that is not an ellipsoid of revolution. This unexpected result stimulated further work by Liouville, Dirichlet, and Riemann amongst others. All of these authors regarded an ellipsoid as stable if there was no effective force acting normally to its surface. The question of what would happen if the ellipsoid was perturbed in some way was mostly investigated only on the assumption that the ellipsoidal shape was retained, an artificial requirement dictated by the simplification it confers on the mathematical analysis. More general perturbations were considered only sketchily.

The first to attempt thorough investigations of the stability question when arbitrary shapes are permitted were Thomson and Tait in the second edition of their *Treatise on Natural Philosophy* of 1879. However, they gave no proofs of their assertions, and nor did Thomson when he published on the same theme separately in 1882. The task of providing a full mathematical analysis was left to Liapunov and Poincaré independently in 1884 and 1885. This near coincidence of publication was matched by a similarity of procedures, inasmuch as the greater generality forced each of them to use approximative methods using Lamé series expansions. It also inspired the Russian to write to Poincaré, and a correspondence ensued.[27]

As Liapunov explained, he wished to start from Thomson and Tait's idea that the motion is stable 'when the energy with a given moment of momentum, is either a minimum or a maximum'. By stable he meant that the shape is

---

[26] For a modern historical treatment of work prior to Liapunov and Poincaré, see J. Lützen 'Joseph Liouville's work on the figures of equilibrium of a rotating mass of fluid', *Archive for History of Exact Sciences* 30(1984):113–66, and *Joseph Liouville 1809–1882, Master of Pure and Applied Mathematics* (New York: Springer Verlag, 1990).

[27] The Liapunov–Poincaré correspondence has been published by V. I. Smirnov and A. P. Youshkevitch in *Cahiers du Séminaire d'Histoire des Mathématiques* 8(1987): 1–18.

restored after it has been perturbed. However, in order to obtain any results he found that he had to preclude certain types of variation arbitrarily: those which involve small fountains. As Poincaré pointed out, these include the cases where a small body of the fluid forms a drop with high velocity and breaks off from the parent body for ever.

Poincaré also started from the definition of stability provided by Thomson and Tait and sought to prove it. He established this result to his own satisfaction, although not to that of Liapunov, who wrote to Poincaré to say that he would have preferred to see a direct demonstration although he found himself unable to offer one.[28] Poincaré then investigated the oscillations of a fluid mass slightly perturbed from an ellipsoidal equilibrium. Some he showed were stable, but others were not. Poincaré also exhibited an infinite family of non-ellipsoidal equilibrium shapes of revolution, which he called 'piriforme' or pear-shaped.

There is, therefore, little overlap between the work of Liapunov and Poincaré on the shape of the Earth on the one hand and Poincaré's work on the stability of the solar system on the other. Not only are the technical conditions for stability different (which is why they have not been described in this paper) but the aim of the research differs as well. In the case of the fluid Earth, the point of interest is the equilibrium shapes that can arise, and their stability. In the dynamical problem it is the evolution of the orbits that is sought.

In 1892 Liapunov published a lengthy paper on the stability question for the dynamics of n bodies in motion, which has gradually and steadily grown in importance since its translation into French by Davaux in 1907. Liapunov explained that it had taken him two years to publish his paper, and that in that time Poincaré had published not only his prize-winning paper but also the first volume of his *Les Méthodes Nouvelles de la Mécanique Céleste*. He had been able to add some notes indicating where his work overlapped with the first of these, but not the second. So Liapunov's paper is independent of Poincaré's later papers, although well-informed concerning the series of earlier papers on differential equations.

Liapunov proposed an alternative definition of stability to the Poisson definition adopted by Poincaré. He regarded a solution to a system of differential equations as stable if solutions that start at a given time close enough to the given solution remain arbitrarily close to it at all later times. So a solution is stable if small changes in the initial conditions produce only small changes in the path. He then set himself the task of investigating when the usual method for solving the equations of motion of a system do give

[28] For the subsequent history of attempts to establish a rigorous stability criterion, see Smirnov and Youshkevitch, *Cahiers*.

reasonable solutions. This method, developed extensively by Thomson and Tait, by Routh, and by the Russian Zukovsky, proceeded by approximating the differential equations in question by keeping only their first-order terms in the dependent variables. The resulting linear equations could then be solved, but, said Liapunov, nothing a priori guaranteed that it was legitimate to replace the original problem by the linear one. He investigated when this was, and for technical reasons restricted his attention to the cases where the coefficients entering the differential equations are either constant or periodic, noting that these conditions are often satisfied in practice.

Liapunov's problem thus differed from Poincaré's in being more general. It is therefore not surprising that the remarkable complexity of the restricted three body problem was not investigated in anything like the global detail attempted by Poincaré. Instead the Russian showed how the characteristic exponents (which he obtained for the perturbed motion as Poincaré had) determined the stability of motion and hence the validity of the linearization procedure. When the coefficients were constants, stability would be ensured if all the characteristic exponents had negative real parts. If however all but two were of that form, but the remaining two were purely imaginary, then stability depended essentially on the existence of a periodic solution to the given equations. He investigated this problem in a way different from Poincaré's, as he observed (p. 392). When the coefficients were periodic, stability again depended on the characteristic exponents: a certain conditional stability (which he defined) held with respect to those coefficients that were of modulus less than one.

It is therefore clear that Liapunov's analysis was almost entirely a local analysis. Liapunov emphasized those orbits which do not stray far from a given, periodic one. His account is more general than Poincaré's in some ways, and more rigorous. Above all it put forward a definition of stability that has been found more natural by most mathematicians since. The analysis given by Poincaré is unmatched in its attention to the global behaviour of the solutions, with all the importance that has for the qualitative theory of systems of differential equations. Moreover, despite the considerable difference that the two authors attached to the term 'stable', something very like Liapunov's idea of stability can be found in Poincaré's idea of asymptotic solutions.

Because Liapunov published in Russian, the immediate response to these ideas was almost entirely directed to Poincaré's presentation. For example, when in 1898 Klein and Sommerfeld devoted half of volume two of their *Über die Theorie des Kreisels*[29] to the stability of motion of a top, Poincaré's ideas were prominently presented, but Liapunov's name is mentioned only once. Klein and Sommerfeld opened with an 'anschauliche' discussion which they

[29]   C. F. Klein and A. Sommerfeld, *Über die Theorie des Kreisels*, 4 vols. (Leipzig: Teubner, 1897–1910).

offered in the spirit of Poincaré's qualitative analysis, and proceeded to consider both the asymptotic behaviour and the stability of motion in Poincaré's meaning of the terms. The book became a highly influential textbook, and in this way Poincaré's approach to differential equations was applied to an important topic he had not himself considered.

Celestial mechanics after Poincaré

To what extent was Poincaré's work successful celestial mechanics, in the sense of enriching our understanding of nature or our power over the world? Its adoption by astronomers could not have been assured in advance. The forceful presentation of the idea that in celestial mechanics practically anything can happen, and that it is practically certain that any given configuration will recur infinitely often almost exactly as it was, might not have been what working astronomers wanted to hear. Weierstrass wrote to Mittag-Leffler that astronomers would not be pleased with it, not only because it offered no help to practical astronomy but also because it shattered illusions about the apparent certainty of certain long-held beliefs. For example, certain configurations of the planets that in Laplace's opinion presaged the eventual destruction of the solar system now seemed more likely to imply its stability.[30] The negative results concerning series solutions would also not have been welcome. Here again, Weierstrass's comments were astute. He observed that the non-existence of convergent trigonometric series solutions was merely asserted to follow, but was not actually deduced, from the non-existence of analytic solutions. Poincaré returned to the point in later writings, without ever resolving it completely. We now know that such convergent series solutions can be found.[31] If astronomers were to prove reluctant to deny power series methods pride of place (witness Sundman's later work, or among mathematicians, Painlevé[32]) might that not be because it is the methods of

[30] See Weierstrass's letter to Mittag-Leffler of February 1889, in G. Mittag-Leffler, 'Zur Biographie von Weierstrass', p. 50.

[31] Weierstrass to Mittag-Leffler, *ibid.*, p. 56, and the comments by Moser on the work of Kolmogorov and Arnold in J. Moser, *Stable and Random Motions in Dynamical Systems* (Princeton University Press, 1973) especially p. 9.

[32] See the references to Sundman and Painlevé in Gilain, 'La théorie géométrique'. Sundman's great achievement (see his 'Recherches sur le problème des trois corps' *Acta Societatis Scientiarum Fennicae* 34 and 'Mémoire sur le problème des trois corps', *Acta Mathematica* 36(1912–13):105–79) was to show that there are power series solutions to the three body problem which converge for all values of time. Unfortunately, the convergence is too slow to be of any practical use. Sundman's work allows for collisions between the three bodies, but the general *n*-body problem continues to defy such analysis. It was one of Painlevé's achievements to show that initial conditions leading to a collision are almost impossible to describe.

power series which apply one way or another to the day to day, month to month, year to year work of locating bodies and plotting their orbits?

In fact, Whittaker was able to find numerous examples of astronomers benefiting from Poincaré's analysis. Several authors were attracted to the question of establishing the existence of periodic solutions, which has both the theoretical importance Poincaré gave it and an obvious observational import. One of the most important writers in this direction was G. H. Darwin, who gave an account of a general method for finding periodic orbits based on extensive numerical calculations of six different families of periodic orbits.[33] Another was Schwarzschild, who in 1898 attempted a proof of Poincaré's conjecture (made in the first volume of the *Nouvelles méthodes*) that the periodic orbits are in fact dense. This idea was first made rigorous by E. Hopf in 1930.[34]

Nor did Poincaré's strictures regarding the non-convergence of Lindstedt's series, the non-existence of single-valued analytic solutions of specified types, and the asymptotic convergence of other series mean that series methods were inappropriate to the problem. Indeed, no others were available for quantitative work, and Poincaré himself was at pains to stress that asymptotic series can be extremely useful in numerical work. The example he gave of Stirling's series proved to be an attractive one and was often quoted by authors (such as Minkowski and Whittaker) seeking to give an account of what Poincaré had done. Whittaker finds many authors, including Poincaré, investigating asymptotic and trigonometric series methods and interpreting their results in a way that shows how much their thinking had been influenced by the ideas of the prize-winning paper.

The conclusion one comes to is that the paper of 1889 was remarkably successful in three ways: in harnessing contemporary innovations in the field of pure mathematics; in changing the range of possible questions and answers in applied mathematics; and in influencing the working practices of astronomers. Yet in the commemorative volume of *Acta Mathematica* written in 1921 there is a suspicious lack of names of those who have followed Poincaré into the field of topological dynamics. The theory of asymptotic curves, stability and the recurrence theorem, seems not to have been taken up. Topological dynamics, it would seem, had to wait until Birkhoff came to it in 1912. Part of the reason for this impression would seem to be that the various

---

[33] G. H. Darwin 'Periodic orbits' *Acta Mathematica* 21(1897):99–242 = *Scientific Papers*, vol. IV, pp. 1–113 (Cambridge University Press, 1911). Darwin worked extensively on the question of the shape of the Earth, where he was also happy to acknowledge the merits of Poincaré's ideas.

[34] K. Schwarzschild, *Astronomische Nachrichten*, 147(1898):17 and 278–9; E. Hopf, 'Zwei Sätze über den wahrscheinlichen Verlauf der Bewegungen dynamischer System', *Mathematische Annalen* 103(1930):710–19.

authors set themselves the narrow, if difficult, task of writing only about what Poincaré had done.[35] For whatever reason, they in any case missed one fascinating way in which Poincaré's ideas were taken up in a quite different domain of physics to memorable effect. This is the case of thermodynamics (see S. G. Brush, *Kinetic Theory*).[36]

In 1896 Zermelo (then a student of Max Planck's) noticed that Poincaré's recurrence theorem could be applied to thermodynamics, an application that, he suggested, Poincaré had missed. Poincaré had in 1892 published a book on thermodynamics, which was heavily criticized by Tait for not presenting the 'true, (i.e. the statistical) basis' of the second law of thermodynamics.[37] Poincaré replied that he had deliberately chosen to avoid all mechanical explanations, which he found most unsatisfactory, a point he alluded to briefly in a later essay, 'Le mécanisme et l'experience', 1893. Zermelo began by giving an alternative derivation of the recurrence theorem for the sake, he said, of physicists who might find Poincaré's original paper too difficult. He deduced that if it is the fate of every system of mass-points to return arbitrarily close to any initial configuration, then there can be no irreversible processes at work. So no mechanical system, and in particular, no gas, can be said to have a continually increasing entropy. There was, Zermelo noted, the possibility that the real states of physical systems were always those that Poincaré had said occurred only with zero probability, but this seemed to him to defy our sense of causality.

Zermelo made no proposals about how Boltzmann's kinetic theory could be saved, so Boltzmann felt compelled to reply. He agreed that Poincaré's

---

[35]   It is amusing to note, however, the neglect of Poincaré's ideas in the textbook tradition. He is not mentioned in E. L. Ince, *Ordinary Differential Equations* (London: Longman, 1926, Dover reprint, 1956), a book good enough still to be in print. He is mentioned but not adequately discussed in F. R. Moulton, *Celestial Mechanics* (New York, 1902, Dover reprint, 1963), and he is marginalized in the six-volume work by A. R. Forsyth *Theory of Differential Equations* (Cambridge University Press, 1900–1906), where attention is concentrated admittedly on the complex theory. But Forsyth was to be heavily criticized for his insistence that a solution of an equation meant an explicit solution and his disregard of qualitative analysis. Times have changed recently. The fashion for chaos has, at least in France, brought Poincaré's analysis of homoclinic points into the scientific best-seller lists, in I. Ekeland's book *Le Calcul, les Figures du Temps de Kepler à Thom* (Paris: Editions du Seuil, 1984), translated (by Ekeland himself) as *Mathematics and the Unexpected* (University of Chicago Press, 1988).

[36]   See S. G. Brush, *The Kind of Motion We call Heat: A History of the Kinetic Theory of Gases in the 19th Century* vol. 2 (Amsterdam: North Holland, 1976) [Studies in Statistical Mechanics, vol. 6] esp. pp. 630–40, upon which this account is based, and the same author's *Kinetic Theory*, vol. 2 (Pergamon Press, 1966) pp. 208–46 for translations of the papers by Zermelo and Boltzmann.

[37]   Tait, review of Poincaré's *Thermodynamique*, *Nature* XLV(1892):246.

recurrence theorem was valid, but disputed its application to the theory of heat. Boltzmann argued for the essentially statistical nature of his theory. Almost all the time entropy would increase, he said, but there would be times when it suddenly and briefly decreased. Such phases of the motion would, however, usually be a long time apart. Poincaré's recurrence theorem would hold, but the recurrence time would usually be very long indeed, so long that there would be no disagreement with experience.

Zermelo replied very quickly, and before the year was out Boltzmann again replied. While many of the details of Boltzmann's views lacked any mathematical justification, and were soundly criticized on those grounds by Zermelo, the major part of his defence of his views rested on the hypothesis that the universe started from a very improbable state and was still in an improbable state. The direction of time, he said, should be defined as that which goes from the less probable state to the more probable one. With this definition, entropy will in general increase, Poincaré's theorem notwithstanding. With that the exchange came to an end. Whatever the later developments of the theory of thermodynamics, one consequence of Poincaré's recurrence theorem seems paradoxically to have been the recognition of the importance of entropy for any physically meaningful definition of time.

Conclusions

The importance of Poincaré's work, like that of Newton's, must be registered at several levels. First, as befits his interest in the motion of the planets, he emphasized the importance of real over complex considerations. The impact of his work did much to direct attention back to the real case, which mathematicians had tended to subsume completely within the complex one. It had this effect because, within this essentially new domain, he established the importance of qualitative geometric considerations and well illustrated the importance of topological concepts. This created a new field of mathematical enquiry, the global theory of the solution curves to a system of differential equations. Moreover, by showing how the ideas he put forward, notably his novel theory of asymptotic expansions, would inevitably have a significant influence on the kinds of solution methods that could be employed, he made his mathematical discoveries significant for theoretical astronomers. Finally, by indicating the new kinds of behaviour that could be described, he prepared the ground for later theories of topological dynamics, which have proved applicable to many other topics than the three body problem.

Index

Printed in the United States
By Bookmasters